CONTENTS

KU-636-601

SECTION 2   DEFORMATION PROCESSES IN PARTICLE – STRENGTHENED ALLOYS

SECTION 3   CREEP FRACTURE PROCESSES

SECTION 4   CREEP AND FRACTURE

OF CERAMICS

SECTION 5   MATERIALS BEHAVIOUR

AT ELEVATED TEMPERATURES

SECTION 6   DESIGN AND PERFORMANCE

OF COMPONENTS AND STRUCTURES

SECTION 1
CREEP MECHANISMS

DISLOCATION CREEP IN SUBGRAIN-FORMING PURE METALS AND ALLOYS

A. S. Argon, F. Prinz, and W. C. Moffatt

Massachusetts Institute of Technology
Cambridge, Massachusetts   02139, U.S.A.

SUMMARY

Steady state creep and structural transients are considered in subgrain-forming pure metals and alloys. The formation of subgrain boundaries by an evolutionary clustering
process is described, together with the process of subgrain
boundary annihilation by coalescence, due to their migration,
leading to a structural steady state. The role of the various
components of the dislocation content of sub-boundaries is
discussed with particular emphasis on the debris content that
plays a key role in the mobility of sub-boundaries under
stress. Finally, internal stresses, measurable by the stress
dip test in such metals, are attributed to the flexing under
stress of sub-boundaries. The calculated magnitudes of such
internal stresses are in very good agreement with reported
measurements. These internal stresses are responsible for
changing the normal third power law of the strain rate stress
relation into one of higher power.

1.   INTRODUCTION

Creep deformation at high temperatures in engineering
structures is of unquestioned technological importance. In
its relatively idealized form in pure metals and single phase
alloys, creep has been widely investigated. The phenomenology of creep and the many attempts to explain it mechanis-
tically have been reviewed recently [1-4]. In spite of the
fact that considerable detailed criticism has been directed
against it, the simple picture of Bailey [5] and Orowan [6]
that high temperature creep combines two competing mechanisms
of strain hardening and thermal recovery has found wide ac-
ceptance and forms the basis of nearly all current research
on the subject. A point of particular interest has been
steady state creep in pure metals and Class II alloys, its

specific functional form, and its basis in fundamental processes of the glide, climb, and clustering of dislocations
into subgrain walls. This will also be the subject of primary interest to us in this communication, where we will attribute the high stress exponent of the steady state creep
rate to long range dynamic internal stresses, and will propose the bowing of subgrain walls under stress as the principal source of this internal stress. Furthermore, we will
consider the mobility of subgrain walls under stress as an
important part of steady state and present an outline development for the production and annihilation of subgrain walls.
We will then proceed and furnish some new experimental observations on the specific dislocation content of sub-boundaries
in creeping alloys and discuss the role of this dislocation
structure in governing the mobility of sub-boundaries. In
all of this we will strive for internal consistency but leave
the more complete mechanistic description of the complex
evolutionary processes of steady state to a future communication.

2.   STEADY STATE CREEP

2.1   Internal Stresses

The analyses by Bird, Mukherjee, and Dorn [1], Mohamed
and Langdon [7], and others [2-4] of many creep studies have
established that the steady state creep rate is given by an
expression of the form

$$\dot{\varepsilon} = A \left(\frac{\mu\Omega}{kT}\right) \left(\frac{\chi}{\mu b}\right)^3 \left(\frac{D}{b^2}\right) \left(\frac{\sigma}{\mu}\right)^m \tag{1}$$

where b , μ , Ω , χ , and D are the interatomic distance,
the shear modulus, the atomic volume, the stacking fault
energy and the self diffusion constant, and where m , the
stress exponent, is often in the range of 5 . Argon and
Takeuchi [8] have shown that a creep rate expression of the
above form but with an exponent of m = 3 is a natural result of a steady state sequence of processes of gliding and
climbing of dislocations. Under a tensile stress σ , the
process starts by the generation of dislocations at sources
after the passage of a characteristic waiting time requiring
the climb of a short segment a critical distance $\lambda \simeq (0.1)\,\mu b/\sigma$,
to become free from surrounding dislocations. The generated
dislocations move by a predominantly glide motion over a storage distance L that is often larger than subgrain dimensions, where they become trapped by other dislocations within
a trapping distance w . There they are eventually annihilated after some further climb toward the trapping dislocations of opposite sign, resulting in a mobile dislocation
density of

$$\rho_m = \frac{C_m}{b^2} \left(\frac{w}{L}\right) \left(\frac{\sigma}{\mu}\right)^2 \left(\frac{\chi}{\mu b}\right) ,$$ (2)

where C_m is a constant of order 500 and χ the stacking fault energy. The model considers that most of the time, τ_a, is spent by a dislocation in the various climb steps leading to eventual annihilation in which the velocity v_c of the dislocation is given by [9]

$$v_c = v_o \frac{\sigma}{\mu} = C_c \frac{D}{b} \left(\frac{\mu\Omega}{kT}\right) c_j \left(\frac{\chi}{\mu b}\right)^2 \left(\frac{\sigma}{\mu}\right) ,$$ (3)

where C_c is a numerical constant of order 10^3, σ is the climb producing normal stress acting across the half plane of the dislocation, and c_j is the jog concentration along the extended dislocation. This leads to an average velocity $\bar{v}$ during the life time of a dislocation:

$$\bar{v} = L/\tau_a \; ; \; \tau_a - \pi(1-\nu)w^2/2bv_o \; ; \; w = \mu b/4\pi\sigma .$$ (4a,b,c)

Apart from the specific dependence of the strain rate on the stacking fault energy given by Eqn.(1), which results in part from the form of the climb velocity of Eqn.(2) and in part from a factor $(\chi/\mu b)$ given in Eqn.(2) that gives a decreased source configuration probability discussed by Argon and Takeuchi [8], the form of Eqn.(1) with $m = 3$ is identical to the one introduced by Weertman [10] as the rational creep law.

Argon and Takeuchi [8] have proposed further that powers higher than 3 in the stress exponent of Eqn.(1) in subgrain forming alloys are due to dynamic internal stresses that re-sult from the bowing of the boundaries of such subgrains as they migrate under stress. According to this model, most sub-boundaries bow out under an applied stress to an ampli-tude y over the sub-boundary facet length of δ, as the sub-boundary nodes, subject to geometrical constraints, drag behind as shown in Fig. 1. Such flexing of sub-boundary

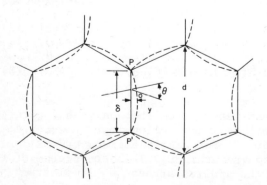

Fig. 1 Idealized hexagonal subgrains with randomly flexed sub-boundaries under stress.

4

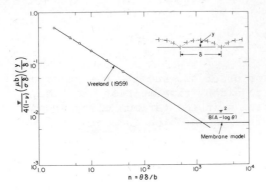

Fig. 2   Amplitude of sub-boundary flexure under stress for sub-boundaries containing small numbers of dis-locations, shown by the slanted line (from Argon and Takeuchi [8]).

facets, treated as a collection of discrete dislocations, has been studied by Vreeland [11], who finds, as shown in Fig. 2, that the amplitude of flexing of a sub-boundary with lattice misorientation θ under an applied stress σ is given by

$$\frac{y}{\delta} = 0.8 \cdot \frac{4(1-\nu)}{\pi} \left(\frac{\sigma\delta}{\mu b} \right) \left(\frac{b}{\theta\delta} \right)^{2/3} .$$ (5)

It is well known that when the plane of a low angle boundary is rotated from its lowest energy orientation, large and long range internal stresses are produced [12]. On this basis, Argon and Takeuchi [8] have associated the long range stresses measured in the stress dip experiments [13] to such rotations of sub-boundary planes and proceeded to calculate the ampli-tude σ_i of these randomly positive and negative stresses by treating the volume enclosed by the flexed lobes of sub-boundaries as if they were partially constrained shear trans-formations with transformation shear strains of θ. This has given

$$\sigma_i = (1/8) \, (\mu\theta) \, (y/\delta) ,$$ (6)

which upon substitution of Eqn.(5) and the use of the steady state relationship between subgrain size d and stress

$$d = 2\delta = K(\mu b/\sigma)$$ (7)

$$\sigma_i/\mu = (C_i(1-\nu)/\pi)(K\theta)^{1/3}(\sigma/\mu)^{2/3}; \quad (C_i = 0.317).$$ (8)

Takeuchi and Argon [3] have shown from the analysis of creep experiments of others that K varies from near 10 for close packed metals to near 60-70 for alkali halides and oxides, with an overall average being near 30. This dependence of the internal stress σ_i on the applies stress σ is plotted

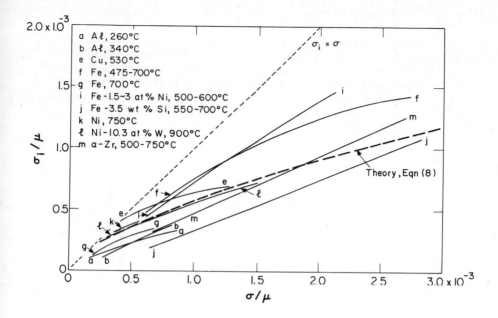

Fig. 3 Comparison of computed dependence of internal stress on applied stress, with the measured dependence for a variety of pure metals and Class II alloys (from Argon and Takeuchi [8]).

in Fig. 3 together with the internal stresses measured with the stress dip test in many pure metals and Class II alloys for a typical lattice rotation of $\theta = 1°$. The agreement between theory and the entire family of experimental results is good.

If the stress in the general expression for steady state creep given by Eqn.(1) with an exponent $m = 3$ is now interpreted to be the effective stress

$$\sigma_e = \sigma - \sigma_i \quad , \tag{9}$$

the overall creep rate should then be given by an expression of

$$\dot{\varepsilon} = A \left(\frac{\mu\Omega}{kT}\right) \left(\frac{\chi}{\mu b}\right)^3 \left(\frac{D}{b^2}\right) \left[\frac{\sigma}{\mu} - \frac{C_i (1-\nu)}{\pi} (K\theta)^{1/3} \left(\frac{\sigma}{\mu}\right)^{2/3}\right]^3 \tag{10}$$

where A is a numerical constant of the order of $(6 \times 10^6) c_j$ [8]. The stress dependence of this equation for $K = 30$ and $\theta = 1°$ is plotted in Fig. 4 over a decade of σ/μ in which most power-law creep experiments are conducted. Clearly,

6

as Eqn.(10) and the plot of Fig. 4 shows, $\dot{\varepsilon}$ exhibits a threshold behavior at a stress that makes the content of the last parenthesis vanish. We dismiss this behavior as unreal and expect that both grain-boundary sliding and Nabarro-Herring creep will add a substantial component of strain rate to the total in this low stress range. That this must be the correct explanation is shown from much of the actual internal stress measurements shown in Fig. 3 which indicate that at these low stress levels the internal stress makes up almost the entire applied stress leaving only the mechanisms that do not involve dislocation mobility to produce inelastic strain. The plot of Fig. 4 shows that in much of the useful range of stress $(10^{-4}-10^{-3})\mu$ the apparent stress exponent is in the neighborhood of 5, going from 6 to 4.5. This is the normal reported behavior. Much above a stress of $2 \times 10^{-3}\mu$, the curve goes to an asymptotic form of a power of 3 as $\sigma_i/\sigma \to 0$. In this range, however, additional low temperature processes involving thermally activated overcoming of slip obstacles can produce increasingly large components of strain rate which have been excluded from the model that gives rise to Eqn. (9). It is this additional component that results in the so-called power-law break-down behavior.

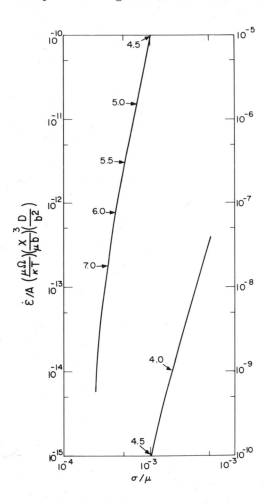

Fig. 4 Computed dependence of steady state creep strain rate on the applied stress. Numbers indicate the level of the local power law creep exponent (from Argon and Takeuchi [8]).

2.2 Structural Steady State

Although it has rarely been demonstrated unambiguously,
it has often been taken for granted that steady state creep
implies an invariant and statistically stationary microstructural state of dislocation density and subgrain size
etc., and that going from one steady state to another by increasing or decreasing the applied stress changes the dislocation structure reversibly to a smaller or larger scale
respectively. It is this reversible connection between subgrain size d and applied stress σ, expressed in Eqn.(7),
which found use in Eqns.(8) and (9), that permits the increase and decrease of the internal stress as well. Holt [15]
has proposed a model both for cell formation at low temperature and subgrain formation during creep that takes as an
initial condition a random collection of parallel dislocations having a high content of interaction energy and considers their segregation without loss or cancellation into
cell or subgrain walls as an analog to a spinodal decomposition. By dimensional analysis and scaling laws, this results
in a subgrain size that is proportional to the initial mean
dislocation spacing and hence proportional to the inverse of the
flow stress that presumably held the dislocations in equilibrium against their mutual interaction. The upper and
lower bound estimates to the lattice misorientation θ that
is consistent with Holt's model are

$$\theta_u = \sqrt{3}\ Kb\ \sqrt{\rho_i}\ ; \qquad \theta_\ell = (3^{1/4}/\sqrt{2})b\ \sqrt{\rho_i}\ , \qquad (11a,b)$$

which are obtained by accumulating all dislocations in a
lozenge shaped region into its short diagonal: first, without any loss of dislocations and assuming them all to be of
the same sign, and second, according to a random-error-sum
allowing for cancellations. For an initial dislocation density $\rho_i = 10^{14}m^{-2}$ b = 2.5 × 10^{-10}m , and K = 30 we estimate θ_u = 7.44° and θ_ℓ = 0.133° . Although these estimates comfortably bracket the actual experimental value of
θ = 1° , we find the basic unidirectional break-down of Holt's
model incompatible with the requirements of a dynamic steady
state. Instead we adopt as a working model the suggestion of
Prinz and Argon [16] that both cell formation at low temperature and subgrain formation during creep is an evolutionary
and statistical dislocation clustering process that should be
a natural ingredient of strain hardening. In the statistical slip plane hardening theory of Kocks [17], e.g., tight
loops are left around impenetrable regions in the slip plane
where the obstacle density is too high. These regions act as
nucleation sites of clusters and eventually as cell or subgrain walls. Thus, we expect that the storage length L is
related to the mean dislocation spacing λ by some

statistical proportionality factor K. At low temperature the cell walls where dislocations cluster are immobilized by the boundary debris consisting of many sessile dislocation segments and compensated prismatical loops that stitch the wall dislocations together [16]. At high temperature, however, the required non-conservative motion of the boundary debris is accomplished by diffusion and the sub-boundaries become mobile. Thus, a dynamic, structural steady state becomes possible by the coalescence and mutual annihilation of mobile sub-boundaries to compensate for the generation of new sub-boundaries by evolutionary clustering in the regions that have been previously cleared of boundaries. Below we give a rough outline of this complex balance, a fuller description of which will be published elsewhere later.

In such a dynamic, structural steady state, the dislocation density that enters into temporary storage at clustering locations a distance L apart, during an increment of strain is:

$$d\rho^+/d\varepsilon = 2/bL ; \quad L = K\lambda ; \quad \lambda = \rho^{-1/2} \quad (12a,b,c)$$

$$d\rho^+/dt = \dot{\varepsilon} \, d\rho^+/d\varepsilon ; \quad \dot{\varepsilon} = b \, v_g/Lw , \quad (13a,b)$$

where v_g is an average glide velocity of dislocations between generation at sources and trapping at eventual annihilation sites. The use of Eqns. (12b), (12c), (13b) and (14c) gives

$$d\rho^+/dt = (8\pi v_g \, \rho/bK^2)(\sigma/\mu) . \quad (14)$$

The rate of storage is counteracted by a mutual annihilation rate of dislocations during storage, given by

$$d\rho^-/dt = -\rho/\tau_a ; \quad \tau_a = 4(1-\nu)/2b \, v_o \rho \quad (15a,b)$$

where v_o is defined in Eqn. (3). Thus, the net dislocation storage rate at cluster nuclei is given by

$$d\rho/dt = \alpha\rho - \beta\rho^2 = (8\pi \, v_g \, \sigma/\mu bK^2) \, \rho - (2bv_o/\pi(1-\nu))\rho^2, \quad (16)$$

where the abbreviations α and β are defined by the equation. We find immediately that the steady state density ρ_{ss} located primarily at clusters is given by

$$\rho_{ss} = \alpha/\beta = (4\pi^2 (1-\nu)/b^2K^2)(v_g/v_o)(\sigma/\mu) . \quad (17)$$

For a lattice misorientation θ this defines immediately the subgrain size

$$d = 2\theta R/b\rho = (R\theta K^2/2\pi^2(1-\nu))(v_o/v_g)(\mu b/\sigma) , \qquad (18)$$

where R is a redundancy factor giving the ratio of the total dislocation line length in the sub-boundary (geometrically necessary plus redundant dislocations and debris) to the line length of the geometrically necessary dislocations responsible for the lattice rotation θ . The solution of Eqn.(16) for the dislocation density $\rho(t)$ for an initial condition of ρ_o at $t = 0$, and a final steady state density given by Eqn.(17) is

$$\rho/\rho_{ss} = [1 + ((\rho_{ss} - \rho_o)/\rho_o) \exp (-\alpha t)]^{-1} , \qquad (19)$$

giving a time constant τ_{bu} for the build-up of a steady state, which is

$$\tau_{bu} = bK^2\mu/8\pi v_g \sigma . \qquad (20)$$

In Eqns.(14-20) for simplicity we assumed that σ is constant and equal to the applied stress. In a more accurate treatment it is necessary to allow for a gradual build-up of an internal stress as the sub-boundaries take shape, flex, and produce the internal stresses that we discussed in Section 2.1 above. For example, the time constant τ_{bu} itself will increase with sub-boundary evolution as $\sigma \rightarrow (\sigma-\sigma_i)$. Thus, the initial sub-boundary evolution will be rapid in regions where sub-boundaries had been previously annihilated but will slow down later until at a structural steady-state τ_{bu} must become equal to the time constant τ_{sba} of sub-boundary annihilation,

$$\tau_{bu} = \tau_{sba} \equiv d/2 v_b \qquad (21)$$

where v_b is the sub-boundary migration velocity under stress. This gives another expression for the sub-grain size

$$d = (bK^2/8\pi)(v_b/v_g)(\mu/\sigma) . \qquad (22)$$

Comparison of this sub-grain diameter with the previous estimate given by Eqn.(18) gives the lattice rotation

$$\theta = (\pi(1-\nu)/4R)(v_b/v_o) . \qquad (23)$$

Finally, Eqn.(22) together with Eqn.(7) gives for the proportionality constant K in Eqns.(7), and (12b)

$$K = 8\pi(v_g/v_b) \quad . \tag{24}$$

Inspection of Eqns.(21-24) shows that the sub-boundary mobility is a key factor in the structural steady state.

In a series of detailed in-situ creep studies of aluminum, Caillard [18] has found that at high temperature, well structured sub-boundaries, having little or no boundary debris, can be displaced by glide alone without any internal rearrangement in a unique direction $\vec{n}$ that is parallel to the line of intersection of the two planes defined by the pairs of Burgers vectors and line vectors, $\vec{b}_1$, $\vec{s}_1$, and $\vec{b}_2$, $\vec{s}_2$ of the sub-boundary dislocations, in the majority of sub-boundaries where the lattice misorientation θ is governed by two sets of geometrically necessary dislocations. This gives

$$\vec{n} \ // \ (\vec{b}_1 \times \vec{s}_1) \times (\vec{b}_2 \times \vec{s}_2) \ , \tag{25}$$

where the vectors $\vec{s}_1$ and $\vec{s}_2$ are the line direction vectors used in the definition of the Burgers vectors $\vec{b}_1$ and $\vec{b}_2$ by the conventional right-handed line integration process. We assume that this is a general possibility, and that the geometrically necessary line content of most sub-boundaries can be displaced in a unique direction by glide alone. In a creeping alloy actual sub-boundaries will always have a substantial debris content that can translate only by climb and which at steady state is replenished as rapidly as it "evaporates". Since this boundary debris is attached to the geometrically necessary and redundant dislocations it needs to be dragged along in the direction $\vec{n}$ prescribed by the geometrically necessary dislocations of the boundary. Thus, for the purpose of computing the net velocity v_b of the sub-boundary it is appropriate to consider that the steady state density of the boundary debris must undergo a virtual translation, largely by climb, in the direction $\vec{n}$ through the forces applicable only to the geometrically necessary dislocations by the applied stress. Thus, we obtain

$$v_b = v_c \ \rho_{gn}/\rho_d \ \sin\psi = v_o(\sigma/\mu \ R_d \ \sin\psi) \tag{26}$$

where ψ is the average angle between the Burgers vector of the debris dislocations in the sub-boundary and the direction $\vec{n}$. In Eqn.(26) R_d is the ratio of the line length of debris dislocations ρ_d , translatable only by climb, to the line length of the geometrically necessary dislocations ρ_{gn} . The line length of redundant dislocations – being compensated plus and minus pairs of the same type as the geometrically necessary dislocations - will not interact with the applied stress but will also have only negligible glide resistance. Incorporation of Eqn.(26) into Eqn.(21), and (23) give for the time constant τ_{sba} for the change of a structural steady state upon change of stress and for the lattice misorientation

$$\tau_{sba} = Kb\ R_d\ \sin\psi/\ 2\ v_o(\sigma/\mu)^2 \qquad (27)$$

$$\theta = \pi(1 - \nu)\sigma/4\mu R\ R_d\ \sin\psi\ . \qquad (28)$$

In Eqn.(27), σ is to be interpreted as the final stress at which steady state is sought, and τ_{sba} is to be taken in this form only if it is shorter than the static recovery time τ_a given by Eqn.(15b).

2.3 Transients

When stress is increased or decreased during creep in a sub-grain forming metal or alloy, prominent transients occur. The transients that have attracted most attention recently are those occurring in reverse creep upon complete removal of the applied stress [19,20], and the structural change transient occurring in the reaching of a new steady state upon partial removal of the applied stress. Gibeling and Nix [19] have observed two different forms of behavior during reverse creep upon complete removal of the stress. During the first phase after removal, the reverse strain rate is a power function of the current remaining component of the total recoverable strain with a power roughly equal to the stress exponent of the preceding steady state creep under stress. In the last stages of reverse creep, however, this dependence becomes linear. We interpret the observations of Gibeling and Nix, that the same time consuming processes of glide and climb must be occurring during the early phases of reverse creep as in the forward creep process, but now under the long range internal back stresses set up by the flexed sub-boundaries. The later stages of reverse creep, linear in the remaining recoverable strain, is then interpreted to be due to the straightening-out of the flexed sub-boundaries, by climb of the attached sub-boundary debris. Thus, we estimate the recoverable non-linear and linear components of the reverse strain to be, respectively,

$$\varepsilon_{rn} \simeq b\ \rho_m\ L = (C_m/4\pi)(\sigma/\mu)(\chi/\mu b)\ , \qquad (29)$$

$$\varepsilon_{r\ell} \simeq \sigma_i/\mu = (C_i(1-\nu)/\pi)(K\theta\mu/\sigma)^{1/3}\ . \qquad (30)$$

For appropriate values of the physical constants, and the numerical constants C_m and C_i given earlier, we find that while $\varepsilon_{rn}/\varepsilon_e$ is 4.12 for Al , 0.64 for Cu , and 0.83 for Pb ; the ratio $\varepsilon_{r\ell}/\varepsilon_e$, due to straightening of flexed sub-boundaries is 0.57 for all alloys [8]. These values are satisfactorily close to those actually measured by Gibeling and Nix.

Another transient of interest is that in reaching a new structural steady state upon partial stress removal. According to the model of subgrain boundary evolution and annihilation described in Section 2.2, we expect that the time constant for reaching the new steady state should be the shorter of τ_a or τ_{sba} . For this purpose, we calculate the ratio of

$$\tau_{sba}/\tau_a = K R_d b^2 \rho \sin\psi / 4(1-\nu)(\sigma/\mu)^2 . \tag{31}$$

Considering that R_d is likely to be considerably less than 0.1, that $K \approx 30 \sin \psi$ 0.5 , and that $\rho (\mu b/\sigma)^2$ is usually around unity [3] in creep, we find the above ratio to be less than 0.5 , and conclude that sub-boundary migration and annihilation should always be the governing mode of recovery in reading a new steady state under partially reduced stress.

2.4   Content of Subgrain Boundaries

As we discussed in Section 2.2 above, considerable knowledge is necessary about the dislocation and debris content of sub-boundaries to determine their mobility under stress. Such determinations are usually quite painstaking. In the procedure that we have developed, sub-boundaries in a crept sample are viewed in a high voltage electron microscope first at room temperature, and preferably with the weak beam technique, to obtain a high resolution record of the entire dislocation content of the boundary in the as-crept state. The same sub-boundary is then observed and photographed at gradually increasing temperatures. This establishes a record and the kinetics of the stepwise annihilation of the redundant dislocation content and additionally reveals the various pinning debris that controls the mobility of the movable sub-boundary dislocations during creep. Finally, after prolonged annealing what remains is the geometrically necessary dislocation content to give the lattice misorientation. A full description of these procedures and results will be published by us elsewhere. A similar series of observations of such in-situ annealing in N_i single crystals deformed at room temperature to a flow stress of 118 MPa is shown in Fig. 5 below. The sequence from a to c shows the stepwise clean-up of the sub-boundary during annealing. Many small tight debris loops visible in frame a have disappeared in frame b. In region C

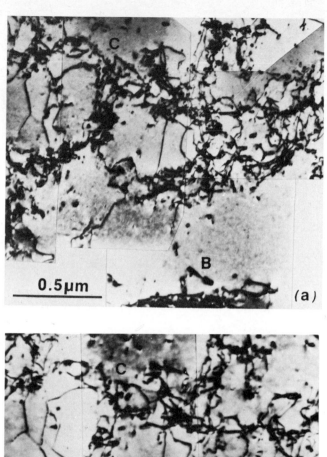

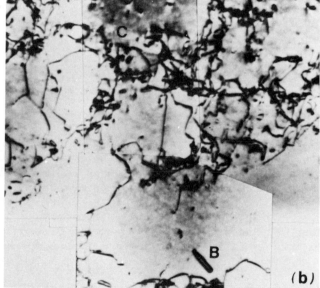

Fig. 5   A sequence of in-situ annealing observations in
a Ni single crystal deformed at RT to a flow
stress of 118 MPa:
a)   total dislocation content of boundaries viewed
at RT;   b) Same area after 1200 sec. at 427° C
$(0.4\ T_m)$,   c)   Same area after an additional
1200 sec at 602°C $(0.5\ T_m)$, $g = <1\ \bar{1}\ \bar{1}>$.

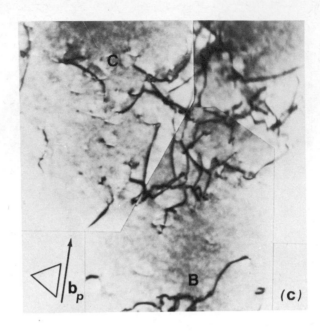

of frames a and b the release of a redundant dislocation
by the disappearance of some pinning debris is clearly visible. This redundant dislocation segment has completely
disappeared in frame c where only the geometrically necessary dislocation content remains. In region B of frames a,
b, and c, a prismatic dislocation loop is seen to disappear.

3. DISCUSSION

We have presented a picture of steady state creep in pure
metals and Class II alloys in which the evolution, migration
and annihilation of sub-boundaries plays an important role.
We have proposed that internal stresses measured by the
stress dip test result from the flexure under stress of sub-
boundaries, and that these internal stresses make a normally
third power dependence of the creep strain rate on stress
appear to have a higher power in the neighborhood between
4 to 6. We find good agreement between the calculated levels
of the internal stress and those measured by the stress dip
tests by many investigators in many close packed pure metals
and Class II solid solution alloys. We have presented, further, an outline of the model of structural steady state in
which sub-boundaries form by an evolutionary clustering process that compensates for the annihilation of sub-boundaries
resulting from their migration and coalescence. We find that
this model gives us reasonable magnitudes for the structural
parameters and satisfactorily predicts the parameters of some
of the important transients. We expect to report on a more
detailed description of these processes in the near future.

Acknowledgement. This research was supported by the U.S. Army Research Office under Contract No. DAAG29-78-C-0014.

References

1. BIRD, J. E., MUKHERJEE, A. K., and DORN, J.E., in "Quantitative Relations Between Properties and Micro-structure", Eds. Brandon, D. G., and Rosen, A., Israel Univ. Press, Jerusalem, 1969, p. 255.
2. SHERBY, O. D., and BURKE, P. M., "Mechanical Behavior of Crystalline Solids at Elevated Temperature," Prog. Mater. Sci., 1967, 13, 325.
3. TAKEUCHI, S. and ARGON, A. S., J. Mater. Sci., 1976, 11, 1542.
4. NIX, W. D., and ILSCHNER, B., in "Strength of Metals and Alloys," Eds. Haasen, P., Gerold, V., and Kostorz, G., Pergamon Press, Oxford, 1980, vol. 3, p. 1503.
5. BAILEY, R. W., J. Inst. Met., 1926, 35, 27.
6. OROWAN, E., J. West Scotl. Iron Steel Inst., 1946-47, 54, 45.
7. MOHAMED, F. A., and LANGDON, T. G., Acta Met., 1974, 22, 779.
8. ARGON, A. S., and TAKEUCHI, S., submitted to Acta Met.
9. ARGON, A. S., and MOFFATT, W. C., Acta Met., in the press.
10. WEERTMAN, J., Trans. ASM., 1968, 61, 680.
11. VREELAND, T., Jr., Acta Met., 1959, 7, 240.
12. OROWAN, E., Nature, 1942, 149, 643.
13. AHLQUIST, C. N., and NIX, W. D., Script. Met., 1969, 3, 679.
14. MOTT, N. F., and NABARRO, F.R.N., in "Report on Strength of Solids," Physical Society, London, 1948, p. 1.
15. HOLT, D. L., J. Appl. Phys. 1970, 41, 3197.
16. PRINZ, F., and ARGON, A. S., Phys. Stat. Sol., 1980, 57, 741.
17. KOCKS, U. F., Phil. Mag., 1966, 13, 541.
18. CAILLARD, D., "Etude de la Microstructure de Fluage, dans l'Aluminium aux Temperatures Moyennes", D. Sc. Thesis, Univ. Paul Sabatier, Toulouse, France, 1980.
19. GIBELING, J. C., and NIX. W. D., submitted to Acta Met.
20. PARKER, J. D., and WILSHIRE, B., private communication, to be published.

A MICROSCOPIC APPROACH OF THE ALUMINIUM CREEP RATE
AT INTERMEDIATE TEMPERATURE

D. Caillard and J.L. Martin[1]

Laboratoire d'Optique Electronique du CNRS, BP4347,

31055 Toulouse-Cedex, France

ABSTRACT

High voltage electron microscopy and weak beam observations at 100 kV are used to analyse the microstructure properties in Aluminium samples after creep in the temperature range
80-200°C. The subboundary properties are described in the load
applied state, as well as their interactions with mobile dislocations. The crossing through subboundaries by gliding defects
appears to be the rate controlling process of creep under these
conditions.

1. INTRODUCTION

In a recent review on the fundamental properties of
dislocations, high voltage electron microscopy (HVEM) is considered as "a new technique for old problems"[1]. Indeed the
higher penetration of electrons through crystals allows more
realistic in situ experiments to be performed, and microscopic
mechanisms directly observed under these conditions are probably more representative of the bulk materials than in 100 kV
microscopes.

The "old problem" investigated here relates to the microstructure properties and its evolution under load, corresponding to the so-called "constant strain rate" stage of creep, at
intermediate temperatures. (In the following, we will refer to
it as stage II). The material investigated is Aluminium in the
temperature range 80°C-200°C. A special attention is paid to
the role and properties of subboundaries after the creep test
and under load, and to their interactions with mobile dislocations, in an attempt to correlate such microscopic observations

(1) Now at Laboratoire de génie atomique, EPFL,
    33, av. de Cour, CH-1007 Lausanne, Switzerland

with the creep rate and its dependence on temperature and stress. The intermediate temperature range is only studied at the moment, since it is usually considered to involve a different strain rate equation with a well-defined activation energy (2) (3).

2. EXPERIMENTAL PROCEDURE

The polycrystals of Aluminium used are 99.3% pure, of a few 100 μm average grain size and have been creep tested in tension between 0.3 and 0.5 T_M, under stresses of 30 to 80 MPa. The test is interrupted in stage II when the subgrain structure is present; and the samples are cooled down under load to prevent as far as possible any rearrangement of the substructure[1]. Rectangular microsamples (3000x1000x50 μm^3) are then cut out of the macroscopic ones with the same tensile axis, and subsequently thinned. We think that the substructure does not change substantially during thin foil preparation, since 1) we never observed any dislocation traces at the foil surfaces in newly prepared samples, 2) subboundaries situated approximately in the foil plane, are often present, 3) dislocation or subboundary motions are only observed during in situ experiments under the action of stress and temperature.

Two different types of experiments are then performed:

2.1 Dynamic observations in the HVEM

These are carried out at the same temperature as that of the macroscopic creep test and at a deformation rate which is as close as possible to the one recorded during stage II. Changes in the microsctructure under load are observed in areas in which the thin edges are parallel to the tensile axis, so that the local stress is approximately a tensile one.

The technique of in situ straining in the HVEM has its own limitations. Although insufficient data are available to estimate them quantitatively, they have been recently analysed (4,5) and consist mainly of radiation damage and surface effects. Nevertheless, since the first investigations of metallurgical problems (6), many studies of crystal plasticity have been successfully performed (7). In the present investigation range, the subgrain size is of a few microns, and comparable with the observable foil thickness, so that interaction of dislocations and subboundaries are more frequently observed than with the foil surfaces. With regard to radiation damage, we have never

(1) The samples were supplied and creep tested by Dr.Myshlyaev, Institute of Solid State Physics, 142432 Chernogolovka, USSR, who is greatfully acknowledged here.

seen evidence of any loop condensation or slowing down of the observed processes as the experiment proceeds. It is believed that, at the temperatures of interest here, mutual recombination of irradiation point defects at foil surfaces is efficient and ensures a concentration of these which is smaller than the thermal equilibrium one.

The deformation stage we used can be employed between 20°C and 700°C and is described in (8). The low deformation rates (10^{-7} s^{-1} < $\dot{\varepsilon}$ < 10^{-4} s^{-1}) are achieved via a magnetostrictive device which controls the sample elongation. Quantitative measurements can be performed locally:

- The strain is estimated through the areas A_i swept by the n_i mobile defects of Burgers vectors b_i using the well-known formula: $\varepsilon = \sum_i n_i A_i b_i/V$, where V is the crystal volume corresponding to the observed area.

- The stresses can roughly be measured through the smallest radii of curvature of the dislocation loops.

In addition, the observation and analysis of dislocation traces at the foil surfaces have proved to be useful in several cases. As shown below, it can tell whether the defect is gliding, cross sliping or climbing. Such traces were noticed in the early thin foil observations (9). They usually exhibit the same visibility conditions as the corresponding dislocation and they can also anneal out, at different rates (depending on the metal and the temperature). When the defect motion is too fast to be recorded (frame rate 25 im.s^{-1}), these traces can be used to identify its passage through the crystal: i) during glide they are straight and parallel to the intersection of the slip plane with the foil plane and the type and indices of the slip plane can be unambiguously determined, ii) If cross slip occurs, the traces are usually of the form shown in Fig.1 a) and b), i.e. are separated by a translation $\vec{B}$ parallel to the Burgers vector; if cross slip is very frequent, they can look like Fig.1 c) and d), where the $\vec{B}$ vector can also be determined, iii) if climb occurs, both traces are wavy again but cannot be connected by a translation: on Fig.1e), a dislocation with Burgers vector parallel to $\vec{B}$ is gliding upwards (slip traces h and b) and cross slips from a {111} plane on to another one, and then suddenly climbs (traces h' and b'). As explained below, we have observed and analysed dislocation traces connected to the motion of individual defects or subboundaries.

2.2 Static observations of the substructure after creep, at 100 and 1000 kV

B

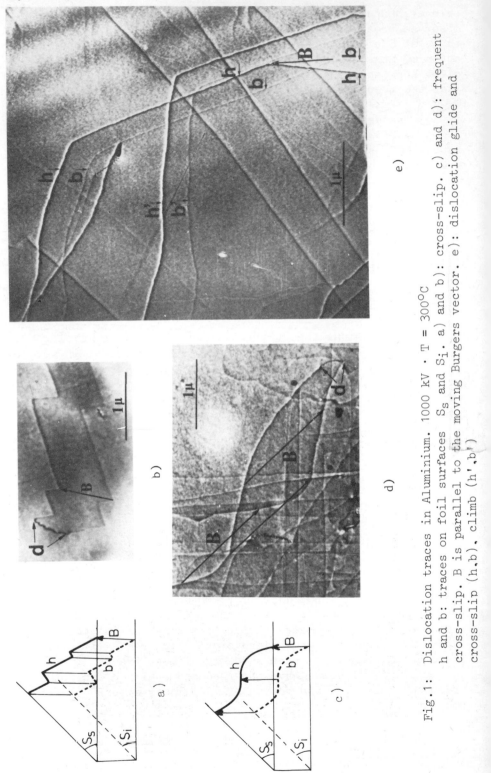

Fig.1: Dislocation traces in Aluminium. 1000 kV · T = 300°C
h and b: traces on foil surfaces S_s and S_i. a) and b): cross-slip. c) and d): frequent
cross-slip. B is parallel to the moving Burgers vector. e): dislocation glide and
cross-slip (h,b), climb (h',b')

2.2 Static observations of the substructure after creep, at 100 and 1000 kV

They are especially useful in the accurate determination of the subboundary geometry. The weak beam technique (10) is used at 100 kV and has proved to be very helpful in the analyses of 100 Å meshes (corresponding misorientation approximately 1°). An image intensifier facilitates focussing. At 1000 kV, thicker foils are observed (up to 3 µm retaining a satisfactory resolution) which provides a better three dimensional picture of the substructure.

3. MAIN OBSERVATIONS

These refer to the movement of individual defects, the properties of subboundaries under stress and the interactions between these two.

3.1 Individual dislocations

A study of the dislocation traces through the subgrains reveals that they move by glide and cross-slip. The latter phenomenon often occurs when the moving defect meets another one, involving probably the Washburn mechanism (11) (12). Unusual slip planes are also frequently observed at such temperatures (13), including {110}, {100} and {112}. An example of cross-slip from {111} on to {112} is illustrated in Fig.2. Apparently, the local stresses are high enough for these processes to occur quite easily and a dislocation moves over the whole subgrain within 1/25 s or less.

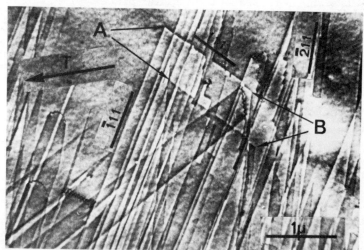

Fig.2: Cross-slip on a 112 plane. Al 300°C, 1000 kV. From {111} on to {111} in A, from {111} on to {112} in B. T: tensile axis

3.2 Subboundaries

a) Static properties

The first detailed observation of subboundaries by electron microscopy can be found in (14), and a description of creep subboundaries in (15). An analysis of their geometry has been undertaken (16) in conventional thin foils using the weak beam technique. Several types of subboundaries are observed and the most frequent ones exhibit three Burgers vectors at 120°. An example is shown on Fig.3; it lies approximately on a

Fig.3: Subboundary with 3 Burgers vectors at 120°. (Al after a creep test at 200°C) 100 kV a) Dark field b) Bright field c) Weak beam picture d) Stereographic projection. x_i, b_i, P_i refer to dislocation directions, Burgers vectors and slip planes for each family respectively.

Fig.3 d)

{100} plane, the Burgers vectors plane being of the {111} type and the respective densities of dislocations X_1, X_2 and X_3 are such that the Frank criterion is fulfilled, within the measurement accuracy. In addition, the dislocation segments in the network lie in their respective glide planes (16).

Another type of subboundaries is shown on Fig.4, consisting of two orthogonal screw dislocation families on a {100} plane (pure twist). Since this is a "thick" foil, subgrain dislocations superimpose their image to the subboundary one.

An analysis of the dislocation densities in the subboundaries, (16)(17), shows that only a few of the six possible Burgers vectors are present (two or three), indicating that the local stress activates only some of the six possible types of sources. (See an example on Fig.5)

All these features limit the number of possible creep subboundaries.

b) Migration of subboundaries

This was first observed by (18) during macroscopic creep tests. It also takes place during in situ experiments, and analysis of the dislocation traces behind the moving sub-

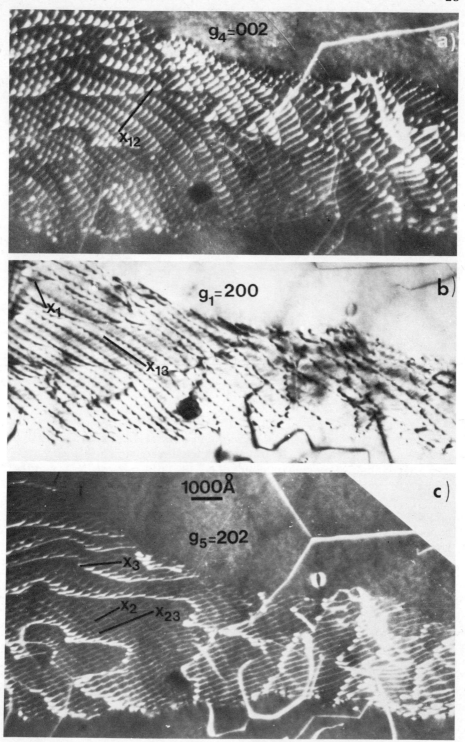

Fig.3: a), b), c)

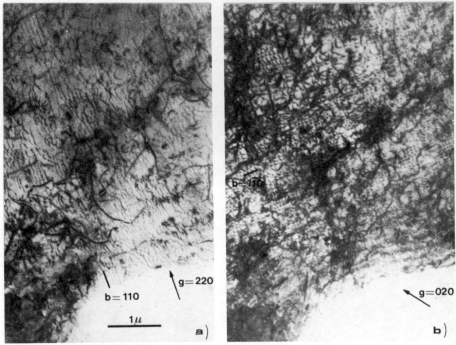

Fig.4: Pure twist subboundary in a {100} plane.
Same experiment as Fig.3. Bright fields.
a) one dislocation family is visible
b) both families are visible

boundary indicates a glide phenomenon. Several examples are shown in (16) and some in (19). In agreement with (18), it was found that the strain connected to this migration is only a small fraction of the total strain.

In addition, the triple junctions at intersecting sub-boundaries can also migrate, as illustrated in Fig.6. This involves a continuous movement of subboundary dislocations

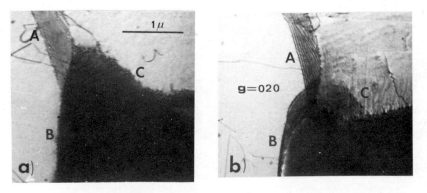

Fig.6: Migration of a triple junction. 200°C. 1000 kV
A,B,C:intersecting subboundaries

Fig.5: Subgrains after a macroscopic creep test
at 200°C (Montage). Weak beam picture
100 kV. Subboundaries have three Burgers
vectors at 120° which are the same all
over the area

through the triple junctions. Consequently, subboundary bowing
under the applied stress does not occur to a significant extent
so that the long range stress field remains small, at least
smaller or equal to the applied stress.

Subboundaries can also be destroyed and rebuilt (19),
but seldom enough to contribute significantly to strain (16).

The phenomena described under this heading probably play
an important role in the substructure evolution during transients on the creep curve.

c) Interaction of subboundaries and dislocations
Mobile dislocations can annihilate with subboundary
dislocations. An example is shown on Fig.7. The subboundary
(Fig.7a) exhibits three dislocation families with Burgers
vectors at 120° (like Fig.3). A short time later (Fig.7b),
dislocations were gliding towards the subboundary (corresponding slip traces are b and h) and A B is at the intersection of
the slip plane and the subboundary is sketched in Fig.7d.

On Fig.7b the meshes in the subboundary plane are distorted along A B and looser than on Fig.7a, i.e. the incident
dislocations have annihilated with network dislocations, having
opposite Burgers vectors. Therefore the Frank criterion is no
longer satisfied in the A B area, where an important long distance stress field is built up. This can however be avoided
by a change in shape of the subboundary which occurs by migration (Fig.7c and sketch on Fig.7d). A step is produced at the
subboundary surface.

Another situation is observed frequently where the
gliding dislocations are crossing through the subboundary.
This process has been described in (17) and especially in (16).
It probably occurs through a complex cross-slip mechanism which
involves significant waiting times, and which appears to be the
rate controlling process of creep, at least under these conditions (16). Since the subboundaries with three Burgers vectors
at 120° are the most frequent ones, a special study of their
interactions with mobile dislocations has been undertaken (16)
(20). (An example of subboundary crossing can be seen on Fig.8)

The frequent observation of subboundary crossing by
mobile dislocations is in agreement with more macroscopic
observations which indicate that the slip line length is larger
than the subgrain size (21).

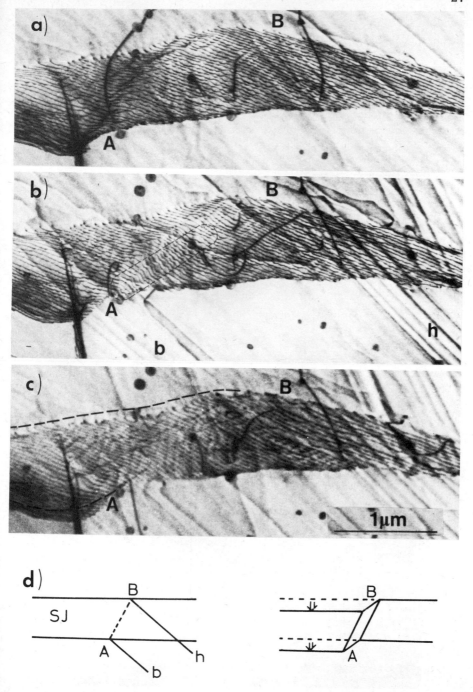

Fig.7: Dislocation annihilation in a subboundary. In situ
       creep test at 200°C. 1000 kV. a) initial state
       b) foil slip traces in b h , sub^y slip trace along A B.
       c) change in sub^y shape d) corresponding schemes

4. CONCLUSION

Electron microscope observations, both conventional and dynamic, have given a realistic picture of creep at intermediate temperatures. The subboundary structure migrates with mobile triple junctions, by glide of their individual dislocation segments. Subboundary migration and construction are the dominant phenomena during transients on the creep curve. Analysis

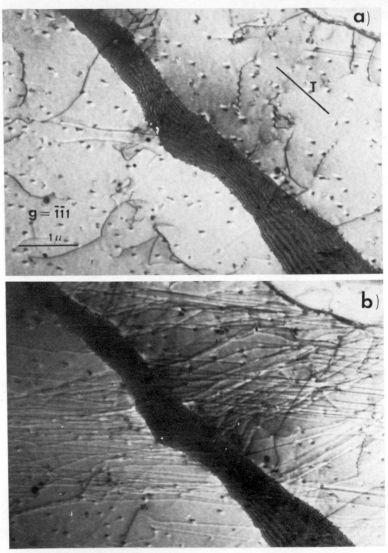

Fig.8: Sequence of dislocation cutting through
subboundaries. In situ creep test at
200°C a) initial state b) About 20 dislocations have moved through the area.
T tensile axis.

of the subboundary geometry reveals no significant long range stresses in the crystal and the networks exhibiting three Burgers vectors at 120° are the most numerous. The movement by glide and cross-slip of individual dislocations accounts for most of the total strain. They can annihilate with sub-boundary defects, but also cross through them. The latter phenomenon also involves cross-slip and is probably the rate controlling mechanism.

REFERENCES

1. FRIEDEL, J. - Dislocations in Solids, Ed. Nabarro, F.R.N., North Holland Publ., 1979, p.26

2. DORN, J.E., SHERBY, O.D. and LYTTON, J.L. - Acta Met., 1957, 5, 219

3. NIX, W.D. and ILSCHNER, B. - Proceedings of ICSMA V, Ed. Haasen P. et al., Pergamon Ltd 1979, 3, p.1503

4. MARTIN, J.L. and KUBIN, L.P. - Ultramicroscopy, 1978, 3, 215

5. MARTIN, J.L. and KUBIN, L.P. - Phys. Stat. Sol.a), 1979, 56, 487

6. SAKA, H., IMURA, T. and YUKAWA, N. - J. Phys. Soc. Japan, 1968, 25, 906

7. KUBIN, L.P. and MARTIN, J.L. - Proceedings of ICSMA V, ibidem, 1979, 3, p.1639

8. VALLE, R. and MARTIN, J.L. - Electron Microscopy 1974, Ed. Sanders, J.V. et al., Austr. Acad. of Sc. Publ. 1974, 1, p.180

9. HIRSCH, P.B. J. of Inst. of Metals, 1959, 87, 406

10. COCKAYNE, D.J.H., RAY, I.L.F. and WHELAN, M.J. - Phil. Mag., 1969, 20, 1265

11. WASHBURN, J. - Appl. Phys. Letters, 1965, 7, 183

12. CAILLARD, D. and MARTIN, J.L. - Proceedings of ICSMA IV, Ed. Champier G. et al., ENSMIM Publ. 1976, 1, p.105

13. CAILLARD, D. and MARTIN, J.L. - Microscopie Electronique à Haute Tension 1975, Ed. Jouffrey, B. and Favard, P., S.F.M.E. Publ., 1976, p.305

14. AMELINCKY, S. and DEKEYSER, W., Solid State Phys., 1959, 8, 325

15. BALL, C.J. and HIRSCH, P.B. - Phil. Mag. 1955, 46, 1343

16. CAILLARD, D. - Doctorate Thesis, Toulouse 1980

17. CAILLARD, D. and MARTIN, J.L. - Proceedings of ICSMA V, Ididem, 1979, $\underline{2}$, p.1323

18. EXELL, S.F. and WARRINGTON, D.H. - Phil. Mag. 1972, $\underline{26}$, 1121

19. MYSHLYAEV, M.M., CAILLARD, D. and MARTIN, J.L. - Scripta Met. 1978, $\underline{12}$, 157

20. To be published

21. PARKER, J.D. and WILSHIRE, B. - Phil. Mag. 1980, $\underline{A41}$, 665

DEFORMATION AND DISLOCATION BEHAVIOUR IN METALS AND SINGLE-PHASE ALLOYS AT ELEVATED TEMPERATURES

S. Karashima

Department of Materials Science, Faculty of Engineering, Tohoku University, Sendai 980, Japan

Summary

Experimental results obtained in our laboratory on the shapes of primary creep curves, dislocation structures, internal stresses and transient behaviours following stress relaxation and stress change tests during creep are reviewed. These results show that it is reasonable to classify the creep behaviours in pure metals and single-phase alloys into pure metal-, Class I alloy-, Class Ia alloy- and Class II alloy-types. Dislocation behaviours during creep and creep mechanisms are discussed. It is made clear that pure metals and Class I alloys show the creep behaviours which are characteristic of the cases where creep is controlled by recovery and by viscous glide of dislocations, respectively. It is also shown that creep behaviours in Class Ia and Class II alloys are intermediate between those in the typical cases mentioned above. The origin of internal stress is also discussed.

I. INTRODUCTION

In high temperature creep the steady-state creep rates, $\dot{\varepsilon}_S$, are usually expressed by the following equation :

$$\dot{\varepsilon}_S \propto (\sigma_c/E)^n \exp(-Q_c/RT) \qquad (1)$$

where σ_c is the creep stress and E the Young's modulus, and R and T have their usual meanings. It is well established that the activation energy for creep, Q_c, is equal to that for diffusion, Q_d, in many metals and alloys. Although this fact suggests that high temperature creep deformation is somehow related to the diffusion of constituent atoms, it is not possible to identify the creep mechanism only by it.

On the other hand, it is widely accepted that the stress exponent, n, is equal to about 5 in pure metals and some alloys, while it is about 3 in the other alloys [1]. Considering that the creep mechanisms are reflected in the values of n, alloys are often classified into Class I and Class II alloys [2].

Quite recently a very excellent review paper has been published on high temperature creep [3]. Creep behaviours are divided into the following three categories in it. Namely, 1) pure metal creep, in which the dislocation substructure plays a dominant role, 2) Class I alloy type creep (for ex. Al-Mg alloys), in which solute effects dominate the creep resistance, and 3) Class II alloy type creep which shows an intermediate behaviour.

As shown later, however, it is more reasonable to classify the alloys which show intermediate behaviours further into two items as follows. Creep deformation of some alloys (for ex. Fe-Mo alloys) is controlled by viscous glide of dislocations dragging solute atoms with them, though dislocation substructure is developed during creep. In the other alloys (for ex. Cu-Al alloys) creep is considered to be controlled by recovery just like in pure metals, in spite of the fact that dislocations move in viscous manner in them. They will tentatively called Class Ia alloy type and Class II alloy type hereafter, respectively.

The present paper aims at summarizing our experimental results obtained mainly on high temperature creep of metals and single-phase alloys. Some characteristic features of various types of creep behaviours in alloys as well as those in pure metals will be described first. Correctly speaking, the classification is related not to materials but to creep behaviours [4]. Nevertheless, the terms such as Class I alloys are used in this paper, for the sake of convenience, according to their creep behaviours shown under typical creep conditions which give the steady-state creep rates of $10^{-4} - 10^{-7}/s$.

2. INSTANTANEOUS PLASTIC STRAIN AND PRIMARY CREEP CURVE

When stress is applied to metallic materials occurs generally instantaneous plastic strain, which is followed by time-dependent creep deformation. However, in Class I and Class Ia alloys in particular the former is usually very small as compared with that in pure metals.

Primary creep curves take various shapes according to materials and creep conditions. In pure metals the initial creep rate is very high and the strain rate falls in the course of primary creep, eventually reaching a constant or steady-state value [5]. This will be called normal (N-type) primary creep curve.

Class I alloys behave in a very different way to show

inverse (I-type) and sigmoidal (S-type) primary creep curves
[6].   In the former, which appears in the case of low stresses,
the creep rate gradually increases to reach a constant value.
On the other hand, at higher stresses obtained is the latter
behaviour, in which it increases at first and then decreases
during primary creep.

As shown in Fig.1 I-type, linear (L-type),S-type and N-
type primary creep curves are obtained in this order as the
creep stress is increased in Fe-Mo (Class Ia) alloys [7], which
were listed in Class I alloys in our previous papers.   In L-
type behaviour the creep curve is apparently linear throughout.
S-type primary creep curves are generally obtained in Cu-Al
alloys with more than 10 at% Al, which belongs to Class II
alloys [8].

The appearance of primary creep curves other than N-type
seems to be characteristic of all alloys.   This feature is
more conspicuous as the creep temperature is increased or the
creep stress is decreased.

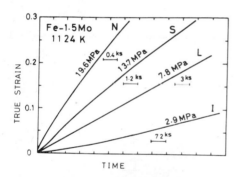

Fig.1 Various types of creep
curves in Fe-15 at% Mo
alloy [7].

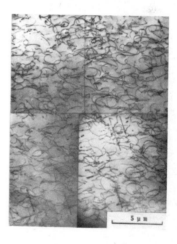

Photo.1 Dislocation
structure at steady-state
creep of Al-5.5 at% Mg
alloy [9].

3. DISLOCATION STRUCTURE

Although dislocations are uniformly distributed during
creep of Class I alloys [9], dislocations generally tend to be
localized to form subgrain and cell structures in pure metals
[10] and in Class II alloys [8], respectively.   Hereafter,
both structures will be called cell structure making no
distinction between them.   In Class Ia alloys cell formation
is also reported [11].

As a typical example of Class I alloys the dislocation
structure after creep in an Al-5.5 at% Mg alloy is reproduced
in Photo.1 [9], which clearly shows that the dislocations are
distributed uniformly even at the steady-state.   In the case
of uniform dislocation arrangement, all the dislocations are
considered mobile.   In fact, the change in dislocation density
corresponds quite well with that in creep rate during primary
creep as shown in Fig.2 [12].

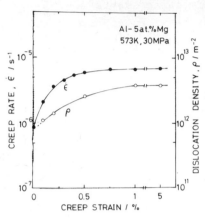

Fig.2 Creep rates and
dislocation densities in
primary creep of Al-5.5
at% Mg alloy [12].

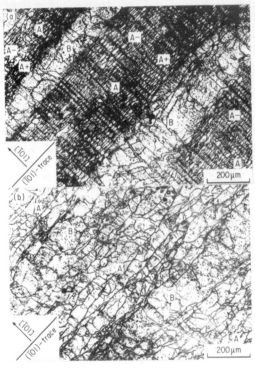

Photo.2   Dislocation structures
in high temperature creep of
copper single crystal [13].

Next, typical example of the substructure formation in
copper single crystals [13] will be described.   A series of
photographs represented in Photo.2 show the substructures in
the same area of a copper single crystal (initial dislocation
density of about $10^8/m^2$) subjected to creep to the initial
stage of primary creep and to the steady-state, respectively.
Substructure observation was made by means of etch-pit technique using Livingston's etchant on the specimen surface which
was parallel to the cross slip plane.   Structure at an early
stage of primary creep (Photo.2a) is substantially the same as
that due to deformation at room temperature.   Regions A in which
small cells elongating along (Ī01)-trace and regions B containing large cells within them are observed alternately with

a period of several microns, their boundaries running parallel
to the plane of deformation band, ($\bar{1}$01). With the progress of
creep deformation, the cell structures in both regions become
similar to each other as seen in Photo.2b. In Fig.3 the cell
sizes in both regions measured along the direction of slip
bands, [$\bar{1}$01], are shown as a function of the creep time. The
mean creep rates which were obtained by measuring the change in
length between two indentations made on the boundaries of A and
B regions using a micro-Vickers hardness tester after certain
creep times. The mean creep rates obtained in regions A and
B as well as those throughout the specimen are given in Fig.4
as a function of creep time. From the results shown above it
may be concluded that creep rate is closely related to substructure formation.

It has been reported that the total dislocation density
ρ_{total}, and the dislocation density in cell walls, ρ_{SB}, increase as the creep strain increases to take a constant value
at the steady-state, while the density of dislocations within
cells, ρ_{SG}, continues to decrease until the steady-state is
reached. The results obtained in alpha-iron polycrystals are
shown in Fig.5 which also shows that the changes in ρ_{SG} and in
creep rate correspond well with each other [14]. Smilar
results were obtained on copper single crystals [10].

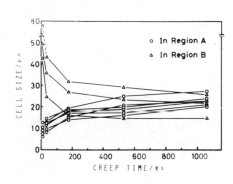

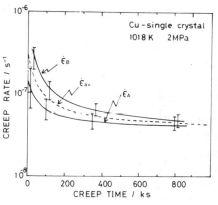

Fig.3 Cell sizes in A and B
regions due to high temperature
creep of copper single crystal
[13].

Fig.4 Mean creep rates in
A and B regions, $\dot{\epsilon}_A$ and $\dot{\epsilon}_B$,
and in whole specimen, $\dot{\epsilon}_{av}$,
in high temperature creep
of copper single crystal [13].

4. INTERNAL STRESS

It is well known that creep strain is not a unique function of creep stress, σ_c. This fact implies that it is necessary to take into account the stress hindering creep deformation, σ_i, which is usually called internal stress and
cosequently that dislocation movement is controlled by effective

or frictional stress, σ_e, which is defined by $\sigma_c = \sigma_e + \sigma_i$.

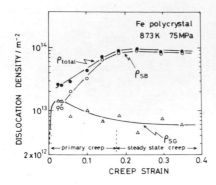

Fig.5 Dislocation densities in high temperature creep of alpha-iron polycrystal [14].

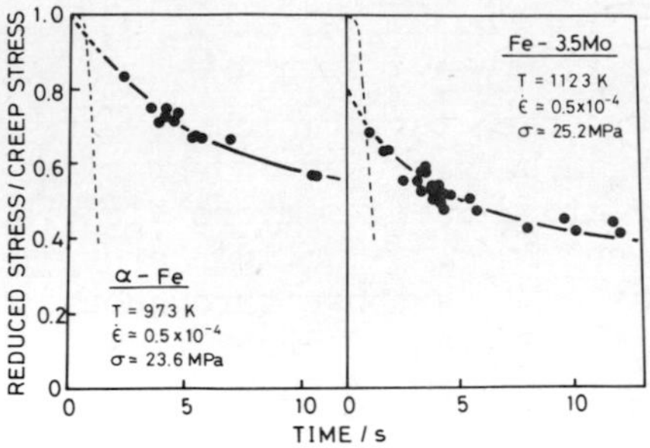

Fig.6 Relation between times at which stagnation of deformation ceases and reduced stress levels, σ'/σ_c, in stress transient dip tests [17].

It is widely accepted that σ_e or σ_i is a measure to discriminate creep behaviours [15]. In the case of recovery controlled creep σ_i is nearly equal to σ_c, indicating that σ_e is almost negligible. On the contrary, an appreciable amount of σ_e exists in Class I alloys in which viscous glide of dislocations dragging solute atoms controls the creep rate.

Although various methods to determine σ_i have been proposed so far, the extrapolation technique in stress transient dip test [16] is believed to give the most reliable value of σ_i.

The results obtained by means of this technique on alpha-iron and an Fe-3.5 at% Mo alloy are represented in Fig.6 [17]. Solid circles in the figure show the time at which ceased the stagnation of deformation at each reduced stress level, σ'/σ_c. The point at which each curve connecting the solid circles intersects the ordinate axis gives the value of σ_i/σ_c. It is clearly seen that $\sigma_i/\sigma_c \simeq 1$ in alpha-iron. On the other hand $\sigma_i/\sigma_c < 1$, indicating that σ_e is not negligible in the case of this alloy, which is considered to belong to Class Ia. The same is true in Al-Mg alloys (Class I alloys) [6]. It is to be noted that in Cu-Al alloys (Class II alloys) with $n \geq 5$ $\sigma_i/\sigma_c \simeq 1$, the primary creep curves showing S-type behaviour [18].

5. TRANSIENT BEHAVIOUR

Transient behaviours appearing after the creep condition is changed during the steady-state creep are also of use to know creep behaviours.

At first, the results obtained by stress relaxation tests during the steady-state creep will be stated. The rate of stress change immediately after the relaxation test started, $\dot{\sigma}_{ro}$, was determined from a tangent of the stress-time curve. This rate was converted into the strain rate, $\dot{\varepsilon}_{ro}$, using a constant, K, which is determined by the rigidity of the testing machine and the specimen. The ratio, $\dot{\varepsilon}_{ro}/\dot{\varepsilon}_c$ ($\dot{\varepsilon}_c$: the steady-state creep rate), is nearly equal to unity when the appreciable amount of σ_e exists, while it is smaller than unity in the case where σ_e is negligibly small [15]. This can be used to judge creep mechanisms. An example of the results obtained on alpha-iron and an Fe-3.5 at% Mo alloy [17] is shown in Table 1, from which derived is the conclusion that in alpha-iron creep deformation is controlled by recovery and that the dislocations move in viscous manner dragging solute atoms in the alloy.

	alpha-iron	Fe-3.5 at% Mo
Temperature/K Creep rate/$10^{-4}s^{-1}$	973	1123
~ 0.5	0.59	0.90
~ 1.3	0.55	1.15
~ 2.5	—-	0.97
~ 5.0	0.35	0.92
Average	0.49 ± 0.10	0.98 ± 0.01

Table 1 Ratio, $\dot{\varepsilon}_{ro}/\dot{\varepsilon}_c$, by stress relaxation test during steady-state creep [17].

The second method is to examine the shape of transient creep curves following a sudden increase or decrease of the creep stress during the steady-state creep [19]. This is occasionally called stress change test. In pure metals N-type transient is usually observed and Class I alloys show I-type transient after the stress is suddenly increased. On the contrary, at a time of stress drop I-type and N-type usually appear in pure metals and in Class I alloys, respectively. This expectation has been confirmed in aluminium and Al-Mg alloys [20] and in an Fe-4.1 at% Mo alloy [21]. However, one should be very careful in examining transient behaviour after stress drop, since it has been reported that anomalous transition appears when the stress drop is very large [22].

The confirmation of the existence of instantaneous plastic strain after a sudden increase of creep stress during the steady-state is another method to judge creep behaviours [23]. If the instantaneous elongation and contraction after the sudden increase and decrease of creep stress are represented by Δl_+ and Δl_- respectively, it is excepted that

$\Delta 1_+ \simeq \Delta 1_-$ in Class I alloys. On the contrary, in pure metals the difference, $\Delta 1_+ - \Delta 1_-$, which corresponds to the instantaneous plastic strain after the stress increase, is anticipated to increase as the increase in stress increment, $\Delta\sigma$, is increased. Typical results are reproduced in Fig.7 [23] which clearly shows that an Al-5.5 at% Mg alloy is ascribed to Class I. It is worth noting that in a Cu-10.4 at% Al alloy $\Delta 1_+ \simeq \Delta 1_-$ [24], indicating that the dislocations move in viscous manner in it in spite of the fact that $n \geqq 5$ [18,25].

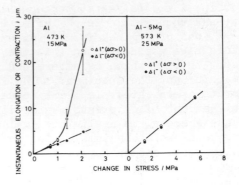

Fig.7 Instantaneous elongation and contraction after stress is respectively increased and decreased by $\Delta\sigma$ during steady-state [23].

6. DISCUSSION

6.1 Internal stress

In Chater 4 it was shown that internal stress is one of the most important factors to identify the creep mechanism. Now, the origin of internal stress will be discussed.

Dislocation theory proposes the following equation to give the flow stress, σ_f, in the specimen with the dislocation density of ρ :

$$\sigma_f = \sigma_0 + \alpha\mu b\sqrt{\rho} \tag{2}$$

where σ_0 is the so-called frictional stress, μ the shear modulus, b the Burgers vector and α a constant. The second term of right hand side of eq. (2) is nothing but the internal stress concerned. According to McLean [26] α is consistently found equal to unity for nickel and for alpha-iron at NPL. However, it has been reported that the observed internal stress is too large to be explained by eq. (2) in Class I alloys (Al-Mg alloys) [20]. Although they have not calculated the value of α, α much larger than unity is obtained when the value of σ_i reported by them is used in eq. (2). Therefore, it seems very urgent to make clear the origin of σ_i in the case of uniform arrangement of dislocations.

On the other hand, from the results on alpha-iron [27] and copper single crystals [10] and a Cu-16 at% Al alloy [8] values of α larger than unity ($\alpha = 1.3 - 2.5$) are obtained except under low σ. as long as the density of dislocations which do not associate with cell walls is used for ρ. Generally the larger is the creep stress, the larger is the α-value. In a Cu-16

at% Al alloy, in only which
the data on total dislocation
density are available [8], α
equal to unity is obtained if
ρ_{total} is used for ρ. It is
unreasonable that α is larger
than unity. The use of ρ_{total}
for ρ also seems illogical.
However, if α is calculated
from the equation in which
$\alpha\mu b/l_o$ is used instead of $\alpha\mu b\sqrt{\rho}$
in eq. (2), more reasonable
values of 0.1 - 0.2 are obtained
for α from the results of alpha-
iron single crystals [28], Fe-
Mo alloys [11,29] and a Cu-16
at% Al alloy [8]. Therefore,
it may be safely said that the
dislocations associated with
cell walls play an important
role in the development of
internal stress at least in the
case where cell structure is
formed during creep. In the
following the experimental
results on a Cu-16 at% Al alloy
[8] to prove the conclusion
mentioned above will be stated.
This alloy belongs to Class II
alloys, showing S-type primary
creep curve as shown in Fig.8a.

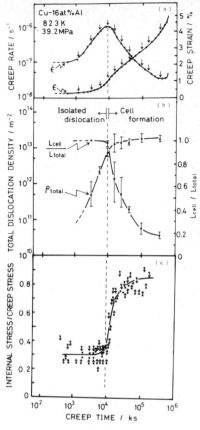

Fig.8 Various parameters
in high temperature creep
of Cu-16 at% Al alloy [8].

It is clearly seen in Fig.
8b that the dislocations are
distributed uniformly and that the total dislocation density
increases in the regime where the creep rate, $\dot{\varepsilon}$, increases.
On the contary, cell structure starts to develop and the
density of dislocations within cells, which are considered to
be mobile, decreases rapidly from the time on when $\dot{\varepsilon}$ begins to
decrease. Notice that the ratio between the length of dislocations existing within cells, L_{cell}, and the total length of
dislocations, L_{total}, including those in cell walls is plotted
in the figure. It is to be noted that σ_i increases drastically
at the same time (Fig.9c). From these results the decrease of
$\dot{\varepsilon}$ may be rationalized considering the two following effects due
to the cell formation : one is the decrease in mobile dislocation density and the other is the generation of large magnitude
of σ_i.

On the other hand, various parameters of the cell structures at the steady-state are reported to be dependent on the
creep stress in many metals and alloys. The dislocation

density within cells, ρ_{SG}, is proportional to σ_c^n, in which n takes the value of $1-2$, and cell sizes to $1/\sigma_c$. The mean dislocation spacing in cell walls, l_o, also depends on $1/\sigma_c$ [5, 30].

In order to make clear which of the three kinds of parameters mentioned above actually acts as the source of the internal stress and consequently determines the creep rate, performed were creep tests of a Cu-0.5 at% Al alloy specimens which had been subjected to prestraining at room temperature to introduce various degrees of cell structures in them [31]. The addition of small amount of aluminium aims at preventing recrystallization during creep. The creep tests were conducted after the dislocation structures due to prestraining had been stabilized by heating the specimens at a temperature a little higher than the creep temperature. Structure observation reveals that the cell sizes at the steady-state or at the stage of minimum creep rate were the same as those obtained after the prestraining. However, it was found that the mean dislocation spacing in cell walls under went a change during creep. The internal stresses as well as the steady-state or minimum creep rates were confirmed to be inversely proportional to l_o. The relation between $\dot{\varepsilon}$ and $1/l_o$ is shown in Fig.9, which infers that the cell formation contributes to hinder the creep deformation through the development of internal stress which in turn is determined by l_o.

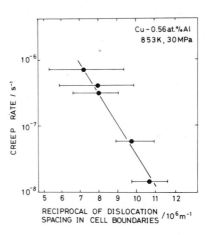

Fig.9 Relation between mean dislocation spacings in cell walls and creep rates in Cu-o.56 at% Al alloy specimens given various prestrains [31].

In conclusion a brief comment on the method to determine internal stress will be given. All our data excepting those of a Cu-16 at% Al alloy were obtained by the extrapolation technique in stress transient dip test [16]. This technique is based on the assumption that creep stagnation periods exist when the creep stress is reduced by certain amounts. However, there are objections against this assumption, maintaining that a true stagnation period for creep at the reduced stress may not exist [3]. Giebling and Nix [32] also deny its existence by comparing the observed value, Δt_{obs}, with the smallest possible recovery delay time, Δt_{min}. In the calculation based on the Bailey-Orowan equation, $\dot{\varepsilon} = r/h$ (r and h are recovery rate and strain hardening cefficient, respectively), used is the assumption that $r = d\sigma/dt \simeq \Delta\sigma/\Delta t$, the validity of

which is very doubtful. Moreover, our above-mentioned results
of internal stress measurement coincide well with those obtained by stress relaxation tests from the point of view of σ_e.
Therefore, it may be concluded possible to estimate σ_i from the
extrapolation technique.

6.2 Dislocation behaviours and creep mechanisms

In this Chapter the controlling mechanisms of creep and
dislocation behaviours during creep in various types of creep
will be discussed based on our above-stated results.

In the first place the experimental facts revealed in an
Al-5.5 at% Mg alloy, which is considered a typical example of
Class I alloys, are reviewed. Considerable magnitude of σ_e
exists, $\dot{\varepsilon}_{ro}/\dot{\varepsilon}_c \simeq 1$ in stress relaxation test, n = 3, and little
or no instantaneous plastic strains were observed at the time
of loading and of stress increase during the steady-state creep,
respectively. All these results indicate that the dislocations
move in viscous manner dragging solute atoms with them and also
that the creep deformation is controlled by their viscous motion. In fact, the effective stress exponent of mean dislocation velocity, m^*, which was obtained by stress relaxation
test, was almost unity [33]. Uniform distribution of dislocations during creep has been expected in the case where $m^* =$
1 from the theoretical consideration [34] based on the assumption that the dislocations arrange themselves so that the strain
rate becomes maximum in creep deformation under a constant
stress. The experimental results agree well with this expectation.

Since the dislocation velocity in this type of alloys is
considered small, the rapid multiplication of dislocations
would not take place even if some dislocations become mobile.
This is why instantaneous plastic strain is hardly observed to
occur, in particular immediately after the stress increase
during creep. By the same reason the probability for the
moving dislocations to tangle with the preceding ones which are
stopped to move against some obstacles may be small. Thus,
uniform distribution of dislocations is produced.

As the creep deformation proceeds the mobile dislocation density
increases due to multiplication, I-type primary creep curve
being obtained. On the other hand, the increase in dislocation
density results in the increase of internal stress (work hardening) due to the decrease in mean spacing of dislocations,
which in turn increases the annihilation rate of dislocations
to promote recovery. The steady-state appears when the work
hardening and recovery are balanced with each other.

Next, summarized are the results obtained on pure metals.
Appreciable amount of instantaneous plastic strains occurred

at the time of stress increase during creep as well as of initial loading, $\dot{\varepsilon}_{ro}/\dot{\varepsilon}_c < 1$ in stress relaxation test, $n \simeq 5$, and σ_i is nearly equal to σ_c, suggesting that σ_e is almost negligible in this case. All the results may support the view that the dislocations move in free flight manner and that the creep deformation is recovery controlled.

Because of the high mobility of dislocations the multiplication rate soon after the application of stress is high enough to produce the considerable amount of instantaneous plastic strain. The dislocations introduced during the instantaneous strain develop into cell structure during primary creep. This fact agrees well with the theoretical expectation cited above [34]. As a result of the cell formation mobile dislocation density decreases and σ_i increases. This is the origin of N-type primary creep curves. While the work hardening increases along with the creep deformation, the decrease of mean dislocation spacing within cell walls promotes the annihilation of dislocations resulting in the more regular distribution of dislocations with larger mean spacing. The unification of cell walls due to their migration also occur concurrently. These factors contribute to recovery. In this case also the balance between the work hardening and recovery brings on the steady-state.

In Class Ia and Class II alloys, however, the situation seems rather complicated. In Fe-Mo alloys no instantaneous plastic strain appeared, S-type transient creep occurred in stress change test, and $\sigma_i/\sigma_c < 1$ [29,35,36]. These facts suggest the viscous dislocation motion in these alloys. However, the cell formation was observed to occur in them [29,36] and in an Fe-1.8 at% Mo alloy n-value was 4 which is appreciably larger than 3 [35]. These features are characteristic of recovery controlled creep. Nevertheless, it is concluded that the creep deformation of these alloys is controlled by viscous dislocation motion.

On the contrary, in spite of the fact that $\Delta l_+ \simeq \Delta l_-$ according to the results of stress change tests [24] to show that the dislocations move in viscous manner, $n \geq 5$ and $\sigma_i/\sigma_c \simeq 1$ [18] in Cu-Al alloys. It is concluded that the creep deformation is recovery controlled in them.

As seen above the results contradicting with each other are often obtained in some alloys. Therefore, collective judgement on the experimental results obtained by various methods is indispensable to made clear the creep behaviours. In particular, the information about σ_e or σ_i obtained from dip tests [16] and from stress relaxation tests [15] seems to be conclusive in identifying the creep mechanism.

7. CONCLUSION

Results on the characteristic features and dislocation behaviours during high temperature creep of pure metals and Class I alloys and discussion based on them are described. The origin of internal stress in the case of uniform dislocation arrangement is left unsolved.

The situation in Class Ia and Class II alloys is somewhat complicated. Although further systematic studies are needed, information on internal stress seems conclusive in identifying the creep mechanism in these alloys.

ACKNOWLEDGEMENT

The present author withes to express his hearty thanks to his co-workers who obtained the valuable results described here through their persistent efforts. He also acknowledges helpful discussions with Professor H. Oikawa during the preparation of this paper.

REFERENCES

1. BIRD, J.E., MUKHERJEE, A.K. and DORN, J.E. - 'Correlations Between High-Temperature Creep Behavior and Structure', Quantitative Relation Between Properties and Microstructure, Ed. Brandon, D.G. and Rosen, A., Israel Univ. Press, 1969, p.255.
2. SHERBY, O.D. and BURKE, P.M. - 'Mechanical Behavior of Crystalline Solids at Elevated Temperature', Prog. Mater. Sci., 1966, 13, 323.
3. NIX, W.D. and ILSCHNER, B. - 'Mechanisms Controlling Creep of Single Phase Metals and Alloys', Strength of Metals and Alloys, Ed. Haasen, P., Gerold, E. and Kostorz, G., Pergamon Press, 1979, p.1503.
4. OIKAWA, H., SUGAWARA, K. and KARASHIMA, S. - 'High Temperature Creep Mechanism in an Al-2 at% Mg Alloy Determined by Stress-Relaxation Tests', Scripta Met., 1976, 10, 885.
5. KARASHIMA, S., IIKUBO, T. and OIKAWA, H. - 'On the High-Temperature Creep Behaviour and Substructures in Alpha-Iron Single Crystal', Trans. Japan Inst. Metals, 1972, 13, 176.
6. OIKAWA, H., SUGAWARA, K. and KARASHIMA, S. - 'Creep Behavior of Al-2.2 at% Mg Alloy at 573 K', ibid., 1978, 19, 611.
7. OIKAWA, H., KURIYAMA, N., MIZUKOSHI, D. and KARASHIMA, S. - 'Effects of Testing Modes on Deformation Behavior at Stages prior to the Steady States at High Temperature in Class I Alloys', Mater. Sci. Eng., 1977, 29, 131.
8. HASEGAWA, T., IKEUCHI, Y. and KARASHIMA, S. - 'Internal Stress and Dislocation Structure during Sigmoidal Transient Creep of a Copper - 16 at% Aluminium Alloy', Metal Sci, J., 1972, 6, 78.

9. MATSUNO, N.,OIKAWA, H. and KARASHIMA, S. - 'Transmission Electron Microscopy of Substructures Formed During High-Temperature Creep in Al-Mg Alloys', J. Japan Inst. Metals, 1974, 38, 1071 (in Japanese).

10. HASEGAWA, T., SATO, H. and KARASHIMA, S. - 'Etch-Pit Studies of the Dislocation Structures Developed during Creep Deformation of Copper Single Crystals', Trans. Japan Inst. Metals, 1970, 11, 101.

11. SAEKI, M.,OIKAWA, H. and KARASHIMA, S. - 'Substructures Developed During Steady-State Creep of Fe-4.1 at% Mo Alloy at High Temperatures', J. Japan Inst. Metals, 1979, 43, 135 (in Japanese).

12. MATSUNO, N. - 'Transmission Electron Microscopy of Substructures Developed During High Temperature Creep of Al-Mg Alloys', Master Thesis, Tohoku Univ., 1974.

13. HASEGAWA, T., KARASHIMA, S. and HASEGAWA, R. - 'Substructure Formation and Nonuniformity in Strain During High Temperature Creep of Copper Single Crystals', Met. Trans., 1971, 2, 1449.

14. ORLOVA, A. and CADEK, J. - 'Some Substructural Aspects of High-Temperature Creep in Metals', Phil. Mag., 1973, 28. 891.

15. ABE, K.,YOSHINAGA, H. and MOROZUMI, S. - 'A Method of Discerning Frictional Stress and Internal Stress by the Stress Relaxation Test', Trans. Japan Inst. Metals, 1977, 18, 479.

16. TOMA, K., YOSHINAGA, H. and MOROZUMI, S. - 'Internal Stress in Al and an Al-Mg Alloy Deforming at High Temperatures', J. Japan Inst. Metals, 1974, 38, 170 (in Japanese).

17. OIKAWA, H., ICHIHASHI, K. and KARASHIMA, S. - 'High-Temperature Creep Mechanisms in Alpha-Iron and An Fe-Mo Alloy Determined by Stress-Relaxation Tests', Scripta Met., 1976, 10, 143.

18. YASUDA, A., OIKAWA, H. and KARASHIMA, S. - 'High-Temperature Creep of Cu-14.5 at% Al Alloy with Special Reference to Recovery and Work Hardening', to be published.

19. FUCHS, A. and ILSCHNER, B. - 'An Analysis of the Creep Behaviour of Iron-Molybdenum Solid Solutions', Acta Met., 1969, 17, 701.

20. HORIUCHI, R. and OTSUKA, M. - 'Mechanism of High Temperature Creep of Aluminium - Magnesium Solid Solution Alloys', 1971, 35, 406 (in Japanese).

21. OIKAWA, H., SAEKI, M. and KARASHIMA, S. - 'Creep Mechanism of Fe-4.1 at% Mo Alloy at High Temperatures', Tetsu to Hagane, 1979, 65, 843 (in Japanese).

22. BLUM, W., HAUSSELT, J. and KÖNIG, G. - 'Transient Creep and Recovery after Stress Reduction During Steady State Creep of AlZn', Acta Met., 1976, 24, 293.

23. OIKAWA, H. and SUGAWARA, K. - 'Instantaneous Plastic Strain Associated with Stress Increments During the Steady State Creep of Al and Al-5.5 at% Mg Alloy', Scripta Met., 1978, 12, 361.

24. OIKAWA, H. - 'The Absence of Instantaneous Plastic Strain upon Stress Changes during Creep of Some Solid Solutions', Phil. Mag., 1978, A37, 707.

25. PAHUTOVA, M., HOSTINSKY, T., CADEK, J. and RYS, P. - 'Creep Mechanism in a Cu-5.5 at% Al Alloy', Phil. Mag., 1969, 20, 975.

26. MAC LEAN, D. - Private Communication, 1980.

27. KARASHIMA, S., IIKUBO, T., WATANABE, T. and OIKAWA, H. - 'Transmission Electron Microscopy of Substructures Developed during High-Temperature Creep in Alpha-Iron', Trans. Japan Inst. Metals, 1971, 12, 369.

28. IIKUBO, T., OIKAWA, H. and KARASHIMA, S. - 'Activation Process in High Temperature Creep of Alpha-Iron', to be published.

29. MIZUKOSHI, D., OIKAWA, H. and KARASHIMA, S. - 'Transmission Electron Microscopy of Substructures Formed by High Temperature Creep in Fe-1.8 at% Mo Alloy', Trans. Iron Steel Inst. Japan, 1978, 18, 696.

30. MUKHERJEE, A.K., BIRD, J.E. and DORN, J.E. - 'Experimental Correlations for High Temperature Creep', Trans. ASM, 1969, 62, 155.

31. HASEGAWA, T., KARASHIMA, S. and IKEUCHI, Y. - 'High-Temperature Creep Rate and Dislocation Structure in a Dilute Copper-Aluminium Alloy', Acta Met., 1973, 21, 887.

32. GIEBLING, J.C. and NIX, W.D. - 'The Existernce of a Friction Stress for High-Temperature Creep', Metal Sci. 1977, 11, 453.

33. OIKAWA, H., KARIYA, J. and KARASHIMA, S. - 'Some Activation Parameters in Steady-State Creep of Aluminium-Magnesium Alloys at High Temperatures', Metal Sci., 1974, 8, 106.

34. KARASHIMA, S., MARUYAMA, K. and ONO, N. - 'An Analysis of Cell Formation due to Plastic Deformation Based on Dislocation Theory', Trans. Japan Inst. Metals, 1974, 15, 265.

35. OIKAWA, H., MIZUKOSHI, D. and KARASHIMA, S. - 'Creep Mechanism in Fe-1.8 at% Mo Alloy at High Temperatures', Met. Trans., 1978, 9A, 1281.

36. OIKAWA, H., SAEKI, M. and KARASHIMA, S. - 'Steady-State Creep of Fe-4.1 at% Mo Alloy at High Temperatures', Trans. Japan Inst. Metals, 1980, 21, 309.

RECOVERY AND WORK HARDENING DURING HIGH TEMPERATURE CREEP OF FCC ALLOYS OF LOW STACKING FAULT ENERGY

Akira YASUDA*, Hiroshi OIKAWA and Seiichi KARASHIMA

Department of Materials Science, Faculty of Engineering,
Tohoku University, Sendai, 980 JAPAN
* On leave from Research Laboratories,
Kawasaki Steel Corporation, Chiba, 260 JAPAN

SUMMARY

   A general differential rate equation is derived for high-temperature deformation of materials, in which the effect of anelasticity cannot be neglected. Variables chosen are total strain-rate, stress, plastic strain and time. The recovery rate, r, and the work hardening rate, h, in these materials are also defined. They are different from those obtained for materials in which dislocations glide in a free-flight manner.
   Values of r' and h' determined experimentally, however, are apparent ones in alloys exhibiting Class II behavior during steady-state creep. They include the contribution of anelasticity. These parameters can be regarded as a measure of the stress necessary for climbing and gliding of dislocations.
   Creep characteristics of Cu-14.5at% Al alloy at 850 K under stresses of 12.5 to 50 MPa are in accord with the present theoretical considerations.

1. INTRODUCTION

   Creep behavior of single phase alloys are frequently classified into either Class I or Class II according to the characteristics during the steady-state stage. Class I behavior is well elucidated by a microcreep which is controlled by the viscous gliding of dislocations. Class II behavior is frequently regarded as a recovery creep which is controlled by the climbing of dislocations. Typical examples of recovery creep, however, are seen not in alloys but in pure metals in which dislocations glide in a free-flight manner.

Many alloys containing a relatively low concentration of solute elements exhibit Class II behavior in the steady-state stage, though dislocations move viscously in these alloys. When the deformation condition is changed at the steady-state, the transient behavior which appears in these alloys is not necessarily the same as that in pure metals, in which dislocations glide in a free-flight manner.

Creep of pure metals and microcreep of alloys (Class I behavior) have been elucidated theoretically to some extents. However, the transient behavior of single-phase alloys which show Class II behavior, or an intermediate behavior between Classes I and II in the steady-state stage, has been investigated only in a few cases[1,2].

In the present paper, transient and steady-state characteristics are analyzed phenomenologically for alloys which show "so-called" Class II behavior during steady-state creep to provide a theoretical framework for discussing creep behavior of single-phase alloys, in which the effect of anelasticity cannot be neglected.

2. THEORETICAL CONSIDERATIONS

When anelasticity is not expected, an equation,

$$d\sigma = \left(\frac{\partial \sigma}{\partial \varepsilon}\right)_t d\varepsilon + \left(\frac{\partial \sigma}{\partial t}\right)_\varepsilon dt \tag{1}$$

is frequently employed for high-temperature deformation and the steady-state creep rates, $\dot{\varepsilon}^s$, is expressed by Bailey-Orowan equation,

$$\dot{\varepsilon}^s = r/h \tag{2}$$

where r is the recovery rate, $r = -(\partial \sigma/\partial t)_\varepsilon$, and h is the work-hardening rate, $h = (\partial \sigma/\partial \varepsilon)_t$.

The validity of these equations has been confirmed in pure metals[3] in which dislocations glide in a free flight manner. Equation(1), however, is not adequate for describing the deformation behavior of alloys which are expected to show obvious anelasticity. In this section a general equation suitable for describing the deformation of such materials is derived.

2.1 Rate Equation

The (total) strain rate, $\dot{\varepsilon}_T$, which can be measured during a test at a given temperature, is uniquely determined when the applied stress, σ, and the information on the structure such as the dislocation density and their arrangements are given. Although the structure parameter, η, cannot be defined uniquely by total (plastic) strain, ε_p, and total time, t, it can be assmued for the first approximation that the infinitisimal change in the parameter, $d\eta$, is determined by infinitisimal change in plastic strain, $d\varepsilon$, and in time, dt. Therefore, an infinitisimal change in total strain-rate, $\dot{\varepsilon}_T$, can be correlated with infinitisimal changes in the plastic strain, ε_p, the applied stress, σ, and the time, t, by the following equation :

$$d\dot{\varepsilon}_T = \left(\frac{\partial \dot{\varepsilon}_T}{\partial \varepsilon_p}\right)_{\sigma,t} d\varepsilon_p + \left(\frac{\partial \dot{\varepsilon}_T}{\partial \sigma}\right)_{t,\varepsilon_p} d\sigma + \left(\frac{\partial \dot{\varepsilon}_T}{\partial t}\right)_{\varepsilon_p,\sigma} dt \qquad (3)$$

Hereafter, the following abbreviations will be used :

$$(\partial \dot{\varepsilon}_T / \partial \varepsilon_p)_{\sigma,t} \equiv L$$

$$(\partial \dot{\varepsilon}_T / \partial \sigma)_{t,\varepsilon_p} \equiv M$$

$$(\partial \dot{\varepsilon}_T / \partial t)_{\varepsilon_p,\sigma} \equiv N$$

Equation(3) has real meanings when ε_p, t, and σ are independent among each other and M, L and N exist.

It is obvious from the results of stress change tests that σ is independent of ε_p and t. No discussions and experimental results have been reported to prove the independence of ε_p and t with each other in the case that dislocations glide viscously.

The variable t represents a factor expressing recovery at a given temperature. When the recovery arises without plastic strain under a given applied stress, the relative amount of stress effective for the glide of dislocations increases. This increase in the effective stress causes an increase in $\dot{\varepsilon}_T$. On the other hand, the plastic strain without recovery shifts the equilibrium position of (mobile) dislocations, which will be discussed later, to increase the stress opposing the movement and the relative amount of stress effective for the glide of dislocations decreases. This decrease in the effective stress causes a decrease in $\dot{\varepsilon}_T$. Therefore, the plastic strain and the recovery can affect on $\dot{\varepsilon}_T$ independently : the variants, t and ε_p, are independent of each other. The presence of derivative terms, M, L and N, can be assumed as far as creep curves and tensile curves at high temperatures are smooth and continuous.

The term $M = \partial \dot{\varepsilon}_T / \partial \sigma$ represents the effect of stress on the total-strain rate. The effect of elasticity is, of course, included in this term. The effect of anelasticity which is proportional to σ for the first approximation can also be included in this term, though anelastic strain is difficult to be distinguished precisely from elastic and plastic (permanent) ones. The terms L and N indicate the effects of plastic strain and recovery on total-strain rates, respectively.

2.2 Recovery Rate and Work-Hardening Rate

In equation(1), $d\sigma = 0$ in constant stress (creep) tests and $d\dot{\varepsilon}_T = 0$ in constant strain-rate (conventional) tests. During the steady-state deformation $d\dot{\varepsilon}_T = 0$ or $d\sigma = 0$ in these tests, respectively. Then, the plastic strain rates, $\dot{\varepsilon}_p$, in both type of tests are expressed by the equation,

$$\dot{\varepsilon}_p = \left(\frac{\partial \varepsilon_p}{\partial t}\right)_\sigma = -\frac{N}{L} . \tag{4}$$

This equation shows that $\dot{\varepsilon}_p$ is expressed as the ratio between infinitisimal plastic strain and infinitisimal recovery. The recovery occurs continuously with a constant rate during the steady-state stage. Therfore, $\dot{\varepsilon}_p$ in equation(4) can be regarded as plastic strain rates of usual meanings. This $\dot{\varepsilon}_p$ is the same as the usual steady-state creep rates, $\dot{\varepsilon}^s$, because both elastic and anelastic strain rates are zero during the steady-state stage.

The parameters corresponding to r and h in equation(2) can be defined based on equation(3) :

$$r = -\left(\frac{\partial \sigma}{\partial t}\right)_{\varepsilon_p} = \frac{N}{M}$$

$$\tag{5}$$

$$h = \left(\frac{\partial \sigma}{\partial \varepsilon_p}\right)_t = -\frac{L}{M}$$

Although r and h defined by equation(5) satisfy equation(1), they have different meanings from those based on equation(2). The parameter r represents the ratio of contributions of recovery to anelasticity on the total strain rate. The parameter h represents the ratio of contributions of plastic strain to anelasticity on the total strain rate. The values of r and h can be defined and equation(4) holds only when $d\sigma = 0$ and $d\dot{\varepsilon}_T = 0$, $i.e.$, under the steady-state conditions. It becomes evident that r and h can be defined and equation(5) holds as long as $M \neq 0$.

3. DETERMINATION OF r AND h

Let us consider the conditions that r and h can be determined experimentally. The deformation behavior after changing deformation conditions can be expressed by the terms L, M and N during the steady-state stage as follows :

$$\frac{1}{M}\left(\frac{d\dot{\varepsilon}_T}{\partial t}\right) = \dot{\sigma} - \frac{L}{M}\dot{\varepsilon}_p + \frac{N}{M} . \tag{6}$$

One can measure only the parameters $\dot{\sigma}$ and $\dot{\varepsilon}_T$ in usual experiments; the values of $r = N/M$ and $h = -L/M$ cannot be obtained even if an assumption that $\dot{\varepsilon}_T \simeq \dot{\varepsilon}_p$ is made.

The time necessary for changing the deformation conditions must be short to minimize the influence on L, M and N. These circumstances lead to that it is usually very diffcult to measure precisely the second derivative with respect to t, $d\dot{\varepsilon}_T/dt$. Therefore, the condition under which the term $(d\dot{\varepsilon}_T/dt)/M$ can be neglected must be employed in measuring $\dot{\varepsilon}_p$ and/or $\dot{\sigma}$ to obtain an appropriate relation between r and h.

3.1 Stress–Relaxation Tests

The stress relaxation is the process of a stress decrease based on the fact that elastic strain changes into plastic strain under the condition $d\varepsilon_T = 0$.

The strain-rate, $\dot{\varepsilon}^{ro}$, immediately after the initiation of stress-relaxation is obtained from equation(3) by replacing $d\varepsilon_p$ with $-d\sigma/K$ (K: the elastic constant of the testing system) :

$$\dot{\varepsilon}_p^{ro} = -\frac{N - \ddot{\varepsilon}_T}{L - KM} .$$

(7)

During stress relaxation $\ddot{\varepsilon}_T = 0$, so that

$$\dot{\varepsilon}^{ro} = -\frac{N}{L - KM} = \frac{r}{h + K}$$

(8)

as long as $M \neq 0$. The (plastic) strain rate immediately after the relaxation (equation (8)) is not equal to that before the relaxation, $\dot{\varepsilon}^s$. The ratio is

$$\dot{\varepsilon}^{ro}/\dot{\varepsilon}^s = \frac{1}{1 + K/h}$$

(9)

Except under the condition $K \ll h$, $\dot{\varepsilon}^{ro} < \dot{\varepsilon}^s$ holds and r and h can be obtained by the following equations :

$$r = \frac{\dot{\varepsilon}^{ro}}{1 - \dot{\varepsilon}^{ro}/\dot{\varepsilon}^s} K$$

(10)

$$h = \frac{\dot{\varepsilon}^{ro}/\dot{\varepsilon}^s}{1 - \dot{\varepsilon}^{ro}/\dot{\varepsilon}^s} K .$$

(11)

These relationships are similar in the forms to those reported by Maruyama and Karashima[4] and Sakurai et al.[5].

When anelasticity exists, K must be replaced by K' which includes the contribution of anelasticity, though such a value of K' cannot be determined. When K is used instead of K', the anelastic strain is included in the plastic strain.

c

3.2 Strain-Rate Change Tests

In strain-rate change tests deformation proceeds under the condition $\ddot{\varepsilon}_T = 0$ except at the moment when the total strain rate is changed. Plastic strain rate immediately after the strain-rate change, $\dot{\varepsilon}_p^t$, is correlated to the changing rate in stress $\dot{\sigma}^t$ by the equation,

$$\dot{\sigma}^t = h\dot{\varepsilon}^t + r .\tag{12}$$

When the relation between $\dot{\varepsilon}^t$ and $\dot{\sigma}^t$ is known, values of h and r can be determined from $\dot{\sigma}^t - \dot{\varepsilon}^t$ curves.

Equation(12), however, holds only when the plastic strain rate is concerned, whereas the observed value includes elastic and anelastic components. Elastic strain contribution can be corrected by the stress and the Young's modulus, but it is very difficult to remove the influence of anelasticity from the observed (total) strain rate. Hence, only the values of r' and h' which include the contribution of anelasticity cannot be neglected, the relation between $\dot{\varepsilon}^t$ and $\dot{\sigma}^t$ cannot be represented by equation(12) : $\dot{\sigma}^t - \dot{\varepsilon}^t$ relation is shown by a curved line and the unique value of r or h cannot be determined.

4. DISCUSSION

In the previous section it is stated that true values of recovery rate and work-hardening rate could be determined only if the relation between anelasticity and stress were known, otherwise the values including the contribution of anelasticity are obtained. We will discuss in the following the meanings of measurable quantities of r' and h' which are affected by anelasticity.

When $\dot{\varepsilon}^{ro} < \dot{\varepsilon}^s$ in a relaxation test, the recovery rate and the work-hardening rate, whatever they are true ones or just apparent ones, can be determined. In the case that disloca- tions glide in a free-flight manner as in pure metals, the stress necessary for gliding can be neglected and the gliding is not necessary to be considered as a thermally activated process. Therefore, the recovery due to climb is regarded as a thermally activated process. The dislocation multiplication, which results in an increase in the dislocation length and/or a decrease in the link-length of dislocation networks, is re- garded as the work-hardening process.

In the case when dislocations glide viscously, that is when the stress necessary for gliding cannot be neglected and the gliding is a themally activated process, no clear boundary exists between the recovery and the work-hardening[4].

Let us consider a dislocation segment, both ends of which are pinned by nodes or some other obstacles. (1)This segment bows out by the action of applied stress to such a degree that the line tension balances with the stress. (2)When the link-length becomes larger than the critical length by recovery, the dislocation becomes free to glide. (3)It glides forward until it is stopped against the foregoing dislocations, reaching the equilibrium position which is determined by the applied stress and the arrangement of other dislocations. These three steps, by which major part of the strain occurs can be regarded as the work-hardening process. (4)With the lapse of time the dislocation moves forward gradually due to the annihilation (steps 5 and 6) of foregoing dislocations. (5)The dislocation itself begins to climb to release the stress. (6)Finally the dislocation is annihilated with another dislocation. These three steps can be regarded as the recovery process.

These two processes occur simultaneously in many places so that it is difficult to measure h and r separately. The parameter $h = (\partial\sigma/\partial\varepsilon_p)_t$, is a measure that represents the work done for gliding of dislocations since most plastic and anelastic strain results from the gliding of dislocations. The parameter, $r = -(\partial\sigma/\partial t)\dot{\varepsilon}_p$, is a measure that represents the work done for other than gliding, that is, the climbing of dislocations.

During a stress-relaxation test, recovery causes a shift of the equilibrium position of dislocations. Dislocations move to keep the equilibrium and some plastic strain occurs. Anelastic strain is regarded as a part of plastic strain in this type of test, when K instead of K' is employed. To keep the equilibrium, the applied stress assists climbing and gliding of dislocations. When the gliding dominates the deformation, the stress effective for the gliding dominates the applied stress. When the climbing dominates the deformation, the stress effective for climbing dominates the applied stress. Therefore, r' and h' measured by a relaxation test, similar to r and h, are parameters corresponding to the work done for climbing and gliding of dislocations, respectively.

The effects of applied stress on r' and h' represent the relative contribution of the gliding and the climbing of dislocations on whole deformation. A large effect of applied stress on r' indicates a large effect of r' on $\dot{\varepsilon}^s$, that is, $\dot{\varepsilon}^s$ is strongly affected by the climbing of dislocations. Whereas a large effect of applied stress on h' indicates a large effect of h' on $\dot{\varepsilon}^s$, that is, $\dot{\varepsilon}^s$ is strongly affected by the gliding of dislocations.

5. SOME EXPERIMENTAL EXAMPLES

Creep of Cu-14.5 at% Al alloy can be regarded as an example for the creep controlled by recovery, though dislocations move viscously in this alloy. Some experimental results

obtained during creep of this alloy of low-stacking fault
energy at 850 K under stresses from 12.5 to 50 MPa will be
shown.

The absence of instantaneous (permanent) strain upon a
sudden increase in stress during steady-state creep[6] indicates clesrly that dislocations move viscously in this material.
The obvious decrease in the apparent Young's modulus with
increasing creep stress (see Fig.1[10]) indicates also the
presence of a large degree of anelasticity in this material
during the steady-state creep.

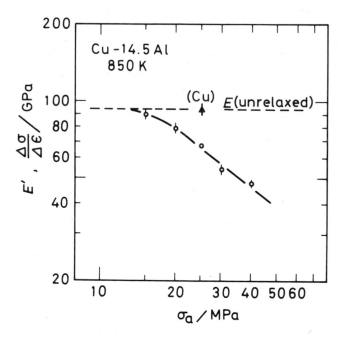

Fig.1 Relaxed Young's modulus as a function of stress during
creep of polycrystalline Cu-14.5 at% Al alloy. The
unrelaxed E value is taken from data given by Pahutova
[11].

The steady-state creep rates are shown in Fig.2[10] as a
function of stress. Under low stresses, the applied stress
exponent, n, is about 6, where no measurable stress effective
for gliding can be detected by dip tests. These results
indicate that creep can be treated as it is controlled solely
by recovery in this stress range.

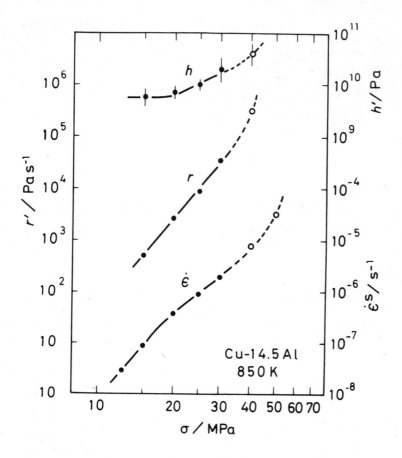

Fig.2 Steady-state creep characteristics of polycrystalline
   Cu-14.5 Al alloy.   Data represented by open marks are
   values obtained before reaching the true steady-state
   stage.   Note the change in slope of the $\dot{\varepsilon}$-line around
   20 MPa.

Under medium to high stresses, n is about 4, where effective stress is obtained by stress dip tests, though it is only
a few percent of the applied stress.   These results indicate
that a role of dislocation glide cannot be neglected in this
stress range.   The presence of a similar transition in steady-
state creep behavior has been reported on Al-Mg alloys[7,8,9].

The apparent r' and h' are determined by the method
decribed in section 3, and are shown in Fig.2.   Stress dependence of r' is insensitive to the stress level, whereas that of
h', n_h, is greatly affected by the stress level.

56

In the low stress region where creep is believed to be solely controlled by recovery, n_h is very small. In the medium to high stress region where the viscous glide motion of dislocations plays an important role, n_h becomes large with increasing stress. These results agree well with the discussion done in section 4.

The relation between $\dot{\sigma}^t$ and $\dot{\varepsilon}^t$ is obtained on this material during creep experiments and is shown in Fig.3[12]. When anelasticity cannot be neglected, theoretical considerations claim that the $\dot{\sigma}^t$-$\dot{\varepsilon}^t$ relation cannot be represented by a straight line. The data for the Cu–Al alloy shown in Fig.3 indicate clearly the trend similar to that expected from the theoretical considerations.

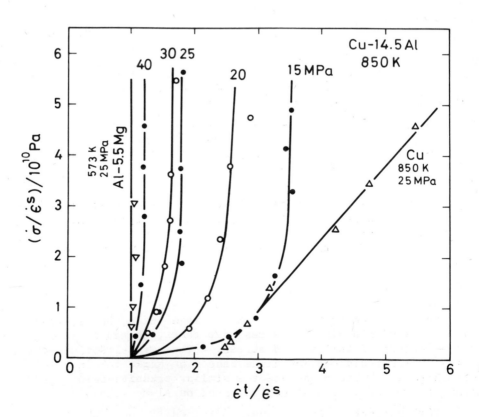

Fig.3 Strain-rate changes upon tensile deformation with a constant-rate increase in stress.

6. CONCLUSIONS

A differential formula among the total strain rate, the plastic strain, the applied stress and time is derived for high-temperature deformation of materials which show obvious anelasticity. The recovery rate, r, and the work-hardening rate, h, in these materials are also defined, which are different from those obtained in pure metals.

Only apparent values of r' and h' can be determined experimentally by stress-relaxation tests and/or strain-rate change tests in alloys exhibiting Class II behavior during steady-state creep. These values include the contribution of anelasticity. These measured but apparent parameters can be regarded as indicators of the stress necessary for climbing and gliding of dislocations. The applied-stress dependence of these apparent r' and h' represents the relative contribution of climbing and of gliding of dislocations to the total strain rate.

ACKNOWLEDMENTS

The authors wish to thank Dr. K. Maruyama for his valuable discussions.

The work was supported in part by a Grant-in-Aid for Scientific Research(C) from the Ministry of Education, and also by Kawasaki Steel Corporations.

REFERENCES

1. POIRIER, J. – 'Microscopic Creep Models and the Interpretation of Stress-Drop Tests during Creep', Acta Met., 1977, 25, 913-917.
2. THORPE, W.R. and SMITH, I.O. – 'A Model for High Temperature Creep Incorporating both Recovery and Thermally Activated Glide', phys. stat. sol. (a), 1979, 52, 487-497.
3. ABE, K., YOSHINAGA, H., and MOROZUMI, S. – 'A Method of Discerning Frictional Stress and Internal Stress by the Stress Relaxation Test', Trans. Japan Inst. Metals, 1977, 18, 479-487.
4. MARUYAMA, K. and KARASHIMA, S. – 'Theoretical Consideration of Measurement of Work-Hardening and Recovery Rates during High Temperature Creep', Trans. Japan Inst. Metals, 1975, 16, 671-678.
5. SAKURAI, S., ABE, K., YOSHINAGA, H. and MOROZOMI, S. – 'Strain Hardening and Recovery in High-Temperature Deformation by Pure-Metal Mode', Nippon Kinzoku Gakkai-shi (J. Japan Inst. Metals), 1978, 42, 432-439.

6. OIKAWA, H. - 'The Absence of Instantaneous Plastic Strain upon Stress Changes during Creep of Some Solid Solutions', Phil. Mag. A, 1978, 37, 707-710.

7. MURTY, K.L., MOHAMED, F.A. and DORN, J.E. - 'Viscous Glide, Dislocation Climb and Newtonian Viscous Deformation Mechanisms of High Tempeaature Creep in Al-3 Mg', Acta Met., 1972, 20, 1009-1018.

8. OIKAWA, H., SUGAWARA, K. and KARASHIMA, S. - 'Creep Behavior of Al-2.2 at% Mg Alloy at 573 K' Trans. Japan Inst. Metals, 1978, 19, 611-616.

9. PAHUTOVA, M. nnd CADEK, J. - 'On Two Types of Creep Behaviour of F.C.C. Solid Solutions', phys. stat. sol. (a), 1979, 56, 305-313.

10. YASUDA, A., OIKAWA, H. and KARASHIMA, S. - 'High Temperature Creep of Cu-14.5 at% Al Alloy with Special Reference to Recovery and Strain Hardening' (to be published).

11. PAHUTOVA, M., private Communications, 1970.

12. YASUDA, A., OIKAWA, H. and KARASHIMA, S. - 'Strain-Rate Change Tests during High Temperature Deformation of Alloys' (to be published).

WORK HARDENING AND RECOVERY RATES OF INTERNAL STRESS IN PURE
METALS AND ALLOYS

H. Yoshinaga

Department of Materials Science and Technology, Kyushu
University, 812 Fukuoka, Japan.

SUMMARY

There are many controversies about the existence of some
appreciable effective stress in deforming pure metals at high
temperatures. However, the method proposed recently by the
present author and his coworkers may be reliable for determining whether the effective stress component is appreciable or
negligible. The results obtained by applying the method to
pure aluminium and Al-5.4at%Mg alloy show that the stress component is negligible in pure aluminium, while it is appreciable in the alloy.

The work hardening and recovery rates of the internal
stress during deformation can be determined by the strain-rate
change test when the effective stress is negligible, while
they can be measured by the applied-stress change test when
the effective stress is appreciable.

The rates measured in the alloy are very different from
those measured in pure aluminium. The difference is discussed
from the effect of solute atmosphere dragging in the alloy.

1.   INTRODUCTION

The great difference observed in high-temperature deformation behaviour between pure fcc and bcc metals and their solution hardened alloys has been inferred by some authors [1]-
[5] to be due to the difference in behaviour of dislocation
motion; the motion is of free-flight type in the pure metals,
while it is viscous in the alloys because the dislocations
drag solute atmospheres. In the former case, the flow stress
is determined by the athermal internal stress, while in the
latter case it is determined by the drag and internal stresses.
The drag stress at high temperatures depends on temperature
and dislocation velocity as the thermal stress does so in gen-

eral.

In the high-temperature deformation, however, the internal stress is hard to be discerned from the thermal effective stress, because the decreasing of the internal stress due to recovery and/or recrystallization occurs simultaneously with its increase due to deformation, and the internal stress will depend on temperature and strain rate in a like manner as the effective stress. And rather many papers reported hitherto [6] -[11] showed that some appreciable effective stress exists in deforming pure metals.

The method proposed recently by the present author and his coworkers [12], which uses the strain-rate change test, may be useful to determine whether the effective stress is appreciable or negligible. The results obtained by applying the method to high-temperature deformation of pure aluminium and Al-5.4at%Mg alloy show that the stress is negligible in pure aluminium, while it is appreciable in the alloy [12].

When the stress component is negligible, the flow stress is determined by the internal stress and the work hardening and recovery rates of the internal stress can be directly measured by the strain-rate change test [13]. When the effective stress is appreciable, however, this stress component must be measured and reduced from the flow stress to determine the internal stress component. The method employed to determine the two rates mentioned above will be described and the rates measured by applying the method to high-temperature deformation of the alloy will be compared with those in pure aluminium. The effect of solute atmosphere dragging on the rates will be discussed.

2. DETERMINATION OF THE TYPE OF DISLOCATION MOTION

When the dislocations move in a free-flight manner, the flow stress will be determined by the internal stress and the effective stress will be negligible. Whether this is the case or not may be determined by the method proposed recently [12]. When this is the case, the flow stress σ will change only through the change of dislocation density and arrangement, and the Bailey-Orowan equation may hold. Then the plastic strain rate $\dot{\varepsilon}$ may be determined by the changing rate of the applied stress $\dot{\sigma}$, recovery rate r and work hardening rate h as

$$\dot{\varepsilon} = (\dot{\sigma}+r)/h \tag{1}$$

The elastic strain rate $\dot{\varepsilon}_e$ is related to $\dot{\sigma}$ as

$$\dot{\varepsilon}_e = \dot{\sigma}/K \tag{2}$$

where K is the apparent Young's modulus which includes the elastic deformation of the machine.

Then the apparent strain rate $\dot{\varepsilon}_a = \dot{\varepsilon} + \dot{\varepsilon}_e$ as

$$\dot{\varepsilon} = \frac{K\dot{\varepsilon}_a + r}{K+h} \tag{3}$$

This equation means that the plastic strain rate will change discontinuously by a sudden change of apparent strain rate.

The plastic strain rate is also expressed as

$$\dot{\varepsilon} = \dot{\varepsilon}_a(1-\frac{1}{K}\frac{d\sigma}{d\varepsilon_a})\qquad(4)$$

When the apparent strain rate is changed suddenly from $\dot{\varepsilon}_{ao}$ to $\dot{\varepsilon}_{af}$ on the way of the steady-state deformation ($d\sigma/d\varepsilon_a=0$) at $\dot{\varepsilon}_{ao}$, the slope of the stress-strain curve immediately after the change may be expressed as

$$(\frac{d\sigma}{d\varepsilon_a})_f = \frac{Kh}{K+h}(1-\frac{\dot{\varepsilon}_{ao}}{\dot{\varepsilon}_{af}})\qquad(5)$$

On the other hand, when the effective stress is appreciable, the plastic strain rate does not change discontinuously, because the strain rate is a continuous function of the applied stress and the internal dislocation structure and these parameters do not change continuously. Then, the plastic strain rate may be almost equal immediately before and after the change. In this case, the equation

$$(\frac{d\sigma}{d\varepsilon_a})_f = K(1-\frac{\dot{\varepsilon}_{ao}}{\dot{\varepsilon}_{af}})\qquad(6)$$

may hold.

The above is the essential part of the theory proposed by the present author and his coworkers [12]. They applied this method to high-temperature deformation of pure aluminium and Al-5.4at%Mg alloy. One of their results is shown in Fig. 1.

It is clear that the effective stress is negligible in pure aluminium, while it is appreciable in the alloy. They also discussed the possible error included in their results and showed that their conclusion may be valid if the data points lie on a straight line passing through the point ($\dot{\varepsilon}_{ao}/\dot{\varepsilon}_{af}=1$, $d\sigma/d\varepsilon_a=0$). The stress change $\delta\sigma$ necessary for the measurement of the slope of the stress-

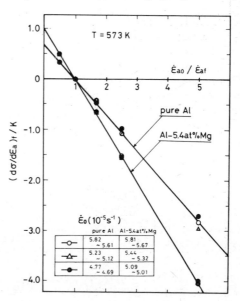

Fig.1 $(d\sigma/d\varepsilon_a)_f/K$ vs. $\dot{\varepsilon}_{ao}/\dot{\varepsilon}_{af}$.

strain curve was as small as 0.5% of the flow stress. Then, in pure aluminium, the possible effective stress may also be so small that it is practically negligible if it exists.

3. DETERMINATION OF THE WORK HADENING AND RECOVERY RATES OF THE INTERNAL STRESS IN PURE METALS

When the effective stress is negligible, the work hardening rate of the internal stress can be directly measured by using the strain-rate change test. Figure 2 is the results obtained by the present author and his coworker [13]. From the slope of the straight line, the work hardening rate can be determined by using equation (5).

Even when the viscous motion of some dislocations such as those on sub-boundaries may also contribute to the plastic deformation, the slope is not affected by the viscous contribution and it is determined solely by the free-flight contribution [14]. However, the estimation of the recovery rate may be affected by the viscous contribution, because the rate was estimated in [13] from the equation

$$r = \dot{\varepsilon}/h \qquad (7)$$

for the steady state deformation. But the error may not be very significant [13].

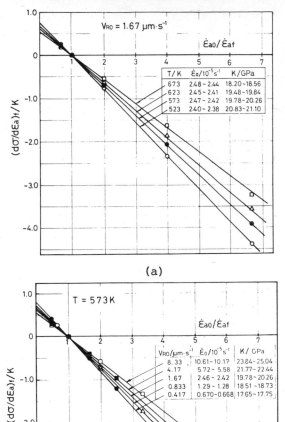

(a)

(b)

Fig.2 $(d\sigma/d\varepsilon_a)_f/K$ vs. $\dot{\varepsilon}_{ao}/\dot{\varepsilon}_{af}$ for pure Al. (a) Temperature dependence. (b) Strain-rate dependence.

4. DETERMINATION OF THE WORK HARDENING AND RECOVERY RATES OF
   THE INTERNAL STRESS IN ALLOYS.

When the effective stress is the dragging stress of the solute atmosphere, the dislocation velocity may be proportional to the effective stress at high temperatures, and the plastic strain rate may be expressed as

$$\dot{\varepsilon} = \alpha\rho bB(\sigma-\overline{\sigma_i}) \tag{8}$$

where α is the geometrical factor, ρ is the mobile dislocation density, b is the Burgers vector, B is the mobility of the dislocation, and $\overline{\sigma_i}$ is the average internal stress.

When the applied stress is suddenly changed by $\Delta\sigma$ on the way of creep deformation, the strain rate may change by

$$\Delta\dot{\varepsilon} = \alpha\rho bB\Delta\sigma \tag{9}$$

Therefore, the internal stress may be obtained from the equation

$$\overline{\sigma_i} = \sigma-\dot{\varepsilon}(\Delta\sigma/\Delta\dot{\varepsilon}) \tag{10}$$

by measuring $\Delta\dot{\varepsilon}$.

Figure 3 is an experimental result obtained in Al-5.7at%Mg alloy under steady-state deformation. The data points may be considered to lie on a straight line.

When the applied stress is changed from σ_0 by $\Delta\sigma$ and the deformation continues under the new applied stress, the internal stress may change. The changing process may be measured by repeating the applied-stress change test at various strains under the new applied stress and returning to the initial steady-state deformation at each time. The process is schematically shown in Fig. 4.

The infinitesimal change of the internal stress during deformation may be expressed as

$$d\overline{\sigma_i} = hd\varepsilon-rdt \tag{11}$$

where $hd\varepsilon$ is the work hardening term and rdt is the recovery term. Then the initial changing rate of the internal stress may be estimated from the initial slope of the internal stress-time curve, as shown in the figure.

Figure 5 is an example of the curves obtained by changing the applied stress to various levels for the same steady-state deformation of the alloy. The initial slope cannot be measured precisely, but rough estimation is possible.

If the values of h and r should not change discontinuously by the applied-stress change, their initial values under the new applied stress would be the same as those in the previous steady-state deformation, and the initial slope would be proportional to the initial plastic strain rate under the new applied-stress. Then the values of h and r would be estimated definitely as shown schematically in Fig.6. But the data points shown in Fig. 7 cannot be considered to lie on a straight line.

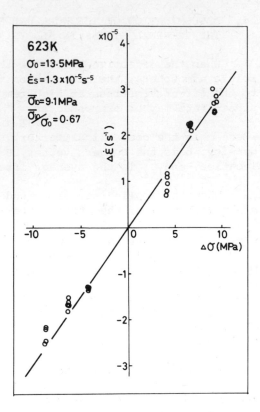

Fig.3 $\Delta\dot{\varepsilon}$ vs. $\Delta\sigma$, obtained by applying the applied-stress change test to the steady-state deformation of Al-5.7at%Mg alloy.

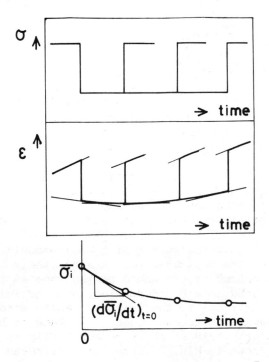

Fig.4 Illustrations for the method proposed to determine the work hardening and recovery rates of the internal stress in deforming alloys at high temperatures.

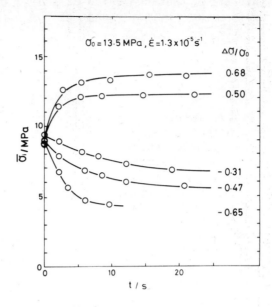

Fig.5 The changing
procedures of the
internal stress
after the change of
applied stress on
the way of steady-
state deformation
of Al-5.7at%Mg
alloy.

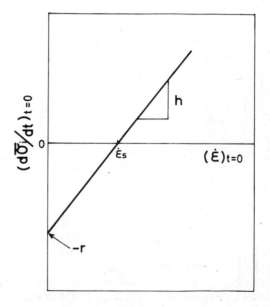

Fig.6 $(d\overline{\sigma_i}/dt)_{t=0}$
vs. $(\varepsilon)_{t=0}$, expected
if the values of h
and r should not
be changed disconti-
nuously by an abrupt
change of applied
stress on the way
of steady-state
deformation of alloys.

This shows that the value of h or r or the both do change discontinuously by the change of the applied stress.

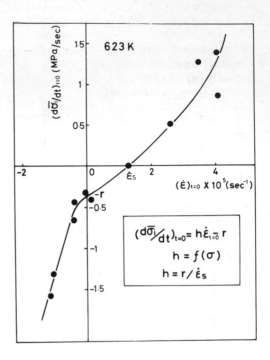

Fig.7 Initial changing rate of the internal stress vs. initial strain rate after the change of applied stress on the way of steady-state deformation of Al-5.7at%Mg alloy.

In this case, the slope of the initial changing rate-initial plastic strain rate curve at the point of $\dot{\varepsilon} = \dot{\varepsilon}_s$, i.e., $\sigma = \sigma_0$ does not give the correct h at σ_0, because

$$\left(\frac{d}{d\varepsilon}\left(\frac{d\overline{\sigma}_i}{dt}\right)_{t=0}\right)_{\sigma=\sigma_0} = \left(\frac{dh}{d\sigma}\right)_{\sigma=\sigma_0}(\sigma_0-\overline{\sigma}_{io})+h(\sigma_0)$$
$$-\frac{1}{\alpha\rho b B}\left(\frac{dr}{d\sigma}\right)_{\sigma=\sigma_0} \quad (12)$$

The slope may be considered to increase as $\dot{\varepsilon}_{t=0}$ increases, i.e., σ increases, though the scatter is large. Thus the value of h is thought to increase by the increase of the applied stress even if the internal dislocation structure is the same, which may be due to the increase of the multiplication rate of dislocations by the increase of the applied stress.

In the present paper, the value of r is estimated from the point $\dot{\varepsilon}_{t=0} = 0$, which means that r is assumed not to change by the decrease of the applied stress to the internal stress, and the first term on the right hand side is assumed to much larger than the third term. If the recovery rate is determined mainly by the internal stress and not affected significantly by the effective stress, the assumption made above may be valid.

The value of h is estimated from the equation

$$h = r/\dot{\varepsilon}_s \quad (13)$$

where $\dot{\varepsilon}_s$ is the steady-state strain rate under the applied stress σ_o. The value of h thus determined is not much different from the value estimated from the slope at $\dot{\varepsilon}=\dot{\varepsilon}_s$, when the large scatter of the data points is considered.

Figure 8 shows the internal stress dependence of r in the alloy at various temperatures. The data measured in pure aluminium [13] are also plotted for comparison. Figure 9 shows the dependence of h.

Fig.8 r/E vs. $\overline{\sigma_{io}}$/E (E; the Young's modulus) at various temperatures, obtained for the steady-state deformation of Al-5.7at% Mg alloy, together with the data for pure aluminium.

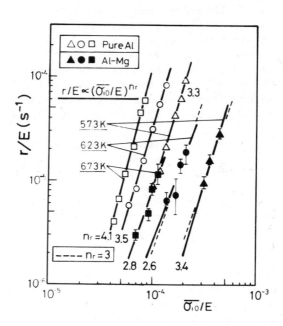

Fig.9 h/E vs. $\overline{\sigma_{io}}$/E at various temperatures, obtained for the steady-state deformation of Al-5.7at% Mg alloy and pure aluminium.

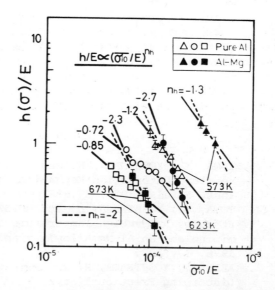

5. DISCUSSION

The value of r is always much lower in Al-5.7at%Mg alloy than in pure aluminium at the same internal stress level, as shown in Fig. 8.

When the internal stress is the same, the dislocation density at places where the stress for the deformation is determined may be assumed to be almost the same between the metal and the alloy. Then, the driving force and the distance which the dislocations must move until they annihilate each other may be almost the same. If this is the case, the experimental result will show that the mobility of the dislocations for the recovery is much lower in the alloy than in the metal. For the recovery the climbing motion of dislocations may be important. Therefore, it is thought that the large dragging effect of solute atmosphere which is well-known for the glide motion of dislocations is also important for the climbing motion.

Contrary to the difference in r, the value of h is larger in the alloy than in the metal in most cases. The difference decreases at higher temperatures and higher internal stresses.

The dislocations tends to segregate to form subboundaries in the metal, while they are distributed rather uniformly in the alloy [15]. The internal stress for the deformation may be determined at subboundaries in the metal, where the dislocation density is deduced to be nearly the same as that of rather uniformly distributed dislocations in the alloy. Then, at the same internal stress, the moving dislocations may be much more in the alloy than in the metal, which may increase the multiplication rate of dislocations in the alloy. This may be the cause of higher in h in the alloy.

As the temperature rises, the mobility of glide dislocations increases, which may be the cause of temperature dependence of the difference in h, but the internal stress dependence cannot be explained.

For the mechanism by which the temperature and internal-stress dependence of h and r requires further investigations.

REFERENCES

1. TOMA, K., YOSHINAGA, H. and MOROZUMI, S. - 'The High-Temperature Deformation Mechanism in Pure Metals'. J.Japan Inst. Metals, 1975, 39, 621.
2. TOMA, K., YOSHINAGA, H. and MOROZUMI, S. - 'Internal Stresses during High-Temperature Deformation of Pure Aluminium and an Al-Mg Alloy'. Trans. Japan Inst. Metals, 1976, 17, 102.
3. ABE, K., YOSHINAGA, H. and MOROZUMI, S. - 'A Method of Discerning Frictional Stress and Internal Stress by the

Stress Relaxation Test'. Trans. Japan Inst. Metals, 1977, <u>18</u>, 479.

4.   OIKAWA, H. and SUGAWARA, K. - 'Instantaneous Plastic Strain Associated with Stress Increments during the Steady State Creep of Al and Al-5.5at%Mg Alloy'. Scripta Met., 1978, <u>12</u>, 85.

5.   OIKAWA, H. - 'The Absence of Instantaneous Plastic Strain upon Stress Changes during Creep of Some Solid Solutions'. Phil. Mag. A, 1978, <u>37</u>, 707.

6.   BOČEK, M. and SCHNEIDER, H. - 'Internal Stress Measurements on Vanadium at Elevated Temperatures'. Scripta Met., 1970, <u>4</u>, 369.

7.   AHLQUIST, C.N. and NIX, W.D. - 'The Measurement of Internal Stresses during Creep of Al and Al-Mg Alloys'. Acta Met., 1971, <u>19</u>, 373.

8.   TOBOLOVÁ, Z. and ČADEK, J. - 'An Interpretation of Steady State Creep'. Phil. Mag., 1972, <u>26</u>, 1419.

9.   KIKUCHI, S., KAJITANI, M., ENJO, T. and ADACHI, M. - 'High Temperature Steady-State Deformation of Cu-Al Solid Solutions'. J.Japan Inst. Metals, 1973, <u>37</u>, 228.

10. OIKAWA, H., MAEDA, M. and KARASHIMA, S. - 'Steady-State Creep Characteristics of Fe-3.5at%Mo Alloy'. J.Japan Inst. Metals, 1973, <u>37</u>, 599.

11. PAHUTOVÁ, M., ORLOVÁ, A., KUCHAŘOVÁ, K. and ČADEK, J. - 'Steady-State Creep in Alpha Iron as Described in Terms of Effective Stress and Dislocation Dynamics'. Phil. Mag., 1973, <u>28</u>, 1099.

12. YOSHINAGA, H., HOLITA, Z. and KURISHITA, H. - 'Determination of High-Temperature Deformation Mechanism in Crystalline Materials by the Strain-Rate Change Test'. Submitted to Acta Met.

13. HORITA, Z. and YOSHINAGA, H. - 'Separate Determination of Work-Hardening and Softening Rates During High-Temperature Deformation of Pure Aluminum'. J.Japan Inst. Metals, 1980, <u>44</u>, 1273.

14. YOSHINAGA, H., KURISHITA, H. and GOTO , S. - 'High-Temperature Deformation Mechanism in Crystalline Materials'. Reports of Graduate School of Engineering Sciences, Kyushu University, 1980, <u>2</u>, 1.

15. HORIUCHI, R. and OTSUKA, M. - 'Mechanism of High Temperature Creep of Aluminum-Magnesium Solid Solution Alloys'. J.Japan Inst. Metals, 1971, <u>35</u>, 406.

THE EFFECT OF INSTANTANEOUS STRAIN ON CREEP MEASUREMENTS
AT APPARENT CONSTANT STRUCTURE

Terence G. Langdon and Parviz Yavari

Departments of Materials Science and Mechanical Engineering,
University of Southern California, Los Angeles,
California 90007, U.S.A.

SUMMARY

Stress increment and decrement experiments were performed
on the Al-5% Mg and Al-5% Zn solid solution alloys. The
results suggest that there is an instantaneous plastic strain
in both materials, except when the stress decrement is less
than ∿25%. It is concluded that an estimate of the stress
exponent at true constant structure requires an extrapolation
to the condition where the instantaneous plastic strain is
zero.

1. INTRODUCTION

Under creep conditions at temperatures above ∿0.4 T_m,
where T_m is the absolute melting temperature of the material,
the steady-state creep rate, $\dot{\varepsilon}_s$, is usually related to the
applied stress, σ, through an equation of the form

$$\dot{\varepsilon}_s = A\sigma^n \tag{1}$$

where A is a constant which incorporates the dependence on
temperature, and n is termed the stress exponent.

The value of n is usually determined for any material by
using several different but nominally identical specimens and
conducting a number of experiments at the same temperature but
over a range of stresses. If the steady-state creep rate is
recorded for each specimen, the value of n is given by the
slope of a logarithmic plot of $\dot{\varepsilon}_s$ versus σ, so that

$$n = \frac{\partial \ln \dot{\varepsilon}_s}{\partial \ln \sigma} \tag{2}$$

Alternatively, if a single specimen is taken into the steady-state stage of creep and then subjected to a change in the applied stress (either incremental or decremental), a new stress exponent may be defined by the relation

$$n^* = \frac{\ln (\dot{\varepsilon}_{s2}/\dot{\varepsilon}_{s1})}{\ln (\sigma_2/\sigma_1)} \tag{3}$$

where $\dot{\varepsilon}_{s1}$ and $\dot{\varepsilon}_{s2}$ are the steady-state creep rates before and after a change in stress from σ_1 to σ_2. In this type of experiment, there is usually a relatively brief transient period immediately following the change in stress, and then the specimen enters steady-state flow. This transient period is generally attributed to an adjustment in the internal substructure to reflect the new stress level, and it follows therefore that n and n* are not strictly the stress exponents under conditions of constant structure.

To overcome this problem, it is possible to define a third stress exponent, designated n', which is given by the relationship

$$n' = \frac{\ln (\dot{\varepsilon}_{i2}/\dot{\varepsilon}_{i1})}{\ln (\sigma_2/\sigma_1)} \tag{4}$$

where $\dot{\varepsilon}_{i1}$ is the instantaneous creep rate immediately before an increase (or a relatively small decrease) in the applied stress from σ_1 to σ_2, and $\dot{\varepsilon}_{i2}$ is the creep rate at any instant after the change.

It follows from equations (3) and (4) that n' → n* when the stress change is performed in the steady-state region and $\dot{\varepsilon}_{i2}$ → $\dot{\varepsilon}_{s2}$. Furthermore, the stress exponent *at apparent constant structure* is obtained by putting $\dot{\varepsilon}_{i2}$ equal to the instantaneous creep rate immediately after the change in stress, so that the measured strain at the new stress level, ε_2, is zero. However, this latter procedure assumes that the *instantaneous* deformation associated with the stress change, $\Delta\varepsilon$, is entirely elastic. Thus, if $\Delta\varepsilon$ is expressed as

$$\Delta\varepsilon = \Delta\varepsilon_e + \Delta\varepsilon_p \tag{5}$$

where $\Delta\varepsilon_e$ and $\Delta\varepsilon_p$ are the instantaneous elastic and plastic strains due to the change in stress, it is required that $\Delta\varepsilon_p = 0$. For the situation where $\Delta\varepsilon_p \neq 0$, it is therefore only possible to estimate the stress exponent *at true constant structure* by extrapolating to determine the value of n' when $\varepsilon_2 + \Delta\varepsilon_p = 0$.

This paper describes the results obtained in a series of experiments which were designed to investigate the nature of

the instantaneous strain following a stress change. Tests were
conducted on two different solid solution alloys, Al-5% Mg and
Al-5% Zn. These two alloys were selected because it is known
that dilute Al-Mg alloys exhibit a stress exponent of $\sim$3 over
a wide stress range at elevated temperatures [1-3] whereas
dilute Al-Zn alloys give a stress exponent of $\sim$4.5 - 5 [4,5].
The latter exponent suggests control by some form of dislocation climb process, whereas the exponent of $\sim$3 indicates
control by glide due to the operation of a viscous drag
process on the moving dislocations [6].

2. EXPERIMENTAL MATERIALS AND PROCEDURE

The materials used in this investigation were Al-5 wt % Mg
and Al-5 wt % Zn having initial grain sizes of 600 and 700 μm,
respectively. All of the creep tests were conducted in double
shear, using specimens cut to the configuration and dimensions
described earlier [7] and with the longitudinal specimen axis
perpendicular to the rolling direction. The experiments were
performed in air under conditions of constant shear stress,
and with the temperature held constant to ±1 K of the reported
values.

The stress change experiments were performed by either
decreasing or increasing the load using an electric jack. For
a stress decrement, a predetermined load, equal in magnitude
to the amount to be removed, was attached with a wire to the
bottom of the load pan, and then this additional load was
carefully removed at the selected strain by raising the
electric jack. For a stress increment, the predetermined load
was placed on the jack, and then the jack was carefully
lowered at the selected strain so that the additional load
became attached to the load pan with a small hook. The time
required to perform a change in stress using this procedure
was less than 1 second. The shear strain was measured continuously throughout the stress change using a linear variable
differential transformer, and monitored on a strip-chart
recorder with an accuracy of 8×10^{-6}.

3. EXPERIMENTAL RESULTS

3.1 Dependence of strain rate on stress and temperature

Initially, experiments were conducted to determine the
value of n from a series of uninterrupted tests at a constant
temperature. The results are shown in Fig. 1, logarithmically
plotted as steady-state shear creep rate, $\dot{\gamma}$, versus shear
stress, τ, for (a) Al-5% Mg and (b) Al-5% Zn. In these two
plots, each experimental datum point represents the result
obtained from a different specimen.

For Al-5% Mg at absolute temperatures, T, of 673 and

723 K, the two sets of data fall along parallel lines with
n = 3.0 ± 0.1. This value is consistent with several reports
on Al-Mg alloys [1-3], although it should be noted that earlier
results on Al-5% Mg show that there is a transition to a
higher value of n (∿4.4) at lower stress levels [3]. For
Al-5% Zn, the datum points at four temperatures in the range
from 573 to 823 K give n = 4.4 ± 0.1, and again this is consistent with other reports on Al-Zn alloys at comparable stress
levels [4,5].

The effect of a change in stress on the macroscopic creep
behavior was examined in detail, and the basic trends were
described earlier [7]. In stress increment experiments, Al-5%
Mg exhibits a brief inverted transient following the stress
change, so that there is a short period in which the creep
rate increases with strain. By contrast, Al-5% Zn shows a
normal transient stage following a stress increase, so that
the creep rate decreases to the final steady-state value.

The situation is more complicated in stress reduction
experiments. In Al-5% Mg, there is a brief normal transient
in which the strain rate decreases to steady-state flow. However, Al-5% Zn exhibits basically an inverted transient stage
following a decrease in stress, although the precise shape of
this transient depends on the magnitude of the reduction.
This is illustrated in Fig. 2 for stress decrement experiments
conducted at 823 K, plotted as shear strain rate versus shear
strain, γ: the stress reduction gives a sharp drop in $\dot{\gamma}$ which
is indicated by the vertical lines. In each case, the initial
shear stress, τ_1, was 0.65 MPa, and the final shear stress
following the reduction, τ_2, varied from 0.40 to 0.21 MPa.
For the smallest decrease, to 0.40 MPa, the upper curve shows
a normal inverted transient which is similar in form to, but
less pronounced than, the transients observed in high purity
Al after stress reductions [8]. For the two larger stress
decrements, the shear creep rate initially decreases, reaches
a minimum at a shear strain which is slightly beyond the point
at which the stress was changed, and then increases again to a
steady-state value. A similar type of transient behavior has
been reported also for Al-11% Zn [9].

3.2 The nature of the instantaneous strain

As discussed in section 1, the primary objective of this
work was to investigate the nature of the instantaneous strain
in stress change experiments. Accordingly, a series of tests
was conducted in which the stress was either increased or
decreased at a single point in essentially steady-state flow.
For both alloys, these tests were performed at a temperature
of 673 K. Initial stresses, τ_1, of 2.88 and 1.69 MPa were
selected for Al-5% Mg and Al-5% Zn, respectively: reference to
Fig. 1(a) and (b) shows that these stress levels at 673 K are

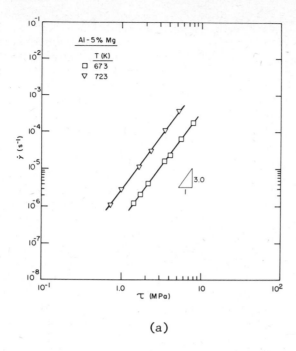

(a)

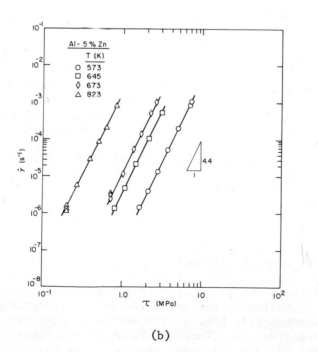

(b)

Fig. 1   Steady-state shear creep rate versus shear stress for
(a) Al-5% Mg and (b) Al-5% Zn.

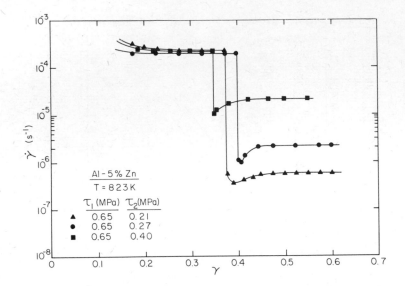

Fig. 2   Shear strain rate versus shear strain for Al-5% Zn,
showing the effect of stress reductions during the
test.

within the linear region of $\dot{\gamma}$ versus τ used to calculate the
values of n.

Figure 3 shows examples of the strip-chart recordings for
Al-5% Mg for (a) a stress increment to a new stress level, τ_2,
of 4.58 MPa and (b) a stress decrement to τ_2 equal to 1.18 MPa:
this is equivalent to changes in shear stress, $\Delta\tau$, of ±0.7, so
that $\Delta\tau/\tau_1$ is either +0.59 [Fig. 3(a)] or -0.59 [Fig. 3(b)],
respectively.

The chart recording is moving from right to left in Fig.
3, and the external notation indicates the scale for time, t,
and strain, γ. In stress increment experiments, the instantaneous strain, $\Delta\gamma$, was measured from the linear displacement
between the point on the forward creep curve at the instant of
stress change and the point on the new forward creep curve
immediately after the change: this is illustrated in Fig. 3(a).
In stress decrement experiments, $\Delta\gamma$ was measured from the
linear displacement between the point at which the stress was
reduced and the first point associated with either forward
creep, an incubation period, or negative (or decreasing) creep:
this is illustrated in Fig. 3(b) where there is a very short
period of negative creep immediately after the stress change.
It was estimated from a large number of tests that the errors
on $\Delta\gamma$ resulting from these procedures were typically about
±15% for stress increments and about ±10% for stress
decrements.

A similar set of chart recordings is shown in Fig. 4 for

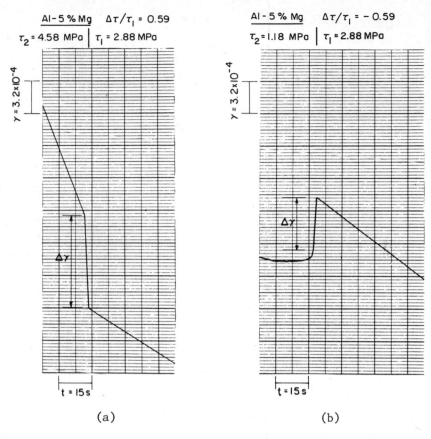

Fig. 3   Strip-chart recordings for Al-5% Mg showing the
effect of (a) a stress increment and (b) a stress
decrement. The procedure used for measuring $\Delta\gamma$
is indicated.

Al-5% Zn for (a) a stress increase to 2.58 MPa and (b) a stress
decrease to 0.63 MPa, and again the recordings indicate the
measured values of $\Delta\gamma$. It is important to note that the scale
for γ differs by a factor of two between Fig. 3(a) and (b), so
that the difference in $\Delta\gamma$ between these two experiments is even
larger than may appear from a comparison of the two recordings.

By conducting a large number of experiments of this type,
it was possible to plot the measured values of $\Delta\gamma$ against the
imposed changes in shear stress, $\Delta\tau$, as shown in Fig. 5 for
(a) Al-5% Mg and (b) Al-5% Zn. Each point in Fig. 5 refers to
a different specimen tested at the same temperature (673 K)
and the same initial stress (either 2.88 MPa for Al-5% Mg or
1.69 MPa for Al-5% Zn). Several tests were performed under
identical testing conditions, and in these cases the individual
datum points in Fig. 5 indicate that there is relatively little
scatter in these experiments.

78

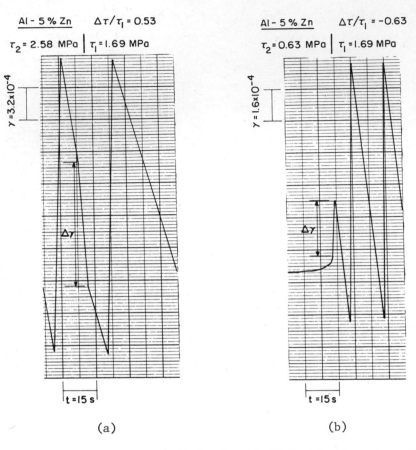

(a)   (b)

Fig. 4   Strip-chart recordings for Al–5% Zn showing the effect
of (a) a stress increment and (b) a stress decrement.
The procedure used for measuring Δγ is indicated.

The solid lines in Fig. 5 indicate the experimental trends
for either a stress increase (upper lines) or a stress decrease
(lower lines). A comparison of Fig. 5(a) and (b) shows that
the instantaneous strains are different for these two alloys,
although in both materials the value of Δγ for any value of Δτ
is larger when making a stress increment. Figure 5(a) shows
that Δγ is relatively insensitive to whether the stress is
increased or decreased in Al–5% Mg, whereas Fig. 5(b) shows
that there is a very significant difference between these two
procedures in Al–5% Zn.

The broken lines superimposed in Fig. 5 represent the
instantaneous elastic deformation of the material, calculated
from the relation

$$\Delta\gamma = \frac{\Delta\tau}{G} \tag{6}$$

where G is the shear modulus. This procedure seems reasonable

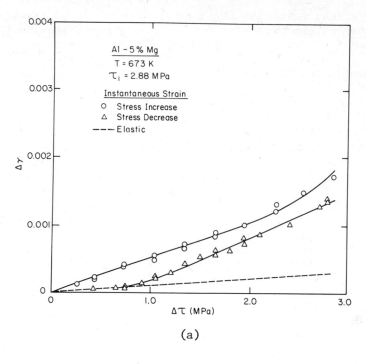

(a)

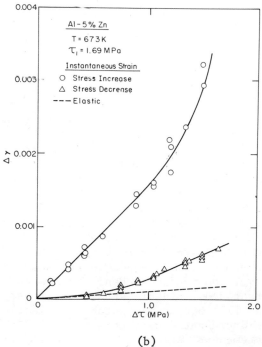

(b)

Fig. 5   Instantaneous shear strain following a stress
         increment or decrement for (a) Al-5% Mg and
         (b) Al-5% Zn. The broken lines indicate the
         instantaneous elastic strain of the material.

because stress reduction tests at room temperature showed that
the values of the instantaneous strains were similar to those
estimated from the elastic modulus at this temperature. The
values of the shear moduli for these two alloys are very close
to the value for pure Al [10,11], so that G was taken as [12]

$$G = (3.022 \times 10^4 \text{ MPa}) - (16.0 \times T) \text{ MPa K}^{-1} \qquad (7)$$

where T = 673 K in Fig. 5. From the positions of these broken
lines, it appears that the instantaneous shear strain generally
contains a plastic contribution in addition to the elastic
strain, except at very small stress reductions ($\Delta\tau$ below about
0.8 and 0.5 MPa for Al-5% Mg and Al-5% Zn, respectively) where
the experimental measurements of $\Delta\gamma$ are essentially equal to
the anticipated instantaneous elastic shear strain, $\Delta\gamma_e$. Under
all other experimental conditions, the separations between the
broken and solid lines in Fig. 5 are equal to the instantaneous
plastic shear strain, $\Delta\gamma_p$. Thus, it is concluded that, for
almost all of the stress change conditions used in this
investigation, $\Delta\gamma_p \neq 0$.

4. DISCUSSION

The results contained in Fig. 5 indicate that there is an
instantaneous plastic shear strain, $\Delta\gamma_p$, in stress increment
and decrement tests, except when the stress reduction is less
than about 25 - 30% of the initial stress level. It is
therefore not correct, in general, to assume that the instantaneous strain is entirely elastic.

There has been considerable discussion concerning the
nature of the instantaneous strain [13], and it is instructive
to compare the present results with two other viewpoints.

On the one hand, early experiments by Wilshire and co-
workers [14,15] demonstrated a large plastic contraction in
stress reduction experiments on Cu single crystals and a small
but significant plastic contraction in polycrystalline Cu.
More recently, a similar effect was reported for high purity Al
[16], although it was noted that the instantaneous contraction
was entirely elastic for stress decrements up to ∿50%. This
result is therefore comparable with the present observation
that a stress reduction of >25 - 30% is necessary in order to
establish $\Delta\gamma > \Delta\gamma_e$. On the other hand, Oikawa and co-workers
reported an instantaneous plastic strain in stress increment
experiments on Al [17] but no instantaneous plastic strain
either in stress decrement experiments on Al [17] or in stress
increment or decrement experiments on Al-5% Mg [17], Cu-4.1% Al
(Cu-10.4 at % Al) [18], Fe-3.0% Mo (Fe-1.8 at % Mo) [18], or
Fe-6.8% Mo (Fe-4.1 at % Mo) [19].

The results obtained by Oikawa and Sugawara [17] are
shown in Fig. 6 for (a) Al-5% Mg and (b) pure Al, plotted as

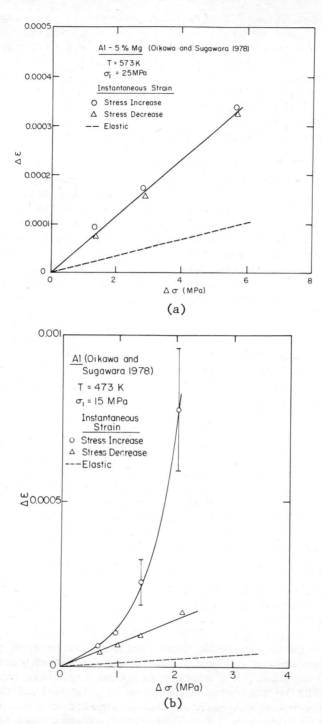

Fig. 6   Instantaneous normal strain following a stress
increment or decrement for (a) Al-5% Mg and
(b) pure Al.   The broken lines indicate the
instantaneous elastic strain of the material.

the instantaneous normal strain, $\Delta\varepsilon$, in tensile experiments versus the change in applied stress, $\Delta\sigma$. The changes in stress were performed from initial stress levels, σ_1, of 25 and 15 MPa, respectively. Since Al has a steady-state stress exponent, n, of ~ 4.5, and this is similar to the behavior of Al-5% Zn [Fig. 1(b)], it would be anticipated that the data shown in Fig. 6(a) and (b) are directly comparable to the present results in Fig. 5(a) and (b). However, a comparison of these two plots reveals two important similarities and a significant difference.

First, the broken lines in Fig. 6 again show the instantaneous elastic deformation of the material, calculated from equations (6) and (7) for pure Al by putting Young's modulus, E, equal to $2G(1 + \nu)$, where ν is Poisson's ratio (0.34 for Al). This procedure gives $E \simeq 6.1 \times 10^4$ MPa at 473 K, which is not significantly different from the value of $E \simeq 6.4 \times 10^4$ MPa quoted by Oikawa and Sugawara [17]. Thus, the positions of these broken lines again suggest that both materials may exhibit an instantaneous plastic strain in stress increment and decrement situations.

Second, the overall trends are essentially the same in both Figs. 5 and 6, with Al-5% Mg exhibiting relatively similar values of $\Delta\gamma$ (or $\Delta\varepsilon$) in stress increment or decrement experiments but Al-5% Zn or pure Al giving much higher values of $\Delta\gamma$ (or $\Delta\varepsilon$) in experiments where the stress is increased.

The significant difference between Figs. 5 and 6 is the possibility of a measurable plastic component in the work of Oikawa and Sugawara [17] even at very low values of $\Delta\sigma$. Furthermore, it should be noted that, whereas the *absolute* values of $\Delta\sigma$ in the experiments on Al-5% Mg in Fig. 6(a) are much higher than in the present investigation [where $\Delta\tau < 2.88$ MPa in Fig. 5(a)], the *percentage* decrements used by Oikawa and Sugawara [17] were very low and never exceeded 25%. In summary, therefore, the present results on Al-5% Zn indicate $\Delta\gamma \simeq \Delta\gamma_e$ in stress decrements up to $\sim 0.25 \, \tau_1$, the results of Parker and Wilshire [16] on high purity Al give $\Delta\varepsilon \simeq \Delta\varepsilon_e$ up to $\sim 0.50 \, \sigma_1$, but the data of Oikawa and Sugawara [17] shown in Fig. 6(b), also on Al, suggest $\Delta\varepsilon \neq \Delta\varepsilon_e$ under all experimental conditions.

There are two important conclusions arising from the present experiments and analysis. First, a determination of the stress exponent, n', at true constant structure requires an extrapolation to the value of n' when $\Delta\varepsilon_p$ is zero. Second, it has been suggested that stress change experiments may be used to distinguish between a recovery creep process when $n \simeq 4.5$ and a viscous glide process when $n \simeq 3$. This argument is based on the concept that the former process leads to an instantaneous plastic strain at a stress increment, whereas the latter process will not give an instantaneous plastic strain since the movement of dislocations is then restrained by their associated clouds of solute atoms [20,21]. However, the present data in Fig. 5, when combined with the results of Oikawa and Sugawara [17] in Fig. 6 where the elastic modulus line is included, indicate that both types of process may lead to some instantaneous plastic strain. Thus, it is suggested that the criterion is better stated in terms of the relative magnitudes of $\Delta\gamma$ (or $\Delta\varepsilon$) in stress change experiments, since recovery creep and n $\simeq$ 4.5 tends to give large differences in the instantaneous strains between stress increments and decrements [Figs. 5(b) and 6(b)] whereas in viscous glide and n $\simeq$ 3 these differences tend to be relatively small [Figs. 5(a) and 6(a)].

5. CONCLUSIONS

(i) There appears to be an instantaneous plastic strain in stress increment and stress decrement experiments on materials exhibiting control by either dislocation climb or viscous glide, except when the stress decrement is less than $\sim$25%.

(ii) Materials deforming by viscous glide exhibit instantaneous strains which are relatively similar in stress increment and decrement experiments, whereas materials deforming by dislocation climb exhibit much larger instantaneous strains when the stress is increased.

(iii) An estimate of the stress exponent at true constant structure requires an extrapolation to the condition where the instantaneous plastic strain is zero.

ACKNOWLEDGMENT

This work was supported by the United States Department of Energy under Contract DE-AM03-76SF00113 PA-DE-AT03-76ER10408.

REFERENCES

1. OIKAWA, H., SUGAWARA, K. and KARASHIMA, S. - "Creep Behavior of Al-2.2 at % Mg Alloy at 573 K," *Trans. Japan Inst. Metals*, 1978, 19, 611.
2. PAHUTOVÁ, M. and ČADEK, J. - "On Two Types of Creep Behaviour of F.C.C. Solid Solution Alloys," *Phys. Stat. Sol. (a)*, 1979, 56, 305.
3. YAVARI, P., MOHAMED, F.A. and LANGDON, T.G. - "Creep and Substructure Formation in an Al-5% Mg Solid Solution Alloy," to be published.
4. ENDO, T., NOMURA, T., ENJYO, T., and ADACHI, M. - "Stress Dependence of Strain Rate during Steady State Creep of Al-Zn Solid Solution Alloys," *J. Japan Inst. Metals*, 1971, 35, 427.
5. HAUSSELT, J. and BLUM, W. - "Dynamic Recovery during and after Steady State Deformation of Al-11 wt % Zn," *Acta Met.*, 1976, 24, 1027.
6. MOHAMED, F.A. and LANGDON, T.G. - "The Transition from Dislocation Climb to Viscous Glide in Creep of Solid

D

Solution Alloys," *Acta Met.*, 1974, 22, 779.

7. LANGDON, T.G., VASTAVA, R.B. and YAVARI, P. - "The Influence of Substructure on Creep Behavior following Stress Changes," *Proceedings of the 5th International Conference on the Strength of Metals and Alloys*, Eds. Haasen, P., Gerold, V. and Kostorz, G., Pergamon Press, Oxford, 1979, vol. 1, p. 271.

8. HORIUCHI, R. and OTSUKA, M. - "Mechanism of High Temperature Creep of Aluminum-Magnesium Solid Solution Alloys," *Trans. Japan Inst. Metals*, 1972, 13, 284.

9. BLUM, W., HAUSSELT, J. and KÖNIG, G. - "Transient Creep and Recovery after Stress Reductions during Steady State Creep of AlZn," *Acta Met.*, 1976, 24, 293.

10. ABE, K., TANJI, Y., YOSHINAGA, H. and MOROZUMI, S. - "Young's Modulus of an Al-Mg Alloy at Elevated Temperatures," *J. Japan Inst. Light Metals*, 1977, 27, 279.

11. KÖSTER, W. and RAUSCHER, W. - "Beziehungen zwischen dem Elastizitätsmodul von Zweistofflegierungen und ihrem Aufbau," *Z. Metallk.*, 1948, 39, 111.

12. SUTTON, P.M. - "The Variation of the Elastic Constants of Crystalline Aluminum with Temperature between 63 K and 773 K," *Phys. Rev.*, 1953, 91, 816.

13. POIRIER, J.P. - "Microscopic Creep Models and the Interpretation of Stress-Drop Tests during Creep," *Acta Met.*, 1977, 25, 913.

14. DAVIES, P.W. and WILSHIRE, B. - "On Internal Stress Measurement and the Mechanism of High Temperature Creep," *Scripta Met.*, 1971, 5, 475.

15. DAVIES, P.W., NELMES, G., WILLIAMS, K.R. and WILSHIRE, B. - "Stress-Change Experiments during High-Temperature Creep of Copper, Iron, and Zinc," *Metal Sci. J.*, 1973, 7, 87.

16. PARKER, J.D. and WILSHIRE, B. - "Rate-controlling Processes during Creep of Super-purity Aluminium," *Phil. Mag. A*, 1980, 41, 665.

17. OIKAWA, H. and SUGAWARA, K. - "Instantaneous Plastic Strain Associated with Stress Increments during the Steady-State Creep of Al and Al-5.5 at. pct. Mg Alloy," *Scripta Met.*, 1978, 12, 85.

18. OIKAWA, H. - "The Absence of Instantaneous Plastic Strain upon Stress Changes during Creep of some Solid Solutions," *Phil. Mag. A*, 1978, 37, 707.

19. OIKAWA, H., SAEKI, M. and KARASHIMA, S. - "Steady-State Creep of Fe-4.1 at % Mo Alloy at High Temperatures," *Trans. Japan Inst. Metals*, 1980, 21, 309.

20. MARUYAMA, K. and KARASHIMA, S. - "Theoretical Consideration of Measurement of Work-Hardening and Recovery Rates during High Temperature Creep," *Trans. Japan Inst. Metals*, 1975, 16, 671.

21. ABE, K., YOSHINAGA, H. and MOROZUMI, S. - "A Method of Discerning Frictional Stress and Internal Stress by the Stress Relaxation Test," *J. Japan Inst. Metals*, 1976, 40, 393.

BERYLLIUM MICRODEFORMATION MECHANISMS

R. S. Polvani, B. W. Christ and E. R. Fuller, Jr.

National Bureau of Standards, Fracture and Deformation

Washington, D.C.   20234

SUMMARY

Microtensile and microcreep behaviors of beryllium were studied using a capacitance type extensometer capable of resolving .1 μm/m over long times. The nature of the dislocation processes responsible for microdeformation are not entirely clear; but surely, the primary difference between micro and conventional deformation is the extent to which the dislocations move and not the nature of the processes. Despite low temperatures, microcreep of Instrument Grade Beryllium appears to be diffusion limited.

1.   INTRODUCTION:

To correct for drift, inertial guidance systems can require frequent realignment. Drift, which is the loss with time of the system's ability to recognize a reference condition, is a serious performance limitation of the system, and is attributable to various causes. One notable cause of drift is dimensional instability of the gyroscope components. This source of instability is important because a shift in the center of mass of less than 25 ρm is specified in current design practice, and advanced designs are predicated on achieving stabilities of 0.25 ρm or less.

To understand the size of the problem, consider that this level of stability is required from an electro-mechanical, multi-component system, and despite stresses in the components to 70 MN/m^2. Further, instability data which report either the inherent capability (Dimensional Instability) or creep at very low stresses and room temperature (Microcreep) are sparse, but more importantly frequently contradictory (1-4). And so, the likelihood of dimensional instability is inferred - by both system designers and the material producers - from micro level tensile measurements. The microyield strength (MYS) is the value of the tensile stress which causes .1 µm/m of plastic strain. Underlying the use of the MYS is the assumption - still unproven - that higher MYS values are quantative indicators of better long term stability. Unquestionably, this is qualitatively true; but surely, the application warrants using microcreep measurements, and more importantly an understanding of the nature of micro deformation.

Our research addresses this need in two ways. One is to develop a microcreep model which reflects the contributions of stress, temperature and material processing. The other is to develop a basic understanding of the dislocation processes that limit dimensional instability. The experimental approach uses correlations between microdeformation measurements and microstructural observations.

The conclusion that metals creep extensively, at low stresses and room temperature is not surprising or new. Chalmers initially coined the term "microplasticity" in the mid 30's to characterize his observations for tin [5]. Much later, Wyatt, studying creep of pure metals at very low temperatures, identified several features of microcreep [6]. He observed that creep at low stresses and temperatures is - understandably - a transient response, and that the strain was either proportional to the logarithm of time at very low applied stresses or a power of the time at still higher stresses. Hughel identified the contribution of several metallurgical features to the micro tensile behavior of beryllium [1]. Again studying a "pure" metal, copper, Lubahn recognized an important experimental hazard with microcreep experiments. "Straightness" of the sample, he felt had something to do with the anomaly that the creep strain can seem to be unrelated to the applied stress [7]. In our laboratories, prior efforts have gone into developing heat treatments which improve the stability of steel gauge blocks to better than 2-3 µm/m/year [8].

2. EXPERIMENTAL

2.1 Material: Beryllium is a commercial product which is available in different purities or "grades"; and beryllium is an especially useful material for constructing space appliances. Two typical uses are for infra-red mirrors, and inertial guidance system gyroscopes. As a construction material beryllium, and especially the Instrument Grade, has clear advantages. The density is so very low that the specific strength and stiffness exceeds other metal systems and compares favorably with composite materials. Unlike composites the mechanical anistropy is very low. Beryllium has a high thermal conductivity, which is comparable to or better than aluminum alloys. Lastly, at temperatures below several hundred degrees there are no known phase instabilities.

Instrument Grade beryllium is the basis for this study. The Instrument Grade has a grain size that is less than 10 μm, is dispersion hardened by at least 4.25 percent BeO, and in addition by much smaller concentrations of several other phases. Instrument Grade is a hot-pressed material which is formulated from recycled materials; and so, the impurity content exceeds 0.62 percent. The most concentrated residuals are iron, aluminum, carbon and silicon, and are the basis for fractional percent concentrations of Be_4 (Al Fe), BeC and Be_{13} (Fe X) particles. All are relatively small sized particles; for example we found the Be_{13} (FeX) particles are of the order of less than .1 μm [9].

The chemical composition of the material used in this study is shown in Table I.

Table 1: Typical Chemical Composition of Instrument Grade Beryllium

Element	Be	BeO	Fe	C	Al	Mg	Si	Other Metallics
wt. %	99.44	7.04	.12	.19	.05	.01	.05	<.10

Understandably, the macro level properties of this complex, recycled formulation are variable. In addition, the surface of beryllium is easily damaged, even when good machining techniques are used. Machining artifacts take several forms including very high - allegedly compressive - residual stresses, twins and frequently microcracks [20]. To minimize anomalies in the mechanical behavior which might result from surface artifacts, three steps were taken. First, to prevent forming surface flaws, all samples were machined using the NMAB recommended practice [10]. Second, to minimize the contribution of the surface, samples with a large cross section (0.953 cm diameter) were used. Third, to minimize the twinning and residual stresses which had formed, a stress relief treatment was used after machining. The annealing conditions were:

- Heat to 788 °C at less than 38 °C/h

- Soak at 788 °C for 1 h

- Cool to room temperature at less than 38 °F/h

This is the MIT treatment which is recognized industrially and is typically used to relieve gyroscope components.

2.2 <u>Apparatus</u>: To obtain creep strains with a usable precision of at least 0.1 μm/m, we are developing - still - two (2) distinctly different test systems. The simpler test system uses an extensometer with a parallel plate capacitor for the strain transducer, and has a resolution of .01 μm/m for several hours or .1 μm/m for periods of many hundreds of hours. The degradation of the resolution is caused by small changes in the laboratory environment - relative humidity, barometric pressure and importantly temperature. A second system is being developed to make much more precise measurements; and this facility is based on an optical extensometer. A polarized laser interferometer is the basis for this system from which a long time resolution of at least .01 μm/m is expected.

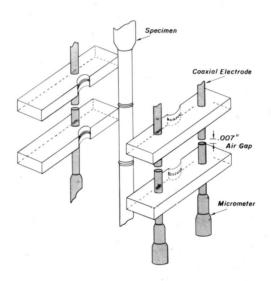

Figure 1. A schematic of the capacitor based extensometer.

The notable feature of the .1 μ/m system is the extensometer which is shown schematically in Figure 1. This gauge consists of three (3) fully independent parallel plate capacitors, which fasten to the test specimen by means of vee shaped ribs. For each sample, the ribs are individually fitted to the gauge. The capacitors are all of the driven ground type and arranged equidistantly about the specimen's circumference. The advantage of using at least three (3) independent elements for the gauge is that both the axial and bending strains in the sample can be monitored. Admittedly, the value of the bending strains were subsequently calculated from either Morrison's Relation [11] or preferably by using vector algebra. The advantage of using vector algebra is that both the magnitude and location of the maximum bending strain are defined. In addition to the extensometer gauge another notable feature is the bridge circuit design [12]. By using an appropriate design for the bridge circuit, the output signal is linearly related to the strain. Typically, conventional capacitance gauging responds inversely to the displacement and is then - in some fashion - corrected. Lastly, this gauge has a resolution of better than 0.9 ρm [13].

To simulate the temperature conditions in an operating gyroscope, tests are performed at 62.67 °C. Using the heat from two banks of light bulbs samples are maintained at 62.67 °C and with a standard deviation of less than 0.01 °C over intervals of several thousand hours. To monitor the temperature and thermal gradients, four (4) thermistor probes are glued to the sample just outside the gauge section. The specimen and load train are isolated from the laboratory using a reflectively lined, foam insulated chamber. An important feature of this chamber is the labyrinth seals through which the load train passes. Specifically, the labyrinth seals - largely - eliminate lateral constraints on the sample and provide surprisingly effective thermal isolation.

A conventional creep frame is used to dead weight load the sample. However, ball-end-joints are used to grip the specimen. Microcreep experiments are critically sensitive to the bending of a misaligned sample. For example, even with reasonable alignment, and especially at low stresses, the bending strains may equal and-more likely-exceed the axial strain. To minimize bending strain several approaches are used. Primarily, the sample terminates in and is gripped through low friction ball joints. Further, additional ball joints are used in the load train and the mass of the loading system has been minimized. The labyrinth seals, which minimize the lateral constraints, have been already mentioned. Lastly, the misalignment is continually monitored, and in the early stages of the loading sequence the indication of misalignment adjusted to a minimum.

3.   RESULTS AND DISCUSSION

Several important features of micro deformation were identified. The following characteristics provide insight into the mechanism which are responsible for microdeformation; and are:

- The MYS (Micro Yield Strength) value and .2 percent yield strengths correlate.

- Grain size, dispersed phase concentration and chemistry follow conventional strengthening roles.

- Microcreep has no steady state region and remains a transient response even after several thousand hours.

- A low sensitivity to the applied stress, and the need to consider an effective stress rather than simply the nominal stress are two characteristics of microcreep.

3.1 A Matter of Scale: The primary difference between deformation on the micro and macro scales seems to be - largely - the extent to which dislocations move about, rather than the kind of dislocation processes operating.

To identify the relationships between both the mechanical behavior of Instrument Grade beryllium and the micro features, we statistically modelled the behavior. The data base for the analyses were the production records of a former domestic beryllium producer (Kaweki Berylco Inc.) and cover different grades of beryllium over several years. The possibility that the MYS might be a downward extrapolation of the .2 percent yield strength was an initial concern of this modelling. The correlation which demonstrates an interdependence between these two strengths is shown in Figure 2. Two measures were used to test the significance of the interdependence. One measure is the value of the standard deviation of the residuals about the regression line; and is ± 7.2 MN/m^2. Notably, the reproducibility of an MYS value is estimated at ± 3.4 MN/m^2. A more recognizable test of significance is R, the linear correlation coefficient. On a scale ranging from 0 to 1, the R for an interdependence is 0.93.

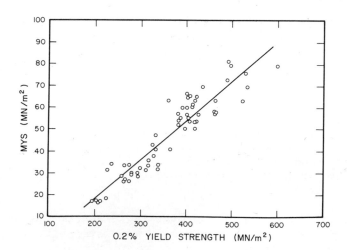

Figure 2. A correlation between the MYS and corresponding 0.2 percent yield strength.

Additional physical support for arguing that micro tensile deformation is no more than a downward extrapolation of the macro behavior is provided by Bonfield's metallographic observations [14] and our statistical analyses of the role microstructure plays. Because beryllium is an hcp metal which deforms by easy glide on basal planes, Bonfield observed slip lines in suitably polished samples. Importantly, he found deformation is a progressive process. Initially there were no slip lines. For deformation of a few μm/m slip lines appeared but only a few, and then only in widely separated grains which were appropriately oriented. However, with increasing deformation both the density of slip lines increased, and the orientation of the grains became less important.

Using the statistical approach rather than metallographic observations we found several microstructural features play a conventional role in determining the MYS value. MYS depends on the inverse square root grain size (R = 0.92). In fact, the performance of the entire range of beryllium grades, which were considered, could be normalized by the Hall Petch plot. Further, the BeO concentration - the primary dispersed phase - is an even stronger determining factor (R = 0.95). Using the BeO concentration as the independent variable, we found that over a range of from 0 to 8.0 percent the Grades all fit onto a common curve. Chemistry has a less clear role except that silicon - clearly - has a negative role. Increasing the silicon concentration reduces the MYS value.

Additional support for arguing a downward extrapolation is valid can be gained from both Bonfield's work and studies at Battelle which probed the nature of the dislocation process [2]. To do this the strain hardening exponent for the micro deformation of Instrument Grade eryllium was measured. The reported value of 0.5 supports the much earlier work of Carnahan, who demonstrated for other metals, that the MYS value corresponds to the stress needed to cause a dislocation to loop past a barrier [15].

And so, in a tensile test the yield strength is a measure of the resistance of the material to dislocation motion; and importantly the value determined for tensile yield strength depends on the extent or scale of the deformation.

3.2 Transient Behavior: Unlike the tensile case, the nature of the rate limiting event for microcreep is far less clear. We have observed several important features or characteristics of microcreep; and one is the transient character of the creep curve. Creep curves typical of our experiments and over a stress range of 20.7 to 138 MN/m^2 are shown in Figure 3. There is no evidence in these curves of a steady state and the curves are best described using a logarithmic time dependence of the form [16].

$$e = e_o + A\ln (1 + t/B) \tag{1}$$

where

e_o = the instantaneous plastic strain

A = parameter #1

B = parameter #2

t = time.

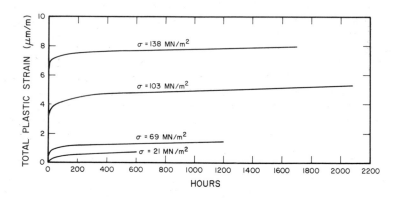

Figure 3. Typical microcreep curves for Instrument Grade Beryllium at 62.67 °C.

Fitting the data to this non-linear equation was performed by first linearizing Eq. (1) using a dummy variable to obtain linear Normal Equations and then using conventional regression analysis. In addition, to considering the log dependence on time, the Andrade Relation was also treated. In all the cases considered – our experiments and the published results of several other studies – the preferred fit is to the log time dependence [2, 17]. Two (2) indexes were used to test the fit which were R, and the mean of the residuals. No less interesting than the time dependence of microcreep is that Eq. (1) applies both well below and above the MYS value.

3.3 The Deformation Process: This transient character of the microcreep curve, together with two (2) other characteristics of the behavior are – at least – indicative of diffusion creep.

Sintered, dispersion hardened metal alloys, which are creeping at high rates and of necessity high temperatures, typically exhibit three (3) features; features which are identifiable characteristics of diffusion creep [18, 19]. One is a transient - rather than a linear - creep curve. Second the creep rate has a low sensitivity to the strain rate; and lastly, for creep to occur the applied stress must exceed a critical or threshold value. The driving force for strain is an effective stress which is the difference between the nominal stress and threshold values.

We found Instrument grade beryllium, which is a sintered, fine grained, dispersion hardened material, exhibits all three (3) characteristics. The creep curve is a transient function. Second, the stress exponent is 1.27 rather than exactly one (1). Because microcreep is transient, there was no steady state creep rate, which could be used to evaluate the stress exponent. Instead, the strain rate at 100 hours was used. Lastly, the applied stress must exceed a threshold value of about 21 MN/m^2. Figure 4 is a plot of the total creep strain at 500 hours for samples with applied stresses of 20.7, 60, 103, 138 MN/m^2. Notably, the plot is linear with respect to stress, has a high correlation coefficient (R = .93), and the intercept is offset from the zero (0) stress level. Positive creep occurs only where the applied stress is greater than 21 MN/m^2.

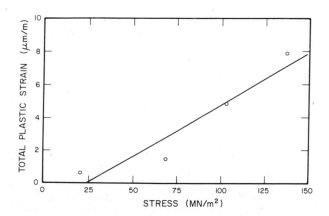

Figure 4. The total plastic strain at 500 hours plotted against the nominal stress.

Despite these three (3) characteristics, which apparently identify the deformation process, even a cursory calculation - using bulk diffusion - shows that the creep rates are too high for diffusion limited creep. To clarify the role of diffusion we are - currently - attempting to determine the role of grain size and temperature to measure the activation energy for creep.

The physical explanation for the observation of the threshold, very likely, is simple. The surface of beryllium is very easily damaged by machining, and this damage can take the form of twins, microcracks, and importantly compressive residual stresses which can approach a large fraction the yield strength [20]. The MIT treatment, which was used to relieve our samples, should be 100% in eliminating twinning and we consistently found no evidence of twinning. But only about 50% effective in eliminating residual stresses [21]. So, the need to exceed a threshold, may very simply be a need for a forward bias of the stress. Tests at levels below 21 MN/m^2 loading are not conclusive, but either no creep or very little negative creep was observed.

4. CONCLUSIONS:

1) Tensile mode microplasticity is the downward extrapolation of the macro behavior, apparently the result of the same dislocation processes.

2) Test methodology is a primary experimental concern.

3) Microcreep - apparently - is diffusion limited.

5. REFERENCES

1. T. J. Hughel: The Metallurgy of Beryllium, Institute of Metals Monograph No. 28, Chapman and Hall, London, 1963, 546-552.

2. C. W. Marschall, R. E. Maringer and F. J. Cepollina: AIAA Paper 72, 325, 13th Structures, Structural Dynamics and Materials Conference, San Antonio, Texas, April 10-12, 1972.

3. J. W. Lyon, Jr.: NASA Report No. TM-X-2046, Cambridge Research Center, Cambridge, Mass. 1971.

4. P. Y. Koo: Report No. TR-75-144-012-MIM-49, Autonetics Division (Rockwell Corporation) Anaheim, CA, 1975.

5. B. Chalmers: Proc. Roy. Soc. (London) 156A, 1936, 427.

6. D. H. Wyatt: Proc. Phys. Soc. (London) Sect. B, 66, 459.

7. H. A. Lequear and J. D. Lubahn: Trans ASME, 79, 1957, 97.

8.   M. R. Meyerson; Metal Treating (December, 1964 - January 1965), 5.

9.   R. S. Polvani, C. J. Bechtoldt, J. Orban and B. W. Christ: Society of Engineering Science Inc. 16th Annual Meeting Evanston, Ill. Sept. 5-7, 1979.

10. National Materials Advisory Board: Rept. No. MAB-205-19, National Research Council, Washington, D.C., 1966.

11. J. M. Morrison: Proc. Inst. Mech. Eng. (London) Vol. 42, 1939, 198.

12. F. K. Harris and R. D. Cutkosky: Inst. Soc. Amer. J., February, 1961, 63.

13. R. S. Polvani and B. W. Christ: SAMPE Quarterly, Vol. 10, No. 4, 1979, 32.

14. W. Bonfield and C. H. Li: Acta Met., Vol. 12, May 1964, 577.

15. R. D. Carnahan: Microplasticity, Rept. No. TDR-269 (4240-10)-14 Aerospace Corp., El Segunido, CA, 1964.

16. F. Garofalo: Fundamentals of Creep and Creep-Rupture in Metals, MacMillan, New York, 1965, 12.

17. P. F. Weihrauch and M. J. Horton: The Dimensional Stability of Selected Alloy Systems, Alloyd General Corporation, Final Report of U.S. Navy Contract N140 (131) 75098B.

18. F. K. Sautter and E. S. Chen: Oxide Dispersion Strengthening ed. C. S. Ansell, Gordon and Breach, New York, 1968, 495.

19. B. Burton, Met. Sci. J., 5, 1971, 11.

20. R. E. Maringer and A. G. Imgram: Trans. ASME, October, 1968, 846.

21. K. A. Leischer, "Minuteman Producibility Study No. 8 - Thermal Treatments for Beryllium" Report C5-996/32 North American Aviation Autonetics, Anaheim, CA, 1965.

6.   ACKNOWLEDGEMENTS

The Office of Naval Research has sponsored and encouraged this project (Contract No. NR-039-146). J. McCarthy and F. Petrie (C. S. Draper Laboratory) provided information about using beryllium for gyroscopes. J. Orban (NBS) performed the statistical analyses.

ON THE POWER-LAW BREAKDOWN DURING HIGH TEMPERATURE CREEP OF FCC METALS

A. Arieli and A. K. Mukherjee.

Division of Materials Science and Engineering,
Department of Mechanical Engineering,
University of California, Davis, CA 95616, U.S.A.

1. SUMMARY

The breakdown of the power-law creep with increasing stress was described as a transition from climb-controlled dislocation creep to obstacle-controlled dislocation glide mechanism. Using the creep experimental data at the transition, the obstacle width for the glide mechanism was calculated for Pb, Cu, Al and Ni. The dependence of the estimated width of obstacles on the temperature and stacking fault energy indeed points to the fact that the obstacles are likely to be the repulsive trees in the dislocation intersection process. The supposed constancy for the diffusivity-normalized strain rate value for various materials at the breakdown region was discussed in the context of the proposed hypothesis.

2. INTRODUCTION

When deformed at high temperatures ($T > 0.4\ T_M$, T_M = absolute melting temperature) most metals and alloys obey a power-law creep, i.e., the strain rate $\dot{\varepsilon}$ is related to stress σ by an expression of the form

$$\dot{\varepsilon} \ \alpha \ \sigma^n \tag{1}$$

where n is the stress sensitivity of the strain rate. However, as pointed out by Sherby and Young [1], Sherby and Burke [2] and Weertman [3], at stress levels higher than about 5×10^{-4} G to 10^{-3} G, where G is the shear modulus, the power law breaks down and the creep rates are higher than that predicted from an extrapolation from lower stress (power-law regions). In Weertman's [4] climb-controlled creep model, such a breakdown is expected to occur when the expression $\sigma\Omega/KT$ (where Ω = atomic volume, K = Boltzman's Constant and T = absolute temperature) in the equation for climb of edge dislocation is of the order of or greater than 1. This leads to a value of stress at which the breakdown should occur of $\sigma \simeq 3 \times 10^{-2}$ G, which is about an order of magnitude higher than observed experimentally. Weertman [3] has suggested that the experimental observation could be reconciled to the prediction of the dislocation creep model if a dislocation pile-up or some sort of stress concentration factor is involved to raise the creep rate to the required value above that predicted by the power law relation.

A somewhat different explanation has been put forward by Sherby and Young [1]. According to their interpretation, the power law breakdown may be associated with two contributing factors: (a) increased contribution from dislocation pipe diffusion and (b) excess vacancy generation. At higher stress levels the increased creep rate is also associated with an activation energy of creep that is about three-fourths of that for lattice diffusion. Robinson and Sherby [5] have suggested that under these conditions diffusion via dislocation core makes a significant contribution to the total diffusivity and an effective diffusion coefficient, D_{eff} that takes into account the relative flux of atoms through the volume and through dislocation core should be taken into account. Recently Luthy, Miller and Sherby [6] have used such effective diffusivity concept to correlate the creep stress and steady state creep rate, $\dot{\varepsilon}_s$, in aluminum in the intermediate temperature range over 21 orders of magnitude in $\dot{\varepsilon}_s/D_{eff}$. The results of torsional creep data for very large strains can be successfully correlated using garafalo-type [7] hyperbolic-sine function and such effective diffusivity terms.

A completely different approach has been suggested by other investigators. Ashby and Frost [8] have suggested that the power-law breakdown regime is the transition from a diffusion-limited dislocation motion to the obstacle-limited glide regime. Recently Nix and Ilschner [9] have put forward a model based on thermally activated glide in the subgrain interior and diffusion controlled recovery at the subgrain walls. The total creep rate is given as an algebraic sum of the individual creep rate due to glide and that due to climb-controlled recovery process. The justification for writing an equation of this form was to rationalize power law break-down as a natural consequence of the onset of low temperature flow processes. Quite reasonable correlation was obtained in describing the creep behavior of five fcc metals over extended ranges of stress and strain rates.

In this paper we have taken a similar approach, i.e., the power law breakdown with increasing stress signifies a transition from diffusion controlled creep to glide controlled low temperature thermally activated dislocation mechanism. The microstructure of deformed creep specimens supports the contention that a change of deformation mechanism occurs at high stresses (or low temperatures). Servi and Grant [10] studying the creep behavior of aluminum, observed a transition from

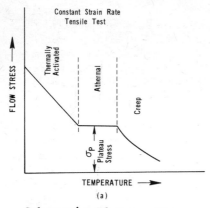

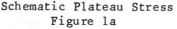

Schematic Plateau Stress
Figure 1a

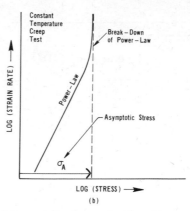

Schematic Asymptotic Stress
Figure 1b

intergranular slip bands at intermediate temperature to clear subgrains at high temperatures. Heard [11] found that sodium chloride displayed a similar transition from "polygonized" microstructure at low stresses to "slip" microstructure at high stresses in elevated temperature compression tests. Other circumstantial evidence lies in the comparison of approximate values of yield stress plateau (σ_p) Fig. 1(a) and creep stress asymptote (σ_A), Fig. 1(b). The comparison is shown in Table 1. It should be emphasized that whereas the normalized values of plateau stress σ_p/G were obtained at yield, that for the stress asymptotes at the power law breakdown region σ_A/G, were obtained from steady state creep stress where significant degree strain hardening has taken place during transient creep. The elevated temperature tensile tests (for σ_p/G) and creep tests (for σ_A/G) were conducted by a different group of workers with different levels of impurity and prehistory for the specimens. The σ_p/G results for Pb and LiF are from single crystals. These factors can be the possible reason for the differences in the values for the plateau stress and the asymptotic stress values for the same material. Even then, these two sets of values for each material are close enough to give some justification for the hypothesis that power law breakdown is due to a transition in deformation mechanisms.

The purpose of this paper is to reanalyze the published creep data on Pb [12], Cu [13], Al [10] and Ni [14] in the context of the approach outlined earlier. These fcc metals are known to deform by dislocation intersection mechanism at low and intermediate temperatures and by climb-controlled creep of edge dislocations at elevated temperatures. In the following section the rate equation for these two mechanisms for deformation will be analyzed in order to ascertain the criterion for the breakdown of power law. The experimentally observed asymptotic creep stress will be used along with the criterion mentioned above for the purpose of correlation.

Material	Yield stress plateau $\sigma_{P/G}$, 10^{-4}	Creep stress asymptote, $\sigma_{A/G}$, 10^{-4}	ref for σ_P	ref for σ_A
Al	0.33	5	a	bcde
Pb	0.45 (single crystal)	2.5	f	g
Ni	2.3	4	h	i
		10		j
Zr	15.4	17.8	k	l
LiF	0.49 (single crystal)	3	m	n

Table 1   Comparison of approximate values of creep stress asymptote and yield stress plateau

References

(a) HOWE, S., LIEBERMAN, B., and LUCKE, K., Acta Met. 1961 9, 625.
(b) SERVI, I.S. and GRANT, N.J., Trans AIME, 1951, 191, 909.
(c) WEERTMAN, J., J. Mech. and Phys. Sol., 1956, 4, 230.
(d) SHERBY, O.D., Trozera, T.A. and Dorn, J.E., Proc. ASTM, 1956, 56, 789.
(e) WEERTMAN, J., J. Appl. Phy., 1956, 27, 832.
(f) WEINBERG, F., Cand. J. Phy., 1967, 45, 1189.
(g) WEERTMAN, J., Trans. AIME, 1960, 218, 207.
(h) SASTRY, D.H. and TANGRI, K., Phil. Mag., 1975, 32, 513.
(i) MONMA, K., SUTO, H., and OIKAWA, H., J. Jap. Inst. Metals, 1964, 28, 253.
(j) WEERTMAN, J., and SHAHNIAN, P., Trans. AIME, 1956, 206, 1223.
(k) HERITIER, B., LUTON, M., JONAS, J.J., Met. Sci., 1974, 8, 41.
(l) PAHUTOVA, M. and CADEK, J., Mat. Sci. Engr., 1973, 11, 151.
(m) HAASEN, P., and HESSE, J., NPL Sympo. No. 15. - 'Relation between strength and structure of metals and alloys', 1963, 1, 138.
(n) CROPPER, D.R. and LANGDON, T.G., Phil. Mag., 1968, 18, 1181.

2.   ANALYSIS

2.1 Thermally activated dislocation glide

At low temperatures ($T \leq 0.3\ T_M$) the plastic flow in crystalline materials is dominated by the glide motion of the dislocations over barriers in crystal. The plastic strain-rate depends on the thermal activation of individual dislocation segments over barriers and the deformation mechanisms are thermally-activated. For constant strain rate the thermally-activated deformation mechanisms are characterized by rapidly decreasing flow stresses with increasing temperature. The maximum temperature, T_c, over which thermally-activated deformation mechanisms control the rate of deformation depends on mechanistic details; it increases with increases in activation energy and strain-rate.

The barriers to dislocation glide motion which can be thermally-activated are in general of two different types: linear barriers and localized barriers. In pure FCC metals, dislocations glide easily past the comparatively weak linear barriers and the rate of strain is controlled by the localized barriers. The average velocity of a dislocation cutting through a localized barrier is given by:

$$v = \frac{\nu b^2}{\ell} \exp\left(\frac{-U_c}{kT}\right) \tag{2}$$

where ν = Debye frequency, b = Burger's vector, ℓ = mean distance between barriers, U_c = free energy of activation for cutting localized barriers, k = Boltzmann's constant, and T = absolute temperature.

The activation energy U_c in Eqn. (2) represents the minimum work needed to activate a dislocation over localized barriers at constant temperature and stress. If we consider that the barriers are arranged in a square array where $\ell = L_s$ on a side [15], we can estimate the free energy of activation for cutting of localized barriers for a rectangular force displacement diagram [16], i.e.

$$U_c = \left[F_m - \sigma^* b L_s\right] d \tag{3}$$

where σ^* = effective stress, d = width of barrier, and F_M is the maximum force.

The maximum force over the distance d can be approximated as:

$$F_m = \alpha \Gamma_o \tag{4}$$

where α = relative barrier strength, and Γ_o = average dislocation line energy for unit length.

Thus, Eqn. (3) becomes,

$$U_c = \left[\alpha \Gamma_o - \sigma^* b L_s\right] d \tag{3a}$$

This activation energy approximates conditions that apply when undissociated glide dislocations produce jogs upon intersecting undissociated repulsive forest dislocations. The plastic strain-rate for the square array is then given by

$$\dot{\varepsilon} = \rho_m \nu \frac{b^3}{\ell} \exp\left[-\frac{1}{kT}\left[\alpha \Gamma_o - \sigma^* b L_s\right] d\right] \tag{5}$$

where ρ_m = density of mobile dislocations. The formulation in Eqn. (5) neglects the reversed motion of the dislocations and uses the Debye frequency instead of the fundamental frequency of vibration of a dislocation string for the attempt frequency as an approximation. The reversed motion of dislocations is important only at small stresses and since the deformation considered in this paper takes place at high stresses it can be neglected here.

The length of a dislocation segment subtended by randomly placed weak obstacles, when there is one obstacle per area L_s^2 on the average on the slip plane, is given by Friedel [17]

$$\ell = \left(\frac{2 \, \Gamma_o \, L_s^2}{\sigma^* \, b} \right)^{1/3} \tag{6}$$

and can be easily incorporated in Eqn. (5), i.e.,

$$\dot{\varepsilon} = \rho_m \, \nu \, b^3 \left(\frac{\sigma^* \, b}{2 \, \Gamma_o \, L_s^2} \right)^{1/3} \exp \left[-\frac{1}{kT} \left(\alpha \, \Gamma_o - \sigma^* \, b \, L_s \right) d \right] \tag{5a}$$

Because of the high stresses and temperatures considered in this paper for pure metals we can assume that the density of mobile dislocations, ρ_m, is equal to that of total dislocations, ρ (i.e. at these temperatures all dislocations are mobile) and that the effective stress, σ^*, is equal to the applied stress, σ (i.e., the internal stress in these fcc metals is negligible).

The density of total dislocations is given by [18]

$$\rho = 0.5 \, [\sigma/(Gb)]^2 \tag{7}$$

where G = shear modulus compensated for temperature and the average dislocation line energy per unit length is given by

$$\Gamma_o = a \, G \, b^2 \tag{8}$$

where a = 0.5.

Introducing Eqns. (7) and (8) into Eqn. (5a) we obtain

$$\dot{\varepsilon} = \frac{\nu}{(2a)^{1/3}} \left(\frac{b}{L_s} \right)^{2/3} \left(\frac{\sigma}{G} \right)^{7/3} \exp \left[-\frac{1}{kT} \left(a\alpha \, G \, b^2 - \sigma \, b \, L_s \right) d \right] \tag{9}$$

In order to retain Eqn. (9) within tractable dimensions, we have retained the parameter L_s in the exponential expression instead of using the more appropriate parameter ℓ from Eq. (6). Calculations show that this retention produces a very small error in the final estimated values for d in the Eqns. (11) and 11(a) to follow. Eqn. (9) will be used throughout this paper to calculate the plastic strain-rate when the deformation is dominated by the thermally-activated glide motion of dislocations over localized barriers.

2.2 Diffusion-controlled dislocation climb creep

At high temperatures ($T \geq 0.5\ T_M$) most metals and alloys deform by diffusion-controlled creep. At the steady-state, a balance must necessarily be obtained between the rate of dislocation generation and their rate of removal by recovery. In this paper we are concerned uniquely with the situations where the climb of dislocations is the rate-controlling recovery process and consequently the rate-controlling step of the creep. The rate of dislocation climb creep depends critically upon the dislocation substructure that is present in the metal. However, at steady-state, both creep rate and the rate-controlling substructural details are essentially constant and independent of previous creep history.

The minimum temperature, T_c, above which the diffusion controlled dislocation climb creep controls the rate of deformation is strain rate sensitive; it increases with increases in strain rate. It depends, also, on the value of activation enthalpy for self-diffusion, H_d [1]; it increases with increases in the value of H_d. It is worth mentioning that the activation energy for creep, $H_c = H_d$, is independent of stress and temperature and will vary only among various metals and alloys, i.e. it is a material property.

Mukherjee et al [19] have shown that the steady-state strain-rate, $\dot{\varepsilon}$, in dislocation climb creep can be related to the applied stress, σ, by the semi-empirical relation

$$\dot{\varepsilon} = A\ \frac{D_o\ G\ b}{kT}\ \left(\frac{\sigma}{G}\right)^n\ \exp\ \left(-\frac{Q}{RT}\right) \tag{10}$$

where A = dimensionless quantity incorporating the effects of the dislocation substructure; n = stress sensitivity; Q = activation energy for creep, equal to that for lattice self-diffusion; D_o = pre-exponential frequency for lattice diffusivity and R = gas constant. When the stress vs strain rate relation in creep can be accurately described by Eqn. (10) it is said that the deformation of these metals obeys a power-law When this equation ceases to describe the relation between the steady-state creep rate and applied stress, and the steady state creep rate increases with stress faster than the increase dictated by Eqn. (10), it is said that the power law breaks down. As was shown by Bird, Mukherjee and Dorn [18], during steady-state the quantities A, n and Q in Eqn. (10) are constant and reproducible. Therefore, Eqn. (10) will be used in this paper to calculate the plastic strain-rate when the deformation is controlled by the climb of dislocations.

---

(1) Since we made the assumption that steady-state dislocation climb creep is diffusion-controlled the discussion is restricted to those cases where the activation enthalpy for creep, H_c, is equal to that for self-diffusion, H_d.

3. THE TRANSITION FROM DISLOCATION CLIMB CREEP TO THERMALLY-ACTIVATED DISLOCATION GLIDE

As was mentioned in the Introduction, the scope of this paper is to explain the power law breakdown phenomenon as a transition between dislocation climb creep and thermally-activated dislocation glide. At constant stress, both mechanisms can be operative simultaneously, but the one which causes the highest strain-rate will be dominant over a certain range of temperatures. For the purpose of this paper we will define that the transition will occur during constant temperature creep test at that stress level where the contribution of the two mechanisms to the strain-rate is equal. Conversely, we will say that at a constant applied strain-rate the transition occurs at that temperature where both mechanisms require the same stress to produce deformation rates equal to the applied strain-rate.

The transition can, therefore, be expressed by equating Eqns. (9) and (10)

$$\frac{\nu}{(2a)^{1/3}} \left(\frac{b}{L_s}\right)^{2/3} \left(\frac{\sigma}{G}\right)^{7/3} \exp\left[-\frac{1}{kT}\left(a\alpha G b^2 - \sigma b L_s\right) d\right]$$

$$= A \frac{D_o G b}{kT} \left(\frac{\sigma}{G}\right)^n \exp\left\{-\frac{Q}{RT}\right\} \tag{11}$$

Normalizing Eqn. (10) in terms of (σ/G) and (T/T_M) and rearranging Eqn (11), we obtain

$$d = \left[\frac{K\, T_M(T/T_M)}{Gb}\right]\left[0.5\,\alpha\,b - \left(\frac{\sigma}{G}\right)L_s\right]^{-1} \times \left[\frac{Q}{RT_M}(T/T_M) - \ln\right.$$

$$\left.\left\{(2a)^{1/3}\frac{A}{\nu}\left(\frac{L_s}{b}\right)^{2/3}\frac{D_o\,G\,b}{K\,T_M(T/T_M)}\left(\frac{\sigma}{G}\right)^{n-\frac{7}{3}}\right\}\right] \tag{11a}$$

In constant temperature creep tests, the power-law breakdown (and in our interpretation the transition from dislocation climb to thermally activated dislocation glide) was observed to occur at a stress level that asymptotically reaches a constant value. At different (but constant) temperature tests one observes different values for such asymptotic transition stress. As one approaches the transition from low-stress climb-controlled region, the creep related structural details at steady state, i.e., A, is substantially constant. At the other side of the transition where dislocation-glide is becoming the dominant mechanism, the essential substructural parameters are L_s, α and d. Since the obstacle density was assumed to be constant (i.e., one obstacle on the average per area L_s^2 on the slip plane) and the strength of the obstacles α is not expected to be very sensitive to temperature and stress, the significant structure parameter

that will vary with variations in test temperature and transition stress value is d, the width of the obstacles. In the
next section, the variation of the obstacle width with temperature at the transition stress will be analyzed.

4. THE VARIATION OF THE OBSTACLE WIDTH WITH TEMPERATURE AT THE TRANSITION STRESS

The variation of the obstacle width, d, for the intermediate temperature thermally activated mechanism was estimated
from available creep data from the literature [10,12-14] on
four fcc metals, Pb, Cu, Al and Ni, near the power-law breakdown region. Eqn. (11a) was solved for d using the material
constants listed in Table 2 and the values for σ(transition)
that were read off directly from the original $\log \sigma - \log \dot{\varepsilon}$
plot for primary data. The values for α and L_s were assumed
constants for all materials and testing conditions and were
taken equal to 0.2 and 100b, respectively. The value of α =
0.2 allows for a relatively weak obstacle and $L_s \simeq 100b$ is
typical of values that are obtained in the analysis of dislocation intersection mechanism in fcc metals [16,20,21].

The significant primary experimental data from the literature and the estimated values for d at several temperatures
are listed in Table 3 for the four metals investigated. The
value of d for each metal decreases with temperature (Fig. 2)
but after normalization as (d/G) all the data become very
nearly independent of temperature (Fig. 3).

One of the premises of this analysis is that the power-law
breakdown regime coincides with transition from a diffusion
limited dislocation mechanism to the obstacle limited glide
mechanism. Hence, approaching the transition point from creep
regim, and using creep data, one should be able to estimate
the significant mechanistic and structural parameters for the
obstacle-limited glide mechanism, i.e., in this case the parameter d. The estimated d values in this analysis involving
pure fcc metals should then relate to the width of the repulsive trees in the forest for dislocation intersection mechanism. They should also depend on the stacking fault energy.
The (d/G) values for different metals from Fig. 3 are replotted as a function of γ/b^2 in Fig. 4, where γ = stacking
fault energy. In fcc metals, the equilibrium separation of
the partial dislocations as a function of stacking fault
energy (SFE) is given (for the prototype case for edge
dislocations) by [22]:

$$d_{SFE} = \frac{Ga_1^2}{48 \pi \gamma} \left(\frac{7}{2}\right) \tag{12}$$

where a_1 is associated with lattice parameter. Hence, if
d_{est} of Table 3 that is calculated from Eq. (11a) is indeed
associated with the width of repulsive trees, it should vary

Metal	Physical Properties			Shear Modulus, G^+		Lattice Self-Diffusion, D^{++}		Dislocation Creep Parameters	
	$b \times 10^8$ (cm)	$\nu \times 10^{-12}$ (s^{-1})	T_M(K)	$G_o \times 10^{-11} \left(\dfrac{\text{dynes}}{\text{cm}^2}\right)$	$\Delta G \times 10^{-8} \left(\dfrac{\text{dynes}}{\text{cm}^2 \text{ K}}\right)$	$D_o \left(\dfrac{\text{cm}^2}{\text{s}}\right)$	$Q\left(\dfrac{\text{kcal}}{\text{mole}}\right)$	A	n
Cu	2.56	7.07	1356	4.74	1.7	0.62	49.6	7.4×10^5	4.8
Ni	2.49	9.51	1728	8.412	2.69	1.9	68.0	8×10^5	4.7
Al	2.86	8.71	933	3.022	1.60	3.5×10^{-2}	28.8	3.4×10^6	4.4
Pb	3.50	1.98	600	0.994	0.883	1.37	26.1	4×10^6	4.4

$+$ $G = G_o - \Delta GT$; $\quad$ $++$ $D = D_o \exp(-Q/RT)$, R = gas constant.

Table 2 Material Parameters

	Pb			Cu			Al			Ni		
T/T_M	0.62	0.7	0.78	0.57	0.61	0.67	0.51	0.57	0.69	0.62	0.68	0.79
T°K	370	420	469	773	823	903	477	533	644	1073	1173	1373
σ (tran) $\times 10^{-7}$ dy/cm^2	3.5	2.9	2.4	24.1	19.3	12.5	13.1	9.99	7.92	50.0	35.0	21.0
$(\sigma/G) \times 10^4$ at tran.	5.25	4.65	4.13	7.03	5.78	3.92	5.8	4.61	3.98	9.04	6.64	4.44
$d_{est} \times 10^7$ cm (Eq. 11a)	2.93	2.64	2.42	3.83	2.80	2.10	1.81	1.59	1.35	3.6	3.2	2.0

Table 3 Primary Experimental Data

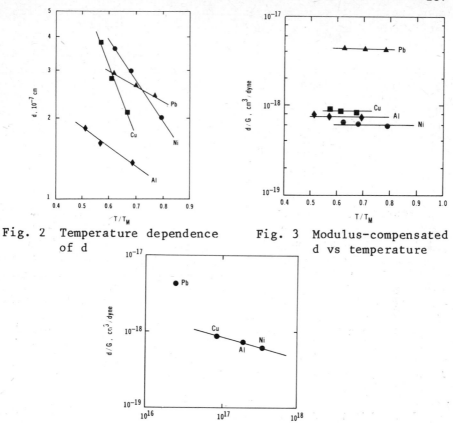

Fig. 2 Temperature dependence of d

Fig. 3 Modulus-compensated d vs temperature

Fig. 4 Variation of d/G with SFE

inversely with SFE as it does for Al, Ni and Cu in Fig. 4. The datum point for Pb does not fall in line and we are unable to explain the deviation at present, although incorrect estimate of SFE for Pb may be one reason. A secondary check will be to compare the estimated values of d from creep data at transition by using Eq. (11a), with values of d for edge dislocation calculated from SFE using Eq. (12). This approximate comparison is shown in Table 4. The two values of d are within one order of magnitude for each metal.

It will be appropriate to summarize the assumptions and approximations that are inherent in this analysis. These are: (a) assumption of a rectangular force-displacement diagram, (b) consideration of only random obstacles, whereas in fact they are more likely to be in the form of cellular or subgrain arrays, (c) consideration of only moderately weak obstacles, (d) undissociated dislocations requiring no energy for constriction in glide regime, (e) mobile dislocation density equal to total density given by Eq. (7), (f) no variation of line energy of dislocation with orientation, (g) negligible or non-existant internal back stress, (h) uncertainties in

Metal	Temp $^\circ K$	d_{est}, cm Eq. (11a)	d_{cal}, cm Eq. (12)	$\left(\dfrac{d_{est}}{d_{cal}}\right)$	value for γ, erg/cm^2	Ref for γ
Al	644	1.35×10^{-7}	2.52×10^{-8}	5.3	150	23
Cu	773	3.83×10^{-7}	9.48×10^{-8}	4.0	55	24
Ni	1073	3.6×10^{-7}	3.8×10^{-8}	9.0	210	25
Pb	370	2.93×10^{-7}	6.32×10^{-8}	4.6	30	26

Table 4   Comparison of estimated d value with distance of separation of partials

the reported values for SFE which are often measured at ambient temperature, but here were used at elevated temperatures, (i) estimation of the separation of partial dislocations for equilibrium condition neglecting the obvious local shear stress that will be present in actual mechanical tests, and (j) difficulty in obtaining the exact asymptotic transition stress from the double logarithmic plot of stress-strain rate data on creep from published literature. However, in spite of these, the approximate analysis does suggest that the power-law breakdown may indeed be correlated with the onset of a low temperature thermally activated glide mechanism, e.g., dislocation intersection process for the fcc metals that are investigated here.

5. SIGNIFICANCE OF $\dot{\varepsilon}/D = 10^{-9}$cm^{-2} FOR POWER LAW BREAKDOWN

Sherby and Burke [2] compiled the data for power law breakdown of several materials (Fig. 5). Surprisingly, the transition seems to be independent of the material investigated, occurring at the normalized value for $\dot{\varepsilon}/D \simeq 10^9$ cm^{-2}. It was suggested by Sherby and Young [1] that in this range, generation of excess vacancies and enhanced contribution from dislocation pipe diffusion may be the reason for this particular value of $\dot{\varepsilon}/D$.

In the context of the present discussion, if the value of $\dot{\varepsilon}/D \simeq 10^9$ cm^{-2} is in general characteristic of the power-law breakdown phenomenon, our analysis should yield a self-consistent explanation for this observation. We suggest that the transition from diffusive climb-controlled mechanism to obstacle-controlled glide mechanism occurs when the probabilities of overcoming the obstacles by each mechanism are equal. Following Conrad et al [27] the probability P_g, that a dislocation will overcome an obstacle by glide is related to the strain rate and structure by the following expression:

$$P_g = \dot{\varepsilon}\left[\rho\, b\,(1/\ell)\, A_1\right]^{-1} \tag{13}$$

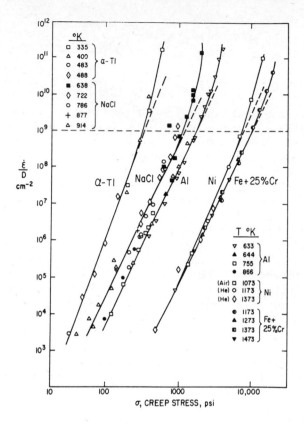

Figure 5   Breakdown of Power-Law Creep at $\dot{\varepsilon}/D = 10^9 \ cm^{-2}$
(after Sherby and Burke, Ref. 2)

where A_1 = activation area for dislocation glide. The probability that the same obstacle will be overcome by dislocation climb is given by:

$$P_c = D/A_2 \qquad (14)$$

where D = lattice self diffusivity and A_2 = activation area for dislocation climb. The transition from dislocation glide to dislocation climb will occur where $P_g = P_c$, i.e.

$$\dot{\varepsilon}/D = \rho b \ (1/\ell)(A_1/A_2) \qquad (15)$$

From Eq. (6), (7) and (8) we obtain

$$(1/\ell) = \left[(\rho^{\frac{1}{2}})/(2^{\frac{1}{2}}a \ L_s^2) \right]^{1/3} \qquad (16)$$

and introducing Eq. (16) into Eq. (15) one obtains

$$\dot{\varepsilon}/D = 0.89 \ \frac{b}{a^{1/3} \ L_s^{2/3}} \ \frac{A_1}{A_2} \ \rho^{7/6} \qquad (17)$$

Taking a = 0.5, L_s = 100 b, b = 3×10^{-8} cm, $A_2 = b^2$ and $A_1 = \beta b^2$,

$$\dot{\varepsilon}/D = 1.62 \times 10^{-4} \beta \; \rho^{7/6} \tag{18}$$

Now if we take $\dot{\varepsilon}/D = 10^9 \text{cm}^2$, then

$$\beta \; \rho^{7/6} = 6.17 \times 10^{12} \tag{19}$$

According to Eq. (19) when $\rho = 10^7 \text{cm}^{-2}$ (the case of a well annealed specimen) and $\rho = 10^9 \text{cm}^{-2}$ (strain hardened specimen), β will be 4.20×10^4 and 1.96×10^2, respectively, i.e., the corresponding activation area for dislocation glide will be $4.20 \times 10^4 b^2$ and $1.96 \times 10^2 b^2$, respectively. If the strain rate corresponding to $\dot{\varepsilon}/D = 10^9 \text{cm}^{-2}$ indeed signify the onset of thermally activated dislocation intersection mechanism in fcc metals, then the activation area in this region should correspond in magnitude to that for intersection process and its value should decrease drastically with strain hardening, i.e., with increase in ρ, the dislocation density. Mitra, Osborne and Dorn [20] and Mitra and Dorn [21] investigated in detail activation parameters for the intersection mechanism of plastic deformation of aluminum and also the nature of strain hardening of fcc metals. Their results indicated that the value of A_1 is indeed in the range of hundreds or thousands of b^2 and that its value decreases sharply with increase in strain hardening. These observations are in agreement with the approximate estimate of the value of A_1 mentioned earlier.

6. CONCLUSIONS

The breakdown of the power-law creep with increasing stress can be visualized as a transition from diffusion-controlled dislocation creep to obstacle-controlled glide mechanism. Using the creep experimental data at the transition, the obstacle width for the glide mechanism was calculated for different test temperatures for Pb, Cu, Al and Ni. The dependence of the estimated width of the obstacles on temperature and stacking fault energy points to the fact that the obstacles are likely to be the repulsive trees in the dislocation intersection process. The estimated activation areas for thermally activated dislocation glide in the transition region and its dependence on dislocation density also adds credence to the proposed hypothesis.

7. ACKNOWLEDGEMENTS

This research supported in part by a grant from the Division of Materials, Metallurgy Program, National Science Foundation (DMR 77-27724). One of the authors (AKM) would like to acknowledge his thanks to Professors Oleg D. Sherby and William D. Nix for many stimulating discussions on their ongoing projects on elevated temperature plasticity of materials and for a very pleasant sabbatical stay at Stanford University. Our thanks are also due to Mr. Roy Arrowood Jr. for his help in compiling the data presented in Table 1.

8. REFERENCES

1. SHERBY, O.D. and YOUNG, C.M. - in Rate Processes in Plastic Deformation of Materials, Eds. J.C.M. Li and A.K. Mukherjee, ASM, 1975, 497.
2. SHERBY, O.D. and BURKE, P.M. - Prog. Mater. Sci., 1967, 13, 325.
3. WEERTMAN, J. - in Rate Processes in Plastic Deformation of Materials, Eds. J.C.M. Li and A.K. Mukherjee, ASM, 1975, 315.
4. WEERTMAN, J. - Trans. Quart. ASM, 1968, 61, 681.
5. ROBINSON, S.L. and SHERBY, O.D. - Acta Met., 1969, 17, 109
6. LUTHY, L., MILLER, A.K., and SHERBY, O.D. - Acta Met., 1980, 28, 169.
7. GAROFALO, F. - Trans. Met. Soc. AIME, 1963, 227, 351.
8. ASHBY, M.F. and FROST, H.J. - in Constitutive Equations in Plasticity, Ed. A.S. Argon, MIT Press, 1975, 117.
9. NIX, W.D., and ILSCHNER, B. - in Strength of Metals and Alloys, Eds. P. Haasen, V. Gerold and G. Kostorz, Pergamon Press, 1980, 3, 1503.
10. SERVI, I.S. and GRANT, N.J. - Trans. AIME, 1951, 191, 909.
11. HEARD, H.C. - Geophys. Mon., 1972, 16, 191.
12. WEERTMAN, J.- Trans. AIME, 1960, 218, 207.
13. FELTHAM, P., and MEAKIN, J.D. - Acta Met., 1959, 7, 614.
14. WEERTMAN, J., and SHAHNIAN - Trans. AIME, 1956, 206, 1223.
15. SEEGER, A. - Phil. Mag., 1955, 46, 1194.
16. BASINSKY, F. - Phil. Mag., 1955, 46, 1194.
17. FRIEDEL, J. - Dislocations, Pergamon Press, Oxford, 1964, 224.
18. BIRD, J.E., MUKHERJEE, A.K. and DORN, J.E. - in Quantitative Relation Between Properties and Microstructure, Eds. D. Brandon and A. Rosen, Israel Univ. Press, Jerusalem, 1969, p. 255.
19. MUKHERJEE, A.K., BIRD, J.E. and DORN, J.E. - ASM Trans. Quart., 1969, 62, 155.
20. MITRA, S.K., OSBORNE, P.W. and DORN, J.E. - Trans. AIME, 1961, 221, 1206.
21. MITRA, S.K. and DORN, J.E. - Trans. AIME, 1962, 214, 1062.
22. DORN, J.E. - in Energetics in Metallurgical Phenomena, Ed. W.M. Mueller, Gordon and Breach, NY, 1964, 1, 241.
23. KANNAN, V.C. and THOMAS, G. - J. A. Phy., 1967, 38, 4076.
24. THORTON, P.R., MITCHELL, T.E. and HIRSCH, P.B. - Phil. Mag., 1962, 7, 1349.
25. BEESTON, B.E.P., DILLAMORE, I.L. and SMALLMAN, R.E. - Met. Sci. J., 1968, 2, 12.
26. BOLLING G.F., HAYS, L.E. and WEIDERSICH, H.W. - Acta Met., 1962, 10, 185.
27. CONRAD, H., deMEESTER, B., YIN, C. and DONER, M. - in Rate Processes in Plastic Deformation of Materials, Ed. J.C.M. Li and A.K. Mukherjee, 175, ASM, Metals Park, Ohio, 1975.

DEFORMATION MECHANISM DIAGRAMS -

MODIFICATIONS FOR ENGINEERING APPLICATIONS

Hiroshi OIKAWA

Department of Materials Science, Faculty of Engineering,
Tohoku University, Sendai, 980 JAPAN

Summary

   *Three-dimensional deformation mechanism diagrams having
the strain-rate axis as a co-ordinate and two-dimensional
sections of these diagrams have been constructed for a model
FCC metal of high stacking-fault energy.   This type of
presentation is useful for recogniqing the general features
of deformation during constant-rate straining at high
temperatures.   Effects of the crystal system and the stacking-
fault energy can be clearly demonstrated on these diagrams.
A guide for choosing good hot-working conditions may be
obtained from this type of diagram.*

1. INTRODUCTION

   Deformation mechanism maps were first introduced by Ashby
[1] in a quantitative manner following the preliminary idea of
creep diagrams proposed by Weertman & Weertman[2].   The maps
have been modified in several ways : for example, the use of
the coordinates σ/G and T_m/T[3] instead of the original
coordinates σ/G and T/T_m, where σ is the tensile stress(N/m^2),
G the shear modulus(N/m^2), T_m the melting temperature(K) and
T the deformation temperature(K).   Maps with the coordinates
σ/G and d/b[4] and with the coordinates d/b and T_m/T[5] have
also been introduced, where d is the grain diameter(m) and b
the atomic diameter(m).   These deformation maps are based on
the equation,

$$\dot{\varepsilon}_s = \dot{\varepsilon}(\sigma, T, d), \tag{1}$$

where $\dot{\varepsilon}_s$ is the steady-state creep rate.   All the variables
are included in the three-dimensional presentation with the
coordinates σ/G, T_m/T and d/b[6].   These diagrams have been
recognized to be useful for discussing *steady-state creep*

(constant-stress straining) at high temperature.

At high temperatures, a steady-state stage can also be obtained in constant-rate straining, such as conventional tensile tests or hot-working. The steady-state deformation in such cases can be expressed by the equation,

$$\sigma_s = \sigma(\dot{\varepsilon}, T, d), \tag{2}$$

where σ_s is the steady-state tensile stress, and $\dot{\varepsilon}$ is the applied strain rate.

Although the physical meaning of equation(2) is the same as that of equation(1), the presentations based on equation(2) are probably more suitable for realizing the general features of high temperature deformation under *constant-rate straining*.

In this paper deformation mechanism diagrams with the co-ordinate $\dot{\varepsilon}$[7,8,9] are presented as examples to show the usefulness of this type of diagram in discussing the deformation under hot-working conditions.

2. CONSTITUTIVE EQUATIONS

Seven deformation mechanisms which are typical in high-temperature deformation are considered in this paper : i) diffusion creep controlled by grain-boundary diffusion, D(b), ii) diffusion creep controlled by lattice diffusion, D(l), iii) Harper-Dorn creep which is believed to be associated with a dislocation mechanism controlled by lattice diffusion, HD, iv) grain-boundary sliding (superplastic flow) controlled by grain-boundary diffusion, S(b), v) grain-boundary sliding (superplastic flow) controlled by lattice diffusion, S(l), vi) power-law creep controlled by dislocation-core diffusion, P(c), and vii) power-law creep controlled by lattice diffusion, P(l).

The strain-rate equation for each mechanism (or behavior) can be expressed by the genralized equation with particular constants[10] :

$$\dot{\varepsilon}_s = AD_o\exp(-Q/RT)(Gb/kT)(b/d)^p(\sigma/G)^n. \tag{1'}$$

The stress in the steady-state stage, σ_s, can also be expressed in the form of equation(2') :

$$\sigma_s/G = A'(kT/Gb)^m(d/b)^q\{\dot{\varepsilon}/D_o\exp(-Q/RT)\}^m. \tag{2'}$$

In these equations D_o and Q are the frequency factor (m^2/s) and the activation energy (J/mol) for diffusion, R and k are the gas and Boltzmann constants, respectively, and A, p, n, A', q, and m are dimensionless constants.

Values of $A' = A^{-1/n}$, $q = p/n$ and $m = 1/n$ for each mechanism are listed in Tables I and II for three model metals, that is, a BCC metal, an FCC metal of high γ (γ : stacking-fault energy), and an FCC metal of low γ. The term kT/Gb is taken as a constant; i.e., 2×10^{-21}m^2, for simplicity in the construction of diagrams. This value can be regarded as a representative value for many cubic metals at high temperatures.

TABLE I
Dimensionless constants in equation (2')
for steady-state deformation at high temperatures

Mechanism	A'	q	m			Ref.
			FCC	FCC high γ	FCC low γ	
D(b)	1.50×10^{-2}	2		1		[11]
D(l)	3.57×10^{-2}	3		1		[11]
HD	5.99×10^{10}	0		1		[12]
S(b)	2.24×10^{-3}	1.5		1/2		[13]
S(l)	3.54×10^{-4}	1		1/2		[13]
P(c)	5.43×10^{-2}	0	1/6.4	1/6.4	1/7.3	[14]
P(l)	3.5×10^{-2}	0	1/4.4	1/4.4	1/5.3	[10]

TABLE II
Diffusion parameters used

Metal	$D_o(l)$	$Q(l)$	$D_o(b), D_o(c)$	$Q(b)$ $Q(c)$
BCC		$17RT_m$		
FCC(high γ)	1×10^{-4} m^2/s	$20RT_m$	1×10^{-4} m^2/s	$9.4RT_m$
FCC(low γ)		$20RT_m$		
Ref.		[15]		[16]

3. DIAGRAMS OF NEW TYPE OF PRESENTATION

3.1 Diagram with the Coordinates $\dot{\varepsilon}$ and T_m/T

An example of a deformation mechanism diagram at a given d for an FCC metal of high γ is shown in Fig.1, in which contour lines of equi-σ_s/G are drawn. In this type of diagram, the gradient of equi-σ_s/G lines indicates the relative effects of $\dot{\varepsilon}$ and T on the σ_s in each domain : a large gradient corresponds to the case in which T exerts a large effect, whereas a small gradient corresponds to the case where $\dot{\varepsilon}$ exerts a large effect. The absolute effect of T or $\dot{\varepsilon}$ can be estimated from the line density of equi-σ_s/G contours measured

E

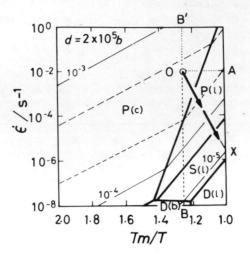

Fig.1 An example of deformation mechanism diagram with
the coordinates $\dot{\varepsilon}$ and T_m/T for $d = 2\times10^5 b$.   The
material is an FCC model metal of high stacking-
fault energy.   Contours are the equi-σ_s/G lines.

along the T_m/T-axis or the $\dot{\varepsilon}$-axis.

From this diagram, we can easily find the dominant mechanism that controls the deformation under consideration.   The
effects of $\dot{\varepsilon}$ and T on σ_s, or the absolute value of σ_s necessary for continuing the deformation at a particular $\dot{\varepsilon}$, can be
determined from the diagram.   The usefulness of this type of
deformation mechanism diagram becomes evident when we are
seeking a more advantageous deformation condition than that
being used.

Let us consider the hot working condition $\dot{\varepsilon} = 10^{-2}s^{-1}$ and
$T = 0.8\,T_m$ (point O in Fig.1).   The value of σ_s decreases with
increasing T along OA, with decreasing $\dot{\varepsilon}$ along OB, or with a
combined change in T and $\dot{\varepsilon}$ along OX.   The value of σ_s can be
lowered by expending heating energy and/or sacrificing the
rate of deformation.   The optimum direction of the change,
i.e., the location of point X, depends on the relative advantages (or disadvantages) associated with changing T and $\dot{\varepsilon}$.
This direction may differ greatly in various cases.

We can determine the new advantageous condition when the
advantages of decreasing σ_s and the disadvantages of changing
T and/or $\dot{\varepsilon}$ are known.   When formability is the most important
problem, we may try to change the deformation condition so
that it is in the S(l) domain where a large strain is allowed.
When the rate of deformation (the productivity) is the most
important factor, we can easily increase $\dot{\varepsilon}$ along OB', because
σ_s does not depend greatly on $\dot{\varepsilon}$ in the domain of P(c).

The capacity of the testing machine necessary for

conventional tests can be readily estimated from this type of
diagram : $\sigma_S \doteqdot 4 \times 10^{-4} G$ under the condition represented by O.
Consistency between the deformation mechanism for various
deformation conditions can also be checked on the diagram :
strain-rate change tests can be readily performed under the
condition represented by O.

3.2 Diagram with the Coordinates $\dot{\varepsilon}$ and d/b

An example of a deformation diagram at $T = 0.8T_m (T_m/T = 1.25)$ is shown in Fig.2. In this type of diagram, the
gradient of the equi-σ_S/G contour lines shows the relative
effects that $\dot{\varepsilon}$ and d exert on σ_S.

Let us consider the deformation conditions $\dot{\varepsilon} = 10^{-2}s^{-1}$
and $d = 2 \times 10^5 b$ (point O in Fig.2). When we wish to reduce
σ_S, $\dot{\varepsilon}$ must be reduced (along OB) as long as the deformation
condition is located within the domains of P(c) or P(l),
though the decrease in σ_S is not significant. Thus we do not
need to know the value of d precisely in order to be able to
control the deformation mechanism and σ_S.

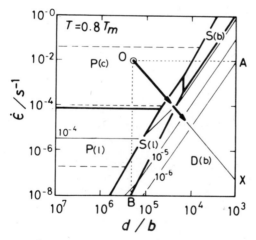

Fig.2 An example of deformation mechanism diagram with
the coordinates $\dot{\varepsilon}$ and d/b at $T = 0.8T_m$.

When the productivity (the rate of deformation) is the
most relevant requirement, we can easily increase $\dot{\varepsilon}$ because σ_S
increases only slightly with an increase in $\dot{\varepsilon}$. When
formability is the most important parameter we should try to
use deformation conditions in the S(b) or S(l) domain where a
large amount of strain can be applied before fracture. The
optimum direction of the change (OX) depends on the relative
disadvantages of a decrease in d and a decrease in $\dot{\varepsilon}$. If the
production of a fine-grained material (changes along OA) is

not difficult, it is worth while trying to use a deformation condition in the D(b) domain where the value of σ_s is a strong function of d (and also of $\dot{\varepsilon}$).

3.3 Diagram with the Coordinates d/b and T_m/T

An example of this type of diagram under a constant $\dot{\varepsilon}$ is shown in Fig.3, in which the gradient of the equi-σ_s/G contour lines indicates the relative effects of d and T on σ_s/G.

Let us consider the deformation condition $d = 2 \times 10^5 b$ and $T = 0.8T_m$ (point O in Fig.3). The mechanism of deformation is expected to be P(c). When we wish to reduce σ_s, T must be raised along OA as long as the deformation is controlled by P(c) or P(l), though the decrease in σ_s is not significant in these domains.

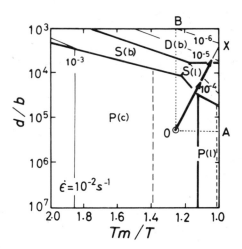

Fig.3   An example of deformation mechanism diagram with the coordinates T_m/T and d/b for $\dot{\varepsilon} = 10^{-2} s^{-1}$.

To decrease σ_s significantly or to deform a material greatly, d must be reduced to ensure that the deformation conditions are located in the domains of S(b) or S(l). If an increase in the temperature does not cause grain growth, a combined change in d and T (along OX) works effectively. The position of X (the direction of the changes) can be determined from a consideration of the relative advantages or disadvantages in changing d and T, which depend on the particular case.

4. THREE-DEMENSIONAL PRESENTATION OF DIAGRAM WITH THE COORDINATE $\dot{\varepsilon}$

The conditions for which the value of σ_s associated with one of the seven mechanisms is lower than for the other mechanisms are shown in Fig.4 as a function of $\dot{\varepsilon}$ ($10^0 \sim 10^{-8} s^{-1}$),

T_m/T (1.0 ~ 2.0) and d/b (10^3 ~ 10^7) for a model FCC metal of high γ. The 3D diagram with the coordinate σ is also shown in Fig.5 for comparison. Seven domains corresponding to all the mechanisms appear on the three-dimensional diagram. Each domain is shown separately in Fig.6 to visualize the boundary conditions for each mechanism. when the deformation conditions are stated, we can easily expect the dominant deformation mechanism from this diagram.

Some domain boundaries are parallel to one coordinate or to two. These characteristics[8] are independent of the crystal system and γ, though the position of some boundaries is greatly affected by these factors.

Most boundary planes in the new type of diagrams such as Fig.4 have steep gradients with respect of the $\dot{\varepsilon}$-axis in comparison with those in σ/G - T_m/T - d/b diagrams (cf. Fig.5). This fact indicates that deformation mechanism is affected mainly by T and by d, as long as the σ_s value required to find $\dot{\varepsilon}$ is not taken into account.

The largest domain is that of P(c) under the considered conditions for three axes. The P(c) is the dominant deformation except when temperature is near the melting point or the grain size is very small.

The second largest domain is that of D(b), although this domain is limited to the relatively small grain size. This mechanism will not appear under practical conditions of high-temperature deformation.

The domains of S(b), S(l), and D(l) appear under conditions of limited ranges of d/b. Narrowness of domains of S(b), S(l) and D(l) is one of the remarkable features of three-dimensional diagrams with the $\dot{\varepsilon}$-axis in comparison with diagrams with the σ/G-axis. The smallest domain is that of HD, which is almost null under the range of present conditions.

5. EFFECTS OF CRYSTAL SYSTEM AND γ

To discuss the effects of crystal system and γ, the $\dot{\varepsilon}$-T_m/T diagrams for three model metals are shown in Fig.7 as examples. General features of this type of 2D presentation have been discussed in 3.1 (Fig.2). The corresponding 2D presentation of the diagram with the coordinates σ/G, T_m/T and d/b are shown in Fig.8 for comparison. Several sets of other 3D- and 2D-diagrams for these three model metals have been presented elsewhere[9].

5.1 Effects of crystal system

Effects of crystal system can be realized by comparing Fig.7(a) for the BCC metal with Fig.7(b) for the FCC metal of high γ. All boundaries in the diagram for the BCC metal are located at lower temperatures and lower $\dot{\varepsilon}$ than in the diagram for the FCC metal. These differences, in turn, are made by

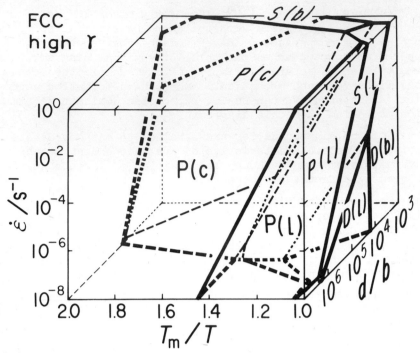

Fig.4 Three-dimensional deformation mechanism diagram with
the coordinate $\dot{\varepsilon}$ for an FCC model metal of high γ.
Thick lines are on the outmost surface, medium lines
are in the interior of the hexahedron.

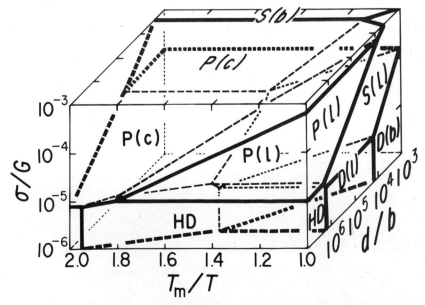

Fig.5 Deformation mechanism diagram with the coordinate σ
for an FCC model metal of high γ.

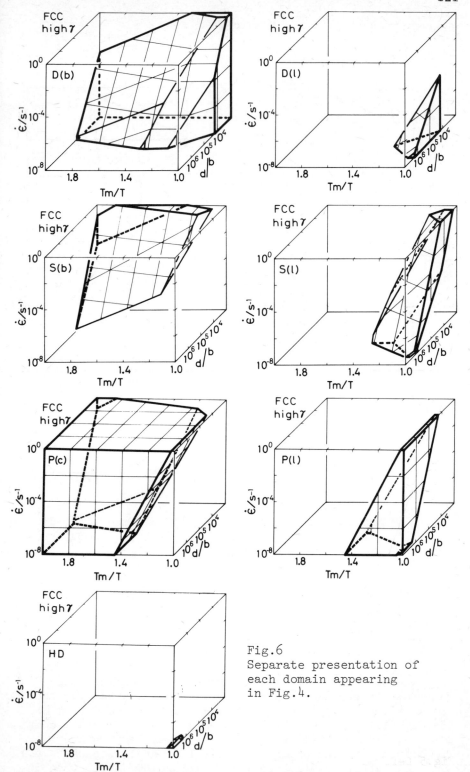

Fig.6
Separate presentation of
each domain appearing
in Fig.4.

a great reduction of the domain of P(c).

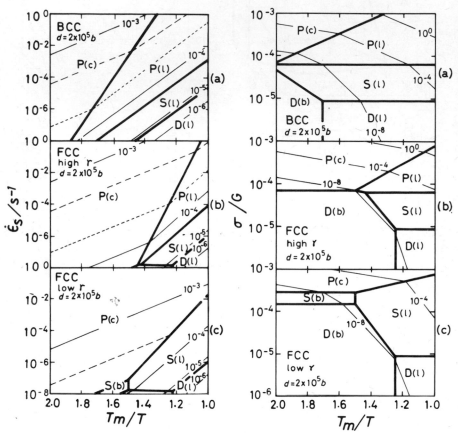

Fig.7 Two-dimensional diagrams with the coordinates $\dot{\varepsilon}$ and T_m/T at $d = 2 \times 10^5 b$ for three model metals. The diagrams correspond to a vertical section of Fig.4.

Fig.8 Two-dimensional diagrams with the coordinates σ and T_m/T (the Ashby type diagram) for three model metals. The diagrams correspond to a vertical section of Fig.5.

One of the most remarkable differences between the 3D diagram with the $\dot{\varepsilon}$-axis for the FCC metal and that for the BCC metal is the range of domains for mechanisms controlled by lattice diffusion. In the diagram for the BCC metal, domains of P(1), S(1) and D(1) extend to the lower temperature in comparison with the diagram for the FCC metal, because Q(b)/Q(1) and Q(c)/Q(1) are larger in the BCC metal than in the FCC metal.

The BCC lattice is favorable for high-temperature working because conventional conditions are likely to be located in the P(l) domain, where the value of σ_s for a given condition is lower than that in an FCC metal of high γ.

5.2 Effects of γ

Effects of γ can be realized by comparing Fig.7(b) with (c). With decreasing γ, the domain of P(c) expands to the higher temperature side, at the same time, the domain of S(l) expands to the lower temperature side. Consequently, the domain of P(l) disappears in the low γ metal.

In the 3D diagrams the domain of P(c) extends to the higher temperature side with decreasing γ, whereas the domain of P(l) is greatly reduced, because the stress exponent, n, for power-law creep increases as γ decreases in FCC metals. The domains of S(l) and S(b) expand with decreasing γ, whereas those of D(l) and D(b) are little affected by γ.

In the FCC metal of low γ, the condition of conventional high-temperature working is likely to be located in the P(c) domain, where an increase in T or a decrease in $\dot{\varepsilon}$ affects σ_s only slightly. The σ_s necessary to attain a given $\dot{\varepsilon}$ increases with decreasing γ in the domain of P(c). These facts indicate clearly that the low value of γ is an unfavorable condition for high-temperature working.

6. DISCUSSION

The deformation mechanism diagrams with the coordinates $\dot{\varepsilon}$, T_m/T, and d/b have the same weaknesses as those with the coordinates σ/G, T_m/T, and d/b. For example, the consideration of deformation mechanisms is limited to those during steady-state deformation, whereas a prolonged transient stage is frequently observed in the domains of P(l) and P(c). In fully-annealed pure metals the stress level increases with increasing strain during the transient stage. Therefore, σ_s values shown in deformation mechanism diagrams with the coordinate $\dot{\varepsilon}$ are the upper limit for the usual hot-working conditions of pure metals. The stress-strain curves of some alloys show a maximum at an early stage of deformation. In these cases the σ_s values for steady-state deformation are the lower limit.

The structure must be stable during deformation, whereas the grain growth occurs during high-temperature long-term deformation, such as superplastic flow and diffusion creep. The increase in d causes an increase in σ_s in order to keep $\dot{\varepsilon}$ constant in the domains of S(l), S(b), D(l) and D(b), or even a change in mechanism may occur. These changes in σ_s or in the mechanism are expected from Figs.2~4. Many metals, especially those of low stacking-fault energy, recrystallize during high-temperature deformations in P(l) or P(c), when the strain exceeds a limiting value. In these cases, the stress-strain curve also has a maximum or shows oscillations. The σ_s

level during continuous recrystallization is a problem to be solved in the future[17].

The deformation mechanisms that we have considered in this paper are limited to the seven well-known processes, though it is possible that an unknown deformation mechanism may play an important role under less common conditions. The Mechanism D(b) dominates under the condition predicted by the present work only when grain boundaries act as perfect efficient sources and sinks for vacancies[18].

In alloys other mechanisms (Class I behavior) can dominate even under the usual conditions[10]. The transition between Class I behavior and Class II behavior (pure metal-like behavior) is not well understood, though some proposals for power-law creep have been made. The effects of solute elements have been little studied in the mechanisms S(l), S(b), and HD.

Effects of γ are taken into accounts only in the cases of P(l) and P(c), because it is well understood that $\dot{\epsilon}$ in these mechanisms increases with increasing γ in FCC metals. The value of γ may also affect $\dot{\epsilon}$ in S(l), S(b), and HD, because the movement of dislocations within grains plays an essential role in these mechanisms. The quantitative estimation of the effect of γ in these mechanisms, however, cannot be done until the details of deformation processes in these mechanisms are revealed.

7. CONCLUSIONS

Although some limitations exist and many improvements are needed, the new type of deformation mechanism diagrams with the coordinate $\dot{\epsilon}$ is useful for considering the general features of the deformation mechanisms which control constant-rate straining of materials. The $\dot{\epsilon}-T_m/T-d/b$ diagrams are useful for finding conditions for easier working, whereas the $\sigma/G-T_m/T-d/b$ diagrams are useful for finding conditions for higher creep-resistance.

The effects of crystal system and γ during constant-rate straining are well demonstrated in the new type of presentation. When the crystal system is changed from BCC to FCC(high γ), domains associated with lattice diffusion; i.e., P(l), S(l), D(l), and HD, decrease greatly because of the difference in lattice diffusion coefficients between BCC lattice and FCC lattice. When γ decreases, the domains of P(l) and P(c) are greatly reduced, and those of S(l), S(b) and HD are obviously increased, because of the increase in n with decreasing γ.

ACKNOWLEDGMENTS

The author would like to record his appreciation of the general interest in this study of his colleagues in the Department of Materials Science, in particular of Professor

S. Karashima. He also wishes to thank Misses Y. Watanabe and
C. Harada for their technical assistance.

References

1. ASHBY, M.F. - 'A First Report on Deformation-Mechanism Maps', Acta Met., 1972, 20, 887-898.
2. WEERTMAN, J. and WEERTMAN, J.R. - 'Mechanical Properties, Strongly Temperature Dependent', Physical Metallurgy, ed. Cahn, R.W., North-Holland Publ., 1965, p.793-819.
3. LANGDON, T.G. and MOHAMED, F.A. - 'A Simple Method of Constructing an Ashby-Type Deformation Mechanism Maps', J. Mater. Sci., 1978, 13, 1282-1290.
4. MOHAMED, F.A. and LANGDON, T.G. - 'Deformation Mechanism Maps Based on Grain Size', Metal. Trans., 1974, 5, 2339-45.
5. LANGDON, T.G. and MOHAMED, F.A. - 'A New Type of Deformation Mechanism Maps for High-Temperature Creep', Mater. Sci. Eng., 1978 32, 103-112.
6. OIKAWA, H. - 'Three-Dimensional Presentation of Deformation Mechanism Diagrams', Scripta Met., 1979, 13, 701-705.
7. OIKAWA, H. - 'Strain Rate-Temperature-Grain Size Diagram of High-Temperature Deformation Mechanisms for Aluminum' Nippon Keikinzoku Gakkai-shi (J. Japan Inst. Light Metals), 1980, 30 376-383.
8. OIKAWA, H. - 'Deformation Mechanism Diagrams with the Strain-Rate Coordinate', Mater. Sci. Eng., 1980, 44, 211-215.
9. OIKAWA, H. - 'Effects of Crystal System and Stacking-Fault Energy on Deformation Mechanism Diagrams', Tech. Rept. Tohoku Univ., 1980, 45, (in press).
10. BIRD J.E., MUKHERJEE, A.K. and DORN, J.E. - 'Correlations between High-Temperature Creep Behavior and Structure', Quantitative Relation between Properties and Microstructure, ed. Brandon D.G. and Rosen, A. Israel Univ. Press, 1969, p.255-342.
11. BURTON, B. - Diffusional Creep of Polycrystalline Materials, TRANS TECH Aedermannsdorf, 1977.
12. YAVARI, P. and LANGDON T.G. - 'The Transition from Nabarro-Herring to Harper-Dorn Creep at Low Stress Levels', Scripta Met., 1977, 11, 863-866.
13. LÜTHY, H., WHITE, R.A. and SHERBY, O.D. - 'Grain Boundary Sliding and Deformation Mechanism Maps', Mater. Sci. Eng., 1979, 39, 211-216.
14. ASHBY M.F. and FROST, H.J - 'The Kinetics of Inelastic Deformation above 0°K', Constitutive Equation in Plasticity, ed. Argon, A.S., MIT Press, 1975, p.117-147.
15. SHERBY, O.D. and SIMNAD, M.T. - 'Prediction of Atomic Mobility in Metallic System', Trans. ASM, 1961, 54, 227-240 & 771-779.
16. HWANG J.C.M. and BALLUFFI, R.W. - 'On a Possible Temperature Dependence of the Activation Energy for Grain Boundary Diffusion in Metals' Scripta Met., 1978, 12, 709-714.

17. SAKAI, T. - 'Recrystallization during High Temperature Deformation', Nippon Kinzoku Gakkai Kaiho (Bull. Japan Inst. Metals), 1978, 17, 195-199.

18. BURTON, B. and REYNOLDS, G.L. - 'Deformation Mechanism Maps : A Cautionary Note', Scripta Met., 1979, 13, 839-841.

THE ROLE OF GRAIN BOUNDARY DISLOCATIONS IN HIGH TEMPERATURE
DEFORMATION

P.R. Howell[+] and G.L. Dunlop[++]

[+]Department of Metallurgy and Materials Science
University of Cambridge, Cambridge, CB2 3QZ England

[++]Department of Physics, Chalmers University of Technology,
412 96 Göteborg, Sweden

ABSTRACT

     It is shown that grain boundary dislocations (GBDs) are
mobile during diffusion creep, power law creep and superplastic deformation. In commercial polycrystalline materials the
main obstacles to the motion of GBDs are intergranular precipitates and triple junctions. These obstacles contribute to
a threshold stress for GBS and diffusion creep. It is considered that the rate at which GBDs can surmount such obstacles
is the major factor influencing a constitutive equation for
GBS.

1. INTRODUCTION

     The role of grain boundary processes has often been neg-
lected in discussions on high temperature creep deformation.
For example, most models of power law creep completely ignore
grain boundary sliding (GBS) although it is well established
that GBS can often make a significant contribution to strain
under these conditions (1, 2). Similarly, it has been disco-
vered that intergranular precipitation or segregation can inhibit diffusion creep at low stresses (3) but it is not widely appreciated that the classical Nabarro-Herring (4, 5)
theory for diffusion creep should be modified to take these
grain boundary sensitive phenomena into account. Another area
where grain boundaries undoubtedly play a vital role is in
superplastic deformation. However, no serious attempt has been
made to relate interfacial microstructure to the flow characteristics of these fine-grained materials.

     Considering this state of affairs, we have attempted over
the past few years to draw a link between the modern concepts
of interfacial structure, the high temperature behaviour of

grain boundaries and the influence of boundaries on the flow
of polycrystals. Interfacial microstructures have been examined by transmission electron microscopy and related to information from mechanical testing. The present paper summarizes
these observations and also attempts to clarify our view of
the micromechanisms for high temperature grain boundary behaviour. It will be seen that, although models based on grain
boundary dislocations (GBDs) provide a good qualitative description of grain boundary properties, considerable more work
is required in order to quantify these general concepts.

2. GRAIN BOUNDARY STRUCTURE IN POLYCRYSTALLINE METALS AND ALLOYS.

Most modern structural theories of grain boundaries are
based on some form of matching between the two crystals (e.g.
the coincidence-site-lattice model of Brandon et al. (6) and
the "O" lattice model of Bollman (7). Deviations from exact
positions of coincidence are considered in terms of periodic
arrays of GBDs; the Burgers vectors of which are characteristic of the coincidence lattice for the two crystals. In this
manner, high angle grain boundaries can be modelled in a way
which is analogous to the approach adopted for low angle grain
boundaries (i.e. in terms of their constituent dislocations).
Considerable support for the coincidence/dislocation type models has been provided by e.g. Schober and Balluffi (8, 9)
who investigated specially prepared gold bicrystals. In these
studies, arrays of dislocations were found in high angle
grain boundaries which were close, in orientation, to a high
density coincidence position.

Support for this dislocation type approach has also been
obtained using commercial polycrystalline metals and alloys.
For example Loberg and Nordén (10) have used both field-ion
microscopy and transmission electron microscopy to image dislocations in high angle grain boundaries in tungsten. Similarly, Howell et al. (11) have presented several examples of high
angle GBDs in an austenitic stainless steel.

The major problem associated with the dislocation models
is their range of applicability (e.g. 12, 13). Limited experimental evidence on this point suggests that a relatively
large proportion of boundaries in commercial materials may be
described in terms of their constituent dislocations. For instance Howell et al. (11, 14) found a high percentage of such
boundaries in both an austenitic stainless steel and in tungsten.

In addition to the dislocations which are geometrically
neccessary (intrinsic GBDs), grain boundaries in polycrystalline materials may contain a variety of other dislocation type defects which are extrinsic in character. These extrinsic
GBDs are created by interactions between crystal lattice dis-

locations and a grain boundary.This interaction process has
been observed in both bicrystal specimens (15) and in polycrystalline materials (16-18). It has been shown that lattice
dislocations can lower their elastic energy by dissociation in
the boundary to form perfect grain boundary dislocations (e.g.
17, 19). Further, it has been suggested that the accomodation
process is limited by climb (e.g. 15) - i.e. the process is
thermally activated.

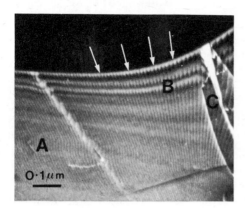 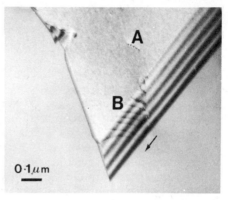

Fig. 1. GBDs in an undeformed Fig. 2. Creation of extrinsic
austenitic stainless steel.   GBDs in a coherent twin boundary.

Figs. 1 and 2 show typical examples of the dislocation
structures observed in boundaries in a 20Cr 30Ni, Ti stabilized stainless steel. In fig. 1, an array of intrinsic dislocations can be seen (arrowed), the spacing of which varies
markedly (from A to B) as the plane of the boundary varies.
The line features exhibiting enhanced contrast (e.g. at C) are
extrinsic dislocations which have interacted with the boundary at room temperature. Fig. 2 illustrates one of the important features of extrinsic dislocations (in this case created
in a primary twin boundary). Interaction of the matrix dislocations in the slip band (A) has produced a series of extrinsic dislocations which have dissociated in the boundary (B)
and have subsequently moved along the interface (in the direction of the arrow). It is this capacity (characteristic
of both intrinsic and extrinsic GBDs) to move in the grain
boundary which allows a grain boundary dislocation description
of such phenomena as sliding and migration to be formulated.
In general, the GBDs must move by a combination of glide and
climb (20). Thus, GBD displacements lead not only to sliding
but also to a diffusive flux. Similarly, most GBDs have a
small step associated with the dislocation core (21) so that
migration of the boundary will normally accompany sliding.

3. GRAIN BOUNDARY SLIDING

Grain boundary sliding (GBS) may be defined as the relative translation of two adjacent grains in a direction pa-

rallel to their mutual interface. GBS becomes a significant
mode of deformation above ∿ 0.4Tm (where Tm is the absolute
melting point). The phenomenon of GBS has been the subject of
a number of studies over the past fifteen years since it is
recognised that it:

(i)   contributes to the overall strain during disloca-
       tion creep (1);

(ii)   is neccessary for diffusion creep (22) (see Sec-
       tion 4); and

(iii) is the major strain-providing mechanism in super-
       plastic deformation (23) (see Section 5).

A number of models for GBS involving extrinsic GBDs have
been suggested (e.g. 20, 21, 24, 25,). Essentially, the models are based on the glide/climb motion of these non-equilibrium defects in grain boundaries and consider dislocation
multiplication in the grain boundary in terms of Frank-Read
and spiral sources. Fig. 3 shows an example of a spiral source operating in a high angle grain boundary in creep deformed
α-brass. A continuous dislocation loop is observed at A and
the distribution of the dislocations at B suggests that a series of such loops has been sectioned by the thin foil. One
possible explanation for this observation is that the source
functioned by the spiralling of the extrinsic dislocations a-
round the point of interaction between a matrix dislocation
and the boundary (19). Subsequent glide of the matrix dislocation, which would involve the concomitant movement of the
point of interaction would then lead to "pinching off" of
each turn of the spiral to produce a series of concentric loops.
It is also of interest to note that groups of dislocations
are seen (e.g. at C). This is consistent with experimental
evidence that GBS is intermittant and occurs in bursts (26,
27).

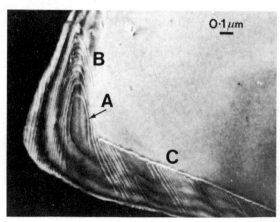

Fig. 3. Operation of spiral GBD source in creep deformed
        α-Brass

Gates (28, 29) has analysed the situation whereby intrinsic dislocations are responsible for GBS. He concluded that good agreement could be achieved between predicted creep rates and those which are experimentally observed. It has also been shown that grain boundary sliding during pure diffusion creep is consistent with an intrinsic GBD description of GBS (30, 31), (see Section 4).

As discussed in some detail by Pond et al. (21, 32), the GBD models can qualitatively explain sliding phenomenology since:

(i)   A threshold stress is often observed for GBS (33, 34). This is consistent with a GBD model since, in a manner which is analogous to matrix slip, a friction stress will have to be overcome. Similarly triple junctions and intergranular precipitates will provide barriers to the motion of GBDs (and see Section 4 - 6).

(ii)   Slide hardening is often found during GBS accompanied by dislocation creep. This will occur when GBD movement is impeded by interactions with other GBDs or pile-ups of crystal lattice dislocations (35).

(iii)   In general, there is a relationship between grain strain and sliding strain. This is a natural consequence of the GBD models involving extrinsic dislocations since matrix slip will generate grain boundary sources (vid. fig. 3).

(iv)   Migration often accompanies sliding. As discussed in Section 2, movement of GBDs leads to simultaneous sliding and migration; the relationship between the two depends on the GBD Burgers vectors and the local boundary plane.

(v)   The role of GBS may be determined by either the diffusive flux required for the climb of GBDs or by the rate of accommodational processes in the adjacent grains. Hence, as experimentally observed, the activation energy for GBS should lie between the extremes for grain boundary and bulk diffusion.

4. DIFFUSION CREEP

Diffusion creep, as predicted by Nabarro (4) Herring (5) and Coble (36), occurs by the transport of vacancies from grain boundaries which experience a tensile stress to those which are under a compressive stress. The vacancy flow can be either intragranular (Nabarro - Herring creep) or intergranular (Coble creep). The occurrence of grain boundary sliding is a requisite for diffusion creep (37, 38) since contiguity between the grains can only be maintained if GBS and

132

diffusive transport occur concurrently.

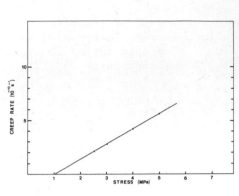

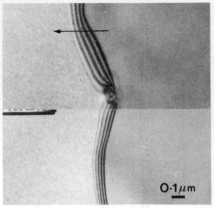

Fig. 4. Linear plot of creep rate versus applied stress for stainless steel.

Fig. 5. Zener drag of a boundary by a TiC precipitate during diffusion creep.

Few microstructural studies have been reported for material which has experienced diffusion creep. Therefore the intrinsic GBD model for GBS has received little experimental support. As shown by Nilsson et al (31) the creep deformation of a 20Cr 30Ni austenitic steel is dominated by Coble creep at stresses below ∿ 5MPa for a testing temperature of 800°C. Fig. 4 is a linear plot of creep rates as a function of applied stress for stresses of 5MPa and below. As may be seen, a threshold stress σ_0 occurs at 1MPa. Considerable evidence for grain boundary migration was observed for stresses below 5MPa (39). Fig. 5 shows an example of a migrating grain boundary in material which had been crept to a strain of 0.1 % at an applied stress of 3MPa. The boundary is migrating in the direction of the arrow and it can be seen that the intergranular precipitate is exerting a strong pinning action on the boundary. It should also be noted that there is no evidence of any matrix dislocation activity in either grain. Three comments can be made concerning fig. 5:

(i)   the migration is consistent with the GBD models of sliding for the reasons detailed in Section 3;

(ii)   the linked processes of sliding/migration will be limited by the pinning action of the intergranular TiC precipitates;

(iii)   the absence of any matrix dislocation activity (40) suggests that GBS must be occurring by the movement of the intrinsic dislocation networks.

Concerning point (ii), use of a simple model for precipitate pinning of a migrating boundary shows that the particles are responsible for a threshold stress (∿ 1MPa) below which migration and hence deformation is wholly suppressed (31).

Detailed examination of the grain boundaries in material crept at σ < 5MPa showed that, in the absence of precipitate particles, the arrays of GBDs tended to be uniform and comparable to those found in undeformed material (and see fig. 1). No evidence was found for interactions of matrix dislocations with grain boundaries (i.e. no extrinsic dislocations were observed). In regions of boundary containing precipitates, the GBDs were invariably distorted. Fig. 6 shows three examples of the interaction between GBDs and intergranular precipitates taken from the same high angle grain boundary. In the three images two resolvable dislocation arrays are present (arrowed in fig. 6 a). One particular set of GBDs has been mobile during the deformation process and many of these dislocations are bowed in the same sense (compare figs. 6 a - c). It is also noticeable that this set of dislocations is often pinned at the intergranular TiC precipitates (e.g. at A,B and C). The regions in front of the bowed dislocation segments contain a very low density of this type of dislocation. These observations suggest that the particles are effective barriers to the movement of the GBDs in the boundary plane - i.e. they inhibit the sliding process. For all cases examined, there was no evidence for the formation of Orowan loops and it is thus concluded that the dislocation bypass mechanism is one of glide/climb of GBDs in particle/matrix interfaces. It is also interesting to note that the dislocations are restricted in spacing both at the leading and trailing edges of the particles. This implies that an attractive interaction exists between the GBDs and the precipitates and indicates that the GBDs can reduce their line energy in the incoherent TiC/matrix interface.

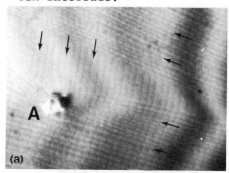

(a)

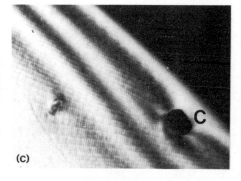

(c)

0.1 μm

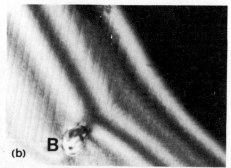

(b)

Fig. 6. Examples of interactions of GBDs with TiC precipitates in stainless steel during diffusion creep.

If we now consider the magnitude of the back stress exerted by the particles on the mobile GBDs, Ashby (41) has estimated that, for precipitate by-pass this back stress, or threshold stress, is given by:

$$\sigma_O = 2E/b\lambda \qquad\qquad 1.$$

where E is the GBD line energy, b is the Burgers vector and λ is the interparticle spacing. Making reasonable approximations for the parameters in this equation (30), a value of $\sigma_O \sim$ 1MPa is obtained which compares well with the experimental value seen in Fig. 4.

The results show that GBS during diffusion creep is consistent with a mechanism involving intrinsic GBDs. The observed threshold stress is consistent with it being determined by the interaction between GBDs and intergranular precipitates. This interaction consists of two parts: firstly the stress necessary to bow dislocations between pinning precipitates and secondly the stress which is necessary to overcome the attractive interaction between precipitates and GBDs. It is also interesting to note that an attractive interaction can also arise between GBDs and intergranular cavities (42).

5. SUPERPLASTICITY

It is well known that the major strain-providing mechanism during micrograin superplastic flow is GBS (23). For the conditions of peak superplastic behaviour, measurements indicate that GBS contributes up to $\sim$ 80 % of the total strain (43). In view of the clear importance of interfacial processes it is somewhat surprising that there have been no reported TEM investigations of interfacial defect structures after superplastic deformation. We have recently studied a superplastic Cu-2·8 % Al-1·8 % Si-0·4 % Co alloy. Rather than being microduplex, as is usual in superplastic alloys, the fine grain structure in this material is maintained by a uniform dispersion of Co-rich precipitates (44, 45). Fig. 7 shows the typical stress/strain-rate behaviour of the particular grain size investigated. Peak superplastic behaviour occurs in region II where the stress sensitivity exponent, n, is a minimum.

Fig. 8 shows a twin boundary in material which had been deformed by 40 % in region III. As may be seen, the boundary contains a high density of extrinsic dislocations in the form of an irregular network (approximate spacing 20nm). As discussed by Howell et al (40), examination of the extrinsic dislocation content in primary twin boundaries allows the extent of matrix dislocation activity to be gauged. The results of measurements of dislocation densities in twin boundaries in the deformed copper alloy are shown in Fig. 9 (46). From this it may be concluded that large numbers of matrix dislocations are mobile during deformation in region III. This number diminishes substantially with decreasing stress in region II and in region I the number of mobile matrix dislocations is virtually zero.

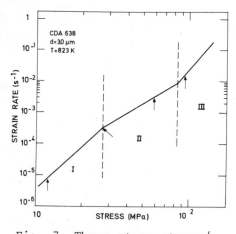

Fig. 7. Three stage stress/ strain-rate behaviour of the Cu-alloy.

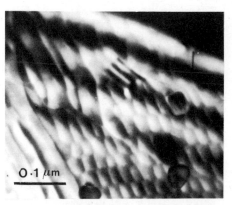

Fig. 8. Extrinsic dislocations in a twin boundary. 40 % strain in region III.

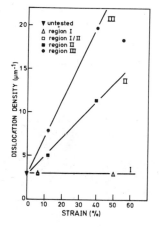

Fig. 9. Densities of dislocations in twin boundaries as a function of strain for regions I - III.

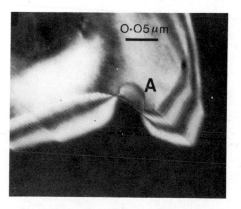

Fig. 10. Pinning of a grain boundary by an intergranular precipitate in the Cu-alloy.

Examination of GBDs in a large number of high-angle grain boundaries in material deformed in regions I to III showed that the overall characteristics for all three regions were very similar and therefore will be considered together in the following discussion. Fig. 10 shows a grain boundary in material which had been strained 12 % in region III. The pinning action which can be seen at precipitate A (compare with fig. 5) leads to the inhibition of grain growth. It also implies the existence of a precipitate induced threshold stress for

grain boundary sliding as discussed in section 4. An array
of intrinsic GBDs can also be seen in this boundary. The spacing and orientation of these GBDs is distorted in the vicinity
of the pinning precipitate. Fig. 11 shows the effect of the
precipitate dispersion on an array of widely spaced GBDs (12
% strain in stage II). In the two beam bright field image of
fig. 11a, a number of weakly diffracting dislocations are seen
together with two intergranular precipitates (arrowed A and B)
The interaction between the dislocations and the particles is
shown in fig. 11b. It can be seen that the dislocations are
strongly pinned at both A and B. A further example of this
type of interaction is seen in fig . 12 (12 % strain stage III).
The precipitate A is exerting a strong pinning effect on the
array of dislocations and considerable bowing of these dislocations can be seen at the "exit" side of the particle. It
would appear that, once unpinning has occurred, there is a
large driving force for continued glide/climb of the GBDs and
a "gap" in the dislocation array arises beyond the forward
edge of the particle. It is also worth noting with respect
to fig. 12, that in common with the observations detailed in
Section 4, no evidence for Orowan looping was observed either
in the matrix or within the boundary itself. Further support
for this is given in fig. 13 where, although considerable dislocation interaction occurred at the particles A, D and E,
no looping is seen. Closer examination of the particle A suggests that the GBDs exist at the particle/matrix interface
(arrowed B). This lends further support to the hypothesis that
by-pass occurs by glide/climb over the particle/matrix interface.

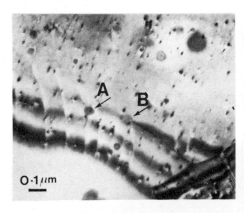
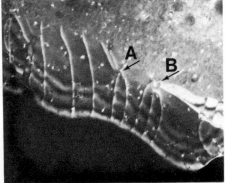

Fig. 11a. Interactions between
GBDs and precipitates during
superplastic deformation.

Fig. 11b. Dark field showing
constriction of GBDs at precipitates.

In addition to the inhibition of the motion of GBDs by
intergranular precipitates, it was found that grain boundary
triple junctions and twin boundary intersections could be

effective barriers to dislocation motion. Fig. 14 shows an example of a pile-up of GBDs against a twin boundary (arrowed); the spacing between individual dislocations changes from ∿ 10 nm at A to ∿ 4 nm at B. Hence, it would appear that the "threshold stress" for grain boundary sliding is dependent, not only on the precipitate spacing, but also on both the grain size and the frequency of twin intersections.

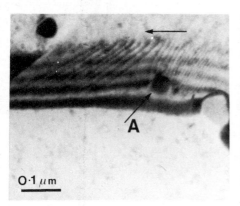

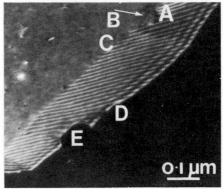

Fig. 12. An array of GBDs pinned at an intergranular precipitate.

Fig. 13. Interactions between GBDs and intergranular precipitates.

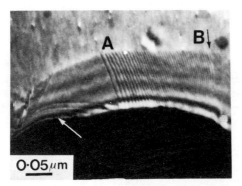

Fig. 14. A pile-up of GBDs against a twin boundary (B). Note also the segments of GBD loops (arrowed). $\varepsilon = 12$ % in region II.

6. CONCLUDING DISCUSSIONS

The observations indicate that grain boundary sliding, whether occurring during diffusion creep, dislocation creep or superplastic flow, can be considered in terms of a GBD type model. It seems that intrinsic GBDs control sliding during diffusion creep and region I deformation of superplastic materials. Extrinsic GBDs may become more effective sliding agents during dislocation creep and regions II and III of

superplastic deformation since the increased matrix dislocation activities will allow the operation of GBD sources of
the type shown in fig. 3.

The inhibition of the motion of GBDs by intergranular
precipitates, noted in section 4 and 5, is consistent with the
work of Horton (33) who showed that the rate of sliding in bicrystals decreases in the presence of intergranular precipitates and that the threshold stress for GBS is dependent upon
particle spacing. Indeed, consideration of the precipitates
alone, allows a reasonable estimate of the threshold stress
to be calculated using either a simple Zener drag model or a
GBD bowing model (section 4). However, the observations of
GBD pile-ups at triple junctions in the superplastic Cu-alloy
(vid. fig. 14) suggest that a further contribution to the
threshold stress can arise at small grain sizes.

Hence, if we adopt the terminology due to Hornbogen (47),
the total threshold stress for grain boundary sliding (and
hence for GBD movement) can be written as:

$$\sigma_0 = \sigma_i + \Delta\sigma_p + \Delta\sigma_{t.j.} \qquad\qquad 2.$$

where σ_i is a friction stress for the movement of GBDs in the
grain boundary (48, 49), $\Delta\sigma_p$ is the contribution due to the
intergranular precipitates, (this appeared to dominate in the
stainless steel) and $\Delta\sigma_{t.j.}$ is the contribution due to dislocation pile-ups at triple junctions etc, (this is likely to
be significant for the small grain sizes of superplastic material).

The rate of GBS in polycrystals will often depend upon the
rate of deformation of the grains which constrain sliding.
However, under certain circumstances (region II superplasticity may well be one) the rate of accommodational strain in
the grains may be sufficiently rapid for GBS to be rate controlling. Whichever is the case, the total rate of deformation of a polycrystalline material at elevated temperatures
will contain a contribution from GBS. Therefore, it is important that a constitutive equation is developed for the
rate of GBS independent of the accommodational strain taking
place within the grains. Since GBS involves creep of the GBD
networks it seems reasonable to draw a parallel to expressions
developed for power law creep involving crystal dislocation
networks. In particular, Ball and Hutchinson (50) have developed an expression for the situation where matrix dislocations traverse a grain and pile-up at a grain boundary. The
pile-up is then relieved through the rate-controlling step of
climb into the boundary. This situation is very similar to
the often observed situation of pile-ups of GBDs at triple
junctions which can be relieved by GBD association or dissociation reactions enabling shear to continue on one or both
of the other two boundaries. A suitable expression for the

rate of GBS would then be of the form:

$$\dot{\varepsilon}_b = \frac{AGb}{kT} \left(\frac{b}{d}\right)^2 \left(\frac{\sigma-\sigma_0}{G}\right)^n D_b \qquad\qquad 3.$$

where A is a constant, G is the shear modulus, b is the magnitude of the Burgers, vector of the mobile GBDs, d is the grain size (related to grain facet length), σ is the applied stress, σ_0 is the threshold stress, n is the stress sensitivity index and D_b is the grain boundary diffusivity.

In Ball and Hutchinson´s (50) expression developed for matrix dislocations the value of n is 2. This is also likely to be the case for GBS. It can be noted that n $\approx$ 2 in region II superplastic flow where GBS is probably rate controlling. Under circumstances where intergranular precipitates provide a more effective barrier to the motion of GBDs than triple junctions, then d in Eqtn. 3 should be replaced by the inter-particle spacing, λ.

7. ACKNOWLEDGEMENTS

Cooperation with T.G. Langdon is gratefully acknowledged as are discussions with H. Nordén. Financial support was received from the Swedish Natural Science Research Council, Carl Tryggers Stiftelse and the Science Research Council, (England).

8. REFERENCES

1. R.L. BELL and T.G. LANGDON, "Interfaces" Butterworths, Melbourne, 1969, 115.
2. H. GLEITER and B. CHALMERS, Prog. Mater. Sci., 1972, 16.
3. B. BURTON, "Diffusional Creep of Polycrystalline Materials" Trans Tech Publications, 1977.
4. F.R.N. NABARRO, Report of Conf on Strength of Solids, Phys. Soc., London, 1948.
5. C. HERRING, J. Appl. Phys., 1950, 21,437.
6. D.G. BRANDON, B. RALPH, S. RANGANATHAN and M.S. WALD, Acta Met., 1964, 12, 813
7. W. BOLLMAN "Crystal Defects and Crystalline Interfaces", Springer Verlag, Berlin, 1970.
8. R. SCHOBER and R.W. BALLUFFI, Phil. Mag., 1970, 21, 109.
9. R. SCHOBER and R.W. BALLUFFI, Phys. Stat Sol. (b), 1971, 44, 103
10. B. LOBERG and H. NORDÉN, Acta. Met., 1973, 21, 21.
11. P.R. HOWELL, A.R. JONES and B. RALPH, J. Matls. Sci., 1975, 10, 35.
12. D.G. BRANDON, Acta Met., 1966, 14, 1479.
13. D.H. WARRINGTON and M. BOON, Acta Met., 1975, 23, 599.
14. P.R. HOWELL, D.E. FLEET, P.I. WELCH and B. RALPH, Acta Met.,1978, 26, 1499.
15. T.P. DARBY, R. SCHINDLER and R.W. BALLUFFI, Phil. Mag. A, 1978, 37, 245

16. P.R. HOWELL, A.R. JONES, A. HORSEWELL and B. RALPH, Phil. Mag., 1976, 33, 21.
17. R.C. POND and D.A. SMITH, Phil. Mag., 1977, 36, 353.
18. P.R. HOWELL and A.R. JONES, J Matls. Sci., 1978, 13, 1830.
19. D.J. DINGLEY and R.C. POND, Acta. Met., 1979, 27, 667.
20. G.R. KEGG, C.A.P. HORTON and J.M.Silcock, Phil. Mag., 1973 27, 1041.
21. R.C. POND, D.A. SMITH and P.W. J. SOUTHERDEN, Phil. Mag. A. 1978, 37, 27.
22. R. RAJ and M.F. ASHBY, Met. Trans., 1971, 2, 1113.
23. D.M. R. TAPLIN, G.L. DUNLOP and T.G. LANGDON, Ann. Rev. Mater. Sci., 1979, 9, 151.
24. Y. ISHIDA and M. HENDERSON BROWN, Acta Met., 1967, 15, 857.
25. D. McLEAN, Phil. Mag., 1971, 23, 467.
26. T.S.KÊ, Phys. Rev., 1947, 71, 533.
27. M. BISCARDI and C. GOUX, Mem. Scient. Rev. Met., 1968, 65,167.
28. R.S. GATES, Acta. Met., 1973, 21, 855.
29. R.S. GATES, Scripta Met., 1974, 8, 55.
30. G.L. DUNLOP, J-O. NILSSON and P.R. HOWELL, J. Microsc., 1979, 116, 115.
31. J-O. NILSSON, P.R. HOWELL and G.L. DUNLOP, Acta Met., 1979, 27, 179.
32. R.C. POND , D.A. SMITH and P.W.J. SOUTHERDEN, Proc. 4th Int. Conf. on Strength of Metals and Alloys, Nancy, 1976, 1, 378.
33. C.A.P. HORTON, Acta. Met., 1972, 20, 477.
34. C.A.P. HORTON, Scripta. Met., 1974, 8, 1.
35. P.M. HAZELDINE and D.A. SMITH, quoted in ref. 21.
36. R.L. COBLE, J. Appl. Phys., 1963, 34, 1679.
37. I.M. LIFSHITZ, Soviet Phys., 1963, 17, 909.
38. G.B. GIBBS, Mem. Sci. Rev. Met., 1965., 17, 781.
39. J-O. NILSSON, P.R. HOWELL and G.L. DUNLOP, Scripta. Met., 1978, 12, 679.
40. P.R. HOWELL, J-O. NILSSON and G.L. DUNLOP, Phil. Mag. A. 1978, 38, 39.
41. M.F. ASHBY, Scripta Met., 1969, 3, 837.
42. L-E. SVENSSON and G.L. DUNLOP, Can. Met. Quart., 1979,18 39.
43. K. MATSUKI and M. YAMADA, J. Jap. Inst. Met. 1973, 37, 448.
44. S.A. SHEI and T.G. LANGDON, Acta Met., 1978, 26, 639.
45. F.A. MOHAMMED, M.M.I. AHMED and T.G. LANGDON, Met. Trans. A., 1977, 8A, 933.
46. L. FALK, G.L. DUNLOP and T.G. LANGDON, "Electron Micro-scopy 1980" Proc. 7th European Cong. on Electron Micro-scopy, The Hague, 1980, 1, 154.
47. E. HORNBOGEN, Proc. 5th Int. Conf. on Strength of Metals and Alloys, Aachen, Pergamon, 1979, 2, 1337.
48. H. GLEITER , Acta Met., 1979, 27, 187.
49. A.H. KING and D.A. SMITH, Met Sci., 1980, 14, 57.
50. A. BALL and M.M. HUTCHINSON, Met. Sci. J., 1969, 3, 1.

CURRENT PROBLEMS IN SUPERPLASTICITY

Terence G. Langdon

Departments of Materials Science and Mechanical Engineering,
University of Southern California, Los Angeles,
California 90007, U.S.A.

SUMMARY

Superplastic metals are capable of very high strains when
pulled in tension. The mechanical behavior of superplastic
materials is reviewed briefly, and then three areas are
described in which a detailed understanding is currently
lacking. These areas are (i) the significance of the decrease
in superplasticity at low strain rates in region I, (ii) the
role of necking and cavitation in fracture, and (iii) the need
for a model of superplasticity which is consistent with all of
the experimental data.

1. INTRODUCTION

Superplasticity refers to the ability of some metals to
exhibit large and essentially neck-free strains when pulled in
tension, so that the elongations at fracture under optimum
conditions are usually in excess of 1000%. Although the
possibility of superplastic deformation was first demonstrated
in the laboratory over forty years ago, primarily in classic
experiments on the Pb-Sn and Bi-Sn eutectic alloys by Pearson
[1], the superplastic process has been subjected to very
extensive investigation and analysis only within the last
fifteen years. This recent interest has arisen because it is
now recognized that superplastic metals have a considerable
potential use in many industrial forming applications,
especially when it is necessary to produce a relatively small
number of identical items (typically, less than ∿1000) of a
complex shape. Any long-term industrial utilization of superplasticity necessitates a reasonable understanding of the
superplastic process and, in particular, it requires a detailed
knowledge of the various parameters which contribute towards
the development of the superplastic characteristics.

This paper has two objectives. First, a short review of

the mechanical behavior of superplastic metals is presented in section 2: this review is necessarily very brief, but more detailed descriptions of superplasticity are given elsewhere [2,3]. Second, the primary role of this paper is to identify important areas in which a detailed understanding is either currently lacking or incomplete, and to describe and summarize the major problems associated with each of these areas. Accordingly, three topics have been selected for detailed review, and these are presented in section 3.

2.   THE MECHANICAL BEHAVIOR OF SUPERPLASTIC METALS

There are two basic requirements for the occurrence of large tensile elongations. First, it is necessary that the grain size of the material is very small and, in addition, that it is relatively stable. Superplastic metals usually have average spatial grain sizes which are less than about 10 µm in size, and they are often of the order of 1 - 3 µm (where the spatial grain diameter, d, is defined as 1.74 × the mean linear intercept grain size). Second, since superplasticity is a diffusion-controlled process, it requires testing temperatures of the order of 0.5 T_m, where T_m is the absolute melting point of the material. In general, these two requirements are incompatible, since pure metals and simple alloys exhibit extensive grain growth at elevated temperatures and it is therefore no longer feasible to retain a very small and stable grain size. This problem may be overcome by using two-phase alloys, such as the Pb-62% Sn eutectic or the Zn-22% Al eutectoid, or by incorporating, as a grain refiner, a very fine dispersion in an essentially single phase alloy. Examples of the latter procedure include CDA 638 (a commercial copper alloy containing 95% Cu, 2.8% Al, 1.8% Si and 0.4% Co, with a dispersed cobalt-rich phase [4]) and Al-6% Cu-0.5% Zr (in which there is a fine dispersion of $ZrAl_3$ [5]).

Superplastic materials are usually tested on tensile machines which impose either a true or approximate constant rate of strain on the sample. The approximation arises because many machines operate at a constant rate of cross-head displacement, which is equivalent to a decreasing effective strain rate throughout the test. In these experiments, the steady-state flow stress, σ, is measured as a function of the imposed strain rate, $\dot{\varepsilon}$, and the data are then logarithmically plotted as σ versus $\dot{\varepsilon}$.

An example of this type of result is shown in the lower part of Fig. 1, where the datum points refer to specimens of the Zn-22% Al eutectoid alloy having a grain size of 2.5 µm and tested at three different absolute temperatures, T, in the range from 423 to 503 K [6,7]. The stress and strain rate are related through the expression

$$\sigma = B\dot{\varepsilon}^m \qquad (1)$$

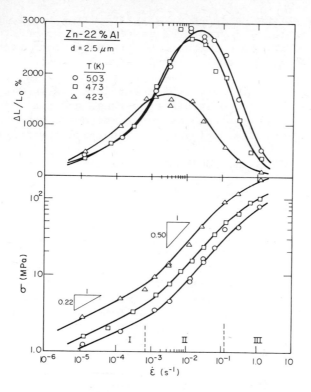

Fig. 1   Elongation to failure (upper) and flow stress (lower)
          versus strain rate for Zn-22% Al [7].

where B is a constant which incorporates the dependence on
temperature and m is termed the strain rate sensitivity. In
Fig. 1, therefore, the data fall into three distinct regions of
flow, designated I, II and III at low, intermediate and high
strain rates, respectively, and there is a distinct and
essentially constant value of m in each of these regions. In
region III, the results in Fig. 1 are not well defined but it
is anticipated that m ≃ 0.2; in regions II and I, the values
of m are 0.50 and 0.22, respectively.

     It is well established on theoretical grounds that there
is a correlation between the measured elongation to failure and
the value of the strain rate sensitivity [8-10]. This correlation is demonstrated in Fig. 1, since each point in the plot of
σ versus ε̇ refers to a different specimen, and these specimens
were pulled to failure and the total elongations, $\Delta L/L_0$%, were
also plotted against ε̇: this plot is shown in the upper part of
Fig. 1, where ΔL is the total increase in length at the point
of failure and L_0 is the initial gauge length of the specimen.

     The experimental results show that, even in a highly
superplastic alloy such as Zn-22% Al, the superplastic condition, with elongations to failure of >1000%, is confined to a

relatively narrow range of strain rates (region II, typically covering about 2 - 3 orders of magnitude of $\dot{\varepsilon}$) and there is a precipitous drop in the fracture strains at both faster (region III) and lower (region I) strain rates. This observation is important from a commercial point of view, because it shows that the optimization of a superplastic forming process will occur in a narrow strain rate range and that, in addition, this range may be centered at rates which are somewhat lower than those generally employed in industrial forming operations.

3. PROBLEM AREAS IN STUDIES OF SUPERPLASTICITY

3.1 The significance of region I

As indicated in Fig. 1, the superplastic region II with m $\simeq$ 0.5 is contained in a relatively narrow range of strain rates between two other regions where the material is no longer highly superplastic and m $\simeq$ 0.2 - 0.3. At high strain rates, where there is a transition to region III, the change in behavior is documented reasonably well and there is no doubt that this represents a transition to normal creep behavior with a stress exponent, n, of $\sim$5 (where n = 1/m, and this is equivalent to m $\simeq$ 0.2). By plotting experimental data on Pb-62% Sn and Zn-22% Al in the form of deformation mechanism maps, it appears that superplastic behavior requires a grain size which is sufficiently small that it precludes the formation of a stable subgrain structure [11].

The transition to region I at the lower strain rates is less easy to explain, and this behavior has been the subject of numerous speculations. An initial problem arising from any consideration of region I is that published data from different laboratories often indicate markedly different types of behavior on the same material at these very low strain rates. This difference is indicated schematically in Fig. 2 for a typical superplastic metal, showing the standard logarithmic plot of σ versus $\dot{\varepsilon}$. In regions III and II, most sets of results are in reasonable agreement, with values of m of $\sim$0.15 and $\sim$0.5, respectively. However, two distinct and separate trends have been reported at low strain rates in region I, as indicated in Fig. 2 by the lines marked A and B. Data falling along line A give m $\simeq$ 0.25 in region I, and this trend is identical to the experimental results shown in Fig. 1; whereas data falling along line B give m $\simeq$ 1.0 in this region, and this is clearly inconsistent with Fig. 1. The confusion is further compounded because both trends have been reported for the same material when tested under nominally identical experimental conditions. For example, a series of experiments at high temperatures on Zn-22% Al gave results lying along line A in the investigations of Mohamed and Langdon [12], Grivas [13] and Vale et al. [14] whereas the datum points fell along line B in the work of Vaidya et al. [15], Misro and Mukherjee [16] and

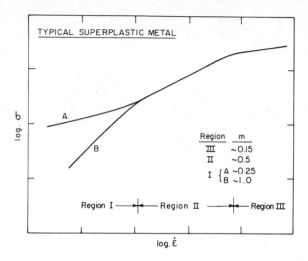

Fig. 2   Schematic illustration of stress versus strain rate,
showing the division into two types of behavior (A
and B) in region I.

Arieli et al. [17]. Furthermore, there is no obvious explanation for this dichotomy in terms of grain size, testing
temperature, or material preparation.

A possible explanation for Fig. 2 is that the datum points
falling on line A are erroneous and arise from concurrent grain
growth [18]. This suggestion is based on the observation by
Rai and Grant [19] on the Al-33% Cu eutectic that the occurrence of grain growth may lead to observations of a false
"region I" with an apparent low m at low strain rates. The
experimental evidence reported by Rai and Grant [19] is shown
in Fig. 3, where specimens were tested at 743 K with initial
grain sizes, d_0, in the range from 1.5 to 10 μm, and the
experimental points show the measured flow stresses at a fixed
strain, ε, of 0.15. In these experiments, there was little or
no grain growth at the larger grain sizes, but there was
extensive grain growth with $d_0 = 1.5$ μm such that, at the
slowest strain rate of 5×10^{-7} s^{-1}, the measured grain size
was 6 μm at $\varepsilon = 0.15$. As a result of this large increase in
grain size, this experimental point was displaced to a higher
stress by a factor of ∿4×, thereby giving the appearance of a
false "region I" with a low strain rate sensitivity.

In practice, a detailed analysis of the various sets of
data shows that this explanation is not able to account for the
experimental points falling on line A for Zn-22% Al [20].
Thus, an alternative explanation has been developed in which it
is argued that the datum points falling on line B in Fig. 2 are
spurious and arise from a failure to allow for the primary
stage of creep. This suggestion is based on the pronounced

146

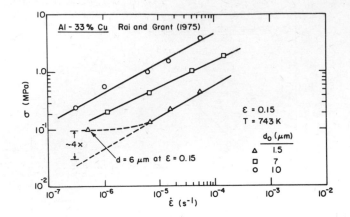

Fig. 3   Stress versus strain rate for Al-33% Cu with three
different initial grain sizes [19].

primary creep which occurs in samples tested in region I, as
shown in Fig. 4 for a specimen of Zn-22% Al with an initial
grain size of 1.9 µm tested at 523 K under a constant shear
stress, τ, of 0.1 MPa. The datum points record the shear
strain, γ, versus testing time, t, and show that the steady-
state shear strain rate, $\dot{\gamma}$, is only achieved in this specimen
at $\gamma > 0.10$ [20]. By contrast, Fig. 5 shows a similar plot of
γ versus t, also for Zn-22% Al, from the experiments of Arieli
et al. [17] purporting to confirm line B in Fig. 2. In this
experiment, the stress was increased during the test to
provide strain rate data at two different stress levels, but
close inspection shows that the slower shear strain rate of
4.4×10^{-7} s^{-1}, which lies in region I, was recorded at a
shear strain of $\gamma \simeq 0.02$. As demonstrated in Fig. 4, this low
level of strain is generally insufficient to establish a true
steady-state flow in region I, and it is therefore concluded
that this measured strain rate, together with all experimental
points on line B, represents an overestimation of the true
strain rate at these very low stress levels.

Having established that the correct trend is shown by
line A in Fig. 2, at least for Zn-22% Al, the problem remains
of explaining the behavior in region I. On theoretical
grounds, it is anticipated that there will be a transition to
diffusion creep and m = 1.0 at sufficiently low stress levels.
Indeed, some evidence for this trend was presented by Vale
et al. [14] and termed region 0. However, calculations of
diffusion creep rates indicate that region I is a distinct
deformation mode which occurs at strain rates which are intermediate between the superplastic region II and the diffusion
creep processes [21]. Two alternative approaches have been
developed in an attempt to explain region I on theoretical
grounds. First, Ashby and Verrall [22] introduced a threshold

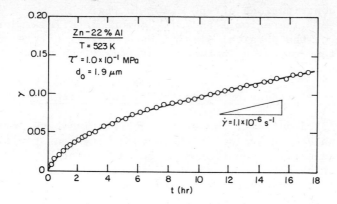

Fig. 4   Shear strain versus time for Zn-22% Al tested in region I [20].

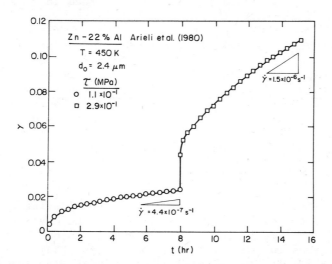

Fig. 5   Shear strain versus time for Zn-22% Al showing a stress change experiment in region I [17].

stress which, it was suggested, arose from fluctuations in the grain boundary area. Second, Gifkins [23] developed a model in which region I was due to grain boundary sliding controlled by barriers inherent in the grain boundary structure. Neither theory is entirely acceptable, as they lead to activation energies which are either equal to or lower than the activation energy for flow in the superplastic region II. This contrasts with the experimental evidence which shows a significant increase in the activation energy when passing from region II to region I [24-27]. Furthermore, the present theories are not able to account for the marked decrease in the importance of grain boundary sliding in region I, since experiments show

F

sliding contributions of ∿50 – 70% in region II but a decrease
to ∿20 – 30% in region I [3,28]. A more detailed understanding of the physical significance of region I is therefore
an important priority in future investigations of
superplasticity.

3.2 The factors governing ductility and fracture processes

As noted earlier, there is a correlation between the total
elongation and the value of the strain rate sensitivity [8-10].
This was clearly demonstrated in an early plot by Woodford
[29] in which data were assembled in the form of m versus the
total elongation to failure. The result is shown in Fig. 6,
where the open symbols are the early data plotted for various
metals and the two closed symbols show additional recent
results on Zn-22% Al (from Fig. 1 [7]) and Pb-62% Sn [30].
The latter material is particularly significant because it
exhibits elongations of up to 4850% without failure, thereby
providing the highest strains so far recorded in a superplastic metal.

Although all of the experimental datum points scatter
about the line in Fig. 6, it has proved difficult to develop a
relationship which permits a reasonable estimate of the
elongation to failure from measurements taken early in the
test. In some cases, a reason for this deficiency is that m
is strain dependent, and it is then necessary to use the
terminal, rather than the initial, value of m [31]. However,
in other materials, where m rapidly reaches an essentially
steady value, it is also difficult to make an accurate
prediction of the overall ductility.

This problem may be illustrated by considering a specific
example. A theory has been presented for the development of
plastic instabilities in tension creep due to macroscopic
irregularities in the specimen cross-section [32]. This
theory, when suitably modified [33], predicts an elongation
to failure which is given by

$$\frac{\Delta L}{L_o} \% = \left\{ \exp \left[\frac{mK}{1 - m} \right] - 1 \right\} \times 100 \qquad (2)$$

where K is a constant equal to ∿2 – 3. Using the data of Fig.
1, and taking m = 0.50 and an upper limit of K = 3, equation
(2) predicts an elongation of 1900% in region II, whereas the
experimental data show a variation from ∿1600% at 423 K to
∿2900% at 503 K. In region I, using m = 0.22, equation (2)
predicts an elongation of ∿130%, which contrasts with the
experimental values of ∿300 – 500%. Since equation (2) is
based on geometrical effects and ignores the possibility of
cavity formation, it should, in principle at least, predict
an upper limit for the total elongation. By contrast, the

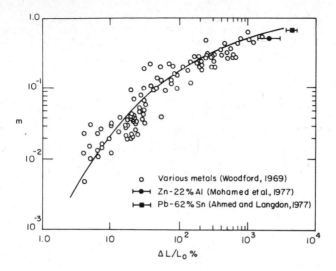

Fig. 6   Strain rate sensitivity versus elongation to failure.

calculations show that the relationship tends to underestimate the overall ductility and, more important, since the value of m in any region is essentially independent of temperature (Fig. 1), it fails to incorporate a specific temperature dependence so that it is unable to account for the higher elongations recorded at the higher temperatures. This suggests that m is a necessary but not sufficient parameter in any theoretical relationship which is developed to predict the total ductility [34].

    An associated problem in studies of superplasticity is to understand the various processes which lead to ultimate fracture. This problem is divisible into two separate, but complementary, parts. First, it is now well established that cavity formation is often important in superplastic metals, not only in materials, such as Cu-based alloys, which exhibit rather modest elongations to failure [35-39] but also in those materials, such as Zn-22% Al, which are highly superplastic [40-42]. Second, it is also clear that, at the macroscopic level, the fracture behavior is connected with flow localization in the form of the development and propagation of external necks.

    Experiments show that flow is essentially uniform in the superplastic region II in Zn-22% Al, and this is indicated by the experimental data shown in Fig. 7 [43]. For these experiments, the gauge lengths of the tensile specimens were divided into 14 segments, each of length ℓ_o, and then the increase in length of each segment, $\Delta\ell$, was measured at selected macroscopic strains using a travelling microscope. Figure 7 plots the percentage strain in each segment, $\Delta\ell/\ell_o\%$, versus the

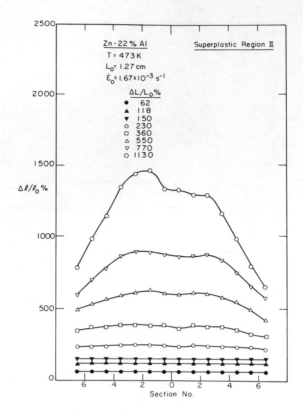

Fig. 7   Variation of local strain along the gauge length for
          Zn-22% Al tested in region II [43].

position along the gauge section (numbered such that zero
corresponds to the mid-point of the specimen). This test was
conducted at 473 K using an initial gauge length, L_O, of 1.27
cm and an initial strain rate, $\dot{\varepsilon}_O$, of 1.67×10^{-3} s^{-1} in
region II. The datum points, taken at various values of $\Delta L/L_O\%$
up to 1130%, show that the flow is entirely uniform up to
∿150%, and thereafter there are only relatively slight perturbations in the measured strains from point to point along the
gauge length. At the higher elongations, the individual values
of $\Delta \ell/\ell_O\%$ tend to increase more rapidly in the center of the
gauge length, but the deformation is quasi-uniform even at the
highest strain (1130%) and there is no evidence of neck
formation.

The behavior is very different in region I, as shown in
Fig. 8 at an initial strain rate of 1.67×10^{-5} s^{-1}. In this
case, the deformation is uniform up to ∿100%, but a neck then
forms near the center of the gauge length and develops very
rapidly to give fracture at the macroscopic elongation of
480%. At the point of failure, the local elongation in the
necked segment is ∿3700%.

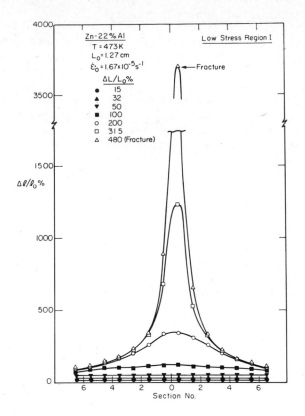

Fig. 8   Variation of local strain along the gauge length for
         Zn-22% Al tested in region I [43].

The results in Figs. 7 and 8 are qualitatively consistent
with the values recorded for m in regions II and I (Fig. 1),
since a low value of m, as in region I, tends to favor the
development and propagation of an external neck whereas necking
is suppressed when m is high. However, it is also clear that,
when these results are combined with the observations of
internal cavitation, it is important to obtain a more detailed
understanding of the fracture characteristics of superplastic
materials, and especially to devote attention to the problems
of internal void growth [44] and interlinkage, the role of
internal necking in inter-cavity ligaments, and the possibility
of developing stable (or unstable) arrays of cavities in materials with high strain rate sensitivities. Some preliminary
work has been conducted to examine the coalescence of
cylindrical holes, drilled prior to testing, in superplastic
tensile specimens [45], but more research of this nature is
clearly needed to provide a more meaningful understanding of
the overall fracture processes.

3.3   The mechanism of superplasticity

Several different mechanisms have been developed theoretically to explain the superplastic process but, without exception, each model is inconsistent with one or more of the established experimental trends [46]. One problem in superplastic deformation is to explain the process by which very large macroscopic strains are produced in the specimen with little or no microscopic changes in the shapes of the individual grains. Extensive studies show that the internal grain structure remains essentially equi-axed, even after specimen elongations of up to several hundreds of percent [2].

All theories of superplasticity are based on grain boundary sliding, and this is consistent with the experimental observations of high sliding contributions in region II. One method of achieving specimen elongation without concomitant grain elongation is shown in Fig. 9(A), based on the theory developed by Ashby and Verrall [22]. A group of four hexagonal grains is shown in (a) at $\varepsilon = 0$. Under the action of an applied stress which is vertical throughout Fig. 9, the grains change their shapes by diffusion to reach the intermediate state at $\varepsilon = 0.275$ where the four central grain boundaries meet at a point. Thereafter, these boundaries split to give two triple junctions separated by a vertical boundary and, ultimately, the configuration shown at $\varepsilon = 0.55$. In this state, the grains have changed their nearest neighbors by sliding over each other, but they have not undergone any local elongation.

At first sight, the sequence shown in Fig. 9(A) is attractive. It leads to a specimen elongation without microscopic strain and, as developed by Ashby and Verrall [22], it predicts a strain rate which is significantly faster than the combination of grain boundary and lattice diffusion creep. Furthermore, direct evidence for this type of grain switching was recorded in Zn-22% Al specimens deformed *in situ* in the high voltage electron microscope [47]. However, close inspection shows that there are four major problems with this sequence. First, the diffusion paths required to change from (a) to (b) are not physically realistic, and the only acceptable boundary movements are uniform translations [48]. Second, the rate equation developed by Ashby and Verrall [22] is incorrect and, when the diffusion paths are changed to take account of this error, the strain rate is approximately equal to that of diffusion creep [49]. Third, the model in Fig. 9(A) gives a specimen elongation without a corresponding increase in the surface area of the material [49]. In a two-dimensional situation, as in the foils examined in electron microscopy [47], the surface area cannot increase because no other grains are available to come to the specimen surface; but in a real polycrystalline sample, pulled in tension, the

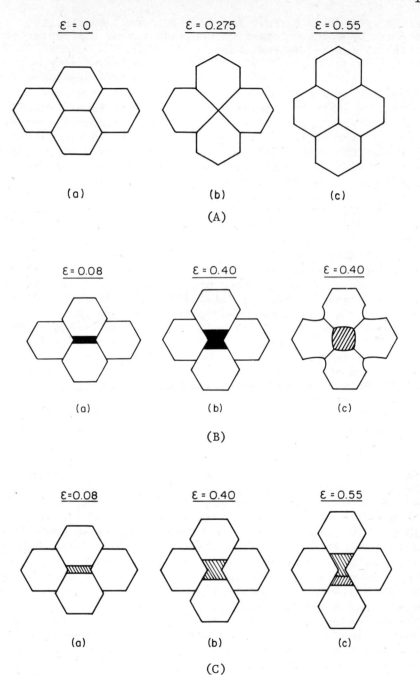

Fig. 9   Examples of grain switching during superplastic flow:
(A) by diffusion [22], (B) by opening up of boundary
fissures [49], and (C) by the exposure of grain
boundary facets [50].

surface area will increase during straining. Fourth, the procedure by which deformation proceeds beyond configuration (c) is not defined in the model.

To overcome these problems, Gifkins [49] presented an alternative sequence, shown in Fig. 9(B). In this case, the grains move apart by sliding and a fissure is formed on the free surface, as at (a) when $\varepsilon = 0.08$. This process continues as at (b), and ultimately a new grain [shaded in Fig. 9(B)] emerges at the surface as shown at (c) when $\varepsilon = 0.40$. At this point, local boundary migration occurs to curve the boundaries, as indicated. The advantage of this sequence is that each position of the group is repeatable as a unit throughout the material, and the process is regenerative to higher elongations. The disadvantage is that it necessitates the opening up of fissures at the surface, and therefore it is again partly two-dimensional because it assumes that the boundaries of the surface grains are essentially perpendicular to the specimen surface. This latter problem may be overcome by using the sequence shown in Fig. 9(C) [50]. At $\varepsilon = 0.08$, the two vertical grains have moved apart exposing a portion of the boundary between them. This boundary facet is further exposed at $\varepsilon = 0.40$, and finally, as indicated at $\varepsilon = 0.55$ in (c), both an exposed grain boundary and an internal grain are visible from the surface. The internal grain then proceeds to the surface with increasing strain, and the boundaries become curved as in Fig. 9(B).

The sequence shown in Fig. 9(C) is capable of explaining the re-arrangements of grains during superplastic flow, but it is now necessary to incorporate this sequence into a model which predicts the various experimental results obtained in region II. In particular, no theory has so far explained the observation that the average amount of sliding varies between the various types of interface in a two-phase superplastic alloy. In Pb-62% Sn, for example, maximum sliding occurs on the Sn-Sn intercrystalline boundaries, the Pb-Sn interphase boundaries are intermediate in behavior, and there is little or no sliding on the Pb-Pb interfaces [28]. It seems likely that these differences in sliding behavior account for the observations of preferential cavity formation at one type of interface in two-phase structures [37], thereby serving to emphasize the intimate relationship between the mechanistic processes of flow and the fracture characteristics.

4. CONCLUSIONS

Three important problem areas have been identified and described in superplasticity. These are (i) the significance of the decrease in superplastic behavior at low strain rates in region I, (ii) the difficulties of predicting the total elongation during superplastic flow and the limited information available on fracture processes, and (iii) the lack of a model of superplasticity which is consistent with all of the experimental data.

ACKNOWLEDGMENT

This work was supported by the National Science Foundation under Grant No. DMR79-25378.

REFERENCES

1. PEARSON, C.E. - *J. Inst. Metals*, 1934, 54, 111.
2. EDINGTON, J.W., MELTON, K.N. and CUTLER, C.P. - *Prog. Mater. Sci.*, 1976, 21, 61.
3. TAPLIN, D.M.R., DUNLOP, G.L. and LANGDON, T.G. - *Ann. Rev. Mater. Sci.*, 1979, 9, 151.
4. SHEI, S.-A. and LANGDON, T.G. - *Acta Met.*, 1978, 26, 639.
5. GRIMES, R., BAKER, C., STOWELL, M.J. and WATTS, B.M. - *Aluminium*, 1975, 51, 720.
6. ISHIKAWA, H., MOHAMED, F.A. and LANGDON, T.G. - *Phil. Mag.*, 1975, 32, 1269.
7. MOHAMED, F.A., AHMED, M.M.I. and LANGDON, T.G. - *Met. Trans. A*, 1977, 8A, 933.
8. HUTCHINSON, J.W. and NEALE, K.W. - *Acta Met.*, 1977, 25, 839.
9. KOCKS, U.F., JONAS, J.J. and MECKING, H. - *Acta Met.*, 1979, 27, 419.
10. NICHOLS, F.A. - *Acta Met.*, 1980, 28, 663.
11. MOHAMED, F.A. and LANGDON, T.G. - *Scripta Met.*, 1976, 10, 759.
12. MOHAMED, F.A. and LANGDON, T.G. - *Acta Met.*, 1975, 23, 117.
13. GRIVAS, D. - *Report No. LBL-7375*, Lawrence Berkeley Laboratory, University of California, Berkeley, 1978.
14. VALE, S.H., EASTGATE, D.J. and HAZZLEDINE, P.M. - *Scripta Met.*, 1979, 13, 1157.
15. VAIDYA, M.L., MURTY, K.L. and DORN, J.E. - *Acta Met.*, 1973, 21, 1615.
16. MISRO, S.C. and MUKHERJEE, A.K. - *Rate Processes in Plastic Deformation of Materials*, Eds. Li, J.C.M. and Mukherjee, A.K., American Society for Metals, Metals Park, Ohio, 1975, p. 434.
17. ARIELI, A., YU, A.K.S. and MUKHERJEE, A.K. - *Met. Trans. A*, 1980, 11A, 181.
18. SUERY, M. and BAUDELET, B. - *Rev. Physique Appl.*, 1978, 13, 53.
19. RAI, G. and GRANT, N.J. - *Met. Trans. A*, 1975, 6A, 385.
20. GRIVAS, D. MORRIS, J.W. and LANGDON, T.G. - "Observations on the Differences Reported in Region I for the Superplastic Zn-22% Al Eutectoid," *Scripta Met.* (in press).

21. LANGDON, T.G. and MOHAMED, F.A. - *Scripta Met.*, 1977, <u>11</u>, 575.
22. ASHBY, M.F. and VERRALL, R.A. - *Acta Met.*, 1973, <u>21</u>, 149.
23. GIFKINS, R.C. - *Met. Trans. A*, 1976, 7A, 1225.
24. MOHAMED, F.A., SHEI, S.-A. and LANGDON, T.G. - *Acta Met.*, 1975, <u>23</u>, 1443.
25. MOHAMED, F.A. and LANGDON, T.G. - *Phil. Mag.*, 1975, <u>32</u>, 697.
26. SHEI, S.-A. and LANGDON, T.G. - *Acta Met.*, 1978, <u>26</u>, 1153.
27. BRICKNELL, R.H. and BENTLEY, A.P. - *J. Mater. Sci.*, 1979, <u>14</u>, 2547.
28. VASTAVA, R.B. and LANGDON, T.G. - *Acta Met.*, 1979, <u>27</u>, 251.
29. WOODFORD, D.A. - *Trans. ASM*, 1969, <u>62</u>, 291.
30. AHMED, M.M.I. and LANGDON, T.G. - *Met. Trans. A*, 1977, <u>8A</u>, 1832.
31. GHOSH, A.K. and AYRES, R.A. - *Met. Trans. A*, 1976, <u>7A</u>, 1589.
32. BURKE, M.A. and NIX, W.D. - *Acta Met.*, 1975, <u>23</u>, 793.
33. MOHAMED, F.A. - *Scripta Met.*, 1979, <u>13</u>, 87.
34. LANGDON, T.G. - *Scripta Met.*, 1977, <u>11</u>, 997.
35. SAGAT, S. and TAPLIN, D.M.R. - *Acta Met.*, 1976, <u>24</u>, 307.
36. SHEI, S.-A. and LANGDON, T.G. - *J. Mater. Sci.*, 1978, <u>13</u>, 1084.
37. CHANDRA, T., JONAS, J.J. and TAPLIN, D.M.R. - *J. Mater. Sci.*, 1978, <u>13</u>, 2380.
38. LIVESEY, D.W. and RIDLEY, N. - *Met. Trans. A*, 1978, <u>9A</u>, 519.
39. HUMPHRIES, C.W. and RIDLEY, N. - *J. Mater. Sci.*, 1978, <u>13</u>, 2477.
40. ISHIKAWA, H., BHAT, D.G., MOHAMED, F.A. and LANGDON, T.G. - *Met. Trans. A*, 1977, <u>8A</u>, 523.
41. MILLER, D.A. and LANGDON, T.G. - *Met. Trans. A*, 1978, <u>9A</u>, 1688.
42. AHMED, M.M.I., MOHAMED, F.A. and LANGDON, T.G. - *J. Mater. Sci.*, 1979, <u>14</u>, 2913.
43. MOHAMED, F.A. and LANGDON, T.G. - "Flow Localization and Neck Formation in a Superplastic Metal," *Acta Met.* (in press).
44. MILLER, D.A. and LANGDON, T.G. - *Met. Trans. A*, 1979, <u>10A</u>, 1869.
45. TAIT, R.A. and TAPLIN, D.M.R. - *Scripta Met.*, 1979, <u>13</u>, 77.
46. GIFKINS, R.C. and LANGDON, T.G. - *Mater. Sci. Eng.*, 1978, <u>36</u>, 27.
47. NAZIRI, H., PEARCE, R., HENDERSON BROWN, M. and HALE, K.F. - *Acta Met.*, 1975, <u>23</u>, 489.
48. SPINGARN, J.R. and NIX, W.D. - *Acta Met.*, 1978, <u>26</u>, 1389.
49. GIFKINS, R.C. - *J. Mater. Sci.*, 1978, <u>13</u>, 1926.
50. LANGDON, T.G. - "The Significance of Grain Boundary Sliding in Creep and Superplasticity," *Metals Forum* (in press).

SECTION 2

DEFORMATION PROCESSES IN PARTICLE – STRENGTHENED ALLOYS

CREEP OF DIRECTIONALLY SOLIDIFIED SUPERALLOYS AND EUTECTIC
COMPOSITES

M McLean and P N Quested

Division of Materials Applications, National Physical
Laboratory, Teddington, Middlesex, TW11 OLW.

Abstract

The effects of solidification conditions on the microstructures and creep properties of both directionally solidified superalloys and in-situ composites have been examined
The creep data of both types of alloys can be described by
a modified power law equation that includes a friction
stress term. A new approach to the measurement of friction
stress has been developed based on modelling of the stress
dip test procedure; the data are shown to be well described
by the model.

Introduction

Over the last decade or more, the major advances in
superalloys have been achieved through developments in
processing technology rather than by alloy development.
Directional solidification has been particularly successfully applied to produce columnar grained structures with
improved creep behaviours over the same conventionally cast
alloys[1] and is now state of the art in the manufacture
of rotor blades for advanced aero-gas-turbines. More
recently directional solidification has been extended to
alloys of near-eutectic composition in an attempt to produce
fibre-, or lamellar-, reinforced composite materials, which
have extraordinary compositional and morphological stability
capable of use at higher temperatures than either conventionally cast or directionally solidified (ds) superalloys[2].
The NPL effort in this area has been concerned with the
effects of the principal parameters of directional
solidification, temperature gradient G and growth rate R,
on the microstructures and properties of both classes of
alloys. In particular, it aims to determine the scope,
through control of the solidification conditions, for
further improvement in the creep behaviour of ds superalloys

over that currently achieved in commercial practice and for
optimisation of the properties of experimental eutectic
composites.

In common with many other microstructurally complex
alloys, the creep data obtained from both ds superalloys and
eutectic composites give unrealistically high values of the
stress exponent and activation energy when described by the
conventional Bailey-Norton equation. The modified Bailey-
Norton equation incorporating a friction stress σ_0,that has
been advocated by Wilshire and co-workers[3,4],has considerable attractions both as an improved method for data extrapolation and as a basis for modelling creep in the heterogeneous materials under consideration. In view of the
practical difficulties in measuring σ_0, normally by interpretation of strain transients following a series of stress
reductions, the procedure has been modelled to provide an
explicit relationship between the test parameters and σ_0 that
allows an objective evaluation of friction stress. This
method has been applied to both types of directionally
solidified materials to provide some insight into the
strengthening mechanisms in such alloys.

Directionally solidified superalloys

The main motivation for the early work on the directional solidification of superalloys appears to have been the
desire to control the grain structures of the castings in
order to reduce (or eliminate) the areas of grain boundaries
normal to the stress axes which were thought to be creep
fracture initiation sites leading to unacceptably low
ductilities in some cast superalloys[5]. This objective
was rapidly achieved and the technology developed to
transfer it to the manufacture of turbine blades[6]; the
imminent introduction of single crystal turbine blades by
Pratt and Whitney is the latest stage in this development[7].
However, directional solidification affects aspects of the
superalloy microstructure other than the grain boundary
morphology. For example, most superalloys solidify with
well defined <100> growth directions in a dendritic
morphology with aligned cells or dendrites parallel to the
direction of solidification. It does not seem clear yet
how these, and other, microstructural parameters influence
the mechanical properties of ds superalloys. The present
work aims to characterise quantitatively the changes in the
various types of microstructural features that result from
changes in the solidification procedures and to determine
their influences on the creep behaviours of the alloys.

Two nickel-base superalloys, IN738LC and Mar M246, that
have been directionally solidified at various rates R between
6 and 3,000 mmh^{-1} and in different temperature gradients G

(between about 4 and 20 Kmm^{-1}), using procedures described elsewhere[8], have been examined in some detail. Various solidification morphologies ranging from near plane front (for low R and high G) through cellular, cellular-dendritic and dendritic to equiaxed structures (for high R) have been obtained[9]. The differences between the dendrite dimensions and forms produced by current commercial ds practice and in the high G and R allowed by advanced laboratory techniques

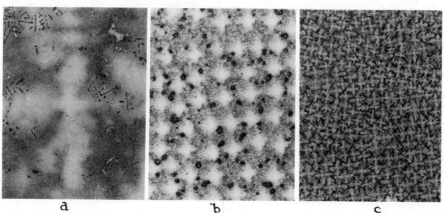

<u>FIG 1</u>. Transverse sections of IN738LC directionally solidi-fied under different conditions and heat treated.
(a) commercial 4 Kmm^{-1}, 1.7 x 10^{-2} mms^{-1}, (b) 13 Kmm^{-1}, 1.7 x 10^{-2} mms^{-1}, (c) 20 Kmm^{-1}, 0.33 mms^{-1}).

are illustrated for IN738LC in Figure 1. The dendrite dimensions in both alloys are correlated with the cooling rate GR during directional solidification in Figure 2; clearly the use of high G and R leads to considerable dendrite refinement. Of course, the dendrite is a manifestation of the different compositions formed as freezing of the alloys proceeds over a finite temperature range. Figure 3 shows the variations in concentrations across the dendrites of various elements that are thought to have important strengthening functions. The dendrite core which being the first solid formed has the highest melting point (the liquidus temperature) is rich in the refractory solid solution strengthener tungsten and in cobalt and chromium, while the last solid to form inter-dendritically (at the solidus temperature) is rich in the γ' formers titanium and aluminium. These compositional differences are also reflected in the form and stability of the γ' precipitate in different parts of the dendrite

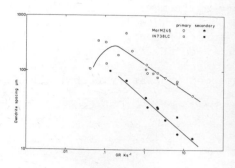

FIG 2. Variation of dendrite
(or cell) spacings with cooling rate GR for IN738LC and
Mar M246.

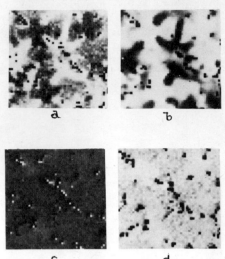

structure. Although, in the
range of dendrite formation,
the solidification conditions
have little influence on the
initial degree of segregation,
this is more easily homogenized by post-solidification
heat treatments of material
with fine dendrites[9]. Any
attempts to rationalise the
mechanical behaviours of these
materials must take into
account these heterogeneities

FIG 3. Elemental distributions
by electron microprobe analysis
of transverse sections of
IN738LC directionally solidified at 0.17 mms^{-1} with
G = 13 Kmm^{-1}
(a) W (b) Ti (c) Al (d) Ta

in composition, structure, and therefore, in strength.

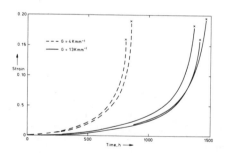

FIG 4. Creep curves for
Mar M246 directionally
solidified at 0.083 mms^{-1} in
commercial and laboratory
conditions and tested at
1123K and 250 MPa.

Typical creep curves for
Mar M246 that has been directionally solidified in commercial and in high G laboratory
rigs and then tested under the
same conditions are shown in
Figure 4. The forms of the
creep curves are similar, but
the laboratory prepared alloys
consistently have lower minimum creep rates and longer
rupture lives than the
commercial castings. The
variations in minimum creep
rate with the rate of solidification, for both Mar M246
and IN738LC directionally
solidified in high temperature

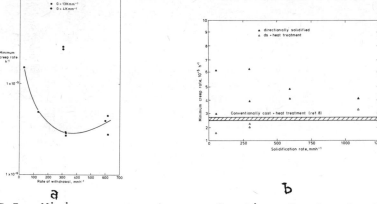

FIG 5. Minimum creep rate as a function of rate of solidification for two directionally solidified superalloys
(G = 13 Kmm^{-1}) (a) Mar M246 at 1123K and 350 MPa, (b) IN738LC
at 1123K and 250 MPa

gradients, are shown in Figures 5a and b respectively; both
alloys show a progressive decrease in creep rate $\dot{\varepsilon}$ with
increasing R and there is a corresponding decrease in rupture
life. These trends can not be attributed to either the
grain morphology or the crystallographic textures of the
alloys which vary little with solidification condition.
Since the high rates of solidification lead to refinement of
the γ' precipitate, in addition to the effects on the dendrite
size described above, the ds alloys were also given a two
stage solutioning plus ageing heat treatment designed to
yield a common γ' precipitate particle size; this changed
the nature of the $\dot{\varepsilon}$ versus R dependence from a rapid decrease
in $\dot{\varepsilon}$ to a slower increase in $\dot{\varepsilon}$ with increasing R (Figure 5b).
This suggests that the γ' size is the major factor in the
strengthening of the ds alloys by the use of high cooling
rates; however, for a constant γ' size the effect of
dendrite refinement appears to be to reduce the creep strength
of the alloy.

The dependence of both rupture life and minimum creep
rate on stress at a constant temperature has been investigated
both for Mar M246 and IN 738LC. The data show an increasing advantage of high G and R solidification as the stress
is decreased relative to both conventionally cast and
commercial directionally solidified material. This is
reflected in higher values of the exponent n in the Bailey-
Norton creep equation.

In-situ composites

The solidification conditions required to produce well
aligned fibre-, or lamellar-, reinforced microstructures
from near eutectic alloys are more stringent than those used

in the directional solidification of superalloys[2].
Consequently, for most <u>in-situ</u> composites there is little
scope for control of the microstructural dimensions while
maintaning a good composite microstructure. Where this is
possible two quite different types of dependence of creep
behaviour on fibre (or lamellar) dimensions are obtained[10].
When the alloys are intrinsically stable with respect to phase
changes, there is a progressive reduction in minimum creep
rate and increase in rupture life with decreasing fibre size
as long as the condition for producing well aligned microstructures is satisfied. The $\gamma-\gamma'-Cr_3C_2$ <u>in-situ</u> composite
that has been developed at NPL[8] can be prepared over a
particularly wide range of solidification conditions and
Figure 6 shows the variation in $\dot{\epsilon}$ with increasing R

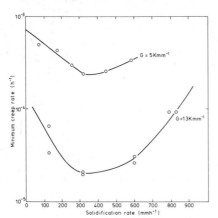

(ie. decreasing fibre size) for
two values of G; the displacement of the two curves can also
be related to the finer microstructures produced in the higher temperature gradient. In
other alloys where one of the
phases is unstable the behaviour
is more complex, but there is an
important regime where the
kinetics of phase transformation
leads to a weakening rather than
strengthening effect of fine
fibres[10].

FIG 6. $\dot{\epsilon}$ vs solidification
rate for the $\gamma-\gamma'-Cr_3 C_2$
composite tested at 1253K
and 122 MPa

Determination of Friction stress

When creep data for both ds
superalloys and <u>in-situ</u> composites are represented by the
conventional Bailey-Norton power law equation, as with many
other engineering alloys, the stress exponent n and activation
energy Q obtained are much larger than are normally associated
with pure metals or solid solutions. Several authors[3,4,11,
12] have proposed that by framing the Bailey-Norton equations
in terms of an effective stress $(\sigma-\sigma_o)$, rather than the
applied stress, "physically realistic" values of n and Q
can be obtained for suitable values of σ_o which is variously
termed a friction-, back-or resisting-stress.

$$\dot{\epsilon} = A(\sigma-\sigma_o)^n \exp(-Q/RT) \qquad (1)$$

Wilshire and co-workers[3,4] have developed and
extensively used a technique for measuring σ_o based on the
interpretation of creep transients following progressive
reductions in stress. By plotting the sum of the delay
periods before creep recommences that are observed after each
stress reduction against the cumulative stress reduction, an

asymptotic value appears to be approached and the stress
remaining is taken to be σ_o. This approach has been
criticised[13,14] because of the arbitrariness of the method
of data analysis. Our attempts to measure σ_o by this
technique, although giving the same type of experimental
data as reported by Wilshire et al[3,4], also point to the
need for a more objective approach to the interpretation of
the data.

The delay periods of zero creep Δt that are observed
after each stress reduction is generally thought to be an
incubation period during which the dislocation substructure
coarsens to the point where yield can occur under the reduced
stress[15,3]. The Wilshire technique essentially follows
the recovery process to very long times in an attempt to
identify the residual stress that can be supported by the
inherent strengthening mechanisms of the alloy. We have
developed a model[16] to describe recovery during the test
sequence of the stress dip technique that leads to an
explicit expression for Δt,

$$\Delta t = \frac{\mu^2 b^2}{K} \left[(\sigma_R - \sigma_o)^{-2} - (\sigma_I - \sigma_o)^{-2} \right] \qquad (2)$$

where μ = shear modulus; b = Burgers' vector; K = recovery
kinetic constant which is highly temperature dependent;
σ_R, σ_I and σ_o = residual, initial and friction stresses
respectively.

By fitting the data produced by the stress dip procedure
to Equation 2 by a least squares analysis, values of
σ_o and of $\frac{\mu^2 b^2}{K}$ can be obtained. Figure 7 shows a typical data
set and the best fit of Equation
2 to the data. (Figure 7 is
identical to the curves obtained
by Wilshire and co-workers[3,4,18] with a simple change of axis
$\sigma_R = \sigma_I - \Sigma_i \, \Delta \, \sigma_i$). The
asymptotic value of stress (σ_o)
indicated by the dashed horizontal line is considerably below
the value that would have been
deduced by inspection.

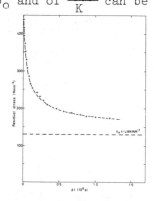

FIG 7. Cumulative incubation
period Δt versus residual
stress σ_R for $\gamma-\gamma'-Cr_3 C_2$
at 973K and an initial
stress of 438.6 MPa.

The form of curves such as
Figure 7 depends on σ_o, $\frac{\mu^2 b^2}{K}$
and σ_I; however, previous
studies have tended to attribute

differences in σ_R versus Δt data for different test conditions only to σ_O and, indeed, to claim significance in the apparent variations of σ_O with the test parameters σ_I and temperature. Since the rate of approach of curves, such as that in Figure 7, to the asymptotic value is determined by K and, to a lesser extent, σ_I it seems just as likely that many apparent variations in σ_O determined by the empirical approach reflect differences in the error in measuring σ_O. McLean[17] has demonstrated that changes in temperature (and therefore in $\frac{\mu^2 b^2}{K}$) and in σ_I give changes in the σ_R versus Δt plots that could be interpretted erroneously as changes in σ_O.

The new numerical method of determining σ_O has been applied to both nickel-base superalloys and to the γ-γ'-Cr_3C_2 eutectic composite to determine both σ_O and the recovery kinetic constant $\frac{\mu^2 b^2}{K}$. The data are well described by equation 2 and the temperature dependence of $\frac{\mu^2 b^2}{K}$ is similar to that for self diffusion of nickel. In addition, the values of σ_O are less dependent on temperature than previously reported[4] being compatible with the theoretical expectation that they should vary as the temperature dependence of μ. All of these factors support the validity of the model. When values of σ_O measured by this technique are compared with the theoretical flow stresses associated with various strengthening mechanisms they indicate that creep deformation in the superalloys considered occurs by dislocation climb around γ' particle, while in the γ-γ'-Cr_3C_2 composite it occurs by a combination of climb around γ' particles and bowing between carbide fibres.

It has been suggested that the absence of a true steady state and the presence of a steadily increasing creep rate in engineering alloys such as superalloys, as shown in Figure 4 is due to a progressive decrease in σ_O due to coarsening of the strengthening precipitate particles[18,19]. This explanation is qualitatively very plausible and appears to be supported by measurements of σ_O. Dyson and McLean[20] have recently shown that, subject to Lifshitz-Slyozov-Wagner coarsening kinetics by volume diffusion, this premise requires that

$$\left(\frac{\dot{\varepsilon}}{\dot{\varepsilon}_{min}}\right)^{\frac{1}{4}} = 1 + \alpha\, t \qquad (3)$$

where $\dot{\varepsilon}$ is the creep rate at time t, $\dot{\varepsilon}_{min}$ is the minimum creep rate and α is a constant that is either independent of applied stress σ if $\sigma_O \propto \sigma$ or decreases with increasing σ if σ_O is independent of σ. Data for ds Mar M246 creep tested with stresses between 320 and 500 MPa at 1123K are analysed

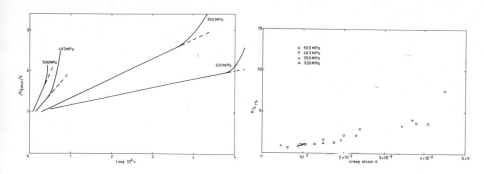

FIG 8. Creep data for Mar M246 ds tested at 1123K with various stresses and replotted as (a) $(\dot{\varepsilon}/\dot{\varepsilon}_{min})^{\frac{1}{4}}$ versus t and (b) $\dot{\varepsilon}/\dot{\varepsilon}_{0.01}$ versus ε.

in terms of Equation 3 in Figure 8a; the opposite trend of increasing α with increasing σ is obtained casting serious doubt on the hypothesis of alloy weakening by particle coarsening. However, when the data are correlated with creep strain ε rather than with elapsed time (Figure 8b) a much better empirical correlation is obtained. Moreover, previous observations by Dyson and Rogers[21] of increased creep rates following prestrain follow the same trends. This suggests that the change in $\dot{\varepsilon}$ may be due to increases in the kinetics of recovery (K) due to damage accumulation. As discussed above this would lead to a more rapid approach of stress dip data to an asymptotic value and therefore to an apparent decrease in σ_O when the data are interpreted by inspection.

Models of creep deformation

A phenomenalogical model of creep deformation in composite materials has been developed[22,23] that incorporates the modified Bailey-Norton description of creep in the super-alloy-type matrices that characterise in-situ composites. Since the fibres are very long, at steady state the matrix and reinforcing phases must extend at the same rate to maintain continuity; this leads to an off-loading of stress from the weaker to the stronger phase until both deform with the same strain rate. The strengthening effect of fine fibres can be accounted for by postulating the existence of a zone of matrix strengthening adjacent to the fibres as illustrated schematically in Figure 9a. By reducing the fibre size at constant volume fraction the volume fraction of the strengthened interface zone increases leading to the observed decrease in minimum creep rate.

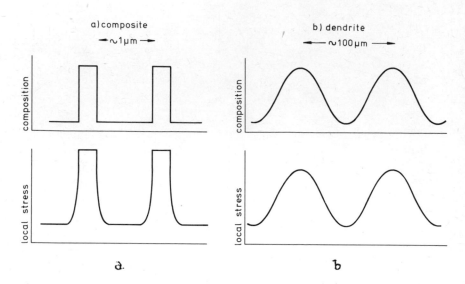

FIG 9. Schematic diagrams showing the relationships between steady state stress distributions and microstructure in (a) fibre reinforced composites and (b) columnar dendritic materials.

Deformation of columnar dendrites can probably be modelled in the same way as shown in Figure 9b. However, there is no sharp interface between the "strong" dendrite cores and the "weak" interdendritic matrix and the larger dimensions involved probably prevent the operation of the interface constraint strengthening. On this basis there should be little effect of dendrite dimensions on creep strength for a constant γ' size. We speculate that the observed decrease in creep strength with decreasing dendrite size after heat treatment (Figure 5) is due to increasing homogenisation reducing the peak to trough stress heterogeneity in the material. It is however important to note that the local stresses in different parts of the dendrite structures, required to give a uniform strain rate, may vary markedly and lead to quite different deformation mechanisms in different parts of the structure.

Conclusions

1.   There is scope for further improvements in the creep behaviours of ds superalloys through control of solidification conditions.

2.   The creep strengths of <u>in-situ</u> composites that are not subject to phase instabilities are increased by reducing the microstructural dimensions by directional solidification in high temperature gradients and at high growth rates.

3.   The creep data for both classes of alloys can be described by the modified Bailey Norton power law equation (Equation 1). A new method of analysing stress dip creep data has been developed to allow the friction stress to be determined objectively.

4.   A phenomenological model of creep of composite materials has been developed that accounts for the observed strengthening due to microstructural refinement; an extension of the model to deal with aligned dendritic structures is indicated.

References

1.   ERIKSON, J.S., SULLIVAN, C.P. and VERSNYDER, F.L. - "High Temperature Materials for Gas Turbines", Ed. Sahm, P.R. and Speidel, M.O., Elsevier, Amsterdam, 1974, p 315.

2.   Proceedings of conference on "In-Situ Composites III", Ginn and Co, Lexington, Mass., 1979.

3.   THREADGILL, P.L and WILSHIRE, B., - Proceedings of Iron and Steel Inst. Conf. on "Creep Strength in Steels", Sheffield, 1972.

4.   PARKER, J.D. and WILSHIRE, B. - Metal Science 1975, 9, 248.

5.   VERSNYDER, F.L and GUARD, R.W. - Trans. ASM 1960, 52, 485.

6.   VERSNYDER, F.L., BARLOW, R.B., SINK, L.W. and PIARCEY, B.J. - Modern Casting 1967, p.68.

7.   GELL, M., DUHL, D.N. and GIAMEI, A.F. - Proceedings of Fourth International Symposium on Superalloys, American Society for Metals, Metals Park, Ohio, 1980.

8.   BULLOCK, E., QUESTED, P.N and McLEAN, M. - Ref. 2, p420.

9.   QUESTED, P.N. and McLEAN, M. - Proceedings of Conference on "Solidification Technology in the Foundry and Cast house", The Metals Society, London - to be published.

10.   QUESTED, P.N., BULLOCK, E. and McLEAN, M. - Proceedings of Conference on "Solidification and Casting of Metals", The Metals Society, London, 1978, p.466.

11. AHLQUIST, C.N. and NIX, W.D. Acta Met. 1971, 19, 373.

12. LAGNEBORG, R. - J. Materials Sci. 1968, 3, 596.

13. BURTON, B. - Metal Science 1975, 9, 297.

14. GIBELING, J.C. and NIX, W.D. - Metal Science 1977, 11, 453.

15. MITRA, S.K. and McLEAN, D. - Proc. Roy. Soc. Lond. 1966, 295A, 288.

16. McLEAN, M. - Proc. Roy. Soc. Lond. 1980, 371A, 279.

17. McLEAN, M. - Ref 7, p 661.

18. BURT, H., DENNISON, J.P. and WILSHIRE, B. - Metal Science 1979, 13, 295.

19. WILLIAMS, K.R. and CANE, B.J. Materials Science and Eng. 1979, 38, 199.

20. DYSON, B.F. and McLEAN, M. - to be published.

21. DYSON, B.F. and ROGERS, M.J. - Metal Science 1974, 8, 261

22. BULLOCK, E., McLEAN, M. and MILES, D.E. - Acta Met. 1977, 25 333.

23. McLEAN, M. - Proc. Roy. Soc. London 1980, 373A, 93.

Acknowldgements
    Parts of this work were supported by Ministry of Defence (Procurement Executive) and others carried out as part of the COST 50 European collaborative programme.

A UNIFIED VIEW OF STEADY-STATE CREEP IN PURE MATERIALS AND
PARTICLE-STRENGTHENED ALLOYS

H.E. Evans and G. Knowles

Central Electricity Generating Board, Berkeley Nuclear

Laboratories, Berkeley, Gloucestershire, UK.

SUMMARY

A recovery model is described for single-phase materials
and compared with a wide range of experimental results on
common metals and ionic solids. Both the Dorn equation and the
Stocker-Ashby correlation can be accounted for satisfactorily.
An extension of the model is made to particle-strengthened alloys
and the existence of threshold and transition stresses identified.

1. INTRODUCTION

One of the most successful means of correlating steady-
state creep rates, $\dot{\varepsilon}_s$, is through the Dorn equation:

$$\dot{\varepsilon}_s = \frac{A\, D_L\, \mu b}{kT}\left(\frac{\sigma}{\mu}\right)^n \tag{1}$$

where A and n are constant, σ is the applied stress, μ is the
shear modulus, b is the Burgers vector, D_L is the self-
diffusion coefficient (of the slower-moving ion in ionic
solids), T is absolute temperature and k is Boltzmann's
constant. The success of the equation is demonstrated on simple
metals by the correlations given by Bird et al. [1] and on
typical ionic solids by the work of Langdon [2,3]. However, a
complication associated with the empirical nature of the
approach is that the constants A and n (equation 1) are not
uniquely defined for any particular material. Thus, Stocker and
Ashby [4] have shown that n varies typically within the range
3 to 6 and that A increases by several orders of magnitude over
this range. A significant feature of this correlation, however,
is that A is of order unity only when n = 3 suggesting that A
then contains no important physical variables and that
equation (1) with these values of the constants has an intrinsic
physical meaning [4,5]. An additional complication though is
that, even with higher values of n, the correlation given by
the equation tends to fail at high stresses leading to gross
underestimates of the creep rate in the worst cases (e.g.
Refs. [1,3]).

It is clear that any theory which presumes to provide a
"unified view" of the creep process must account for these
observations and should be able to predict creep rates to small
error without the use of disposable parameters. Finally, it
should be recognised in the model that, although the activation
energy for the creep rate is that for lattice diffusion at high
homologous temperatures, there is a decrease to about 2/3 rd of
this value at $T/T_m \sim 0.6$, where T_m is the absolute melting
temperature. This change is a characteristic of all systems in
which a wide-enough temperature range has been used.

Various theories have been developed (e.g. Refs. [6-10])
which account for some of these observations and comparison
with this work will be made where appropriate. The purpose of
the present paper however is to concentrate on the authors'
recovery creep theory, expanded elsewhere [11-15] in detail,
and to show that it can meet the requirements cited above.
Particular attention will be paid to the case of pure metals
and simple ionic solids but it will be shown towards the end of
the paper how the basic concepts can also be applied to
dispersion-strengthened alloys.

2. OUTLINE OF THE MODEL

2.1 Basic Concepts

The aim of this outline is to describe the model in its
simplest form which shows how the Dorn equation (equation 1)
has physical significance with a value of $n \sim 3$ and A of order
unity.

During steady-state creep, the dislocations which are
potentially mobile exist as a well-defined, 3-dimensional network of average mesh size, x. The creep process entails the
gradual rearrangement of link lengths within any mesh of this
network at a rate determined by thermally-activated climb. This
rearrangement entails a redistribution of link lengths such that
some links lengthen and others shorten; the process is
essentially symmetrical as shown schematically in Figure 1
where link 1 has grown at the expense of link 2. Since random
behaviour is assumed,
each value of link
length within its
distribution will be
equally likely and
the form of this
distribution will be
simply rectangular
as shown in Figure 2.

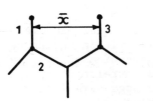 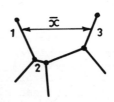

Figure 1: Redistribution of link lengths

The upper and lower limits of the distribution are defined by
the conditions required to make the mesh unstable and to
release a dislocation loop. This release occurs by the
operation of a Frank-Read source but the critical length for

this can be achieved either by continued growth (e.g. link 1
of Figure 1) or by initial skrinkage (link 2 of Figure 1),
removal of the attractive
junction and immediate
local rearrangement to
produce a link length
exceeding critical size.
Once such a link exists
release of a loop is
instantaneous as is its
glide through the lattice.
After a certain slip
distance, s, the loop
again encounters network
dislocations and becomes
re-absorbed randomly into
the distribution function
shown in Figure 2. These

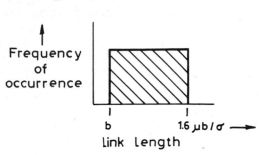

Figure 2 : Distribution of link
lengths in steady-state.

arguments lead to a lower limit of this function of order b
and an upper limit slightly less than the critical source
length, say 1.6 µb/σ.

This general pattern of behaviour in which creep is controlled by the rate of release of dislocation loops is common
to most other recovery creep theories [6-9, 16] but differs
conceptually from that of Nabarro [10] in that here creep
strain results from the glide of the released loop rather than
from the climbing action of the network dislocations. It has
previously been shown [11] that this contribution is several
orders too small and can be ignored. In addition, the assumption that the slip process is very rapid precludes application
of the theory to those materials in which lattice friction or
solute drag forces are high [e.g. 17].

2.2 Effects of Stress Changes

It is implicit in the present theory that all link lengths
within the distribution are potentially capable of release and
all can grow or shrink progressively by repeated steps of
thermal activation to the limits of the distribution function.
It will be shown later that this view allows a ready summation
of individual component forces to provide the average climb
force but it also represents a difference to the model of
Lagneborg [7] who considered individual links to be broken only
by single activated events and identified a large fraction of
links as permanently immobile. Bearing in mind that, in the
real case, a network is continuously being modified by slip and
climb, the present approach seems more appropriate.

With this model it is easy to make an approximate estimate
[18] of the effect of increasing the applied stress by an
increment, $\Delta\sigma$, typical of tests designed to evaluate the
hardening coefficient, h [19]. The average mesh spacing, $\bar{x}$

(Figure 1), is of order $\rho^{-\frac{1}{2}}$ where ρ is the dislocation line length comprising the 3-dimensional network and is related to stress, σ, by the Frank-Read relationship [e.g. 20,21]:

$$\rho^{-\frac{1}{2}} \sim \bar{x} = \frac{\alpha\mu b}{\sigma} \qquad (2)$$

where α is a constant or order unity.

The number, $\Delta\rho$, of dislocations instantaneously released by the stress increment, $\Delta\sigma$, can be obtained directly from equation (2) as

$$\Delta\rho = \frac{2\Delta\sigma \, \rho^{\frac{1}{2}}}{\alpha\mu \, b} \quad , \quad \Delta\sigma \to o. \qquad (3)$$

The tensile plastic strain, $\Delta\varepsilon_p$, produced is then

$$\Delta\varepsilon_p \sim \Delta\rho.s.b \qquad (4)$$

where, as before, s is the average slip distance. Since the model envisages a well-defined, 3-dimensional network, it is most unlikely that s will be appreciably greater than the mesh size since a gliding loop should readily interact with the surrounding links producing lengths which will be less than the critical value. The most direct approach, therefore, is to take $s = \bar{x} \sim \rho^{-\frac{1}{2}}$. Substituting into equation 4 and evaluating the hardening coefficient h gives

$$h = \left(\frac{\Delta\sigma}{\Delta\varepsilon_p}\right) = \frac{\alpha\mu}{2} \quad , \quad \Delta\sigma \to o \qquad (5)$$

if, for purposes of convention, the elastic component of the strain is excluded. Clearly, then, values of h numerically similar to those found experimentally can be derived by considering large numbers of mobile dislocations but small slip distances. It is reiterated that this behaviour is the most reasonable to be expected from a well-developed network and, indeed, the corollary that a poorly defined network leads to large slip distances and low values of h seems to be substantiated by measurements in the early stages of primary creep [18,22].

The effect of a stress decrement will not be considered quantitatively here but it can be appreciated from Figure 2 that the upper limit on the link-length distribution applicable to the new applied stress $(\sigma - \Delta\sigma)$ is greater than existed at the previous stress σ. The effect of this is to reduce greatly the number of links that can be released through the Frank-Read mechanism for a time, Δt, until recovery (i.e. climb) processes can enlarge the distribution to its new steady state. The observed effect on strain is to produce an incubation period in which the creep rate is much less than expected for the new, reduced stress.

2.3 The Climb Rate

It is clear from Figure 1 that migration of the links

entails movement of the nodes also and, although little is known of the node geometry, there is no reason to suppose that they cannot exhibit climb mobility. This is a fundamental difference to the views of Spingarn et al. [9] and will be returned to when discussing threshold-stress effects.

It has been shown [11] that even though the whole of the link length is capable of acting as a vacancy source, the climb velocity of the nodes is still expected to exceed that of the links so that the rate of climb of the latter controls the creep process. The diffusion situation, thus, is one of transfer of vacancies from one link length (e.g. 1 of Figure 1) to another (e.g. 3 of Figure 1) such that some dislocations climb by vacancy emission and others by vacancy absorption. Assuming for the present that the dominant vacancy path is through the lattice (so that the observed activation energy for creep is that for lattice self-diffusion) the diffusion field can be considered to be cylindrical. The vacancy concentration, c, at a radial distance r from the dislocation line is then given by

$$C = A' + B' \ln r \tag{6}$$

where A' and B' are constants. At the dislocation line, the vacancy concentration C_b, is maintained at the level given by:

$$C_b = C_o \exp\left(\frac{F\Omega}{bkT}\right) \underset{\sim}{} C_o\left[\frac{F\Omega}{bkT} + 1\right] \tag{7}$$

where C_o is the equilibrium concentration in the lattice, Ω is the atomic volume, F is the climb force and the first-order approximation holds when $F\Omega/bkT \ll 1$. The vacancy concentration is taken to fall to C_o and to be maintained at this value midway between dislocations, i.e. $r = \bar{x}/2$, so that noting that the flux, j_ℓ, from a link of length ℓ is

$$j_\ell = - 2\pi b \ell D_v \left(\frac{dc}{dr}\right)_{r=b} \tag{8}$$

gives, finally from (7):

$$j_\ell = \frac{2\pi \ell D_v C_o \Omega F}{b kT \ln\left(\dfrac{\bar{x}}{2 b}\right)} \tag{8}$$

Here D_v is the lattice diffusion coefficient for vacancies such that $D_v C_o \Omega = D_L$. The climb velocity, V_ℓ, follows:

$$V_\ell = \frac{j_\ell b^2}{\ell} = \frac{2\pi D_L F b}{kT \ln\left(\dfrac{\bar{x}}{2 b}\right)} \tag{9}$$

It is not possible to give a detailed description of the rearrangements occurring within the network because of the immense complexity of the problem. Local values of the climb force will show large variations as, of course, will the directions of climb and the situation is not unlike the

apparent random motion of individual molecules in a gas. As
in that case, however, it is possible to predict the overall
behaviour of very large numbers of components by using an
averaging approach. To do this in the present case, it is
assumed that the direction of climb is always such as to allow
the average climb force, F, to do work. This, in turn, has
three components, f_i, which by the nature of this approach,
can simply be summed arithmetically to give F. Thus, the climb
force, f_1, due to the applied stress σ is

$$f_1 = \sigma_n \, b \sim \frac{\sigma}{2} \, b \tag{10}$$

where σ_n is the normal component. The elastic interaction
between links provides the second force, f_2:

$$f_2 = \frac{\mu \, b^2}{2 \, (1 - \nu)\bar{x}} \tag{11}$$

where ν is Poisson's ratio. Finally, the network will
experience a tendency to coarsen, thereby increasing $\bar{x}$ and
reducing overall line length. The average force, f_3, due to
this effect is [6]:

$$f_3 = \frac{2\Gamma}{\bar{x}} = \frac{\mu \, b^2}{\bar{x}} \tag{12}$$

where Γ is the dislocation line energy taken as $\frac{1}{2} \mu \, b^2$.
Summing the individual force components for $\nu = 0.33$ and
inserting in equation (9) gives the climb velocity of a
dislocation link as

$$v_1 = \frac{2\pi \, D_L \, b}{kT \, \ln\left(\frac{\bar{x}}{2b}\right)} \cdot \left[\frac{\sigma b}{2} + 1.2 \, \frac{\mu \, b^2}{\bar{x}}\right] \tag{13}$$

In practice, the interaction force, f_2, contributes only $\sim 10\%$
to F.

2.4 The Creep Rate

The uniaxial, steady-state creep rate, $\dot{\varepsilon}_s$, is the product
of the rate of release, $\dot{N}$, of dislocation loops per unit
volume and the slip produced by each:

$$\dot{\varepsilon}_s = 0.7 \, \dot{N} \, \phi \, b \tag{14}$$

where ϕ is the area swept by a released loop ($\sim \bar{x}^2$) and the
factor 0.7 resolves the shear strain to uniaxial. Since each
link climbs in a direction allowing the force to do work, the
rate of release is:

$$\dot{N} = 2 \left(\frac{v_\ell}{\bar{\ell}}\right) \frac{3}{\bar{x}^3} \tag{15}$$

where $3/\bar{x}^3$ is the total number of links per unit volume and the
factor 2 recognises that links can be released from either side
of the distribution of Figure 2. Noting that $\bar{\ell} = \bar{x}/\sqrt{3}$ for a

hexagonal mesh gives from equations 13, (14) and (15) with $\bar{x}$ from equation (2):

$$\dot{\varepsilon}_s = \frac{A \, D_L \, \mu \, b}{kT} \left(\frac{\sigma}{\mu}\right)^3 \tag{16}$$

where A is essentially a constant, varying only logarithmically with σ:

$$A = \frac{4.2 \sqrt{3} \, \pi}{\alpha^2 \ln\left(\frac{\alpha\mu}{2\sigma}\right)} \, b \left\{ 1 + \frac{2}{\alpha} \left[1 + \frac{1}{2\pi \, (1 - \nu)} \right] \right\} \tag{17}$$

For a typical value of α = 1.6, ν = 0.33 and σ/μ = 10^{-4}, A has a value of ∿ 2. It is clear from this section, therefore, that the Dorn equation (equations 1 and 16) with n = 3 and A of order unity has physical significance. The following section will extend the treatment to consider higher values of n and to show how A must increase accordingly.

Before proceeding, it may be of interest to see how the theoretical equations (16) and (17) are capable of predicting creep rates for those conditions under which they apply, i.e. n = 3 and $Q_c = Q_L$. A comparison with data on uranium dioxide is given in Figure 3 from which it can be seen that the predictive error is small.

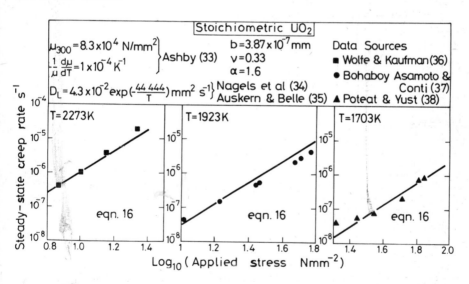

Figure 3 : Comparison with data on uranium dioxide.

3. GENERALISATION OF THE MODEL FOR PURE MATERIALS

3.1 The Slip Distance

Although the direct approach used in Section 2 assumes that the slip distance s = $\bar{x}$, it must be recognised that in many materials such an idealised situation may not apply. In particular, depatures are likely to occur when features extraneous to the 3-dimensional network, e.g. sub-cell walls, twin interfaces, grain boundaries, or in thin sections, the specimen surface exist at distances comparable with, but still greater than, the network mesh size. Under these conditions it seems more reasonable to take s equal to a constant, independent of $\bar{x}$. The value used in all previous comparisons (e.g. [11] for s has been 2 μm.

At somewhat larger values of $\bar{x}$ in those materials which form sub-boundaries, the network size will correspond to the sub-cell size and the present model will not apply. Under these conditions, the release of a dislocation loop will depend on recovery processes occurring within the sub-boundary [23,24] so that the subsequent slip distance will correspond to the sub-cell size and exhibit the same stress dependence [25]. Even with this proviso, however, the most frequent observation (cf. the compilation for common metals given in Ref. [17]) is that the sub-cell size is appreciably greater than that of the network so that the present model is expected to be applicable.

3.2 The Transport Path

In Section 2, the climb rate was assessed intentionally only for the case in which the transport of matter occurred through the lattice and so produced an activation energy for creep, Q_c, equal to that for lattice diffusion, Q_L. At intermediate temperatures however, Q_c decreases to $\sim 0.6\ Q_L$ and the present model can account for this by a change in diffusion path to dislocation cores; the associated activation energy, of course, being that for pipe diffusion.

The vacancy concentration difference between climbing links is again of order (cf. equation 7)

$$\frac{2F\ C_o\ \Omega}{b\ kT}$$

So that the diffusion flux, j_p, from one link to another a distance $2\bar{\ell}$ away is:

$$j_p \sim \frac{2\pi\ b\ D_p\ F}{\bar{\ell}\ kT} \tag{18}$$

where the dislocation pipe is taken to have a radius b and a transfer area of $2\pi\ b^2$. The climb velocity of the link follows directly from equation (9):

$$V_\ell = \frac{j_p b^2}{\bar{\ell}} = \frac{2\pi\ b^3\ D_p\ F}{\bar{\ell}^2\ kT} \tag{19}$$

3.3 The Overall Creep Equation

Since the climb rate contributions from the two diffusion paths are likely to be independent of each other, the general creep equation can be obtained by summing the respective contributions. Thus, using equations (9) and (19) for V_ℓ, (10), (11) and (12) for F, equation (2) for $\bar{x}$ together with equations (14) and (15) gives for the direct case that $s = \bar{x}$:

$$\dot{\varepsilon}_s = \frac{A'\mu b}{kT} \left(\frac{\sigma}{\mu}\right)^3 \left[\frac{D_L}{\ln\left(\frac{\alpha\mu}{2\sigma}\right)} + \frac{3\,D_p\,\sigma^2}{\sigma^2\mu^2}\right] \tag{20}$$

where the constant A' is $A\ln(\alpha\mu/2\sigma)$ (equation 17).

For the case that the slip distance = s, a constant, a similar exercise yields

$$\dot{\varepsilon}_s = \frac{A's}{\alpha b^2}\,\frac{\mu b}{kT} \left(\frac{\sigma}{\mu}\right)^4 \left[\frac{D_L}{\ln\left(\frac{\alpha\mu}{2\sigma}\right)} + \frac{3\,D_p\,\sigma^2}{\sigma^2\mu^2}\right] \tag{21}$$

It can be seen that the stress exponent n of creep rate is predicted to increase above the expected value of 3 such that: n = 4 occurs when lattice diffusion pre-dominates but the slip distance is invariant; n = 5 occurs when pipe diffusion pre-dominates and $s = \bar{x}$; n = 6 occurs again when pipe diffusion pre-dominates but the slip distance is invariant. In both equations (20) and (21) the ratio of the terms in the square brackets gives the diffusion path parameter P:

$$P = \frac{\alpha^2\,\mu^2\,D_L}{3\,\sigma^2\,D_p\,\ln\left(\frac{\alpha\mu}{2\sigma}\right)} \tag{22}$$

values of P >> 1 indicate that lattice diffusion predominates whereas for P << 1 pipe diffusion is dominant. The transition between the two occurs when P = 1.

A comparison of these equations with published creep data has been given previously for the case of pure metals [11,15], simple ionic solids [12] and zirconium and its alloys [13] and shown to give good quantitative agreement. The usefulness of equation (21) is demonstrated for zircaloy-2 in Figure 4 for the temperature range 1523-1773K at which lattice diffusion is taken to be dominant. For the comparison s was taken at the standardised value of 2 μm and the diffusion coefficient from Kidson [26]. The predictive error is again small, typically a factor of 2 to 3.

3.4 The Stocker-Ashby Correlation

As was mentioned in Section 1, Stocker and Ashby [4] have shown in their survey of published data how the apparent value of n in equation 1 affects the required value of A. Their correlation for pure materials is demonstrated in Figure 5 and it is the present purpose to show that the general creep equations derived here can generate such a relationship.

G

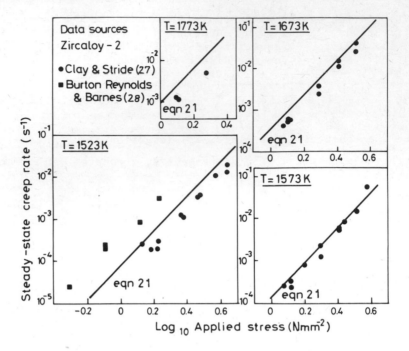

Figure 4 : Comparison with data on zircaloy-2.

It has already been shown how equation (20) with P >> 1 reduces to the exact Dorn equation with n = 3 and A = 2 and this point is located in Figure 5. The same equation with P >> 1 reduces to

$$\dot{\varepsilon} = \frac{3\ A'}{\alpha^2} \cdot D_p \cdot \left(\frac{\sigma}{\mu}\right)^5 \left(\frac{\mu b}{kT}\right) \tag{23}$$

which can be rewritten in the Dorn equation format as

$$\dot{\varepsilon} = \left\{\frac{3\ A'}{\alpha^2} \cdot \frac{D_p}{D_L}\right\} \left(\frac{\mu b}{kT}\right) D_L \left(\frac{\sigma}{\mu}\right)^5 = A \left(\frac{\mu b}{kT}\right) D_L \left(\frac{\sigma}{\mu}\right)^5 \tag{24}$$

The artificial constant A has been evaluated for $\alpha = 1.6$, $\sigma/\mu = 10^{-4}$ with archetypal values [29] of D_p and D_L at 0.6 T_m to give the appropriate point for n = 5 in Figure 5. Similar computations have been made for values of n = 4 and n = 6 from equation (21) and the values of the apparent constants A are again given in the Figure. The values shown for the observed range of creep index of 3 to 6 lie closely to the mean line deduced by Stocker and Ashby.

The present theory indicates, therefore, that values of n > 4 are associated with pipe diffusion control and that the Dorn equation using lattice diffusion coefficients will lead to erroneous correlations. In fact equation (23) shows that, with the appropriate diffusion coefficients, a simple but physically meaningful equation does exist for this regime with n = 5 and $A = 3\ A'/\alpha^2$.

3.5 Breakdown of the Dorn equation at High Stresses

Examples of the gross underestimates produced by equation (1) at high stresses are readily available in the literature [1-3] for a variety of materials but the attempt by the present authors to rationalise this behaviour [14] used lead, nickel and α-iron as typical examples. The case of lead is duplicated in Figure 6a and this example is used here simply because the departure from the Dorn equation shown (equation (1) with n = 3 and A = 1) clearly manifests itself as a linear relationship of higher n value. Furthermore, these data points at high stresses were obtained at the lowest test temperatures where the creep process would be expected to be controlled by pipe diffusion. The present model was used previously [14] to account for the breakdown by a change in transport path and the arguments used are summarised below.

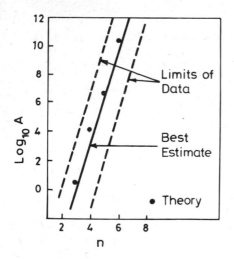

Figure 5 : Prediction of the Stocker-Ashby correlation

Consideration of the general equations (20) and (21) shows that a modified form of the Dorn equation exists, viz

$$\dot{\varepsilon}_s = B\ D* \left(\frac{\mu b}{kT}\right) \left(\frac{\sigma}{\mu}\right)^m \tag{25}$$

where D* is an effective "diffusion coefficient" which makes allowance for the change in diffusion path with test conditions and is defined as

$$D* = \left[\frac{D_L}{\ln\left(\frac{\alpha\mu}{2\sigma}\right)} + \frac{3\ D_p\ \sigma^2}{\alpha^2\mu^2}\right] \tag{26}$$

B is a constant given either by A'b from equations (20) and (17) or by A'/αb² from equations (21) and (17). For the case of lead considered here, a previous comparison [11] was made using equation (21) and so the results of Figure 6a have been replotted in Figure 6b using the constants appropriate to equation (21) with elastic constants and diffusion coefficients from the literature [32,39,40]. It is obvious that the poor correlation has now been greatly improved indicating that the so-called 'breakdown' can indeed be attributed simply to a change in diffusion path. Furthermore, it is clear that the theoretical line offers a good description of the data over the whole experimental range.

4. APPLICATION TO PARTICLE-STRENGTHENED ALLOYS

4.1 Threshold Stress, σ_0

It is not possible here to give a detailed account of creep in such materials but some of the concepts developed earlier can be used to account for widely-observed effects. The first to be considered is the threshold stress σ_0 [e.g. 41] determined either as the stress asymptote at which the steady-state creep rate becomes zero [42] or from stress-decrement tests at which the incubation period becomes 'infinitely' long [43]. Since steady-state creep is not possible below such a stress level, it follows that climb-induced recovery and the release of dislocation loops is also prevented. Although, as for single-phase materials, the potentially mobile dislocations should exist as a

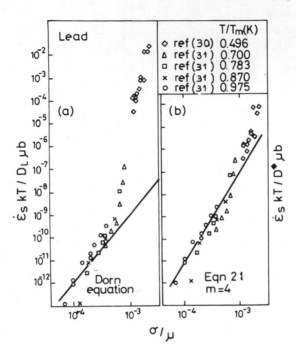

Figure 6 : The correlation of creep data on lead according to (a) the Dorn equation and (b) the modified equation.

3-dimensional network, at low applied stresses the average link length will exceed the interparticle spacing λ (Figure 7). In determining the creep rate it is then not sufficient to consider just the release of a loop from the network but rather the rate of climb around particles. This problem has received appreciable attention in recent years [41,42,44-48]. A critical comparison of the various models is given elsewhere [41] as are the details of the present approach outlined below.

Figure 7 shows a link of the network bowed between two particles under a shear stress τ and experiencing a normal climb force

$$F_n = \Gamma (\cos \Theta_1 + \cos \Theta_2)$$

along the segment in contact with the front face of particle 1. Here Γ is the dislocation line energy and, from geometry,

$$\Theta_1 = 90 - \sin^{-1}\left[\frac{\tau b \lambda}{4\Gamma}\right], \quad \Theta_2 = 90 - \sin^{-1}\left[\frac{\tau b \lambda}{4\Gamma}\right]$$

so that

$$F_n = \frac{3}{4} \cdot \tau b \lambda \tag{27}$$

A critical concept is that the node at x (Figure 7) is fixed in space through interaction of the associated links with particles. This contrasts with the situation for single-phase

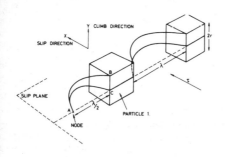

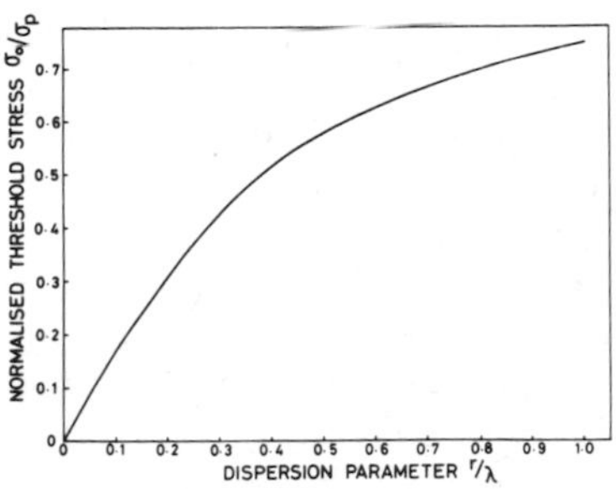

Figure 7 : Detail of climb around particles.

Figure 8 : Variation of $\frac{\sigma_o}{\sigma_p}$ with $\frac{r}{\lambda}$.

materials in which it seems more reasonable to assume climb mobility of the nodes. This difference is crucial since the dislocation segment between the node and particle 1 must increase its length by an amount determined by λ and the particle 'radius' r in order to surmount the particle. Without such by-pass extensive creep is not possible and, thus, the increase in line length and, hence, total energy, to achieve the critical climb configuration can be equated to the work done by the applied force (equation 27) to obtain an estimate of σ_o. It must be emphasised that climb mobility of the node would allow by-pass without necessarily increasing line length and no threshold stress could exist.

Assuming that the dislocation segment as it climbs follows the surface of a cylinder, it can be shown quite simply [41] that the threshold stress, σ_o, is:

$$\sigma_o = \frac{8\Gamma r}{3\lambda b}\left\{r^2 + \left[\frac{4\Gamma}{\sigma b}\sin^{-1}\left(\frac{\sigma b \lambda}{8\Gamma}\right)\right]^2\right\}^{-\frac{1}{2}} \tag{28}$$

It is apparent from this equation that for small arguments there will be only a slight dependence of σ_o on applied stress σ which would not be detected experimentally. There is a strong dependence on the dispersion parameters, however, as shown in Figure 8 where the normalised stress σ_o/σ_p, where σ_p is the Orowan stress = 1.6 $\mu b/\lambda$, increases markedly with the ratio r/λ. The maximum value is likely to be ∿ 0.75 and applicable to a densely-packed dispersion in which r = λ. Thus, to a good approximation, a constant dispersion will have a unique value of threshold stress with a temperature dependence of that of the shear modulus. Systematic variations in σ_0 with exposure time or temperature then reflect changes in the dispersion due to thermal ageing, solution or precipitation effects only.

4.2 The Transition Stress, σ_p

The above analysis has considered creep under low stresses where the average link size $\bar{\ell} > \lambda$. It has been shown that under these conditions the driving force for climb arises from the applied stress (equation 27) and does not contain a contribution associated with the concurrent increase of mesh spacing (cf. Section 2.3). This is understandable since the local climb of a dislocation around the particles cannot produce a decrease in the total line length of the specimen as a whole – in fact, as has been shown the reverse occurs and the line length increases. With increasing applied stress, however, $\bar{\ell}$ decreases to a stage at which $\bar{\ell} = \lambda$ at some critical stress σ_p. This configuration, of course, corresponds to that of athermal yield and σ_p is identified as the Orowan stress. At still higher values of applied stress, $\bar{\ell} < \lambda$ and the creep process approaches that for single-phase materials. In particular, climb and subsequent glide of links can now occur without interference from particles and so the climb force increases to F (equations 10-12), the value applicable to single-phase structures. Again, this is understandable since local decreases of dislocation density can now occur during the climb process.

A consequence of this transition at σ_p is that the creep rate again becomes a strong function of the network size $\bar{x}$ (cf. equations 13 and 15) which in turn is a strong function of applied stress [e.g. 49] in this range. This qualitative picture, thus, anticipates a marked change in the stress sensitivity of creep at a stress equal to the yield stress and this, indeed, is a common observation [47,49,50].

5. CONCLUDING REMARKS

Most of the paper has been concerned with the presentation of a theory of creep for pure metals and simple, ionic compounds. The model used envisages creep strain to arise as the result of glide of a dislocation loop from one attractive junction to another and the release of this loop to occur by climb. It is envisaged that all links are potentially mobile and can change their length by repeated thermal activation, i.e. progressive climb. The use of a rectangular link-length distribution then seems entirely consistent with this view and certainly leads to a natural explanation of the observed high values of the strain hardening coefficient and of incubation periods on a stress reduction. Furthermore, the ability of the theory to provide good quantitative agreement with observed creep rates over a wide range of materials and testing conditions without the use of disposable parameters constitutes a thorough test of the model employed. As part of this it is shown how the basic equations can reproduce the simple Dorn equation with n = 3 and A ∿ 1 and moreover can account for the Stocker-Ashby correlation [4] between the constant A and higher values of n. In addition, it is demonstrated that departure from the Dorn equation at high values of σ/μ simply reflect a transition in the transport path from lattice diffusion to pipe. A modified form of the equation exists, however, which accounts for this change and this should be used for correlation purposes where the transport path alters over the range of experimental conditions.

It is recognised that the model is not applicable under conditions in which the network size may be greater than sub-grain or indeed grain diameter nor should it be used for those alloys in which solute drag effects significantly reduce the rate of dislocation glide.

The model is capable of explaining stress dependence to a value 6 by invoking pipe diffusion as the controlling step in climb and a fixed slip distance of the released loop. It is not necessary to use the concept of a threshold stress in pure materials and it is, in fact, shown that threshold effects will not occur unless dislocation nodes are sessile in climb. This is considered unlikely for such materials and the present model thereby differs significantly from that of Spingarn et al. [9] which does assume the nodes to be fixed. On the other hand, such a situation probably pertains in particle-strengthened alloys and, for these, threshold effects are important. The calculations made in this paper show that it is permissible to consider the threshold stress to be a unique parameter for a given dispersion, a result which should simplify the modelling of behaviour in engineering alloys and extrapolation over service conditions.

ACKNOWLEDGEMENT

This paper is published by permission of the Central Electricity Generating Board.

REFERENCES

1. BIRD, J.E., MUKHERJEE, A.K. and DORN, J.E., 'Correlations
    between High-Temperature Creep Behaviour and Structure',
    Quantitative Relation between Properties and Microstructure,
    Eds. Brandon, D.G. and Rosen, A., Israel Universities Press,
    Jerusalem, 1969, p. 255.

2. LANGDON, T.G., 'Creep Mechanisms in Ice', Physics and
    Chemistry of Ice, Eds. Whalley, E., Jones, S.J. and
    Gold, L.W., Royal Soc. Canada, Ottawa, 1973, p. 356.

3. CROPPER, D.R. and LANGDON, T.G., 'Creep of Polycrystalline
    Lithium Fluoride', Philos, Mag., 1968, 18, 1181.

4. STOCKER, R.L. and ASHBY, M.F., 'On the Empirical Constants
    in the Dorn Equation', Scripta Met., 1973, 7, 115.

5. WEERTMAN, J., 'High Temperature Creep Produced by
    Dislocation Motion', Rate Processes in Plastic Deformation
    of Materials, Eds. Li, J.C.M. and Mukherjee, A.A., Amer.
    Soc. for Metals, 1975, p. 315.

6. MCLEAN, D., 'Resistance to Hot Deformation', Trans. AIME,
    1968, 242, 1193.

7. LAGNEBORG, R., 'A Modified Recovery-Creep Model and its
    Evaluation', Metal Sci., 1972, 6, 127.

8. GITTUS, J.H., 'Creep, Viscoelasticity and Creep Fracture in
    Solids', Applied Science Publishers, London, 1975.

9. SPINGARN, J.R., BARNETT, D.M. and NIX, W.D., 'Theoretical
    Descriptions of Climb Controlled Steady State Creep at High
    and Intermediate Temperatures', Acta Met., 1979, 27, 1549.

10. NABARRO, F.R.N., 'Steady-State Diffusional Creep', Philos.
    Mag., 1967, 16, 231.

11. EVANS, H.E. and KNOWLES, G., 'A Model of Creep in Pure
    Materials', Acta Met., 1977, 25, 963.

12. EVANS, H.E. and KNOWLES, G., 'Dislocation Creep in Non-
    Metallic Materials', Acta Met., 1978, 26, 141.

13. EVANS, H.E. and KNOWLES, G., 'On the Creep Characteristics
    of β-Zircaloy', J. Nucl. Mtls., 1978, 78, 43.

14. EVANS, H.E. and KNOWLES, G., 'The Importance of Pipe
    Diffusion in Correlating Creep Data', Strength of Metals
    and Alloys: Proc. 5th Intnl. Conf., Eds. Haasen, P.,
    Gerold, V. and Kostorz, G., Aachen, 1979, p. 301.

15. EVANS, H.E. and KNOWLES, G., 'Dislocation Creep in Nickel
    and Copper', Metal Sci., 1980, 14, 152.

16. DAVIES, P.W. and WILSHIRE, B., 'On Internal Stress
    Measurement and the Mechanism of High Temperature Creep',
    Scripta Met., 1971, 5, 475.

17. MOHAMED, F.A., 'Creep Behaviour of Solid Solution Alloys', Mtls. Sci. and Eng., 1979, 38, 73.

18. EVANS, H.E., 'A Model of Strain Hardening during High-Temperature Creep', Philos, Mag., 1973, 28, 277.

19. DAVIES, P.W., NELMES, G., WILLIAMS, K.R. and WILSHIRE, B., 'Stress-Change Experiments during High-Temperature Creep of Cu, Fe and Zn', Met. Sci. J., 1973, 7, 87.

20. ISHIDA, Y. and MCLEAN, D., 'Effect of N and Mn on Recovery Rate and Friction Stress During Creep of Fe', J. Iron and Steel Inst., 1967, 205, 88.

21. BARRETT, C.R. and NIX, W.D., 'A Model for Steady State Creep based on the Motion of Jogged Screw Dislocation', Acta Met., 1965, 13, 1247.

22. MITRA, S.K. and MCLEAN, D., 'Work Hardening and Recovery in Creep' Proc. Roy. Soc., 1966, A295, 288.

23. GITTUS, J.H., 'Theoretical Equation for Steady-State Dislocation Creep. Effects of Jog Drag and Cell Formation', Philos. Mag., 1976, 34, 401.

24. MARUYAMA, K., KARASHIMA, S. and OIKAWA, H., 'Analysis of Steady-State Creep of Fe-Mo Alloys from the View Point of Recovery', as Ref. 14, p. 283.

25. EVANS, W.J., 'A Model for Strain Hardening during Creep', Metal Sci., 1976, 10, 170.

26. KIDSON, G.V., 'A Review of Diffusion Processes in Zirconium and its Alloys', Electrochem, Techn., 1966, 4, 193.

27. CLAY, B.D. and STRIDE, R., 'The Primary and Secondary Creep Properties of β-Phase Zircaloy-2 in the Region 1200-1500°C', CEGB Report RD/B/N3950, 1977.

28. BURTON, B., REYNOLDS, G.R. and BARNES, J.P., 'Tensile Creep of Beta-Phase Zircaloy-2', J. Nucl. Mtls., 1978, 73, 70.

29. BROWN, A.M. and ASHBY, M.F., 'Correlations for Diffusion Constants', Acta Met., 1980, 28, 1085.

30. FELTHAM, P., 'On the Mechanisms of High-Temperature Creep in Metals with Special Reference to Polycrystalline Lead', Proc. Phys. Soc., 1956, B69, 1173.

31. WEERTMAN, J., 'Creep of Indium, Lead and some of their Alloys with various Metals', Trans. AIME, 1960, 218, 207.

32. HUNTINGTON, J.B., 'The Elastic Constants of Crystals', Solid State Phys., 1958, 7, 213.

33. ASHBY, M.F., 'A First Report on Deformation-Mechanism Maps', Acta Met., 1972, 20, 887.

34. NAGELS, P., VAN LIENDE, W., DE BATIST, R., DENAYER, N., DE JONGHE, L. and GEVERS, R., 'IAEA Symp. Thermodynamics, Vienna, 1966, Paper AM-66/46.

35. AUSKERN, A.N. and BELLE, J., 'Uranium Ion Self Diffusion in UO$_2$', J. Nucl. Mtls., 1961, 33, 311.

36. WOLFE, R.A. and KAUFMAN, S.F., 'Mechanical Properties of Oxide Fuels', WAPD-TM-587, 1967.

37. BOHABOY. P.E., ASAMOTO, R.R. and CONTI, A.E., 'Compressive Creep Characteristics of Stoichiometric UO$_2$' G.E. Report, GEAP 10054, 1969.

38. POTEAT, L.E. and YUST, C.S., 'Grain Boundary Reactions during Deformation', Ceramic Microstructures, Eds. Fulrath, R.M. and Park, J.A., Wiley, New York, 1968, p. 646.

39. PETERSON, N.L., 'Diffusion in Metals', Solid State Phys., 1968, 22, 409.

40. OKKERSE, B., 'Self-Diffusion in Lead', Acta Met., 1954, 2, 551.

41. EVANS, H.E. and KNOWLES, G., 'Threshold Stress for Creep in Dispersion-Strengthened Alloys', Metal Sci., 1980, 14, 262.

42. HAUSSELT, J.H. and NIX, W.D., 'A Model for High Temperature Deformation of Dispersion Strengthened Metals Based on Sub-Structural Observations in Ni-20Cr-2ThO$_2$', Acta Met., 1977, 25, 1491.

43. PARKER, J.D. and WILSHIRE, B., 'The Effect of a Dispersion of Cobalt Particles on High Temperature Creep of Copper', Metal Sci., 1975, 9, 248.

44. GITTUS, J.H., 'Theoretical Equation for Steady-State Dislocation Creep in Material Having a Threshold Stress', Proc. Roy. Soc., 1975, A342, 279.

45. BROWN, L.M. and HAM, R.K., 'Dislocation-Particle Interactions Strengthening Methods in Crystals, Eds. Kelly, A. and Nicholson, R.B., Elsevier, Amsterdam, 1971, p. 9.

46. LAGNEBORG, R., 'Bypassing of Dislocations Past Particles by a Climb Mechanism', Scripta Met., 1973, 7, 605.

47. EVANS, H.E. and KNOWLES, G., 'Critical Stress Parameters in the Creep of Dispersion Strengthened Alloys', The Strength of Metals and Alloys Vol. 2, 854, Nancy, 1976.

48. SHEWFELT, R.S.W. and BROWN, L.M., 'High Temperature Strength of Dispersion Hardened Single Crystals. II Theory', Philos. Mag., 1977, 35, 945.

49. KNOWLES, G., 'The Creep Strength of a 20%Cr/25%Ni/Nb Steel Containing Controlled Particle Dispersions', Metal Sci., 1977, 11, 117.

50. WILLIAMS, K.R. and WILSHIRE, B., 'On the Stress and Temperature-Dependence of Creep of Nimonic 80A', Met. Sci. J., 1973, 7, 176.

A THEORETICAL CONSIDERATION OF THE MICROSTRUCTURAL ORIGINS OF
FRICTION STRESS IN A CAST γ'-STRENGTHENED SUPERALLOY

R A Stevens and P E J Flewitt

Central Electricity Generating Board (SER),
Scientific Services Department, Canal Road, Gravesend,
Kent, DA12 2 SW, England.

SUMMARY

The role of microstructure on the creep properties of the cast
nickel-base superalloy IN738, containing a bimodal
distribution of γ' precipitates, is described. A theoretical
model is developed which predicts the contribution to the
friction stress of back stress resulting from the dislocation
climb process, allowing for both the effective precipitate
spacing as a function of applied stress and the compensating
decrease in line length of the inter-precipitate dislocation
segment. The experimentally observed increasing strain rate
which follows the transient stage of creep in this alloy is
explained in terms of the decrease in friction stress
resulting from precipitate coarsening.

1.   INTRODUCTION

The rates of deformation of a wide range of materials at high
temperatures are commonly described by the empirical
relationship [1,2]:

$$\dot{\varepsilon}_s = A \sigma^n \exp (-Q/RT) \qquad \qquad ...(1)$$

where $\dot{\varepsilon}_s$ is the secondary creep rate, σ is the applied stress,
A is a constant, Q is the activation energy for creep, R is
the gas constant and T is temperature. At stresses and
temperatures where the deformation mode is recovery-controlled
dislocation creep, various models have been put forward to
give theoretical justification to this relationship. The
value of the stress exponent, n, predicted by these models is
generally either 3 (e.g. [3]) or 4 (e.g. [4,5]) although
apparent exponents of 4 may reduce to 3 if the mean free path
of mobile dislocations is assumed to be equal or proportional
to the size of the dislocation recovery network, rather than
to be constant [6]. In recent studies Evans and Knowles [7,8]

choose to leave this question open and point out that creep data for some pure materials such as lead and α-iron conform to a value of n = 4, but, significantly from the point of view of superalloys, data for pure nickel are best fitted to a value of n = 3, indicative of a dislocation mean free path proportional to the network size. Furthermore these authors show that the large measured exponents (up to n = 6 [9]) for many pure materials and single-phase alloys, particularly at higher stresses, may be explained by assuming a stress dependence of the overall diffusion coefficient, which becomes more dependent on pipe diffusion contributions at high stresses.

In general, solid solution and multi-phase alloys have stress exponents as high as 40 [9] and activation energies for creep which are considerably larger than those observed for pure materials. Also, the value of n in these materials is frequently observed to increase dramatically with increasing stress, far beyond the factors predicted by Evans and Knowles [7] in their analysis. Several authors (e.g. [9,10]) have sought to rationalise this behaviour by proposing that creep in complex alloys takes place under an effective stress which differs from the applied stress by an amount σ_o, generally termed the friction stress, the magnitude of which may depend on such parameters as applied stress, temperature, solute-dislocation interactions and the presence of second phase particles [11]. By invoking this concept creep data for single [12] and multiphase alloys [9,13,14] can be fitted to a modified form of equation (1):

$$\dot{\varepsilon}_s = A^* (\sigma - \sigma_o)^P \exp^{-Q^*/RT} \qquad \ldots (2)$$

where the activation energy Q^* approximates to that for bulk self-diffusion and the exponent of effective stress, p, approaches the theoretical value of 3 to 4. For example, Evans and Harrison [15] have analysed data for a wide range of high-temperature materials, including nickel-base superalloys, and give the best empirical fit as:

$$\dot{\varepsilon}_s \propto \left[(\sigma - \sigma_o) / \sigma_{0.05} \right]^{3.5} \qquad \ldots (3)$$

where $\sigma_{0.05}$ is the 0.05% proof stress.

The creep rates of materials for which the exponent of applied stress, n, increases progressively with increasing stress, shown schematically in Fig.1 (a), can therefore be rationalised by a constant value for friction stress, σ_o, in equation (2). However, a wealth of published data exists in which the creep rates of multiphase alloys are found to fall into two [9,11,14,18] or even three [19,20] distinct regimes of stress dependence. This effect is shown schematically in Fig.1 (a), supported by the experimental results of Harrison and Evans [20] for Nimonic 90, which are reproduced in Fig.1(b).

Furthermore, experimentally determined values of σ_0 [9,21]
frequently fall into two or more distinct regimes as a
function of applied stress [9,11,14,15,19,20,22], Figs.1 (c)
and (d). Consequently, various authors [17,18,20] have
proposed that the transitions in stress sensitivity of both
the strain rate and the friction stress correspond to
transitions in the creep deformation process.

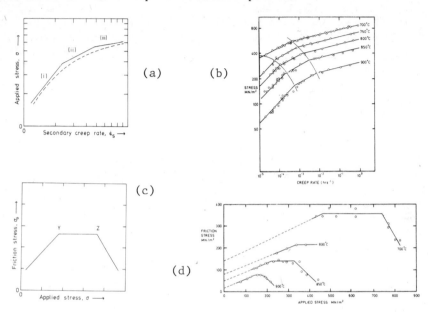

Fig.1. (a) Schematic representation of the dependence of
secondary creep rate on applied stress in various
two-phase alloys. (i) and (ii): [9,14-18]
(i), (ii) and (iii): [19,20]. The dotted line
shows the dependence predicted by equation (2)
if σ_0 is a constant.

(b) Specific experimental results of Harrison and
Evans [20] for Nimonic 90.

(c) Schematic representation of the types of
transition in σ_0 reported for various γ'-
strengthened alloys. Y: [11,15,19,20]
Z: [19,20,22].

(d) Transitions reported by Harrison and Evans [20]
for Nimonic 90.

In this contribution we study the cast nickel-base superalloy
IN-738 and examine ways in which physically reasonable values
of friction stress account for the observed stress-dependence
of the minimum creep rate. The microstructural origins of
friction stress are discussed, with specific reference to
the role of the bimodal distribution of precipitates of the
ordered (L1$_2$) γ' phase.

2. EXPERIMENTAL TECHNIQUES

IN-738 alloy with the composition given in Table 1 and grain size $\sim$ 1mm (mean linear intercept) was supplied in the form of cast carrot-shaped ingots of approximate length 100mm and diameter 20mm, tapering to 10mm.

	Ni	C	Co	Cr	Mo	W	Ta	Nb	Al	Ti	B	Zr
Wt.%	Bal	0.17	8.5	16.0	1.75	2.6	1.75	0.9	3.4	3.4	0.01	0.1

TABLE 1. Nominal Composition of IN-738

Round section creep specimens (5.05mm diameter x 27.5mm gauge length, $\frac{3}{8}$" BSF threads) were machined from each ingot and subjected to the standard manufacturing heat treatment of 2h at 1393K (1120°C), air cool, 16h at 1118K (845°C), air cool, to produce a bimodal distribution of γ' spheriods, $\sim$ 90nm diameter, and γ' cuboids, $\sim$ 1µm cube edge. High sensitivity, constant load, uniaxial creep rupture tests were performed in air at 1123K, with the temperature controlled to within $\pm$ 2 degrees along the specimen gauge.

The total volume fraction of γ' precipitates was established by an electrolytic extraction technique [23]. Using the transmission electron microscope (TEM), γ' precipitate sizes were determined by extraction replication [24] and the volume fraction of γ' cuboidal precipitates was measured from carbon surface replicas, the volume fraction of γ' spheroids then being given by difference. Such measurements involved sampling from a large number of areas to minimise the effect of variation of γ' dispersion parameters within the cast structure.

Thin foils were prepared from ruptured creep specimens by slicing 3mm diameter discs from the specimen gauges, normal to the stress axis. These discs were mechanically thinned to $\sim$ 0.08mm prior to jet polishing at 80V in a 2% perchloric acid/ethanol electrolyte maintained at $\sim$ 230K. Dislocation structures were then studied at both 100 and 120 kV in a JEOL JEM 100-CX transmission electron microscope equipped with a double tilt eucentric stage.

3. RESULTS

The variation of size, square lattice spacing and volume fraction of γ' spheroidal and cuboidal precipitates during the term of the creep tests at 1123K is summarised in Fig.2. Previous studies [25] have shown the change in γ' dispersion parameters due to classical diffusion - controlled Ostwald ripening at this test temperature (1123K) to be essentially independent of applied stress.

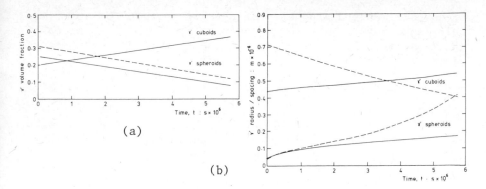

Fig.2. (a) Change in volume fraction of γ' cuboids and
spheroids during exposure at 1123K. The dotted
line shows the effective spheroid volume fraction
in the available matrix, i.e. after allowing for
the finite volume occupied by cuboids.

(b) Change in radius (full lines) and spacing (dotted
lines) of γ' precipitates during exposure at 1123K.

Creep curves for IN-738 in the stress range 200 to 400 MPa at
1123K are shown in Fig.3. The most noteworthy feature is that
with the exception of the tests at the highest stresses
(σ > 350 MPa), all the curves display 'pseudo-tertiary'
behaviour throughout, i.e. a progressively increasing strain
rate following the transient stage. When the minimum creep
rates, $\dot{\varepsilon}_m$, are plotted as a function of stress, Fig.4 (a), two
distinct regimes are apparent, with a transition at a stress

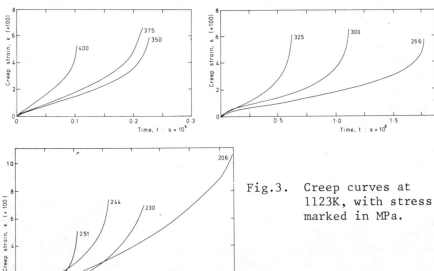

Fig.3. Creep curves at
1123K, with stress
marked in MPa.

of σ ≈ 315 MPa and exponents of n = 6.1 and 9.6 for stresses
below and above this point respectively. Consequently, for
the reasons outlined in Section 1 the results are presented as
plots of $\dot{\varepsilon}_m^{1/3}$ Vs. σ and $\dot{\varepsilon}_m^{1/4}$ Vs. σ in Figs. 4 (b) and (c), and
again two distinct regimes are apparent.

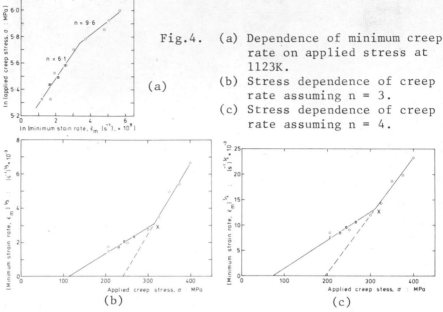

(a)

Fig.4. (a) Dependence of minimum creep
rate on applied stress at
1123K.
(b) Stress dependence of creep
rate assuming n = 3.
(c) Stress dependence of creep
rate assuming n = 4.

(b)   (c)

The arrangements of dislocations in thin foils taken from
transverse sections of creep specimens tested in both 'high'
and 'low' stress regimes were examined in the TEM under a
range of imaging conditions including high resolution centred
dark field and two-beam diffraction such that g.b ≠ 0 for the
Burgers vector of possible dislocations.

In the 'high' stress regime, Fig.4, (σ > 315 MPa) dislocations
overcome γ' precipitates by a cutting process as shown in the
centred dark field micrograph Fig.5 (a) (σ = 400 MPa) where
the foil normal is close to [110] and g is set to 002.
Glissile dislocations cannot shear the ordered γ'
precipitates without introducing high energy surfaces of
antiphase boundary or complex fault domains. However, groups
of partial dislocations can shear the γ' precipitates
resulting in overlapping fringe contrast due to the creation
of superlattice intrinsic or extrinsic stacking faults, which
at the low strain rates associated with creep deformation can
be relaxed by vacancy-dislocation reactions [26] giving
extended faults within both cuboidal and spheroidal γ'
precipitates, Fig.5 (a). In the 'low' stress regime,
σ < 315 MPa, no such evidence of precipitate shearing was
detected and here dislocations overcome the spheroidal γ'
precipitates by a combined process of slip and climb, Fig.5(b).

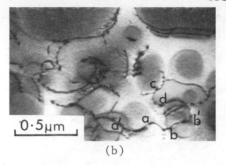

(a)   (b)

Fig.5.   (a) Dislocation cutting of γ' precipitates in specimen
             tested to rupture at 400 MPa, 1123K. Dark field
             image, foil normal [110], g = 002.
         (b) Dislocation arrangement in specimen tested to
             rupture at 206 MPa, 1123K.

Referring to this Figure, the dislocation lines ab, cd, are
bowed between spheroids but, significantly, no Orowan loops
were observed. The dislocation line aa' is at an early stage
of climb, whereas the line bb' is at a more advanced stage of
by-pass. Thus the stress of ∿ 315 MPa, Fig.4, defines a
transition between the by-pass of γ' precipitates by
dislocation climb and cutting.

4.   DISCUSSION

The generally observed transitions in friction stress for γ'
strengthened alloys, Fig.1 (c), fall into two basic types;
either from σ_0 increasing with stress to constant ('Y'), or
from σ_0 constant to decreasing with stress ('Z'). One of the
most rigorous studies on superalloys to date is that of
Harrison and Evans [20], who evaluated friction stresses in the
nickel-base alloy Nimonic 90 over a wide range of temperature
and applied stress, and observed both types of transition,
Figs. 1 (b) and (d). These results led to the formulation of
a creep deformation map for which it was suggested that the
transition in stress sensitivity of both creep rate and
friction stress, Y, corresponds to that from creep by
recovery-controlled climb to a regime in which precipitate
shear occurs during the initial stages of creep. However, due
to the build up of immobile dislocation tangles during primary
creep the friction stress increases, reducing the effective
stress on mobile dislocations such that cutting can no longer
occur and secondary creep is again by diffusion-controlled
climb. The second transition, Z, was then suggested to be
the transition from this regime to one in which the applied
stress is sufficient for precipitate cutting to occur throughout the creep life. Since precipitate cutting is observed
only in the 'high' stress regime in IN-738, Fig.5, the
transition X in stress sensitivity of creep rate, Fig.4, could
correspond to either of the above types of transition depending

on whether or not the degree of cutting during primary creep is sufficient to be detected in thin foils. We shall therefore investigate the feasibility of both possibilities.

If the transition corresponds to Z in Fig.1 (c), then the constant value of σ_0 in the 'low' stress regime is given by the intercepts in Fig.4 as either $\sim$ 115 MPa (n = 3) or $\sim$ 80 MPa (n = 4). The gradients below and above the transition point would then mean that σ_0 goes to zero at either $\sigma \simeq$ 370 MPa (n = 3) or $\sigma \simeq$ 400 MPa (n = 4). However, once γ' precipitates are cut the associated creation of antiphase boundary must make a significant contribution to friction stress. It is therefore unreasonable for σ_0 to be zero at 400 MPa and we consider instead the transition Y of Fig.1 (c). On this basis σ_0 has constant values of either $\sim$ 240 MPa (n = 3) or $\sim$ 195 MPa (n = 4) in the 'high' stress regime and decreases with decreasing stress in the 'low' stress regime.

Origins of stress-dependent components of friction stress from two sources have been proposed. Firtly, Ajaja et al. [27] concluded that solid solution effects within the matrix of superalloys give rise to a significant, stress dependent, contribution to the friction stress. Secondly, Evans and Harrison [28] expressed the total friction stress as:

$$\sigma_o = \left[\sigma_o\right]_{IN} + c\sigma \qquad \ldots(4)$$

where the stress-dependent term, $c\sigma$, was considered to arise from the contribution of immobile dislocations and $\left[\sigma_o\right]_{IN}$, the "inherent friction stress", was equated to the back-stress for dislocation climb over γ' precipitates. The same authors suggest that $\left[\sigma_o\right]_{IN}$ can be determined by extrapolation of σ_0 to zero applied stress, Figure 1 (d) [20], and related to the climb back-stress through the model of Shewfelt and Brown [29], the value being temperature dependent due to changes in γ' dispersion parameters with temperature.

However, whatever the origins of the other components of σ_0, it is only the component due to γ' precipitates that is likely to change significantly during high-temperature exposure, as a result of precipitate coarsening. Before evaluating this component it is necessary to establish the size of active dislocation links, r_s, which is given by:

$$\sigma - \sigma_o = \alpha\mu b/r_s \qquad \ldots(5)$$

where μ is the shear modulus, b is the Burgers vector and α is a constant with values in the range 0.6 to 1.2 [17]. From Fig.4 (b) we find that r_s varies from $\sim$ 280 - 560 nm at 200 MPa applied stress, to 130 - 260 nm at 300 MPa. Thus in this stress regime mobile dislocation links are large relative to the spheroidal γ' precipitates ($\sim$ 100nm diameter) but small relative to the cuboidal precipitates ($\sim$ 1µm cube

edge). Dislocation links in close proximity to cuboidal
precipitates may therefore be regarded as immobile and we need
only consider the back stress resulting from dislocation by-
pass of spheroidal γ' precipitates, which as shown in Fig.5
is by a climb mechanism in the stress range 200 to 300 MPa.
To evaluate the back stress we adopt a modified form of the
Shewfelt and Brown analysis [29], which takes account both of
the effective spacing of precipitates of finite size as a
function of stress and of the balance between the increase in
length of the climbing dislocation segment and the decrease in
length of the inter-precipitate segment. Details of the
model are given in the Appendix and it is of note that the
assumed geometry of a climbing dislocation corresponds to the
dislocation line a'a b b' shown in Fig.5 (b). This model
enables the back stress due to dislocation climb to be
calculated from the γ' coarsening kinetics, Fig.2, as a
function of time at 1123K. As shown in Fig.6, the value of
back stress associated with the initial γ' dispersion
parameters is $\sim$ 40 MPa but the rate of decrease thereafter is
dependent on the shear stress acting on the mobile
dislocation. This effective shear stress will be less than
the applied shear stress due to the other components of σ_0,
and will therefore be in the range 20 to 50 MPa for applied
tensile stresses of 200 to 300 MPa, Fig.4 (b).

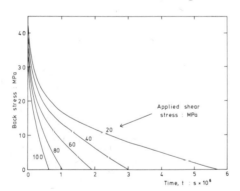

Fig.6. Back stress (tensile) for dislocation climb over γ'
spheroidal precipitates as a function of applied shear
stress and aging time at 1123K.

From Fig.6 the decrease in friction stress during creep tests
can therefore be predicted. For the creep tests in the stress
range 200 to 300 MPa it is assumed that but for microstructural
degradation the classical tertiary stage would commence at a
time of approximately two thirds of the rupture life, this
being the time of onset of tertiary creep due to grain
boundary cavitation in a variety of microstructurally stable
materials [30]. It is also the earliest time at which
surface cracking has been detected in a range of nickel-base
alloys [31]. Fig.7 therefore shows the calculated decrease in
friction stress required to account for the increase in creep

rate, $\dot{\varepsilon}_s$, from the end of transient creep to two thirds of the
rupture life, for creep tests in the stress range 200 to 300
MPa at 1123K.   Here we assume that σ_O is constant for
$\sigma > 315$ MPa and that the exponent of applied stress is n = 3,
as in Fig.4 (b).   Furthermore, since the creep tests were
performed at constant load, the applied stress was determined
through consideration of specimen extension, and hence net
sectional area, as a function of time.   Fig.7 shows a good
correlation in both the absolute magnitude of, and the average
gradient of, the decrease in friction stress required to
explain the increase in creep rate with time and that
predicted by the dislocation climb back-stress model.   The
weakest correlation is at the lowest stress studied, 206 MPa,
but it should be noted that in the duration of this long term
test coarsening and increase in volume fraction of γ' cuboids,
Fig.2, will lead to a larger number of immobilised
dislocation links and hence to a compensating increase in
other contributions to σ_O, a factor not taken into account in
Fig.7.   It is also possible that this relatively low stress
is approaching another transition in deformation mechanism,
e.g. to a regime in which diffusional creep makes a
significant contribution [20].

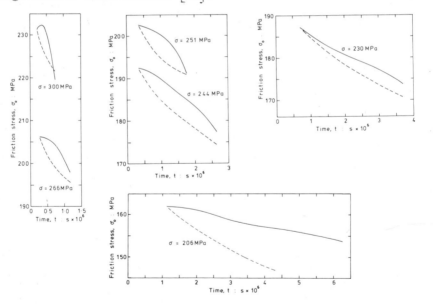

Fig.7.   Variation in σ_O during creep tests at 1123K, from
the end of primary creep to two thirds of the rupture
life.   Full lines: as required to account for the
observed increases in creep rate through equation (2),
with n = 3.   Dotted lines: the predicted decrease
in the back stress due to dislocation climb.

5.   ACKNOWLEDGEMENT. This paper is published with the permission of the Director General, CEGB, South Eastern Region.

6.   REFERENCES

1.   McLean, D. Rep. Prog. Phys., 1966, 29, 1.
2.   Mukherjee, A.K., Bird, J.E. and Dorn, J.E. Trans.Am.
           Soc. Metals, 1969, 62, 155.
3.   Gittus, J.H. Acta Met., 1974, 22,789.
4.   Gibbs, G.B. Phil. Mag., 1966, 13, 317.
5.   Lagneborg, R. Met. Sci. J., 1972, 6, 127.
6.   Evans, H.E. and Williams, K.R. Phil. Mag., 1973, 28,227.
7.   Evans, H.E. and Knowles, G. - 'Strength of Metals and
           Alloys',(ICSMA5, Aachen), Ed. Haasen, P.,
           Gerold, V. and Kostorz, G. Pergamon Press, 1979,
           p.301.
8.   Evans, H.E. and Knowles, G. Acta Met., 1977, 25, 963.
9.   Threadgill, P.L. and Wilshire, B. - 'Creep Strength in
           Steel and High Temperature Alloys', ISI, London,
           1972, p.8.
10.   Lagneborg, R. J. Mat. Sci., 1968, 3, 506.
11.   Dennison, J.P., Holmes, P.D. and Wilshire, B.
           Mat. Sci. Eng., 1978, 33, 35.
12.   Nelmes, G. and Wilshire, B. Scripta Met., 1976, 10,697.
13.   Williams, K.R. and Wilshire, B. Met. Sci. J., 1973, 7,
           176.
14.   Parker, J.D. and Wilshire, B. Met. Sci., 1975, 9, 248.
15.   Evans, W.J. and Harrison, G.F. Met. Sci., 1976, 10,307.
16.   Davies, P.W. and Dennison, J.P. Met. Sci., 1975, 9, 319.
17.   Bergman, B. Scand. J. Metall., 1975, 4, 97.
18.   Lagneborg, R. Scripta Met., 1973, 7, 605.
19.   Evans, W.J. and Harrison, G.F. Met. Sci., 1979, 13, 346.
20.   Harrison, G.F. and Evans, W.J. - 'Engineering Aspects
           of Creep', (IME, Sheffield), 1980, to be published.
21.   Davies, P.W., Nelmes, G., Williams, K.R. and Wilshire, B.
           Met. Sci. J., 1973, 7, 87.
22.   Sidey, D. Met. Trans., 1976, 7A, 1785.
23.   Kriege, O.H. and Baris, J.M. Trans. ASM., 1969, 62, 195.
24.   Stevens, R.A. and Flewitt, P.E.J. Metallography, 1978,
           11, 475.
25.   Stevens, R.A. and Flewitt, P.E.J. Mat. Sci. Eng., 1979,
           37, 237.
26.   Kear, B.H., Leverant, G.R. and Oblak, J.M. Trans. ASM.,
           1969, 62, 639.
27.   Ajaja, O., Howson. T.E., Purushothaman, S. and Tien, J.K.
           Mat. Sci. Eng., 1980, 44, 165.
28.   Evans, W.J. and Harrison, G.F. Met. Sci., 1979, 13, 641.
29.   Shewfelt, R.S.W. and Brown, J.M. Phil. Mag., 1977,35,945
30.   Dennison, J.P. and Wilshire, B. - 'Advances in Research
           on the Strength and Fracture of Materials', (ICF4,
           Waterloo, Canada), Ed. Taplin, D.M.R. Pergamon Press
           1977, p.635.

31. Tipler, H.R., Lindblom, Y. and Davidson, J.H. - 'High
        Temperature Alloys for Gas Turbines', Ed.
        Coutsouradis, D. et.al. Applied Science Publishers
        (London), 1978, p. 359.

APPENDIX. THE CLIMB OF DISLOCATIONS OVER PRECIPITATES.

First we must calculate the effective spacing, L, of
precipitates of finite size as a function of stress. The
average centre-to-centre spacing of precipitates of radius R
encountered by a straight dislocation line (i.e. at zero
applied stress) is $4R/3f$ where f is the volume fraction of
precipitates [1]. A small shear stress $\Delta\tau$ will cause the
dislocation to bow out to a radius of curvature $R_{\Delta\tau}$, Fig.A1.:

$$R_{\Delta\tau} = \frac{\Gamma}{\Delta\tau b} = \frac{\mu b}{2\Delta\tau} \qquad \qquad ...(1)$$

where Γ ($\simeq 1/2\ \mu b^2$) [1] is the line energy of the dislocation.
The average radius of a precipitate section in the glide plane
is $R_S = (2/3)^{\frac{1}{2}} R$ [1] hence from Fig.A1 the area of matrix, ΔA,
swept by the dislocation is approximately given by the shaded
area:

$$\Delta A = \frac{2R}{3f} \left\{ R_s + \frac{\mu b}{2\Delta\tau} - \left[(\frac{\mu b}{2\Delta\tau} + R_s)^2 - (\frac{2R}{3f})^2 \right]^{\frac{1}{2}} \right\} \qquad ...(2)$$

The area of matrix, A_m, in the glide plane associated with
each precipitate can be deduced from the square lattice
spacing of precipitates:

$$A_m = (\frac{\pi}{f}) R_s^2 - \pi R_s^2 \qquad \qquad ...(3)$$

Thus by considering the total area of matrix swept by the
individual segments of a long, initially straight, dislocation
line under the action of a shear stress $\Delta\tau$, the new effective
precipitate spacing, $L_{\Delta\tau}$, can be deduced from the number of
extra precipitates encountered:

$$L_{\Delta\tau} = \frac{4R}{3f} \left[1 + \frac{f\Delta A}{2\pi R_s^2 (1-f)} \right]^{-1} \qquad \qquad ...(4)$$

The effective spacing L at any applied stress τ can therefore
be computed numerically from the cumulative effect of
successive increments in the applied shear stress.

Fig.A2 shows a dislocation in an intermediate stage of climb
over two such precipitates. The segments AA' and BB' take the
shortest paths over the precipitate surfaces and the segment
AB bows to a radius of curvature $R_c = \mu b/2\tau$. The dislocation
will climb over or under a given precipitate depending on the
position of the point of incidence relative to the precipitate
equator. It can then be shown geometrically that:

Length of arc AB $= \dfrac{2\Gamma}{\tau b} \sin^{-1}\left(\dfrac{\lambda/2 + R_s - x}{\Gamma/\tau b}\right)$

Length of arc AA' $= 2R \sin^{-1}\left(\dfrac{x}{R}\right)$
$$\quad\dots(5)$$

where $\lambda = L - 2R_s$ is the surface-to-surface precipitate spacing and the chord length x is related to the 'climb height' z, by:

$$x = R\left\{1 - \frac{1}{3}\left[\frac{1}{z + (\frac{1}{3})^{\frac{1}{2}} R}\right]^2\right\}^{\frac{1}{2}} \qquad \dots(6)$$

The total length of dislocation line between points P and Q is therefore $\ell = $ arc AB + arc AA'. From this the rate of change of line length as the dislocation climbs, $d\ell/dz$, is computed. The back force on this section is then given by:

$$F_{(back)} = \Gamma\ \frac{d\ell}{dz} = \frac{\mu b^2}{2}\ \frac{d\ell}{dz} \qquad \dots(7)$$

and the corresponding back stress acting over the length PQ = L is:

$$\tau_{(back)} = \frac{F_{back}}{bL} = \frac{\mu b}{2L}\frac{d\ell}{dz} \qquad \dots(8)$$

Thus the threshold shear stress for climb is:

$$\tau_3 = \frac{\mu b}{2L}\left(\frac{d\ell}{dz}\right)\ max \qquad \dots(9)$$

where the maximum value of $d\ell/dz$ is computed by evaluating $d\ell/dz$ from the point where the dislocation first touches the precipitates (making a tangent at the points ST) to the point where AA', BB' lie normal to the glide plane. Unlike previous analyses [2,3] this maximum back stress does not necessarily occur when the dislocation has reached the top of the precipitate.

REFERENCES:

1.   Brown, L.M. and Ham, R.K. - Strengthening Mechanisms in Crystals, Ed. Kelly, A. and Nicholson, R.B. Elsevier, Amsterdam, 1971, p.9.
2.   Lagneborg, R. Scripta Met., 1973, 7 605.
3.   Evans, H.E. and Knowles, G. Met. Sci., 1980, 14, 262.

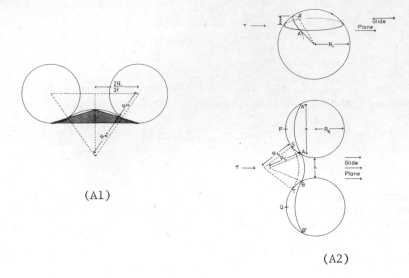

(A1)

(A2)

FIGS. A1 and A2.   Models used for calculating climb
back stress.

THE ROLE OF AGING IN THE MODELING OF ELEVATED TEMPERATURE DEFORMATION

Erhard Krempl

Department of Mechanical Engineering, Aeronautical Engineering, and Mechanics, Rensselaer Polytechnic Institute, Troy, New York 12181

ABSTRACT

In elevated temperature service diffusion induced metallurgical changes (aging) can occur which can significantly affect the deformation behavior. When aging is found to be present it should be accounted for in the material model similar to the practice in concrete technology. It is shown that aging can significantly affect the creep behavior and that it might be the cause of creep-plasticity interaction. The theory of viscoplasticity based on total strain and overstress is introduced and it is indicated how the two material functions of the theory might be identified in the presence of aging. The overstress is shown to be similar to the effective stress used in materials science. The proposed identification procedure could serve as an alternate way of determining the friction stress which may be equivalent to the equilibrium stress used in the viscoplasticity theory based on total strain and overstress.

INTRODUCTION

Ample metallurgical and metallographic studies demonstrate that microstructural changes can occur when an engineering alloy is subjected to extended elevated temperature service (see for example [1-4]). The chemical reactions that take place and the metallurgical changes within the alloy are frequently well known. They can affect the deformation behavior significantly and should therefore be accounted for in phenomenological theories which represent the elevated temperature deformation behavior and which are used to predict the performance of components by inelastic analysis. These changes and those caused by aggressive environments are presently not explicitly recognized in engineering analysis.

While details of the microstructural changes need not be accounted for in a phenomenological approach it is necessary to model those microstructural processes which have a significant effect on the deformation and fracture behavior.

It appears that there are three important basic phenomena which may occur at elevated temperature:

1) Diffusion processes with chemical reactions which occur in the absence of deformation and which proceed independently of it. They may be caused by aggressive environments or by internal reactions alone. These processes are referred to as aging.

2) Diffusion processes with chemical reactions which are initiated by deformation. They are referred to as strain aging.

3) Diffusion processes without chemical reactions which undo the effect of prior plastic deformation. These processes can return the material to its original (virgin) condition and are generally known as annealing or recovery.

It is of course clear that in reality all three of these diffusion processes can occur simultaneously. For the purpose of their identification, however, it is necessary to separate these effects especially since they are physically distinct and therefore must have separable repositories in constitutive equations or material models.

These effects have not been adequately dealt with in materials modeling. The purpose of this contribution is to discuss only the role of aging on the deformation behavior, the implications for the identification of elevated temperature deformation phenomena and to discuss the modeling of aging within the theory of viscoplasticity based on total strain and overstress.

IDENTIFICATION OF AGING

In a phenomenological approach aging is identified by performing the same test (tensile or creep or relaxation test or any other mechanical deformation test) with specimens of different age, i.e., specimens which were exposed (without mechanical loading) to the given elevated temperature environment for different periods of time. If aging is significant then the results of the test will depend on age [5-6], the time spent in the elevated temperature environment before the start of the mechanical test.

This assessment of the influence of age on deformation is well known in concrete technology [7,8] and it is practice to

account for the age of the material in general and in computerized stress analysis, e.g. [9,10].

IMPORTANCE OF AGING FOR METAL DEFORMATION

The practice of concrete technology (where aging is
caused by chemical reactions at room temperature) is not at
all shared in elevated temperature metals technology. There
are recent studies [1] which suggest that creep rate and ductility are affected by age. But the bulk of mechanical testing, specifically creep testing, ignores the influence of
aging completely. Indeed, aging as defined here is not mentioned at all in the textbooks on creep [11-13].

The tests reported in [1,14] strongly suggest that aging
phenomena have a significant influence on mechanical deformation and should therefore be determined by separate experiments. Figure 1 taken from [14] shows the significant influence of aging on the creep behavior of Type 304 stainless
steel. The creep strain at the same stress level at 100 hrs
differs almost by a factor of five due to the influence of
different aging times. In this case aging reduces creep; it
is conceivable that it may enhance creep under other
conditions.

Suppose now that a regular creep test is started right
after the specimen reaches 593°C. In the usual creep tests
the measured creep strain is assumed to be caused by the load
alone. Figure 1, however, shows that age contributes also.
The normally measured creep strain and therefore the shape
of the creep curve is a function of time under load and of
age. The age dependence can be significant and must be determined by separate tests.

AGING AND CREEP-PLASTICITY INTERACTION

The presence of aging may also cause a wrong interpretation of some deformation phenomena such as creep-plasticity
interaction.

Currently recommended constitutive equations for elevated
temperature stress analysis of nuclear components [15] formu-
late creep and plasticity separately and creep-plasticity
interactions cannot be represented [5]. This fact is now
acknowledged [16] and "unified theories" are under development [16-26]. Aging is, however, only explicitly considered
in [26].

An example of what is usually considered creep-plasticity
interaction is given in Fig.2. After creep-testing at 86.2 MPa

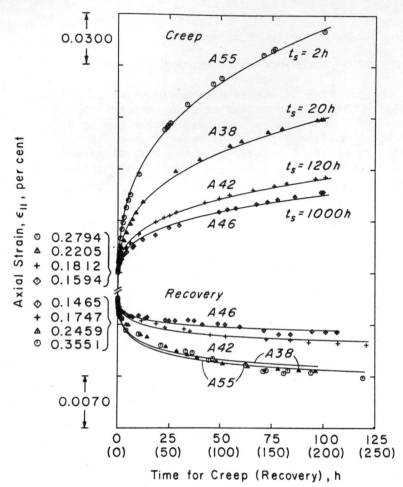

Fig.1 The Influence of Prior Aging at 593°C (t$_s$ is the Aging Time) on the Creep Behavior at 593°C at a Stress Level of 86.2 MPa and on the Subsequent Recovery. Type 304 Stainless Steel from [14].

for 2000 hrs at 649°C the specimen was unloaded and a tensile test was performed at room temperature. The stress-strain diagram of this specimen (CD – residual tensile curve) is above the one obtained with a virgin specimen (OB – typical initial tensile curve). Prior creep is said to have "hardened" the material and therefore creep-plasticity interaction has taken place.

This author provides a different interpretation of the same results. During the 2000-hour creep test the material aged and its yield strength increased due to aging. (The evidence presented in [1] and [14] suggests that aging makes 304 SS "harder.") The subsequent tensile test reveals this fact.

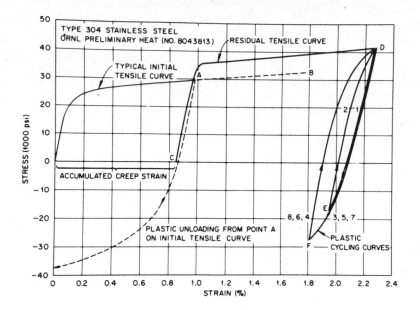

Fig.2 The Stress-Strain Behavior at Room Temperature
of a Virgin Specimen (OAB) and of a Specimen
Subjected to a Prior Creep Test at 86.2 MPa
and 1200°F (649°C) for 2000 hrs (CAD). Type
304 Stainless Steel from [15].

It is therefore suggested that the flow stress increase
is not due to creep but due to aging. This assertion can,
however, only be proven by an additional tensile test performed
on a companion specimen which was exposed for 2000 hrs to the
649°C environment without mechanical loading. If this companion test would exhibit the flow stress increase as the
curve CD in Fig.2 this author's assertion would be correct.

This example may suffice to demonstrate that aging should
be given more attention in elevated temperature materials
modeling. Its effect can seriously cloud the interpretation
of many deformation phenomena.

IDENTIFICATION OF THE MATERIAL FUNCTIONS OF THE THEORY OF
VISCOPLASTICITY BASED ON TOTAL STRAIN AND OVERSTRESS IN
THE PRESENCE OF AGING

In [26], we investigated the general possibility of the
above-mentioned theory to reproduce classical creep curves
($\sigma_o > 0$, $\dot{\varepsilon} \geq 0$; $\ddot{\varepsilon} \leq 0$) and nonclassical creep curves ($\sigma_o > 0$;
$\dot{\varepsilon} > 0$, $\ddot{\varepsilon} > 0$) in the presence of aging. To represent this
phenomenon the two material functions of the theory were made
to depend on the age of the material. We have [24-26]:

$$\dot{\varepsilon} = \dot{\sigma}/E + \frac{\sigma - g[\varepsilon,t]}{Ek[\sigma - g[\varepsilon,t],t]} \tag{1}$$

where ε is the engineering strain, σ is the true stress, E is the modulus of elasticity, $g[\varepsilon,t]$ is the "equilibrium stress-strain curve[i] odd in ε, and k is the positive viscosity function. A superposed dot denotes differentiation with respect to time [26, Eq. (9)]. The quantity $\sigma - g$ is called the overstress. Square brackets denote "function of."

The dependence of the equilibrium stress-strain curve and of the viscosity function on age t says that they can change with age and deformation.

We assume in the following that aging proceeds independently of deformation and that the dependence of g and k on age is slow so that they change insignificantly during the duration of a typical short-time mechanical test.

For creep ($\sigma = \sigma_o = $ const) we get from (1)

$$\dot{\varepsilon} = \frac{\sigma_o - g[\varepsilon,t]}{Ek[\sigma_o - g[\varepsilon,t],t]} \, . \tag{2}$$

We see that the creep strain rate depends on the two material functions which in turn depend on age. It is shown in [26] that the age variation of these two functions can significantly affect the creep behavior. It is therefore of interest to know how they change with age. The creep test itself is not suitable for answering this question. The age influence must be determined separately.

For large strains Eq.(1) admits an asymptotic solution [27]

$$\{\sigma - g[\varepsilon,t]\} = (E - g')k[\{\sigma - g[\varepsilon,t]\},t]\dot{\varepsilon} \quad \text{[ii]} \tag{3}$$

where $g' = \dfrac{\partial g}{\partial \varepsilon} + \dfrac{\partial g}{\partial t}\dfrac{1}{\dot{\varepsilon}} \ll E$ since g' represents the tangent modulus in the plastic range. In the above we have used $\{\ \}$ to indicate the limiting overstress. Equation (3) can in principle be inverted (usually done numerically [25]) to get

$$\{\sigma - g[\varepsilon,t]\} = f[\dot{\varepsilon},t] \, . \tag{4}$$

---

[i] Strictly speaking $g[\varepsilon, t = \infty]$ is the equilibrium stress-strain curve.

[ii] Mathematically (3) is only valid for infinite strain and therefore infinite time; numerical experiments [24-26] show that (3) applies in the plastic range where the tangent modulus is small compared to E and changes very little.

To determine the influence of age on g[ ] and k[ ] we expose a batch of specimens for different periods of time to the
elevated temperature environment in the absence of mechanical
loading. Then identical tests using a servocontrolled testing
machine with an extensometer on the specimen gage length are
to be performed on the specimens with different age. The
tests are to be done at a constant temperature and involve
initially a constant strain rate followed by step changes of
the rate in the plastic range as shown in Fig.3.

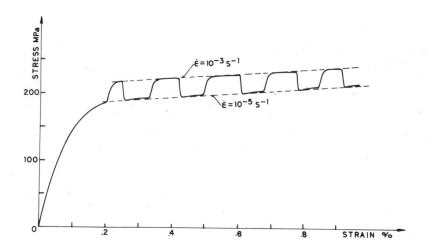

Fig.3   The Room Temperature Response of Type 304 Stainless Steel to Step Changes in Strain Rate

If aging is important we can expect the following results
as a function age:

1) At a given constant strain rate an increase or a
   decrease in the flow stress (dashed curves in
   Fig.3).

2) A change in the spacing of the flow stresses obtained at different strain rates (more than two
   strain rates can be used).

If 1) is found then aging has an effect on work hardening;
aging affects the rate dependence if 2) is found.

With the information gained from such tests we can in
principle determine k[ ] and g[ ] as a function of age using
the methods of [24,25]. (In the evaluation we assume that the
influence of age is negligible during a particular test. The
test duration is therefore limited.) With the functions g[ ]

and k[ ] known the creep curve can be calculated from (2) and compared with experiment.

RELATION OF PRESENT THEORY TO THE EFFECTIVE (FRICTION, REDUCED) STRESS CONCEPT USED IN MATERIALS SCIENCE

Equation (2) says that the creep rate depends only on the overstress. Our theory does not agree with engineering creep theories [11-12] but resembles recent creep equations presented in the materials science literature. In these papers, [27-31] to name just a few, creep rate is dependent on an effective stress which is defined as the applied stress minus the friction stress which is assumed to be constant. Roughly speaking the present g[ ] corresponds to the friction stress.

Phenomenologically the overstress is necessary since negative creep rates can be observed at positive stresses following unloading [28], p.87. Equation (2) reproduces this fact if $\sigma < g[~]^{(i)}$. (In [28,29] creep rate is proportional to the fourth power of the effective stress; no negative creep rates can be reproduced by this expression. The effective stress is raised to the 3.5 power in Eq.(3) of [27]; again difficulties arise when negative effective stresses are used.) Also since k is a strongly decreasing function of the argument [25] the theory predicts that it is very difficult to determine g[ ] by "stress dip" or "strain dip" experiments. This prediction is confirmed by the results cited in the literature.

An alternate possibility of determining the equilibrium stress-strain curve g[ε] is by strain rate change experiments [25]. It might be possible to apply this technique for the determination of the friction stress. Also the tests reported in Fig.5 of [25] can be used to decide whether the friction stress is constant or a function of deformation.

ACKNOWLEDGEMENT

This research was supported by grants from the National Science Foundation and from the Office of Naval Research.

(i) Presently contemplated augmentations of g[ ] in (1) for cyclic loading [32] predict a smaller creep rate for $0 \le \sigma_o \le g[~]$ than for $\sigma_o > g[~]$ for equal values of $|\sigma - g[~]|$.

REFERENCES

1. Sikka, V.K., Brinkman, C.R. and McCoy, H.E., "Effect of Thermal Aging on Tensile and Creep Properties of Types 304 and 316 Stainless Steels," in Structural Materials for Service at Elevated Temperatures in Nuclear Power Generation, Schaefer, A.O., ed., MPC-1, ASME, New York, 1975.

2. Van Leeuwan, H.P. and Schra, L., "Parametric Analysis of the Effects of Prolonged Thermal Exposure on Material Strength," Proceedings, Creep and Fatigue in Elevated Temperature Applications, Vol.1, Paper C134/73, 134.1-134.9, Institution of Mechanical Engineers, 1975.

3. Swindeman, R.W., Pugh, C.E., "The Influence of Cyclic Loadings on Creep Behavior of Austenitic Stainless Steels," ORNL 5429, Oak Ridge National Laboratory, 1978.

4. Private communications by C.R. Brinkman, G.V. Smith, R.W. Swindeman, D.A. Woodford.

5. Krempl, E., "Cyclic Creep. An Interpretive Literature Survey," Welding Research Council Bulletin 195, Welding Research Council, New York, NY, June 1974.

6. Krempl, E., "On the Interaction of Rate and History Dependence in Structural Metals," Acta Mechanica, 22, 53-80 (1975).

7. Neville, A.M., Properties of Concrete, Wiley, New York, 1963.

8. Orchard, D.F., Concrete Technology, Vol.1, Wiley, New York, 1973.

9. Argyris, J.H., Pister, K.S. and William, K.J., "Thermomechanical Creep of Aging Concrete - A Unified Approach," Assoc. Int. des Ponts et Charpents, Memoirs, Vol.36, 1976, pp.23-57.

10. Bazant, Z.P., "Theory of Creep and Shrinkage in Concrete Structures: A Precis of Recent Developments," Mechanics Today, 2, S. Nemat-Nasser, editor, Pergamon Press, 1975.

11. Rabotnov, Yu.N., Creep Problems in Structural Members, North Holland, 1969.

12. Hult, J., Creep in Engineering Structures, Ginn/Blaisdell, 1966.

13. Gittus, J.H., Creep, Viscoplasticity and Creep Fracture in Solids, Halsted Press, 1975.

H

14. Cho, U.W. and W.N. Findley, "Creep and Creep Recovery of 304 Stainless Steel at Low Stresses and Effects of Aging," Engineering Materials Research Laboratory Report EMRL-75, Brown University, Providence, RI, July 1980.

15. Corum, J.M. et al., "Interim Guidelines for Detailed Inelastic Analysis of High-Temperature Reactor System Components," ORNL-5014, Oak Ridge National Laboratory, December 1974.

16. Corum, J.M., "Future Needs for Inelastic Analysis in Design of High-Temperature Nuclear Plant Components," Computers and Structures, 13, 231-240 (1981); also in Computational Methods in Nonlinear Structural and Solid Mechanics, A.K. Noor, H.G. McComb, Jr., editors, Pergamon Press, 1980.

17. Hart, E.W., "Constitutive Relations for the Nonelastic Deformation of Metals," Trans. ASME, Journal of Engineering Materials and Technology, 98, 193-202, 1976.

18. Laflen, J.H. and Stouffer, D.C., "An Analysis of High Temperature Metal Creep: Part I - Experimental Definition of an Alloy," and "Part II - A Constitutive Formulation and Verification," Trans. ASME, Journal of Engineering Materials and Technology, 100, 363-380 (1978).

19. Lee, D. and Zaverl, F., Jr., "A Generalized Strain Rate Dependent Constitutive Equation for Anisotropic Metals," Acta Metallurgica, 26, 1771-1780 (1978).

20. Miller, A.K., "An Inelastic Constitutive Model for Monotonic, Cyclic and Creep Deformation: Part I - Equations Development and Analytical Procedures," and Part II - Application to Type 304 Stainless Steel," Trans. ASME, Journal of Engineering Materials and Technology, 98, 97-113 (1976).

21. Chaboche, J.L., "Viscoplastic Constitutive Equations for the Description of Cyclic and Anisotropic Behavior of Metals," Bulletin de L'Academie Polonaise des Sciences, Série des Sciences Techniques, Vol.XXV, No.1, pp.33-43, 1977.

22. Krieg, R.D., Swearengen, J.C. and Rhode, R.W., "A Physically-Based Internal Variable Model for Rate-Dependent Plasticity," in "Inelastic Behavior of Pressure Vessel and Piping Components," PVP-PB-028, American Society of Mechanical Engineers, 1978, pp.15-28.

23. Walker, K.P., "Representation of Hastalloy-X Behavior at Elevated Temperature with a Functional Theory of Visco-plasticity," presented at ASME/PVP Century 2 Emerging Technology Conference, San Francisco, California, August 1980.

24. Liu, M.C.M. and Krempl, E., "A Uniaxial Viscoplastic Model Based on Total Strain and Overstress," J. Mech. Physics of Solids, 27, 363-375 (1979).

25. Krempl, E., "The Role of Servocontrolled Testing in the Development of the Theory of Viscoplasticity Based on Total Strain and Overstress," presented at ASTM Symposium on Mechanical Testing for Deformation Model Development, Bal Harbor, Florida, November 1980 and submitted for publication in the Proceedings.

26. Krempl, E., "Viscoplasticity Based on Total Strain. The Modeling of Creep with Special Considerations of Initial Strain and Aging," Trans. ASME, Journal of Engineering Materials and Technology, 101, 380-386 (1979).

27. McLean, M., "Friction Stress and Recovery During High-Temperature Creep: Interpretation of Creep Transients Following a Stress Reduction," Proc. R. Soc. Lond. A371, 279-294 (1980).

28. Davies, P.W., Nelmes, G., Williams, K.R. and Wilshire, B., "Stress Change Experiments During High-Temperature Creep of Copper, Iron and Zinc," Metal Science Journal, 7, 87-92 (1973).

29. Williams, K.R. and Wilshire, B., "On the Stress and Temperature Dependence of Creep of Nimonic 80A," Metal Science Journal, 7, 176-179 (1973).

30. Ahlquist, C.N. and Nix, W.D., "The Measurement of Internal Stresses During Creep of Al and Al-Mg Alloys," Acta Metallurgica, 19, 373-385 (1971).

31. Gittus, J.H., Ref.13, p.70 ff.

32. Krempl, E., forthcoming.

FACTORS CONTROLLING THE CREEP BEHAVIOR OF A NICKEL-BASE SUPERALLOY

D. D. Pearson, B. H. Kear and F. D. Lemkey

United Technologies Research Center

East Hartford, CT   06108, U.S.A.

SUMMARY

The temperature and orientation dependence of the creep properties of single crystals of a Ni-13Al-9Mo-2Ta (at.pct.) alloy have been investigated. The alloy exhibits substantial creep anisotropy over the temperature range studied (1033-1311K), with the <111> orientation exhibiting the lowest creep rate at all temperatures. The existence of precipitates in addition to the γ' below 1073K produces two distinct regimes of creep behavior. At high temperatures, the creep strength is controlled by the γ' morphology. In the <100> orientation, for example, the decisive factor is the formation of γ' platelets, or rafts, perpendicular to the applied stress axis during primary creep. This leads to a marked reduction in steady-state creep rate. Similar effects occur in other orientations, although the situation is complicated by the sensitivity of the γ' morphology to the direction and sense of the applied stress. At low temperatures an additional contribution to the creep strength is obtained from the impedance to dislocation motion by the formation of additional precipitates on a fine scale in the γ matrix. The precipitates form initially as metastable DO_{22} Ni_3Mo, which transform on aging to a mixture of Ni_4Mo and Ni_2Mo while retaining some coarsened DO_{22} Ni_3Mo.

1.0 INTRODUCTION

In a previous article[1], it was shown that the high temperature creep properties of single crystals of a Ni-13Al-9Mo-2Ta (at.pct.) alloy are strongly influenced by changes in the morphology of the γ' phase. In the <100> orientation, for example, the effect of 'rafting' of the γ' phase is to increase the creep rupture life by a factor of four. This was attributed to more effective impedance to matrix dislocation motion offered by plates or rafts of the γ' phase. This paper provides further documentation on this phenomenon, and attempts to define limits with respect to temperature and orientation. As will be shown, the benefits to be derived from γ' rafting are greatest in the high temperature range where diffusive slip in γ' becomes the rate controlling creep mechanism. At lower temperatures, fine-scale precipitation of Ni_xMo phases in the γ matrix provides an additional source of creep strengthening in the alloy. Thus, in this alloy there are two distinct regimes of creep, where different creep strengthening mechanisms are operative.

2. EXPERIMENTAL PROCEDURE

Single crystals of the alloy, Ni-13Al-9Mo-2Ta (at.pct.) were prepared by directional solidification using the seeded Bridgman technique. The residual microsegregation of the crystals due to dendritic growth was removed by a solution heat treatment of 16 hours at 1593K. Microprobe line traces confirmed that the crystals were completely homogenized. Some crystals were creep tested in this condition, which will be referred to hereafter as the solutioned/air cooled condition. Other crystals were given an additional heat treatment of 4 hours, 1353K plus 16 hours, 1144K which simulates a typical coating cycle heat treatment. This will be referred to as the standard heat treated condition. Creep specimens with raised ridges defining the gage length were prepared by low stress grinding. The extensometry consisted of alumina rods extending from clamps on the specimen with creep extension measured using paired LVDT's. Creep displacement was recorded continuously on a strip chart recorder. All tests were conducted in air using constant load machines with temperature controlled to $\pm$ 1K using Pt thermocouples.

3. RESULTS

3.1 Microstructural Observations

Thin foils were examined in TEM to characterize the structure resulting from the two heat treatments prior to creep testing. The solutioned/air cooled specimens consisted of a high volume fraction of fine γ' precipitates, Fig. 1a. Also produced on cooling was an extremely fine DO_{22} Ni_3Mo precipitate residing in the γ matrix. These were imaged using $\frac{1}{2}10$ reflections in the diffraction pattern, Fig. 1b, and imaged in dark field. Fig. 2a shows an example of the resulting microstructure after the additional standard heat treatment. The γ' has coarsened considerably and has developed a high density of misfit dislocations. This feature is unusual for superalloys because most are designed for minimal misfit to reduce the γ' coarsening rate. Also observed after this heat treatment was the additional DO_{22} Ni_3Mo precipitates, Fig. 2b. The misfit of the alloy was determined to be -0.78% from X-ray diffractometer studies of the 420 peak obtained from carefully aligned single crystal slices, Fig. 3. The sense of the misfit was also confirmed from TEM diffraction patterns in which the 040 spots were split with the γ' spot lying outside the γ spot indicating $a_o^\gamma > a_o^{\gamma'}$. The additional DO_{22} precipitates in the matrix were recognized as a potential source of strengthening and the temperature range of its occurrence was unexpectedly revealed during differential thermal expansion studies of the alloy. Using pure Ni as a standard, a sudden expansion of the alloy was observed on heating in the temperature range of 1045 to 1073K, Fig. 4. When specimens were subsequently aged at 1033K, the DO_{22} phase was observed to coarsen initially but within a few hours Ni_4Mo and Ni_2Mo were also present. This phase sequence occurs in binary Ni-Mo alloys[2,3] and has also been observed previously in Ni-Al-Mo alloys[4,5]. These phases were remarkably stable, as can be seen in the micrograph taken from an unstressed portion of a creep specimen tested for 1430 hours at 1033K, Fig. 5.

3.2 Creep Behavior

Results of creep tests on [100], [211] , [111] and [110] oriented specimens are listed in Table I. As previously noted [1], the alloy displays substantial creep anisotropy at elevated temperatures and the morphology of the γ' is

216

Fig.1(a) Bright Field

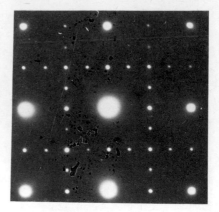

Fig.1(b) 100 D.P.

X60,000

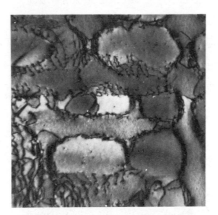

Fig.2(a) Bright Field X30,000

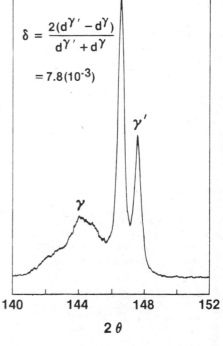

$$\delta = \frac{2(d^{\gamma'} - d^{\gamma})}{d^{\gamma'} + d^{\gamma}}$$

$$= 7.8(10^{-3})$$

γ'

γ

140   144   148   152

2θ

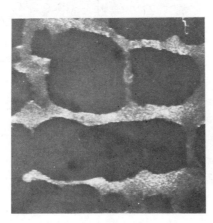

Fig.2(b) Dark Field
Showing DO$_{22}$ X30,000

Fig.3 Diffractometer Scan

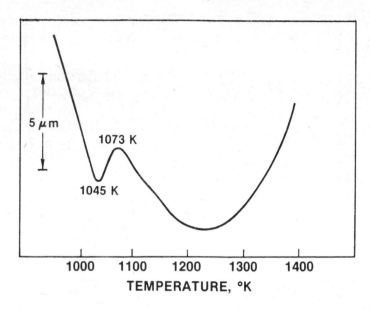

Fig.4 Differential Thermal Expansion Curve

Fig.5 Bright Field
(a)   X120,000

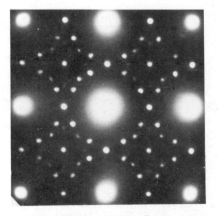

Fig.5(b) 100 D. P.

Fig.5(c) Dark Field
X120,000

218

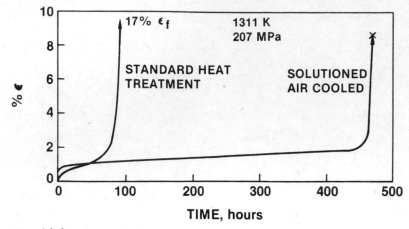

Fig.6(a)

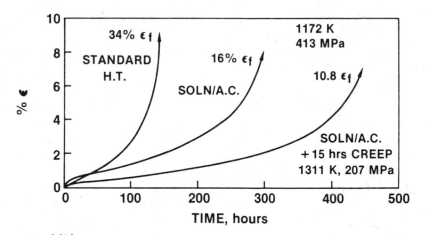

Fig.6(b)

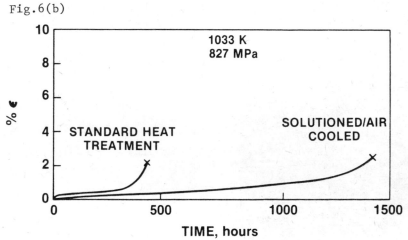

Fig.6(c) [100] Typical Creep Curves

important in determining creep behavior. This is clearly evident in the [100] orientation in which a platelet or rafted γ' structure develops in the solutioned/air cooled specimens during 1311K creep. Representative creep curves for [100] specimens for the two heat treatments at 1033, 1172 and 1311K are shown in Fig. 6. At each temperature the solutioned/air cooled specimens exhibit lower creep rates and correspondingly longer rupture lives. The increased margin in life over the standard heat treated conditions of 3 times at 1033K and 2 times at 1172K would be expected due to the finer γ' size. However, at 1311K the thermal coarsening rate of the γ' would mask any differences in the initial γ' size if the γ' coarsened isotropically. Instead, an increase in life of about 5 times is observed with a drop in the minimum creep rate of about 10 times. This is a consequence of the fine γ' rafts formed during creep of the solutioned/air cooled specimens, which are effective barriers to dislocation circumvention of the γ' by climb. Similar directional coarsening of the γ' occurred in the standard heat treated condition, but the rafts were not nearly as fine or as extensive laterally, as can be seen by comparing the two microstructures in Fig. 7. The effectiveness of the rafted γ' in retarding creep in the solutioned/air cooled specimens was made clear from the following experiement. It was observed that the rafting was much more extensive at 1311K than at 1172K. This suggested that benefits might be obtained at 1172K after "pre-rafting" at 1311K if γ' shearing was precluded at the lower temperature. The experiment had exactly the desired effect, as seen in Fig. 6b.

Improvements in the solutioned/air cooled specimens for [211] and [111] crystals also occur at 1172K, but only the [211] orientation shows an improvement in creep properties at 1311K. Platelet γ' was observed in the [211] crystal at 1311K, but no obvious directional coarsening of the γ' was observed in the [111] orientation. Instead, a more or less saw tooth γ' structure emerged as the γ' appeared to link together in [100] directions.

Comparison of the [100] creep properties with directionally solidified (D.S.) Mar-M200 indicates two distinct regimes of creep behavior, Fig. 8. These two areas are divided by the Ni_xMo solvus temperature, 1073K. Below this temperature, the alloy exhibits substantial stress capability over D.S. Mar-M200. However, at 1089K, 16K above the Ni_xMo solvus, no improvement is evident and the existence of the additional γ' precipitates at the lower temperature is most likely re-

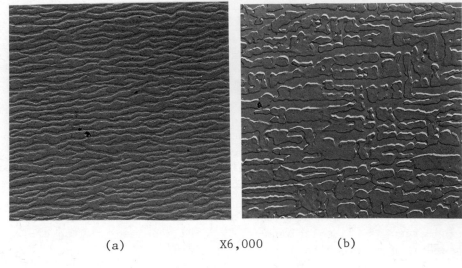

(a)   X6,000   (b)

Fig.7 Rafting of γ' (a) S/AC (b) SHT

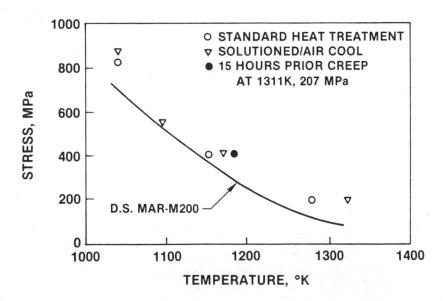

Fig.8 300 Hour Stress Rupture Strength Comparison

sponsible for the observed increased strength. Above 1088K
the alloy shows steady improvement in creep properties with
temperature reaching almost double the stress capability at
1311K. The high temperature improvement reflects retardation
of climb processes in the alloy.

4. DISCUSSION

A characteristic feature of the stress coarsened structure in this alloy is the presence of networks of a/2 [110]
dislocations at the γ/γ' interfaces. These dislocations are
true misfit dislocations, since their Burgers vectors are
appropriate for relaxing the internal stresses due to the
relatively large (-0.78 pct.) negative γ/γ' lattice misfit.
Typically the dislocations are composed of pure edge, or mixed
edge and screw segments, arranged in fairly regular networks[6].
Weatherly and Nicholson[7] have previously observed similar
networks at γ/γ' interfaces in a two-phase Ni-Al binary alloy
after prolonged aging to coarsen the γ' particles. It was
shown that the interface dislocations are a consequence of
dislocation multiplication from relatively few sources.   The
driving force is the internal stress associated with the γ/γ'
misfit. Thus, dislocations are pulled out of sub-boundaries
and then proceed to wrap around the γ' particles until the
equilibrium array of misfit dislocations is established.
Something similar to this must happen in the Ni-Al-Mo-Ta
alloy, because networks of γ/γ' interface dislocations are
observed only in the heat treated alloy. The solutioned/
air cooled alloy contains fine γ' particles and very few dislocations.

Tien and Gamble[8] have shown that changes in the γ' morphology due to stress annealing produce significant changes
in the yielding behavior of [100] oriented U-700 crystals,
at least at temperatures below 1033K. The formation of γ'
rods or plates increases the yield strength over that of
γ' cuboids, with γ' plates producing the largest effect. The
present results show that similar stress-induced changes in
γ' morphology in the Ni-Al-Mo-Ta alloy can also exert a strong
influence on creep behavior at high temperatures. Thus, in
the [100] orientation, the formation of γ' plates, or 'rafts'
in an orientation perpendicular to the applied tensile stress
causes a marked reduction in the creep rate. Such behavior
evidently reflects fundamental changes in the rate controlling
creep mechanism in the alloy. A plausible explanation follows

from the known behavior of conventional γ' precipitation hardened nickel-base alloys[9,10]. Under low stresses at high temperatures, the operative creep mechanism in conventional γ' strengthened alloys involves dislocation motion primarily in the γ matrix, with the mobile dislocations circumventing, or by-passing the γ' particles by combined glide and climb processes. On the other hand, under high stresses at high temperatures, γ' particle by-passing tends to be replaced by γ' particle shearing as the prevailing creep mode. This involves some rather complex dislocation motions, including diffusive slip of loosely coupled superlattice dislocation pairs, Fig. 9. As indicated, the mobility of such dislocations is controlled by the climb rate of the coupling antiphase boundaries (APB's), irrespective of whether they are of edge, screw or mixed character[11,12]. Such diffusive, or viscous slip processes in γ' tend to be more sluggish than the corresponding processes in γ, due to the lower diffusivities in the ordered phase. Thus, the observed reduction in creep rate due to γ' rafting may be attributed to the suppression of γ' particle by-passing as the effective rate controlling creep mechanism, and its replacement by the more sluggish process of diffusive slip in the γ' phase.

Another interesting aspect of diffusive slip in γ' is the possible influence of alloying on dislocation mobility. As shown in Fig. 10, solute segregation to the diffusively roughened APB's could increase the dislocation drag stress. A plausible explanation is the reduction in the number of thermal vacancies associated with the APB's. Such an effect should increase with the solute content in the γ' phase, which perhaps explains the creep strength maximum, corresponding with saturation of this[13] and a related[14] alloy in Mo; at saturation the γ' phase contains about 5 at.pct. Mo. This segregation model seems reasonable because diffusively roughened APB's are known to be favorable sites for phase transformations in ordered alloys[15].

The effectiveness of γ' rafting in inhibiting matrix dislocation motion will depend on the perfection of the γ' platelet structure. In an ideal situation, with γ' platelets extending from one side of the crystal to the other in a perfectly regular manner, by-passing of the γ' phase by matrix dislocation motion is precluded. Thus, a threshold stress for significant creep can be defined, which is the stress required to extract dislocation sources from the γ/γ'

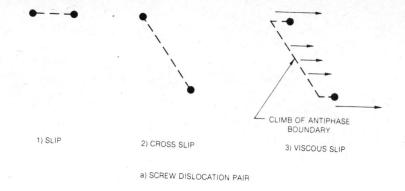

1) SLIP 2) CROSS SLIP 3) VISCOUS SLIP

CLIMB OF ANTIPHASE BOUNDARY

a) SCREW DISLOCATION PAIR

1) SLIP 2) CLIMB 3) VISCOUS SLIP

CLIMB OF ANTIPHASE BOUNDARY

b) EDGE DISLOCATION PAIR

Fig. 9 – Schematic contrasting slip and viscous slip motions of APB-coupled dislocation pairs in the γ' phase[11,12]. Note that viscous slip involves climb of APB's, irrespective of dislocation character.

DIFFUSIVELY ROUGHENED APB

SOLUTE SEGREGATION

a) LOW DRAG STRESS b) HIGH DRAG STRESS

Fig. 10 – Schematic illustrating possible influence of solute segregation on dislocation solubility in γ'.

interfacial dislocation networks and push them through the
γ' phase. In this model, the creep rate is determined primarily by the nucleation rate for dislocation sources, the
mean slip distance in the γ' phase, and the average dislocation velocity.

At a given temperature, stress is the driving force controlling the nucleation and propagation of dislocation pairs
into and through the γ' phase. The sources for γ' slip
probably have their origin in interactions between mobile
dislocations in the γ matrix and the pre-existing array of
γ/γ' interface dislocations. Chance encounters between
mobile dislocations and interface dislocations form small
segments of appropriately paired dislocations having the same
Burgers vector. Under a sufficiently high stress, the paired
segments will become active sources for propagating slip into the γ' phase. After passing through the γ', the dislocation pairs should be absorbed by the networks on the
other side of the γ' particles, since these dislocations are
of opposite sign. Thus, it can be seen that each nucleation
event for γ' slip is unique, and that there is little tendency for propagation of mobile dislocations from one γ'
lamella to the next adjacent one in the fully rafted structure. In other words, the mean slip distance for mobile
dislocations tends to be the width of a single γ' plate.
With regard to the stress required to activate sources at
the γ/γ' interfaces, this appears to be determined primarily
by the magnitude of the APB energy, and the degree of γ/γ'
misfit. The former decreases rapidly with temperature in
the high temperature range, due to interactions with thermal
vacancies. The latter, on the other hand, increases with
temperature, i.e. the γ/γ' misfit becomes more negative,
due to differences in the thermal expansion of the two
phases. It follows that the higher the temperature the more
dominant becomes the γ/γ' misfit interaction stress in controlling the source activation stress.

Several studies[16-18] have shown that stress annealing
of nickel base alloys can produce significant changes in the
morphology of the γ' phase. Fig.11 presents a summary of
the principal findings of an investigation performed on U-
700 crystals[16]. In <100> orientations, both γ' plates
and γ' rods are observed depending on the sense of the
applied stress. Thus, tensile annealing produces plates,
whereas compressive annealing yields rods. In <110> orientations, the effect of stress annealing is the opposite of

that in < 100> orientations, in that rods are formed in tension and plates in compression, albeit in different orientations with respect to the stress axis. In <111> orientations,
the γ' particles remain virtually unchanged in shape during
stress annealing, irrespective of the sense of the applied
stress. The present observations are in agreement with Fig
11, except for minor differences in the response of the
<111> orientation to stress annealing. In this orientation,
all three favorably oriented {100} plate orientations develop
to some extent in different regions of the crystal, such
that the general appearance of the structure is that of
a three-dimensional inter-connecting network of γ' plates.
In the nearby <211> orientation, only one plate orientation
is observed in tensile annealing, presumably because the
shift in the stress axis away from the symmetrical < 111 >
axis is sufficient to favor one plate orientation over the
two. The normal to the favored (100) plate orientation is
inclined at 35° to the [211] tensile axis. A comparison
of observations from various studies[16-18] shows that the
sign of the γ/γ' lattice misfit also exerts a strong influence on stress coarsening behavior, in agreement with theoretical predictions[19]. Thus, alloys with negative misfit
(here defined as $a_o^{\gamma} > a_o^{\gamma'}$) respond to stress annealing in the
manner depicted in Fig. 11, whereas alloys with positive
misfit ($a_o^{\gamma} < a_o^{\gamma'}$) show just the opposite effect. Note that
an apparent contradiction in the data presently exists
for U-700. In this case, a small negative misfit ($a_o^{\gamma} > a_o^{\gamma'}$)
is required for a consistent picture, whereas in fact a
small positive misfit is claimed for this alloy. Perhaps
there is a change in the sign of the misfit with increasing
temperature, due to differential thermal expansion of the
two phases, in which case the stated correlation is upheld.
Another parameter of some significance in stress coarsening
of γ' is the number of active glide systems[17]. Thus,
tensile annealing in < 103 > and < 115 > orientations of a positive misfit alloy produces plates lying parallel to the
stress axis, rather than rods, due to interactions between
glide dislocations and the growing γ' particles. Such behavior has not been found in this study, apparently because
the observations were restricted to high symmetry orientations. However, it has been found that < 1.0 percent creep
strain is associated with the formation of a fully rafted,
or plate-like structure in both <100> and <112> orientations.
This deformation presumably enhances the kinetics of the γ'

226

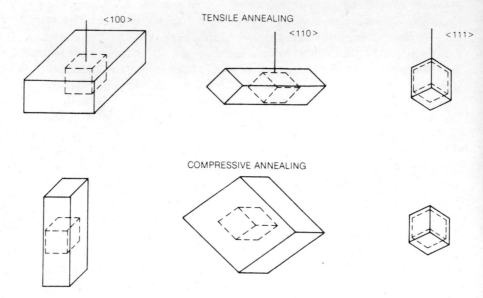

Fig. 11 – Schematic representation of changes in stress annealed shape of γ' precipitates for different directions and sense of the applied stress (after Tien and Copley[16])

coarsening process. A puzzling aspect of γ' plate formation
is that the observed plate orientations are those in which
the misfit dislocations can be viewed as the products of reactions involving mobile dislocation segments in the active
glide systems. For example, in the [211] orientation, hexagonal edge dislocation networks formed by reactions between
glide dislocations in two active systems are appropriate misfit dislocations for the specific (100) plates developed in
this orientation. The significance of this correlation is
not yet understood. However, it does seem reasonable that
the formation of Lomer-Cottrell sessile dislocations by junction reactions between mobile dislocations will stabilize
the networks, and in turn, these networks will stabilize the
γ/γ' interfaces.

The observed changes in γ' morphology with stress direction provide a simple basis for explaining the orientaion
dependence of the creep properties. Thus, both <100> and
<211> orientations are unusually strong in creep because
the formation of γ' rafts early in the creep process effectively impedes matrix dislocation motion. On the other
hand, the <110> orientation is much weaker in creep because
of its inability to develop plates, at least in tensile
creep. In this case, only γ' rods are formed, which are much
less effective obstacles to matrix dislocation motion. In
fact, matrix deformation can be sustained, provided that
the applied stress is capable of pushing the dislocations
between the γ' rods. The <111> orientation is something
of an anomaly. This orientation is strong in creep, despite
the fact that it does not develop a regular γ' rafted structure. However, as noted earlier, the appearance of the
stress annealed structure is that of a three-dimensional inter-
connecting network of small γ' plates. Perhaps, this represents sufficient continuity in the γ' phase to make it difficult for matrix dislocations to circumvent the γ' particles
by glide and climb processes, so that diffusive slip through
the γ' phase becomes the rate controlling creep mechanism.

Previous studies of Mar-M200[20], U-700[21], Mar-M247[22]
and Re-doped Mar-M200[23] crystals have shown a strong dependence of creep strength on crystallographic orientation,
at least for temperatures below 1255K. In general, crystals
with <100> and <111> orientations display higher creep strengths
than other orientations in the stereographic triangle. The

<110> orientation appears to be a special case in that it is
much weaker than the <100> orientation in tensile creep[20],
but can be somewhat stronger in compressive creep[21]. Whatever the circumstances, however, it seems clear that the <111>
orientation is the most creep resistant orientation in
nickel base alloys. This is easily understood in terms of
the low Schmid factor for {111} a/2 <110> slip, and the
strong work hardening due to interactions between mobile dislocations gliding in several intersecting systems[20]. At
temperatures above 1255K, the creep anisotropy in single
crystals is much reduced, apparently because of the onset
of slip in other systems, e.g. {100} a/2 <110> slip in the
<111> orientation. Since thermally activated deformation
processes, such as dislocation cross slip and climb, are
so facile at such high temperature, perhaps a more realistic
way of viewing the deformation process is that of dislocations
with unique a/2 <110> Burgers vector following paths of essentially maximum driving force, with respect to both glide and
climb. Contours of constant Schmid factor, assuming that
mobile a/2 <110> dislocations follow paths of maximum shear
stress show little variation with orientation. A small
benefit in Schmid factor is realized in orientations approaching <111> , but obviously much less than that assuming
only {111} a/2 <110> slip. This explains the relatively
small advantage in yielding and creep behavior observed in
<111> orientations at high temperatures in conventional
alloys. In contrast, the present alloy shows a persistence
in creep anisotropy up to at least 1311K and possibly even
to 1376K. As emphasized in the preceding discussion, this
may be attributed to the influence of changes in the γ'
morphology on the creep properties of the alloy, which in
specific orientations precludes γ matrix deformation, independent of γ' precipitate deformation. In the absence of
γ' rafting, or some other form of interconnected γ' phase,
the creep strength should be more or less independent of
orientation, as in the conventional alloys.

It has been noted that at temperatures below 1033K, the
matrix phase of the Ni-Al-Mo-Ta alloy precipitates metastable γ''-Ni_3Mo (DO_{22} structure). When the alloy is in the
pre-rafted condition, precipitation of γ'' occurs primarily
on the planar γ/γ' interfaces. The DO_{22}-γ'' phase is coherent
with the $L1_2$-γ' phase, and its c-axis is normal to the cube
axis of the γ' plates. In other words, decoration of the
γ/γ' interfaces in the rafted structure involves, at least

initially, fine scale precipitation of the one crystallographic variant of γ'' that offers the best lattice match
with the γ' phase. Upon further aging, the γ'' phase decom-
poses into the equilibrium phase mixture of Ni_4Mo and Ni_2Mo.
The initial precipitation of γ'' is accompanied by a predict-
able elongation of the specimen during cooling down under
a tensile load through the transformation temperature range.
In the conventionally heat treated alloys, containing a uniform distribtuion of γ' cuboids, precipitation of metastable
γ'' occurs in three different crystallographic variants,
coresponding with the best lattice match on three different
cube faces of the γ' particles. Such a specific orientation
relationship between γ' and γ'' has previously been reported
for INCONEL 718 alloys[24,25]. In certain compositions[25]
a 'compact' γ'/γ'' morphology has been observed, such that
the γ' cuboids are surrounded on all sides by thin γ''
platelets in appropriate matching orientations. The present
case seems to be somewhat unique in that the effect has
been observed in a high γ' volume fraction (~75 pct.) alloy.
The INCONEL-type alloys typically contain about 10-15 pct.
γ'. Another difference is that phase instability in this
alloy is due to supersaturation of the γ matrix with respect
to Mo, as opposed to Nb in the INCONEL-type alloys. In solu-
tioned/air cooled alloys, the effects of aging at low temperatures are complex and not yet properly understood. However, it appears that the fine multi-phase Ni_xMo structures
developed are quite resistant to coarsening up to 1033K.

With regard to the influence of these low temperature
phase transformations on creep properties, the most significant finding has been the strong dependence of properties
on the scale of the microstructures. The very fine scale
microstructures developed in quenched alloys exhibit the
highest creep strengths. For example, in the <100> orientation at 1033K solutioned/air cooled specimens exhibit 100
hr. rupture lives at 140 ksi. Under the same conditions
Mar-M200 crystals exhibit 100hr rupture lives at 100 ksi.
Such dramatic improvements in creep strength, however, are
not found in the coarser structures, produced by prior stress
annealing and conventional aging treatments. On the contrary,
in these cases the creep strengths are comparable with those
of Mar-M200. While a qualitative explanation for this fine
scale structural effect on creep properties can be given
in terms of available theories of γ' and γ'' particle
strengthening in nickel base alloys it seems prudent to de-
fer this discussion until tests have been performed on more

stable low temperature microstructures. To this end,
attempts are now being made through compositional modifications to preserve the γ'' phase indefinitely. It should be
possible then to determine the influences of various γ' and
γ'/γ'' morphologies on low temperature strengthening of
these alloys, without complications due to unwanted phase
instabilities.

CONCLUSIONS

1. Homogeneous crystals of the nickel-base alloy Ni-
13Al-9Mo-2Ta (at.pct.) exhibit two distinct regimes of creep
behavior at elevated temperatures. At temperatures below
1073K, the alloy exhibits about a 20 pct. improvement in
stress capability over directionally solidified Mar-M200.
At temperatures above 1073K, the alloy shows an increase
in stress capability with temperature reaching almost twice
that of Mar-M200 at 1311K.

2. The striking differences in creep behavior in
these two regimes are due to fundamental differences in the
operative creep mechanisms. In the high temperature regime,
the decisive factor is the change in the γ' morphology that
occurs under stress, whereas in the low temperature regime,
it is the fine scale precipitation of additional Ni_xMo
phases.

3. In certain orientations, for example <100> and
<211>, the development of continuous plates or rafts of the
γ' phase under stress is responsible for the high temperature creep strengthening effect. The γ' rafting is particularly effective in inhibiting γ matrix deformation, independent of γ' deformation. Thus, the improvement in high
temperature creep behavior is explained in terms of the
difficulties associated with the nucleation and propagation
of diffusively roughened APB-coupled dislocation pairs into
and through the γ' plates.

4. A large negative γ/γ' misfit in this alloy provides
a strong driving force for the formation of γ/γ' interfacial dislocation networks. These networks develop upon
high temperature exposures, with or without the benefit of
an external applied stress. In agreement with previous
work the effect of stress is to create specific γ' morphologies and associated interfacial dislocation networks, depending on the direction and sense of the applied stress.

Table I

Single Crystal Creep Results

Orientation	Heat Treatment	Temp. $^\circ$K	Stress MPa	$\dot{\varepsilon}$ min. $(10^{-5})hr^{-1}$	Rupture Life Hours	Elongation %
[100]	S/AC	1311	207	1.2	458	9.0
	SHT	1311	207	13.0	88	17.0
	S/AC	1172	413	10.8	309	16.0
	SHT	1172	413	20.0	153.2	34.0
	S/AC*	1172	413	5.2	470	10.8
	S/AC	1089	551	2.0	255	15.7
	S/AC	1033	827	0.28	1430	2.2
	SHT	1033	827	0.88	419	3.7
[211]	S/AC	1311	207	0.34	908	3.5
	SHT	1311	207	3.20	230	3.4
	S/AC	1172	413	5.30	341	25.0
	SHT	1172	413	12.0	189	24.2
[111]	S/AC	1311	207	0.47	**	--
	SHT	1311	207	0.50	> 1000	3.5
	S/AC	1172	413	4.40	736.9	10.0
	SHT	1172	413	7.00	418	14.1
[110]	S/AC	1311	207	2.50	125	5.0
	S/AC	1172	413	47.00	18.3	2.0

*   + 15 hrs prior creep 1311K/207MPa
**   Test stopped at 350 hours.

REFERENCES

1. PEARSON, D. D., LEMKEY, F. D. and KEAR, B. H. - Proc. 4th Int. Sym. on Superalloys, ASM, Eds. J. K. Tien et al, 1980, p. 513.

2. OKAMOTO, P. R. and THOMAS, G. - Acta Met. 1971, 19, 825.

3. VAN-TENDELOO, G., DERIDDER, R. and AMELINCKX, S. - Phys. Stat. Sol., 1975, 27, 457.

4. SNOW, D. B. - Proc. 38th Annual Meeting EMSA, Claitor's Publ. Div., Baton Rouge, 1980, p. 334.

5. KERSKER, M. M., AIGELTINGER, E. A. and HREN, J. J. - Proc. 38th Annual Meeting EMSA, Claitor's Publ. Div., Baton Rouge, 1980, p. 158.

6. FRASER, H. L. and PEARSON, D. D. - to be published.

7. WEATHERLY, G. C. and NICHOLSON, R. B. - Phil. Mag. 1968, 17, 801.

8. TIEN, J. K. and GAMBLE, R. P. - Met. Trans. 1972, 3, 2157.

9. CARRY, C. and STRUDEL, J. L. - Acta Met. 1977, 25, 767; 1978, 26, 859.

10. LEVERANT, G. R., KEAR, B. H. and OBLAK, J. M. - Met. Trans. 1973, 4, 355.

11. FLINN, P. A. - Trans. Met. Soc. AIME, 1960, 218, 145.

12. KEAR, B. H. and OBLAK, J. M. - J. DePhysique Colloque C-7, 1974, p. 35.

13. PEARSON, D. D. - private communication.

14. MAXWELL, D. - U.S. Patent 3,617,397, Nov. 2, 1971.

15. LEAMY, H. J. SCHWELLINGER, P. and WARLIMONT, H. - Acta Met. 1970, 18, 31.

16. TIEN, J. K. and COPLEY, S. M. - Met. Trans 1971, 2, 543.

17. CARRY, C. and STRUDEL, J. L. - ICSMA4 , Ed. Laboratoire de Physique du Solide, Nancy, France, 1976, 1, 324.

18. MIYAZAKI, T., NAKAMURA, K, and MORI, H. - J. Mats. Sci. 1979, 14, 1827.

19. PINEAU, A. - Acta Met. 1976, 24, 559.

20. KEAR, B. H. and PIEARCEY, B. J. - Trans. TMS-AIME, 1967, 239, 1209.

21. STRUTT, P. R., KHOBAIB, M., POLVANI, R.S. and KEAR, B. H. - ICSMA4, Ed. Laboratoire de Physique du Solide, 1976, 1, 314.

22. MACKAY, R. A., DRESHFIELD, R. L. and MAIER, R. D. - Proc. 4th Int. Symp. on Superalloys, ASM, Eds. J. K. Tien et al, 1980, p. 385.

23. LIN, L. S., GIAMEI, A. F. and DOIRON, R. E. - Proc. 38th Annual Meeting EMSA, Claitor's Publ. Div., Baton Rouge, 1980, p. 330.

24. PAULONIS, D. F., OBLAK, J. M. and DUVALL, D. S. - Trans. ASM, 1969, 62, 611.

25. COZAR, R. and PINEAU, A. - Met. Trans. 1973, 4, 47.

SECTION 3
CREEP FRACTURE PROCESSES

A UNIFYING VIEW OF THE KINETICS OF CREEP CAVITY GROWTH

B.F. Dyson.

Division of Materials Applications,
National Physical Laboratory, Teddington, TW11 0LW, England.

SUMMARY

The last five years have witnessed considerable theoretical activity in the area of creep cavity growth. Two quite distinct micromechanisms are now thought to occur. One, continuum growth, involves matter transport by time-dependent plastic deformation. The other, growth by stress-directed, vacancy absorption, involves atom transport by diffusion to sinks at suitably stressed grain boundaries. It is the kinetics of the latter mechanism which have caused some confusion, since assumptions of earlier models, which lead to diffusion-controlled growth, are inapplicable to polycrystalline metals. Diffusion-controlled growth is now seen as occurring, at best, only over a very restricted stress/temperature range. At higher stresses, growth rate is enhanced by power-law creep, while at lower stresses, it is slowed down by constraints imposed by grain geometry or grain boundary particles. Some growth rate measurements are presented which are in agreement with this unified view.

1. NOMENCLATURE

r	cavity radius	D_B	grain boundary diffusivity
λ	cavity spacing	T	absolute temperature (K)
$\dot{V}$	volumetric cavity growth rate	γ	surface energy
		σ	uniaxial stress
d	grain size	$\sigma_1 \sigma_2 \sigma_3$	principal stresses
L	particle spacing		
p	particle diameter	σ_m	mean stress $\dfrac{\sigma_1+\sigma_2+\sigma_3}{3}$
z	denuded zone width		

$\bar{\sigma}$ effective stress $\dfrac{1}{\sqrt{2}}\left[(\sigma_1-\sigma_2)^2 + (\sigma_2-\sigma_3)^2 + (\sigma_3-\sigma_1)^2\right]^{\frac{1}{2}}$

$\bar{\varepsilon}$ effective strain $\dfrac{\sqrt{2}}{3}\left[(\varepsilon_1-\varepsilon_2)^2 + (\varepsilon_2-\varepsilon_3)^2 + (\varepsilon_3-\varepsilon_1)^2\right]^{\frac{1}{2}}$

2. INTRODUCTION

There are only a few measurements of creep cavity growth rates reported in the literature [1-3], even though such data are vital for a complete understanding of cavitation fracture. The reasons are easy to understand; apart from the intrinsic difficulties of making these measurements, interpretation is ambiguous due to continuous cavity nucleation occurring in most materials of interest. In contrast, *theories* of growth are in profusion and urgently require experimental validation, although recent papers and reviews have provided a reasonably unified picture.[4-7] This paper mainly emphasises those growth models which the author believes are of most interest to the *users* of engineering materials and provides evidence for this view with some recent measurements of growth rates made without the complications of continuous nucleation.

3. CAVITY GROWTH MODELS

Holes in solids can grow by two quite distinct micro-mechanisms viz:-

3.1 Continuum growth

Cavity volume can increase through the action of the hydrostatic component of stress changing the direction of matter flow during plastic deformation in the continuum adjacent to the cavity. This mechanism was introduced first by McClintock [8], followed by Rice and Tracey [9], to account for hole growth during *time-independent* plastic deformation. For a rigid plastic matrix Rice and Tracey [9] showed that:-

$$\dot{r}_A = r_A B' \dot{\bar{\varepsilon}} \tag{1}$$

where $\dot{r}_A/r_A$ is the specific rate of *dilatory* extension in the direction of principal axis A of a remote uniform effective strain rate field $\dot{\bar{\varepsilon}}$. B' represents the stress-state sensitivity of this process and is given by:-

$$B' = 0.56 \sinh\left[\frac{3\sigma_m}{2\bar{\sigma}}\right] \tag{2}$$

Hellan [10] later showed that equations similar to (1) exist for strain-rate-sensitive deformation, at least for ratios of $\sigma_m/\bar{\sigma}$ greater than approximately unity. In the particular case of power-law creeping solids, Hellan found that B' is additionally a function of the stress exponent n, although an analytical solution was not found. Recent papers [7,11] have analysed the case most appropriate to creep fracture, in which cavities are arranged only at grain boundaries while the surrounding matrix deforms by power-law creep. Needleman and Rice [11] computed growth rates by finite element methods while Cocks and Ashby [7] presented an approximate analytical solution:-

$$\dot{r} = \frac{\beta\lambda^2}{8r} \left[\left[1 - \left(\frac{2r}{\lambda} \right)^2 \right]^{-n} - \left[1 - \left(\frac{2r}{\lambda} \right)^2 \right] \right] \dot{\epsilon} \tag{3}$$

which, for the usual case where $(2r/\lambda)^2 \ll 1$, reduces to:-

$$\dot{r} = \frac{\beta(n+1)}{2} r\dot{\epsilon} \tag{4}$$

β is similar in form to B'of (2) and, for a material with $n = 5$ under uniaxial tension, has a value of 0.57 and therefore:-

$$\dot{r} = 1.7 \, r\dot{\epsilon} \tag{5}$$

compared with Rice and Tracey's expression (from (1)) of:-

$$\dot{r} = 0.29 \, r\dot{\epsilon} \tag{6}$$

Dyson and Taplin [12] used the expression:-

$$\dot{r} = r\dot{\epsilon} \tag{7}$$

due to Hancock [13] to calculate the values of $2r/\lambda$ at the transition between diffusion controlled vacancy growth (section 3.2) and continuum growth as a function of temperature and strain rate. Typical values of volume and grain boundary diffusivity were used, which are close to the values recently recommended by Brown and Ashby [14] from a statistical analysis of published data. Assuming a cavity spacing of 10 μm, the resulting transitional loci for various strain rates are shown in Fig. 1. Most engineering materials of interest are used at temperatures greater than 0.5 T_m and inspection of Fig. 1 reveals that continuum growth will be unimportant at strain rates less than 10^{-4} s^{-1}, given that fracture intervenes at about $2r/\lambda = 0.7$. The effect of imposing a highly tensile triaxial stress state (similar to that around the tip of a crack) will reduce this value by only one or two orders of magnitude. It is concluded therefore that this micromechanism is of no importance in steady load service applications where average strain rates of 10^{-9}–10^{-10} s^{-1} are required. Its importance under transient conditions due to thermal loadings imposed by 'start-ups' and 'shut-downs' where high strain rates occur for short times, is not so clear and specific judgements are then needed.

3.2 Stress-directed, vacancy absorption

Growth by this mechanism is achieved by the cavities acting as vacancy sinks. If the cavity is of radius r and a perfect vacancy sink and σ_n is the normal stress on a perfect vacancy source, then the only restriction on growth by vacancy

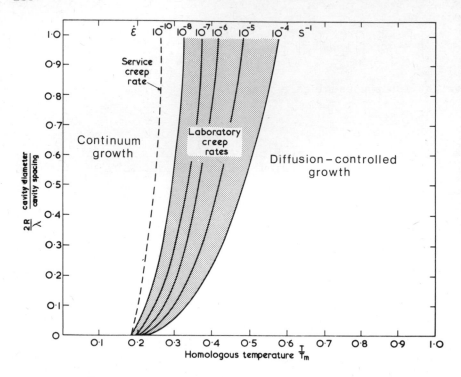

Fig. 1: Showing the strain-rate dependent transition loci
between continuum hole growth and diffusion
controlled growth.   Taken from Ref (12).

absorption is that:-

$$r > \frac{2\gamma}{\sigma_n}$$

Grain boundaries are commonly regarded as being perfect vacancy sources and so cavities on boundaries transverse to an
applied tensile stress will be most stable.   Fig. 2(a) illustrates schematically a set of cavities placed, by some means,
on boundaries of all orientations.   If the specimen were
crept under a stress σ_1, as in Fig. 2(b), then only those
cavities on boundaries normal to σ_1 would grow, likewise in
Fig. 2(c) only those on boundaries normal to σ_2 grow.   This
directional character of cavity growth by vacancy absorption
is in marked contrast to the scalar nature of growth by the
continuum mechanism which results in cavities growing equally
on all boundaries.   That this is not the case has been demonstrated with the following experiments.   Small cavities were
introduced onto all boundaries of Nimonic 80A by prestraining
at room temperature and then heating to 750 °C for a short
time [15].   Creeping one specimen in torsion in a clockwise
manner and creeping another in an anticlockwise manner resulted
in the cavity distributions shown in Fig. 3(a) and (b).   This

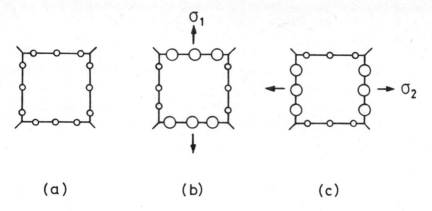

Fig. 2: (a) shows a uniform array of grain boundary holes
which grow by stress-directed vacancy absorption only
on those boundaries subjected to a tensile stress
(b) σ_1 or (c) σ_2.

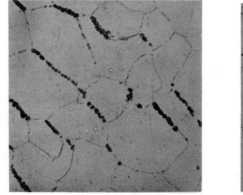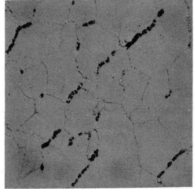

Fig. 3: (a) The creep tensile axis of the torsional stress
state is perpendicular to the cavitated boundary.
(b) The tensile axis has been rotated 90°.
Compare with the schematic in Fig. 2.

procedure overcomes the usual objections raised by continuous
nucleation and the results are a striking confirmation of
growth by vacancy absorption.

Until quite recently, the theoretical effort in analysing
cavity growth *kinetics* by vacancy absorption was directed towards growth being controlled by volume, grain boundary or
surface diffusion [16-21].   There are three principal

I

assumptions in these models, in addition to an assumed constancy of cavity number density, viz:-

(a) that the grains do not deform plastically.

(b) that the grain boundaries are perfect vacancy sources and the cavities, perfect vacancy sinks.

(c) that the *load* on each cavitated grain facet remains constant and equal to the applied load.

When surface diffusion is fast compared with grain boundary or volume diffusion, these assumptions lead to an equation for the volumetric cavity growth rate $\dot{V}_D$, given by:-

$$\dot{V}_D = A\left[\sigma_1 - \frac{2\gamma}{r}\right] \tag{8}$$

where $A = A(\lambda, T)$ but is microstructure insensitive.

Relaxing assumption (a) above [11, 22, 23] leads to the more realistic case of cavities growing in a power-law creeping matrix, where, to a first approximation, the volumetric cavity growth rate, $\dot{V}_{c/D}$, is given by:-

$$\dot{V}_{c/D} = \dot{V}_D + \dot{V}_c \tag{9}$$

where $\dot{V}_c$ is the volumetric equivalent of (7)

ie $\quad \dot{V}_c = 3V\dot{\bar{\varepsilon}} \tag{10}$

The important point to note about coupled diffusion and continuum growth is that it leads to shorter times to fracture than diffusion alone.

The perfect vacancy source/sink capability of grain boundaries has been contentious for some time but without much objective evidence being presented. There seems little doubt however that particles inhibit diffusion creep rates [24] and Harris [25] suggested that they might inhibit cavity growth rates as well. Harris postulated that the interface between a grain boundary particle and the matrix does not behave as a vacancy source/sink but only considered the case of a plastic matrix. Dyson [4] relaxed this to the more realistic case of a power-law creeping matrix. The physical situation is illustrated in Fig. 4(a), where a representative section of a boundary between two cavities with a spacing λ contains particles of diameter p and spacing L. As cavities grow, they deposit matter of thickness z. Since atoms cannot be plated on the particles, stress redistribution must occur to maintain compatibility and this is modelled with the mechanical 2-bar analogue shown in Fig 4(b). Below a transition stress given in [4], the particle-inhibited volumetric growth rate, $\dot{V}_p$, is given by:-

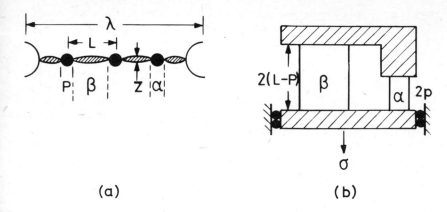

Fig. 4: (a) Physical model of particle-inhibition of growth.
       (b) Mechanical analogue: α represents region of
           stress enhancement around particle: β represents
           region of stress attenuation around boundary.

$$\dot{V}_p = (L-p) \ \lambda^2 \ (1-x) \ \dot{\varepsilon}_L \tag{11}$$

where $x = \left(\dfrac{p}{L}\right)^2$ is the particle area fraction and $\dot{\varepsilon}_L$ the local
creep rate whose upper bound value is given by:-

$$\dot{\varepsilon}_L = \frac{p}{L-p} \ B \ \left[\frac{\sigma}{x}\right]^n \tag{12}$$

$B\sigma^n$ is the dislocation creep rate in remote areas. Combining
(11) with (12) gives:-

$$\dot{V}_p = \frac{p\lambda^2(1-x)}{x^n} \ B\sigma^n \tag{13}$$

Thus, in contrast to diffusion-controlled growth, $\dot{V}_p$ is a power
function of stress.

For values of $x < \left[\dfrac{p}{d}\right]^{1/n}$, $\dot{\varepsilon} \simeq \dfrac{L-p}{d} \ \dot{\varepsilon}_L$

where $\dot{\varepsilon}$ is the specimen strain rate and therefore:-

$$\dot{V}_p \simeq d \ \lambda^2(1-x) \ \dot{\varepsilon} \tag{14}$$

For values of $x > \left[\dfrac{p}{d}\right]^{1/n}$, $\dfrac{L-p}{d} \ \dot{\varepsilon}_L < \dot{\varepsilon} \simeq B\sigma^n$

$$\therefore \dot{V}_p \simeq \frac{p\lambda^2(1-x)}{x^n} \dot{\varepsilon} \tag{15}$$

Even without the presence of particles on grain boundaries, diffusion-controlled growth will be slowed down, again below some transition stress [4], due to compatibility requirements imposed by a non-uniform distribution of cavities. Essentially, the assumption of constancy of load on a cavitated facet is now relaxed. A number of analyses have now been published [4, 7, 26-29] and it is clear that there is some confusion over whether the inhomogeneous distribution is in terms of cavitated boundaries or cavitated grains. Originally [26], it was thought that if every grain were cavitated, then grain boundary sliding could accommodate the dilation displacements and therefore no geometric constraint would occur. Now Raj [28] has shown that this is only true for the regular two-dimensional hexagonal grain structure used by Dyson [26]. For all 'normal' grain geometries, grain boundary sliding *cannot* accommodate dilation displacements and therefore creep is always a compatibility requirement and constrained growth will occur even when every grain is cavitated. Figs 5(a) and (b) illustrate the physical situation when all grains are cavitated. Fig 5(a) is a plane section through the grain structure and Fig 5(b), a plan view of a cavitated facet. The mechanical 2-bar analogue is shown in Fig 5(c). Below a transition stress [4, 7], the upper bound to geometrically constrained cavity growth rate, $\dot{V}_g$ will be given by:-

$$\dot{V}_g = \lambda^2 d B \left[\frac{\sigma}{1-A_c} \right]^n \tag{16}$$

where A_c is the area fraction of cavitated boundaries.

$$\text{Or:-} \quad \dot{V}_g = \lambda^2 d \dot{\varepsilon} \tag{17}$$

where $\dot{\varepsilon}$ is again the specimen strain rate.

It has recently been shown [30] that this model has a stress-state dependence and that when a hydrostatic pressure, P, is superimposed on an axial stress, σ_u, the growth rate is given by:-

$$\dot{V}_g = \lambda^2 d B \left[\frac{\sigma_u}{1-A_c} \right]^n \left[1 - \frac{P}{\sigma_u} A_c \right]^n \tag{18}$$

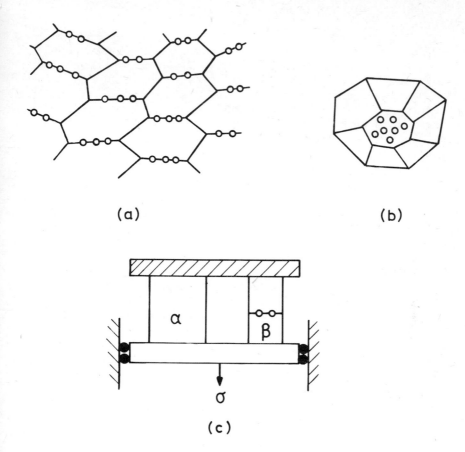

Fig. 5: (a) Plane section through a cavitated microstructure.
(b) Plan view of a cavitated facet.
(c) Mechanical analogue: α represents region of stress
enhancement in cavity free areas: β represents
region of stress attenuation in cavitated areas.

4.   CAVITY GROWTH RATE MEASUREMENTS

A few of the results of a much larger study of cavity
growth rates [31] will now be given.   The material used was
Nimonic 80A and, prior to creep, each specimen was plastically
strained 15% at room temperature and then annealed for 2 h at
750 °C to provide an array of cavities of constant mean
spacing, λ, at the start of the test.   Since this spacing did
not decrease during creep, the material fulfilled one of the
most demanding requirements of the theories presented above.
Some tests were performed in torsion but most were in uniaxial
tension.   Fig 6 shows tension and torsion growth rates at
750 °C plotted against maximum principal stress, σ_1.   The
data disagree with diffusion-controlled growth (8) on two

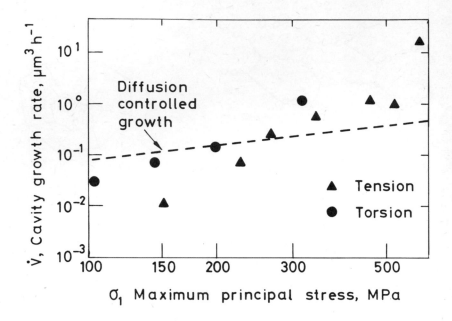

Fig. 6: Volumetric cavity growth rates in tension and torsion
as a function of maximum principal stress.
Temperature, 750 °C

counts.   First, the tension and torsion points clearly do not
superimpose on one another and second, there is a power-law
dependency on stress.   The dotted line was calculated from
(8) using material property data given in [4].   It would
appear that growth rates can be an order of magnitude lower
than those predicted by diffusion control.   Fig 7 shows the
same data plotted as a function of effective stress, $\bar{\sigma}$.   The
torsion and tension data now agree very well and there is even
a small stress-state dependency, as predicted by (18), although
the scatter is large.   Again the diffusion-controlled model
is in total disagreement.   In Fig 8, $\dot{v}$ is plotted as a function of temperature for a constant stress of 460 MPa.   The
activation energy is 320 kJ mol^{-1} compared with an expected
(from diffusion-control) grain boundary activation energy of
170 kJ mol^{-1}.

5.   DISCUSSION

The view expressed here is that cavity growth in materials of interest to the engineer is by a stress-directed,
vacancy absorption mechanism.   Theoretically, under steady-
load service stressing, this mechanism would usually be particle-inhibited or geometrically-constrained, if cavitation

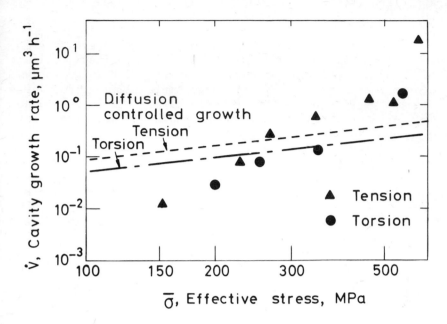

Fig. 7: Same data as in Fig 6 plotted against effective stress.

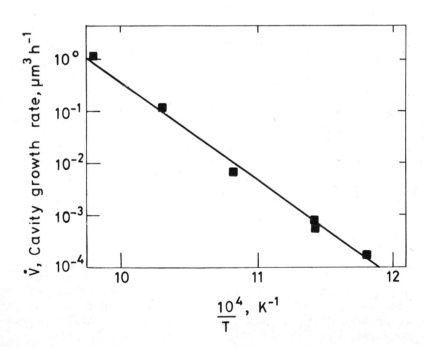

Fig. 8: Volumetric cavity growth rates in tension as in
function of $1/T$.   Stress, 460 MPa

is a fracture problem in the particular material, and so extrapolation from laboratory data needs to be done with caution.

The recent vacancy absorption models are interesting on two counts. They now support the strong experimental view, expressed in the sixties and early seventies, that growth was "deformation-controlled" (see, for example [32]) without dismissing a perfectly sound micro-mechanism. The models also indicate interactions of various strengths between creep rate and amount of cavitation which Cocks and Ashby [7] have used to draw analogies with the Kachanov empirical approach to "damage accumulation". Cocks and Ashby used power-law assisted vacancy absorption, which has only a weak interaction, and found inconsistencies between the micro-mechanism and empirical approaches. This author believes these may be overcome when growth is geometrically-constrained since then the the upper-bound strain rate is given by:-

$$\dot{\varepsilon} = B\left[\frac{\sigma}{1-A_c}\right]^n \tag{19}$$

and is strongly coupled to the fraction of boundaries cavitated. One way in which A_c can change is by continuous cavity nucleation, which again is strain controlled, and is likely to have the following form:-

$$A_c = \kappa\varepsilon \quad \text{where } \kappa \text{ is a constant}$$

$$\therefore \quad \dot{\varepsilon} = B\left[\frac{\sigma}{1-\kappa\varepsilon}\right]^n$$

$$\text{or} \quad \dot{\varepsilon} = B\sigma^n(1 + n\kappa\varepsilon) \quad \text{for} \quad \kappa\varepsilon \ll 1$$

$$\therefore \quad \dot{\varepsilon} = B\sigma^n \exp^{n\kappa\varepsilon} \tag{20}$$

(19) is a Kachanov type of equation while (20) does, in this author's experience, represent the secondary and tertiary shapes of many creep curves very well. That the two types of equation are in fact related does not seem to be appreciated.

The cavity growth results presented in Figs 6 and 7 support power-law enhancement of diffusion-controlled growth at the higher stresses. Its inhibition at lower stresses may be a consequence of either particles or grain geometry. The activation energy derived from Fig 8 is near to that for creep in Nimonic 80A and gives further support to the ideas of constrained or inhibited growth.

6.   CONCLUSIONS

Models of stress-directed, vacancy absorption, developed over the last few years suggest that particles at grain boundaries or inhomogeneous cavity distributions lead to much slower cavity growth rates at engineering stress levels than are predicted from diffusion-controlled models.   New measurements of cavity growth rates, at constant spacing λ, using Nimonic 80A substantiate this and further support the concept of power-law assisted growth at higher stress levels.

7.   REFERENCES

1.   CANE, B.J. and GREENWOOD, G.W. - Met. Sci. 1975, 9, 55.
2.   NEEDHAM, N.G. and GLADMAN, T. - Met. Sci. 1980, 14, 64.
3.   SVENSSON, L-E and DUNLOP, G.L. - to be published.
4.   DYSON, B.F. - Can. Met. Quart, 1979, 18, 31.
5.   SVENSSON, L-E and DUNLOP, G.L. - Met. Reviews, to appear.
6.   BEERÉ, W. - GEGB Report No. RD/B/N4886 July, 1980.
7.   COCKS, A.C.F. and ASHBY, M.F. - Cambridge University
        Report No. CUED/C/Mats/TR 6.7 July, 1980.
8.   McCLINTOCK, F.A. - J. Appl. Mech. 1968, 4, 363.
9.   RICE, J.R. and TRACEY, D.M. - J. Mech. Phys. Solids, 1969,
        17, 201.
10.   HELLAN, K. - Int. J. Mech. Sci. 1975, 17, 369.
11.   NEEDLEMAN, A. and RICE, J.R. - Acta Met, 1980, 28, 1315.
12.   DYSON, B.F. and TAPLIN, D.M.R. - 'Grain Boundaries',
        Proceedings of Institution of Metallurgists meeting,
        Jersey 1976, E13.
13.   HANCOCK, J.W. - Met. Sci. 1976, 10, 319.
14.   BROWN, A.M. and ASHBY, M.F. - Acta Met. 1980, 28, 1085.
15.   DYSON, B.F., LOVEDAY, M.S. and RODGERS, Mary Jo. - Proc.
        Roy. Soc. 1976, A349, 245.
16.   HULL, D. and RIMMER, D.E. - Phil. Mag. 1959, 4, 673.
17.   SPEIGHT, M.V. and HARRIS, J.E. - Met. Sci. J. 1967, 1, 83.
18.   WEERTMAN, J. - Met. Trans. 1974, 5, 1743.
19.   SPEIGHT, M.V. and BEERÉ, W. - Met. Sci. 1975, 9, 190
20.   RAJ, R. and ASHBY, M.F. - Acta Met. 1975, 23, 653.
21.   CHUANG, T-J, KAGAWA, K.I., RICE, J.R. and SILLS, L.B. -
        Acta. Met. 1979, 27, 265.
22.   BEERÉ, W. and SPEIGHT, M.V. - Met. Sci. 1978, 12, 172.
23.   EDWARD, G.M. and ASHBY, M.F. - Acta. Met. 1979, 27, 1505.
24.   BURTON, B. - 'Diffusional Creep of Polycrystalline Materials', Trans. Tech. Publications, Ohio, 1977.
25.   HARRIS, J.E. - J. Nucl. Matls. 1976, 59, 303.
26.   DYSON, B.F. - Met. Sci. 1976, 10, 349.
27.   BEERE, W. - Acta Met. 1980, 28, 143.
28.   RAJ, R. - Met. Trans., to appear.
29.   RICE, J.R. - Acta Met. 1981, to appear.
30.   DYSON, B.F., VERMA, A.K. and SZKOPIAK. Submitted to Acta Met.
31.   DYSON, B.F., LOVEDAY, M.S. and PARTRIDGE, Margaret, - to be
        published.
32.   DAVIES, P.W. and WILSHIRE, B. - Structural Processes
        in Creep 151 LONDON 1961.

CAVITATION IN NICKEL DURING OXIDATION AND CREEP

R.H. Bricknell and D.A. Woodford

General Electric Corporate Research and Development

P.O. Box 8, Schenectady, NY 12308, U.S.A.

ABSTRACT

The mechanisms underlying the formation of cavities in nickel when oxidized at high temperatures are reviewed, and vacancy injection and creep cavitation discounted by the results of a simple experiment. The evidence presented suggests that the cavities are gas bubbles formed by a carbon-oxygen reaction, and that they may be prevented by removal of the carbon.

The effect of carbon on the creep behaviour of nickel in air at 800°C is then investigated at three stress levels. At all stress levels, but especially at the two lower, carbon removal by hydrogen annealing leads to striking enhancements in creep lives and reductions in cavitation rates. Various possibilities, including environmental interactions are then considered to explain these results.

1. INTRODUCTION

Despite the fact that the oxidation of nickel has been studied for a number of years, it was not until about 1965 that attention was drawn to the existence of grain boundary voids in the metal following oxidation (1-3). Subsequent studies have reported the occurrence of these voids in various grades of pure nickel (4-10). As it is known that NiO is a metal-deficient oxide (11), oxide growth is expected to occur at the oxide-air interface by the outward diffusion of Ni^{2+} ions. This must be counterbalanced by a corresponding inward flux of vacancies, and, if these are not eliminated at the metal-oxide interface, they will be "injected" into the metal and may condense at grain boundaries to form voids. As this process had been reported to occur in certain metals during TEM observations (12,13), a majority of the studies assumed this to be the case in nickel (3,4,6,8,10), and thus, it became the generally accepted mechanism for the formation of cavities during oxidation.

Recently, this view has been challenged in an elegantly presented argument by Harris (14), who considers the cavities to be creep voids resulting from the accommodation of stresses generated by the growth of the oxide. This concept of growth stresses was first raised by Pilling and Bedworth (15), discussed in the specific case of nickel by Rhines and Wolf (16), and has been reviewed by Stringer (17). In his paper, Harris (14) summarizes the evidence for vacancy injection and re-interprets it in terms of stress-induced cavity growth. Such an approach is supported by several reports of similar grain boundary cavitation occurring in nickel, or very dilute nickel alloys, following creep (18-23). Hence, a case can be made for either of these two postulated mechanisms.

One further suggestion has been advanced to explain the cavitation observed, and involves the internal oxidation of carbon segregated to grain boundaries to produce CO bubbles (5,7,17). This mechanism is supported by the finding of Caplan et al (10) that nickel in which the carbon content was reduced did not cavitate. However, these authors interpreted the role of carbon to be that of a nucleating agent for grain boundary cavities, which were considered to be formed by the condensation of vacancies generated by creep of the nickel due to oxide growth stresses.

The current work reports the preliminary results of a study investigating the formation of grain boundary cavities in nickel, and assesses the possible contribution of the three suggested mechanisms: vacancy injection, creep due to growth stresses, and gas formation. This separation of contributions is made possible by some recent observations of the embrittlement of various grades of pure nickel following air exposure (24). This embrittlement was shown to proceed by grain boundary penetration of oxygen, and furthermore, to occur at oxygen partial pressures below those required for the formation of an external oxide on nickel. Hence, here we have a case where vacancy injection and creep due to oxide growth stresses can be discounted as mechanisms leading to any cavitation observed. Carbon monoxide and/or dioxide formation at grain boundaries will remain a viable mechanism.

The results gained from this study appeared to have some pertinence with respect to the cavitation observed during the creep of nickel, and their possible relevance is emphasized by the rupture results reported at the end of this paper.

2. EXPERIMENTAL

The current study was conducted on samples machined from Ni270 bar supplied by the International Nickel Company, and of composition given in Table I. Cavitation behaviour was studied on cylindrical pins of 0.25 cm diameter after exposure in flowing air and Ni/NiO packs. For the pack exposures, specimens were sealed in a quartz tube, adjacent to, but not in contact with, a 50:50 mixture of

Ni:NiO. This tube was evacuated to 10^{-3} Pa at room temperature before sealing-off. All specimens were degreased before exposure. Certain samples were analyzed for carbon contents in a Leco combustion apparatus, following the removal by grinding of any surface oxide.

The specimens for the creep study also had a 0.25 cm gauge diameter, with an effective gauge length of 1.14 cm.

TABLE I

Composition of the Ni270 Studied (wt. ppm)

Ni	C	Mg	Mn	Ti	Fe	Cu	S	Cr	Si	Co
bal.	130	10	<10	<10	<10	<10	1	<10	10	<10

3. RESULTS AND DISCUSSION

The microstructure of the Ni270 following exposure in flowing air at 1000°C is shown in Figure 1. Below the external oxide, extensive grain boundary cavitation can be seen in this unetched sample. The cavities are typically 10-15μm in diameter and extend throughout the metal, although their density is greater near the edges.

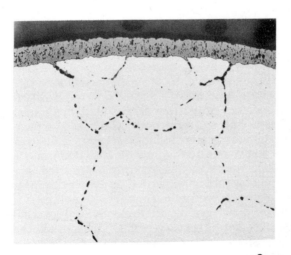

Figure 1:   Ni270 oxidized for 200 hours at 1000°C in flowing air. Unetched. X75.

Following the pack exposure with the 50:50 Ni/NiO powder ($PO_2 \approx 5 \times 10^{-6}$ Pa (25)), the surfaces of the nickel were bright and shiny, and showed no evidence of oxide or oxide film formation. Sectioning of the sample revealed the same extensive grain boundary cavitation, again extending to the centre of the specimen, as was observed after exposure in air, see Figure 2. In contrast, a sample in a sealed silica tube, evacuated to 10^{-3} Pa and baked-out for one hour at 350°C before sealing-off and exposure at 1000°C for 200 hours, led to no visible cavitation. The difference between a test conducted in an evacuated, sealed tube where the available oxygen is rapidly consumed, and that conducted in a vacuum furnace, flowing gas or Rhines' pack where the supply is unlimited, must be stressed.

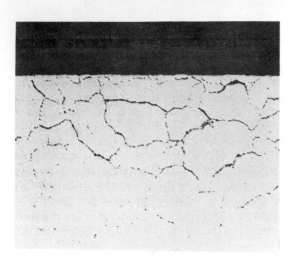

Figure 2:   Ni270 exposed to Ni:NiO pack for 100 hours at 1000°C. Unetched. X50.

The occurrence of copious grain boundary cavitation in Ni270 on pack exposure where no external oxide is formed discounts both creep cavitation and vacancy injection as possible mechanisms, and leaves gas formation as a viable alternative. Carbon is known to be strongly segregated at nickel grain boundaries (5), and oxygen has been detected at the same location following oxidation at 1000°C by Auger spectroscopy (26), hence, a gas-forming reaction appears feasible. External NiO scales would be expected to block the inward diffusion of oxygen (11), but the tracer experiments of Atkinson et al (27) have shown permeation down grain boundaries or microcracks in NiO. In the current study, the effect of carbon was explored by annealing samples for 100 hours in hydrogen at 1100°C. Such treatments are known to be effective in removing carbon from nickel (28-30). These low carbon samples were then exposed in air at 1000°C for 200 hours, and their microstructures examined, see Figure 3. Grain boundary cavitation has been entirely suppressed. A similar effect was obtained by remelting Ni270 with .25 wt% NiO powder, which produced an alloy effectively free of carbon. The results of the carbon analyses are shown in Table II.

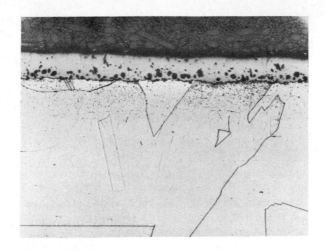

Figure 3: Low carbon Ni270 oxidized for 200 hours in flowing air. Etched to reveal internal oxides, but note lack of cavitation. X150.

TABLE II

Effect of Starting Carbon Level on Cavitation
Observed After Oxidation in Air at 1000°C

Material and Prior Treatment	C Content (wt. ppm)	Grain Boundary Cavitation
Ni270, as received	135, 133	Yes
Ni270, remelted with 0.25% NiO	Not detected.	No
Ni270 + 100 hrs 1000°C H$_2$	3, 7	No

The effect of carbon level in the alloy is seen to be crucial in determining whether cavitation is observed on oxidation (or exposure to oxygen partial pressures). The critical carbon level is unknown, and may vary with other impurity contents, but appears to be in the range 20-100 ppm. Caplan et al (10) interpreted the role of carbon as providing nucleation centres at grain boundaries for voids formed by grain boundary sliding in response to oxide growth stresses. This argument cannot pertain when no external oxide exists, and furthermore, oxide particles should be more effective nuclei for grain boundary creep voids than dissolved solute (31). It could be argued, in the absence of the results of the pack exposure, that hydrogen annealing suppresses cavity formation in Ni270 by reduction of grain boundary NiO particles. However, the alloy with the excess NiO does not cavitate on oxidation. Hence, the formation of grain boundary cavities on the exposure of nickel to oxygen at high temperatures is seen to be a totally environmental effect, and dependent on the carbon present. This strongly supports gas formation as the mechanism of cavity growth, and experiments are in hand to identify any gases present in nickel pores by sensitive mass spectroscopy.

4. EFFECT ON CREEP CAVITATION

The similarities between the cavities seen on oxidation and after creep of nickel led to the speculation that those seen on oxidation were caused by creep. Now it has been shown that the voids after oxidation are a result of environmental reactions, it is instructive to ask the reverse question, and explore the role of environment on the formation of creep cavities in nickel. In the studies described above, cavity formation during oxidation was prevented by testing low-carbon samples. It was, therefore, decided to explore whether a similar effect could be obtained in creep. Hence, a comparison was made on samples annealed for 200 hours at $1000^{\circ}C$ in hydrogen, and samples similarly exposed in evacuated tubes, both of which developed identical grain sizes. These specimens were tested in air at $800^{\circ}C$ at three stress levels. The experimental details and resulting lives are listed in Table III.

TABLE III

$800^{\circ}C$ Stress Rupture Results

Stress Level (MPa)	Prior Annealing Atmosphere	Test Atmosphere	Life (hrs.)
15.80	vacuum	air	23.6
	H_2	air	> 500
17.56	vacuum	air	41.9
	H_2	air	> 1,000
22.95	vacuum	air	4.0
	H_2	air	9.7
15.80	vacuum	argon	13.1

The striking increase in life obtained at the lower stress levels for the low carbon specimens is immediately obvious. Figure 4 shows the creep curves for the samples deformed at the lowest stress level, and demonstrates the dramatic reduction in creep rate after hydrogen exposure.

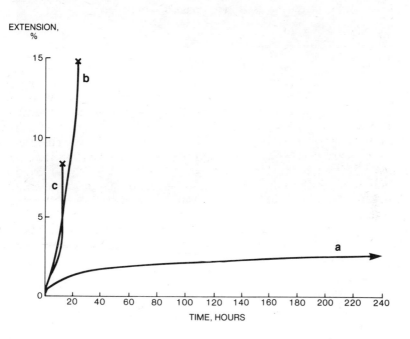

Figure 4: Time-extension plots for 800°C tests, 15.8 MPa.
a.   Low carbon Ni270, in air.
b.   Standard Ni270, in air.
c.   Standard Ni270, in argon.

Scanning electron microscopy reveals the standard Ni270 to have failed intergranularly, after testing at the lowest stress level, see Figure 5. The heavily oxidized surface masks any details of cavitation. Optical metallography of the same sample, Figure 6a, demonstrates the catastrophic damage at grain boundaries throughout the gauge section. Every boundary oriented normally to the applied stress can be seen to have parted and to be filled with oxide. In contrast, only minor cavitation can be seen on other boundaries, and none was found in the head region. The grain size was comparable in both the head and gauge sections at about 2×10^{-4} m. The manner in which the failure of the grain boundaries in the gauge region has occurred is revealed by an examination of the shoulder area, Figure 6b. Here, strings of grain boundary cavities on perpendicular boundaries can be seen interlinking to form the grain boundary cracks which then become filled with oxide.

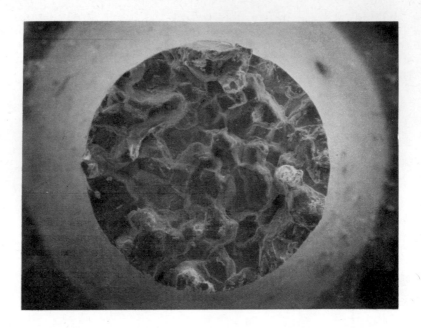

Figure 5:   SEM   micrograph   of   intergranular   creep   fracture   of
standard Ni270 in air at 800°C, 15.8 MPa. X25.

     The microstructure of the low carbon Ni270, unloaded after 500
hours at the same stress level, is shown in Figure 6c. Some cavitation
of perpendicular boundaries is seen in certain areas, but only minimal
interlinkage is seen even after this extended testing period. The grain
size is comparable to that in the standard Ni270, but one or two larger
grains are seen. Again, no cavitation was observed in the head.

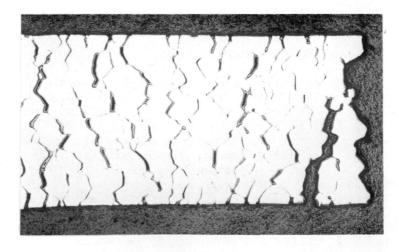

Figure 6a: Standard Ni270 after failure (23 hours) at 15.8 MPa, 800°C
in air. X20.

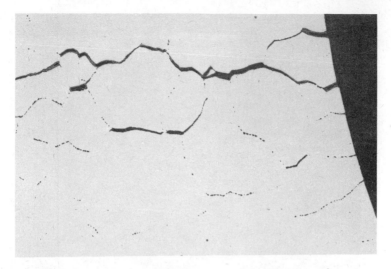

Figure 6b: Detail of shoulder region showing development of grain boundary cracks by cavity interlinkage. X50.

Figure 6c: Low carbon Ni270 unloaded after 500 hours at similar conditions, etched. X20. Slight cavitation visible.

The situation at the highest stress allows far less time for any possible environmental influence on cavity nucleation and growth during the course of testing, but even so, a doubling of rupture life is obtained, and is accompanied by a transition from totally intergranular failure to one displaying predominantly transgranular features, see Figure 7. Heavy oxidation of the fracture surface again prevented identification of grain boundary cavitation.

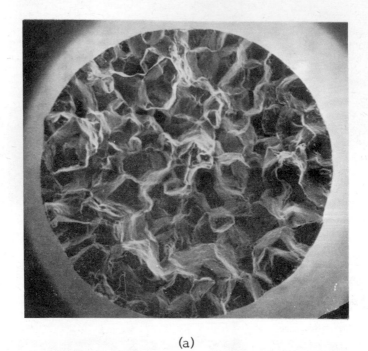

(a)

(b)

Figure 7: Transition from (a) intergranular to (b) predominantly transgranular creep failure in standard to low carbon Ni270 at 800°C, 23 MPa. X30.

To explore further the effect of environment on creep behaviour, a vacuum annealed sample was tested at 15.8 MPa in laboratory grade argon. Rapid intergranular failure occurred in 13 hours, see Figure 4 and Table III, and was accompanied by massive cavitation, see Figure 8.

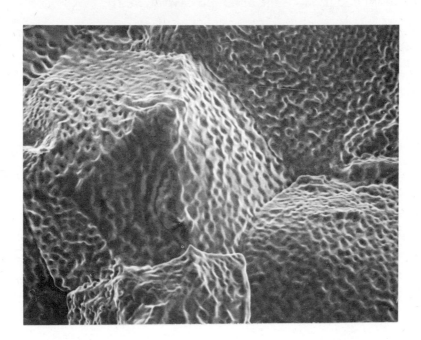

Figure 8: Grain boundary voiding in specimen of standard Ni270 failed after testing in argon at 800°C. X300.

In analysing the above results, many possibilities emerge, of which two appear the most probable: either the presence of carbon on nickel grain boundaries leads to a vast augmentation in the rate of cavity nucleation or growth irrespective of any environmental reaction, or gas formation due to oxygen penetration of grain boundaries has the same effect. The former explanation was shown not to pertain in the case of cavitation during oxidation at 1000°C, but could still be relevant in creep cavitation at 800°C. For the latter to be the case, then the oxygen partial pressure in the laboratory argon would have to be sufficient to produce the reaction. This argon contains 3 to 10 ppm O_2, and would give an oxygen partial pressure of 3×10^{-1} to 1.0 Pa, i.e., far in excess of that of the Ni:NiO pack. Also, unlike air exposure, here no thick external oxide would be produced to impede oxygen ingress. As no cavitation was observed in any of the head sections, it can be concluded that if an environmental reaction is responsible for the enhanced cavitation, then it, of itself, cannot nucleate cavities at 800°C, but acts by stabilizing small creep voids of below the critical size.

One further possibility to be considered is that a gas-forming reaction between carbon and oxygen may result from oxygen already dissolved in the matrix or present as NiO particles, rather than by oxygen penetration of grain boundaries from an external atmosphere. Here too, however, the results would require a stress/strain contribution to produce the observed cavitation.

This work has demonstrated that the removal of even the limited amounts of carbon found in Ni270 leads to spectacular increases in creep life and reductions in cavitation rates. Whilst the mechanism has not been established, similarities with the situation on oxidation suggest the serious consideration of this being an environmental interaction effect.

5.   CONCLUSIONS

1.   The grain boundary cavitation observed on the oxidation of nickel is not due to either vacancy injection or creep cavitation from oxide growth stresses.

2.   The cavitation is shown to be totally dependent on the carbon level of the material, and strongly supports gas formation for the mechanism of cavitation.

3.   Striking effects are also observed in the creep lives and cavitation rates of Ni270 when tested in air after the carbon level is reduced. No improvement in the standard Ni270 is obtained on testing in argon.

ACKNOWLEDGEMENTS

The authors wish to thank Craig Robertson (SEM), Paul Dupree (Creep Tests) and the members of the Metallography Group for valuable experimental assistance.

REFERENCES

1.   Jones, D.A. and Westerman, R.E., Corrosion, 1965, 21, 295.

2.   Wood, G.C., Wright, I.G., and Ferguson, J.M., Corrosion Sci., 1965, 4, 645.

3.   Hancock, P. and Fletcher, R., Metallurgie, 1966, VI, 1.

4.   Douglass, D.L., Mat. Sci. Eng., 1968-9, 3, 255.

5.   Wolf, J.S., NASA TN D-5266, 1969.

6.   Hales, R. and Hill, A.C., Corrosion Sci., 1972, 12, 843.

7.   Lowell, C.E., Grisaffe, S.J. and Deadmore, D.L., Oxid. of Metals, 1972, 4, 91.

8.   Hancock, P., "Vacancies '76", The Metals Society, London, 1976, p.215.

9.   Palmer, L.D. and Cocking, J.L., Microstructural Sci., 1979, $\underline{7}$, 193.

10. Caplan, D., Hussey, R.J., Sproule, G.I., and Graham, M.J., Oxid. of Metals, 1980, $\underline{14}$, 279.

11. Kofstad, P., "Nonstoichiometry, Diffusion and Electrical Conductivity in Binary Metal Oxides", pub. Wiley, New York, 1972, p.246.

12. Dobson, P.B. and Smallman, R.E., Proc. Roy. Soc. (A), 1966, $\underline{293}$, 423.

13. Hales, R., Smallman, R.E., and Dobson, P.S., Proc. Roy. Soc. (A) 1968, $\underline{307}$, 71.

14. Harris, J.E., Acta Met, 1978, $\underline{26}$, 1033.

15. Pilling, N.B. and Bedworth, R.E., J. Inst. Met., 1923, $\underline{29}$, 529.

16. Rhines, F.N. and Wolf, J.S., Met. Trans., 1970, $\underline{1}$, 1701.

17. Stringer, J., Corrosion Sci., 1970, $\underline{10}$, 513.

18. Davis, P.W. and Wilshire, B., J. Inst. Met., 1961-2, $\underline{90}$, 470.

19. Dennison, J.P. and Wilshire, B., J. Inst. Met., 1962-3, $\underline{91}$, 343.

20. Davis, P.W. and Wilshire, B., Phil. Mag., 1965, $\underline{11}$, 189.

21. Bowring, P., Davis, P.W., and Wilshire, B., Metal Sc. J., 1968, $\underline{2}$, 168.

22. Woodford, D.A., Metal Sci. J., 1969, $\underline{3}$, 234.

23. Badiyan, E.E., and Sirenko, A.F., J. Mat. Sci., 1971, 1479.

24. Bricknell, R.H. and Woodford, D.A., Met. Trans., 1981, $\underline{12A}$, in press.

25. Darken, L.S. and Gurry, R.W., "Physical Chemistry of Metals", pub. McGraw-Hill, New York, 1953.

26. Mulford, R.A. and Bricknell, R.H., General Electric Co., Schenectady, NY, unpublished research, 1979.

27. Atkinson, A., Taylor, R.I., and Goode, P.D., Oxid. of Metals, 1979, $\underline{13}$, 519.

28. Sukharov, V.F., Aleksandrov, N.A. and Kudryavtseva, L.S., Phys. Met. Metallogr., 1962, 14, 2.

29. Sukharov, V.F. and Popov, L. Ye., Phys. Met. Metallogr., 1964, 17, 107.

30. Sonon, D.E. and Smith, G.V., T.A.I.M.E., 1968, 242, 1527.

31. Perry, A.J., J. Mat. Sci., 1974, 9, 1016.

GRAIN BOUNDARY SLIDING AND FRACTURE OF METAL BICRYSTALS AT HIGH TEMPERATURES

Tadao Watanabe

Department of Materials Science, Faculty of Engineering,
Tohoku University, Sendai, Japan.

SUMMARY

Some systematic investigations of grain boundary sliding and fracture at high temperatures were made on orientation-controlled bicrystals of zinc, aluminium, magnesium and copper in order to reveal the mechanism of sliding and the effect of grain boundary structure on sliding. The contribution of grain boundary sliding to intergranular fracture at high temperatures was also studied. It was found that grain boundary sliding is closely related to crystal deformation and the sliding behaviour strongly depends on the grain boundary misorietation. A characteristic saturation of sliding was observed under stress cycling, which is also grain boundary structure-dependent. It was made clear that grain boundary sliding is essential for nucleation and growth of cavities and deformation ledges produced on the grain boundary by intersection with crystal slips are potential sites of nucleation of cavities.

1. Introduction

Grain boundary sliding is an important metallurgical phenomenon which can contribute to plastic deformation and fracture of polycrystalline metallic and ceramic materials at high temperatures. Much work has been previously done in order to study fundamental behaviour and mechanism of grain boundary sliding. It is known to be structure-sensitive phenomenon which depends on various structural parameters relating to grain boundary structure and mechanical testing conditions. The experimental results on sliding and intergranular fracture in polycrystalline materials often seem difficult to be fully explained because of complication of the condition under which sliding and fracture are actually going. Moreover, it is not easy to study the effect of grain boundary structure on sliding which is now well accepted, bacause of difficulty in dtermination of the character of many grain boundaries in the materials.

Bicrystals are very useful for basic study of grain boundary sliding because the experiment can be made under simple

geometrical and stress conditions. The character of grain
boundary contained in a bicrystal specimen can be easily determined by crystallographic analysis. Therefore, the study of
grain boundary sliding on systematically orientation-controlled
bicrystals can provide useful basic knowledge about the sliding
behaviour and the effect of grain boundary structure on sliding.
The followings are still remain as subjects to be solved,
(1) relation between crystal deformation and sliding,
(2) the effect of grain boundary structure on sliding,
(3) mechanism of sliding,
(4) relation between sliding and intergranular fracture at high
    temperatures.
The present paper is an overview of the recent works by
the present author and others in order to obtain basic knowledge concerning the above four subjects, on orientation-
controlled bicrystals of some pure metals and alloy.

2. The Relation between Crystal Deformation and Sliding

It seems important to make sure whether crystal deformation
is essential to occurrence of sliding, in order to reveal the
mechanism of sliding. Some relationship between crystal
deformation and sliding was reported early [1-3]. Recent
transmission electron microscopy works on slid grain boundaries
have provided another solid experimental supports for the close
relation between crystal deformation and sliding [4-6].
However, we have not yet fully understood the details about the
way in which crystal deformation contributes to sliding.

A systematic investigation was made on $<10\bar{1}0>$ and $<11\bar{2}0>$
symmetric tilt zinc bicrystals of high purity, to study the
interrelation between crystal deformation and sliding, and the
effect of grain boundary misorientation on sliding [7,8].
The reason why zinc bicrystals with $<10\bar{1}0>$ and $<11\bar{2}0>$ symmetric
tilt boundaries of various grain boundary misorientations were
used, is that zinc has only one prominent slip system (basal
slip) and therefore the simplest interaction between crystal
deformation and the grain boundary was hoped. The details of
the experimental procedure and results are not described here.

Figure 1 shows the creep curves of $<10\bar{1}0>$ tilt zinc bicrystals with different grain boundary misorientations. Creep
deformation becomes difficult with increasing the misorientation mainly because of the difference in the average shear
stress on basal planes in these bicrystals when creep test
is conducted at constant tensile stress and at constant grain
boundary shear stress for 45° inclined grain boundary with
respect to the tensile axis.
Figure 2 shows the sliding curves for the same $<10\bar{1}0>$ tilt
zinc bicrystal specimens as those shown in Fig.1. The rate
of sliding gradually decreases with displacement, i.e."slide-
hardening" occurred. The rate of sliding strongly depends on
the grain boundary misorientation at constant temperature and
grain boundary shear stress. Small angle boundaries appear to

have difficulty in sliding.

A comparison between crystal deformation and sliding for <10$\bar{1}$0> tilt zinc bicrystals is shown in Fig.3. It is clearly showed that a large amount of crystal deformation does not always provide a large amount of sliding among the bicrystals, but sliding depends on the grain boundary misorientation. The 73°<10$\bar{1}$0> tilt bicrystal which showed difficult creep deformation, is most strain-sensitive among these. This figure also shows that the amount of sliding is not linearly related to the amount of crystal deformation, and that sliding becomes less strain-sensitive as sliding is going on. This fact is considered to be related to the origin of slide-hardening. The slide-hardening seems to originate from grain boundary itself not from the localized crystal deformation near grain boundary and others.

Photograph 1 shows an electron channeling pattern (ECP) taken from the slid grain boundary in the 32.5°<10$\bar{1}$0> symmetric tilt zinc bicrystal. The displacement of the scratches was due to sliding. It is clear that ECP lines are not diffused near the grain boundary but sharp as in the grain interior. This clearly means that any localized crystal deformation did not take place near the grain boundary and the bicrystal homogeneously deformed at high temperature.

The contribution of sliding to the total creep deformation of <10$\bar{1}$0> tilt zinc bicrystals was evaluated. The result is shown in Fig.4 which clearly indicates that the contribution $\gamma(=\varepsilon_{gb}/\varepsilon_t)$ decreases with increasing the total creep strain for each bicrystal specimen and the level of γ depends on the grain boundary misorientation. For small angle boundaries below 20° misorientation the contribution is quite small. Among high angle boundaries a slightly-off coincidence boundary (61°<10$\bar{1}$0> tilt) showed rather small contribution.

Another experimental evidence which supports the close relation between crystal deformation and sliding was obtained in the study of sliding in <11$\bar{2}$0> symmetric tilt zinc bicrystals [7].

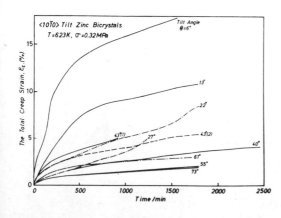

Fig.1.

Creep curves of <10$\bar{1}$0> tilt zinc bicrystals with different grain boundary misorientations.

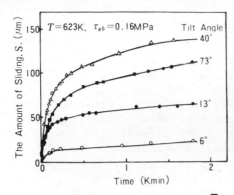

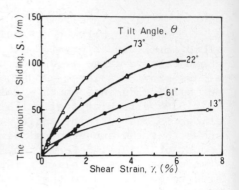

Fig.2. Sliding curves of <10$\bar{1}$0> tilt zinc bicrystals with different misorientations.

Fig.3. Shear strain dependence of sliding in <10$\bar{1}$0> tilt zinc bicrystals.

The sliding took place along <11$\bar{2}$0> direction which is the rotation axis of the tilt boundary. The amount of sliding did not vary so significantly along the straight boundary in the case where no grain boundary migration took place. The sliding curves obtained from the observations on the front and back surfaces of a bicrystal specimen are shown in Fig.5. It is surprising that the amount of sliding is quite different between the surfaces A and B. The amount of sliding on the surface A was 3 - 7 times larger than that on the surface B of the same specimen. This interesting finding was well explained in the consideration of interaction between crystal deformation and sliding, using a dislocation model of sliding specifically proposed for <11$\bar{2}$0> symmetric tilt zinc bicrystals. The schematic representation of the model is also shown in Fig.5. According to this model screw dislocations from the grain interior move along the grain boundary to produce sliding on the surface A where displacement between two crystals due to screw dislocation movement is accumulated. The sliding on the

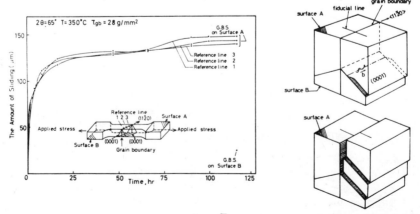

Fig.5. Sliding curve of a 65°<11$\bar{2}$0> symmetric tilt zinc bicrystal and the schematic representation of a model of sliding for this type of grain boundaries.

Photo.1
Electron-channeling-
pattern(ECP) taken from
the slid grain boundary
in a 32.5°<10$\bar{1}$0> symmetric tilt zinc bicrystal.

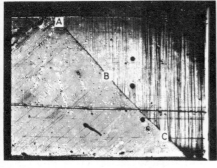

A
S=8.4 μm

B
S=13.8 μm

C
S=14.8 μm

Photo.2
Variation in the amount
of sliding along the
boundary intersected by
crystal slip bands in a
copper bicrystal.

surface B can not be interpreted by this mechanism.   It may be
attributed to the movement of grain boundary structural dislocations originally inbeded in the boundary or to some crystal
deformation-independent mechanism like diffusional mechanism
of sliding [9,10].
    A direct experimental evidence for the close relation
between crystal deformation and sliding was obtained by microscopic observations of sliding in copper bicrystals [11].
A large amount of sliding was observed at the position where
slip bands were more densely intersecting with the boundary,
as seen in Photo.2.

3. The Effect of Grain Boundary Structure on Sliding

3.1. The Misorientation Dependence of Sliding

    The effect of grain boundary structure on sliding is one of
the most interesting subjects of sliding study.   Any type of
grain boundary behaviour is supposed to be more or less grain
boundary structure dependent, as expected from the recent
knowledge about grain boundary structure and properties [12].
Previous works by others clearly revealed a strong effect of
grain boundary structure on sliding in metal bicrystals of
cubic structure [13-16].   However, this sort of work had not
been undertaken in hcp crystals although a recent theory of
grain boundary structure for hcp crystals by Bruggeman, Bishop
and Hartt [17] pointed out the possibility of the existence of
special ordered boundaries,i.e. the near-coincidence boundaries.
Therefore, a study of misorientation dependence of sliding in
hcp mtals was hoped to be able to provide some experimental

evidence for the prediction of the existence of the near-coinci
dence boundaries which were derived by an extension of the
coincidnce site lattice theory of grain boundary structure, from
cubic to hcp crystals.

Figure 6 shows the misorientation dependence of the total
amount of sliding observed 30 hrs after the onset of testing in
<10$\bar{1}$0> tilt zinc bicrystals. It was found that the amount of
sliding very strongly depended on the grain boundary misorientation. A very deep cusp was observed near 55° of the
tilt angle, covering over 30°. In a bicrystal with the 55°
<10$\bar{1}$0> tilt boundary, grain boundary migration markedly took
place during creep deformation, but sliding hardly took place.
The cusp found near Θ=55° is considered to be attributed to the
existence of the 56.6°<10$\bar{1}$0>/Σ9 near-coincidence boundary which
was theoretically predited by Bruggeman et al [17].

Another experimental evidence for the existence of the near-
coincidence boundary was obtained in a study of misorientation
dependence of sliding in <11$\bar{2}$0> tilt zinc bicrystals [7].
A cusp near 50° tilt angle was found which supports the existence of the 49.5°<11$\bar{2}$0>/Σ17 near-coincidence boundary.

The misorientation dependence of sliding for random boundaries which have both tilt and twist components of grain boundary misorientation, was studied in magnesium, aluminium and
iron-silicon alloy bicrystals by the present author and others.
The result on misorientation dependence of sliding in magnesium
bicrystals is shown in Fig.7.   The amount of sliding strongly
depends on the tilt angle, showing a maximum almost around 90°
but not obviously on the twist angle (for five specimens among
seven, the twist angle was almost remained constant (20°±3°).

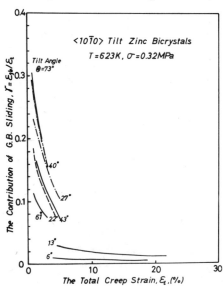

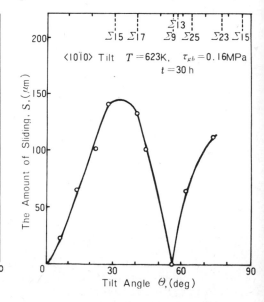

Fig.4. The contribution of
sliding to the total creep
deformation of <10$\bar{1}$0> tilt
zinc bicrystals.

Fig.6. Misorientation dependence
of the total amount of sliding
in <10$\bar{1}$0> tilt zinc bicrystals.

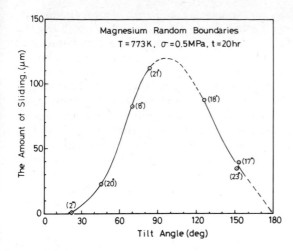

Fig.7. Misorientation
dependence of the total
amount of sliding of
random moundaries in
magnesium bicrystals.

However, both the tilt and twist components seem to be important
for sliding of random boundaries of small angle regime in
iron-silicon alloy bicrystals [18].

Transmission electron microscopy of slid aluminium grain
boundaries have provided some important results on the effect
of grain boundary structure on sliding [19].
Transmission electron microscopy revealed a very distinct
difference in dislocation structure after sliding between
coincidence (strictly speaking, slightly off-coincidence)
boundaries which were rather difficult to slide, and random
high angle boundaries which could easily slide. High density
of lattice dislocations were observed in the coincidence
boundaries, while no lattice dislocation was observed in the
random high angle boundaries. The observed difference in
dislocation structures between the slid coincidence and random
boundaries was well interpreted by considering the absorption
of lattice dislocations into the grain boundaries with different
structures. The absorption of lattice dislocations which was
taken as elementary process of sliding,was supposed to be slow
in the coincidence boundaries and rapid in the random high
angle boundaries from the results of the analysis based on
energetics of dissociation reaction of lattice dislocation in
grain boundaries with different structures [19].

3.2. The Origin of the Misorientation Dependence of Sliding
The misorientation dependence of sliding in <$10\bar{1}0$> tilt
zinc bicrystals shown in Fig.6 has been interpreted by a
dislocation mechanism, originally proposed by McLean [20]
and later modified by the present author and others [8].

The lattice dislocation of Burgers vector b can be regarded
as being dissociated into two components, one with Burgers
vector b_n = b sinθ normal to the grain boundary and the other
with Burgers vector of b_p = b cosθ parallel to the boundary.

The geometrical arrangement of basal slip planes and the grain boundary with respect to the stress axis is schematically shown in Fig.8 for a $<10\bar{1}0>$ tilt bicrystal. In general, the Burgers vector of lattice dislocations is not in the grain boundary plane so that their movement along the boundary involves a combination of climb and glide in the boundary. The resulting sliding is due to the component of Burgers vector b_p parallel to the boundary. When a lattice dislocation dissociate and component dislocations are assumed to move together, the whole dislocation motion along the grain boundary is governed by the net climb force F on the dislocation.

$$F = \sigma_n b_n \pm \tau_{gb} b_p = \sigma b [\cos^2\phi \sin\theta \pm \sin\phi \sin\phi \cos\theta]. \quad (1)$$

The rate of sliding $\dot{S}$ in real crystals should be proportional to the linear density of lattice dislocations ρ_{gb} introduced from the grain interior and later moving in the grain boundary. Therefore, $\dot{S}$ is given by

$$\dot{S} \propto F \cdot b_p \cdot \rho_{gb}. \quad (2)$$

Under an reasonable assumption for ρ_{gb} [8], the sliding rate $\dot{S}$ can be given by

$$\dot{S} \propto \frac{1}{8} \sigma^3 b^2 L \ [\cos^2\phi \sin2\theta \pm \sin2\phi \cos^2\theta] \sin2\phi\sin\theta$$
$$\times \cos2\theta \sin2(\phi \pm \theta) \quad (3).$$

where σ is the applied stress, L the length of the grain boundary in the specimen.
Eq.(3) is rewritten in a simpler form as

$$\dot{S} = K \cdot M. \quad (4)$$

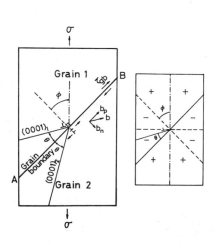

Fig.8. Schematic illustration of a dislocation model for sliding in $<10\bar{1}0>$ tilt bicrystals.

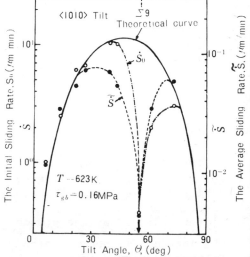

Fig.9. Misorientation dependence of the sliding rate calculated for $<10\bar{1}0>$ tilt bicrystals.

by using the orientation factor M,

$$M = [\cos^2\phi \, \sin2\theta \pm \sin2\phi\cos^2\theta]\sin2\phi\sin\theta\cos2\theta\sin2(\phi\pm\theta) \qquad (5).$$

At constant stress and temperature, the misorientation dependence of sliding is considered to be attributed to the misorientation dependence of the sum of the factor M for each of the two component crystals of a bicrystal specimen.

Figure 9 shows the misorientation dependence of sliding rate which was theoretically predicted using the above equation. The theoretical curve has a maximum near $\theta=45°$ and minima at $\theta=0°$ and $90°$. The theoretical curve well fits the plots of experimental data of the initial sliding rate $\mathring{S}$ and the average sliding rate $\widetilde{S}$ calculated from the data shown in Fig.6, except those within the range $\theta=40°-70°$, when we take the value of constant K as $K=8.1 \times 10^1$ (μm/min) for $\mathring{S}$ and 1.1×10^0(μm/min) for $\widetilde{S}$. The cusp observed over a range from $\theta=40°\sim70°$ is considered to arise from the strong effect of grain boundary structure on sliding in exact or slightly off-coincidence boundaries which has already been mentioned in 3.1.

4. The Effect of Stress Cycling on Sliding

A study of sliding under stress cycling was undertaken using orientation controlled $<11\bar{2}0>$, $<10\bar{1}0>$ and $<41\bar{5}0>$ tilt zinc bicrystals with different grain boundary misorientations [21]. Figure 10 shows a typical sliding/time curve under stress cycling for a $31°<11\bar{2}0>$ tilt zinc bicrystal. Stress cycling was made by converting tensile stress to compressive one, or vice versa at constant time interval, to reverse the direction of shear stress component along the grain boundary.

During the first tension cycle creep deformation, the sliding rate gradually decreased with time due to slide hardening. Upon the stress reversal the sliding rate in the reverse direction became more rapid than just before the reversal. It was found that the grain boundary sliding produced in the first tension cycle was not completely recoverable in the next compression cycle. The amount of sliding produced by each

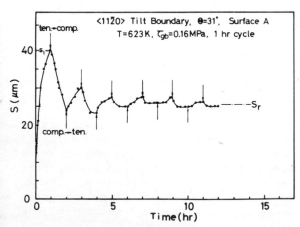

Fig.10. Sliding/time curve under 1 hr stress cycling for a $31°<11\bar{2}0>$ tilt boundary.

K

stress cycling rapidly diminished to give eventually a residual grain boundary sliding, S_r. The value of residual sliding was found to depend on the grain boundary misorientation in <11$\bar{2}$0> symmetric tilt zinc boundaries as reported on the misorientation dependence of sliding under static unidirectional stress condition [7]. Different sliding behaviour was observed among the different types of grain boundaries. Figure 11 shows a sliding time curve for a 27°<10$\bar{1}$0> tilt boundary. The dimimution in sliding rate by stress cycling was less pronounced than that observed for <11$\bar{2}$0> tilt boundaries.

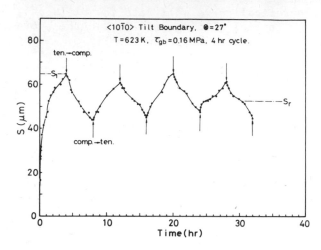

Fig.11. Sliding/time curve under 4 hr stress cycling for a 27°<10$\bar{1}$0> tilt boundary in zinc.

The residual sliding has been proved to be one of the important factors in high temperature-low cycle fatigue intergranular fracture. A recent work on high temperature fatigue intergranular fracture in austenitic stainless steel has demonstrated that the length of fatigue cracks in grain boundaries depends on the residual sliding [22]. Therefore the present investigation on zinc bicrystals would suggest that intergranular fracture due to high temperature-low cycle fatigue will possibly depend on the type and misorientation of grain boundaries.

5. Intergranular Fracture Caused by Sliding

Although much work of fracture under creep condition has been undertaken, thorough understanding of the behaviour and mechanism of intergranular fracture in crystalline materials has not been reached. Most of the previous works have been done with polycrystalline materials. In polycrystals, the type and structure of grain boundaries varies considerably. The grain boundary planes are arranged along various directions with respect to the stress axis so that the type and extent of interaction between crystal deformation and grain boundary may vary as well as the stress condition at the boundary. These would make the intergranular fracture process in polycrystals more complicated and understanding of fracture mechanism more

difficult. These difficulties can be avoided in studying grain
boundary sliding and fracture of bicrystals. The type and
misorientation of grain boundary in the specimens to be tested
can easily be defined and controlled. Observations of grain
boundary sliding, interaction between crystal deformation and
grain boundary, and of intergranular fracture surfaces can also
be easily made under simple geometrical conditions.

From the view point mentioned above, a basic study of
interrelation of grain boundary sliding, crystal deformation
and intergranular fracture at high temperatures was made in
orientation-controlled copper bicrystals [23], aiming at
understanding the mechanism of intergranular fracture at high
temperatures. A recent technique of scanning electron microscope fractography was applied to examine the detailed appearance
of the fracture surfaces. The observed fracture patterns on
the surfaces were crystallographically analysed to obtain some
basic information about the microscopic process of intergranular
fracture in the metal.

The sliding behaviour of copper bicrystals was in principle
similar to that observed in zinc, magnesium and aluminium
bicrystals; the sliding depended on the grain boundary misorientation, and was closely related to crystal deformation. The
slide-hardening was always observed. Copper bicrystals used
fractured in typically intergranular manner by creep deformation.
Random high angle boundaries easily slid and broke while small
angle boundaries were very difficult to slide and would not
break even at large creep strain.

Figure 12 shows the relation
between grain boundary sliding
and fracture of copper bicrystals. The contribution of
sliding to the total creep
defprmation γ was plotted as
function of the total creep
strain. The contribution of
sliding decreased with the
total creep strain for each
bicrystal specimen. The bicrystal specimens finally fractured
and the value of γ_f at fracture
are plotted by the simbol x
in the figure. As seen from
this figure all the data of γ_f
are plotted on a single curve.
This curve almost satisfies the
following relation between γ_f
and the fracture strain ε_f.

$$\gamma_f \; \varepsilon_f = \varepsilon_{gb} = \text{constant.}$$

This means that the intergranular
fracture takes place at a constant

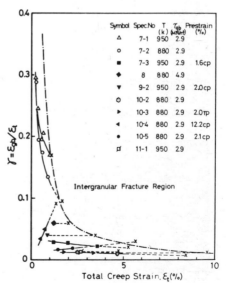

Fig.12.
Grain boundary sliding-
fracture diagram for copper
bicrystals with different
types of grain boundaries.

274

value of sliding.   It is worth while noting that the condition
for final fracture for bicrystal specimens of different types of
grain boundary and for prestrained ones can be predicted from a
single sliding-fracture curve.   It is interesting to see that
fracture will not occur in a bicrystal which has a low level of
γ until γ-ε_t curve crosses the sliding-fracture curve at comaratively large creep strain.   This is the case of bicrystals
containing small angle or coincidence (slightly off-coincidence)
boundaries.   The basic mechanism of the intergranular fracture
is supposed to be the same for grain boundaries of different
types and different misorientations.   Therefore, the difference
in fracture behaviour among different specimens will result from
the difference in sliding behaviour which in general strongly
depends on the type and the misorientation of grain boundaries.

A direct evidence for the importance of sliding and crystal
deformation for intergranular fracture at high temperatures was
obtained by SEM fractography of fractured copper bicrystals[23].
Photographs 3 and 4 show the typical intergranular fracture
pattern observed on fracture surfaces of copper bicrystals.

The pattern has several specific orientations.   The important
crystallographic directions determined by X-ray analysis are
included in the photographs.   The specific orientations are
related to the direction of grain boundary sliding and to the
primary slip traces in the grain boundary plane.   The length

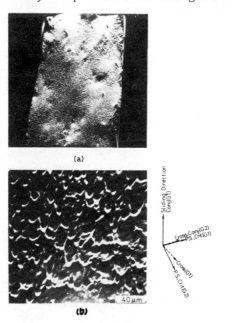

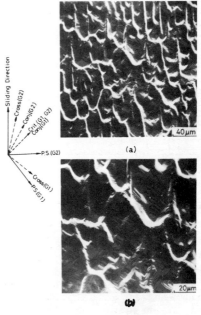

Photo.3.
Scanning electron micrograph
taken from a fractured copper
bicrystal.

Photo.4.
Fracture surface of a copper
bicrystal specimen pre-
strained by 12.2% under
compression before creep test.

of the traces along the sliding direction is somewhat different among the specimens, but the longest one almost corresponds to the amount of sliding at fracture, S_f of each specimen. The traces which were observed along primary slip plane, are ledges formed by intersection of grain boundary with crystal slip. These observations support the mechanism of intergranular fracture in which ledges at the grain boundary formed by crystal deformation become potential sites of cavity nucleation and the nuclei grow due to sliding and link together at the later stage of creep deformation to result in final intergranular fracturing of the specimen.

6. Conclusions

(1). Grain boundary sliding of metal bicrystals was found to strongly depend on the grain boundary misorientation. It is difficult in the small angle boundary and in the exact or slightly off-coincidence boundaries.

(2). Crystal deformation is essential for grain boundary sliding markedly to take place. The dislocation mechanism of sliding well explains the experimental results on sliding behaviour and of dislocation structures in slid grain boundaries of metal bicrystals.

(3). Grain boundary sliding and crystal deformation play the important roles in intergranular fracture of copper bicrystals at high temperatures. The fracture behaviour of copper bicrystals depends on the type and the mis-orientation of grain boundary possibly through grain boundary sliding. The small angle, exact- or slightly off-coincidence boundaries are resistant to intergranular fracture at high temperatures.

Acknowledgements

The present author should like to express his sincere thanks in loving memory to the late Dr. P.W.Davies who properly brought the present author into the field of grain boundary research, in Swansea at the time of 1970. He also thanks Dr.B.Wilshire, Professor S.Karashima and Professor Y.Ishida for their warm-hearted encouragements and invaluable advices.

References

[1]   McLean,D., Creep and Fracture of Metals at High Temper-
      atures, National Physical Laboratory, 1956, p.73.
[2]   McLean,D.,and Farmer,M.H.,J.Inst.Metals,1956/57,85,41.
[3]   Horton,C.A.P.,Scripta Met.,1969,3, 253, Acta Met.,1970,
      18, 1159.
[4]   Ishida,Y. and Henderson-Brown,M.,Acta Met.,1967,15,857.
[5]   Kegg,G.R.,Horton,C.A.P. and Silcock,J.M.,Phil.Mag.,1973,
      27, 1041.
[6]   Pond,R.C.,Smith,D. and Southerden,P.W., Phil.Mag.,1978,
      A37, 27.

[7]   Watanabe,T.,Kuriyama,N.,and Karashima,S.,Proceedings of the
      Fourth International Conference on the Strength of Metals
      and Alloys, Nancy, 1976, p.383.
[8]   Watanabe,T.,Yamada,M.,Shima,S.,and Karashima,S.,Phil.Mag.,
      1979,A40, 667.
[9]   Gifkins,R.C. and Snowden,K.U.,Nature,1966,212,916.
[10] Ashby,M.F.,Raj,R.,and Gifkins,R.C.,Scripta Met.,1970,4,737.
[11] Watanabe,T.,and Davies,P.W., Unpublished work.
[12] Grain Boundary Structure and Properties, Academic Press,
      1976.
[13] Biscondi,M.,and Goux,C.,Mem.Sci.Rev.Metall.,1968,65, 167.
[14] Lagarde,P.,and Biscondi,M.,Can.Met.Quart.,1974,13, 245.
[15] Lagarde,P.,and Biscondi,M.,Mem.Sci.Rev.Metall.,1974,71,121.
[16] Michaut,B.,Silvent,A.,and Sainfort,G.,Mem.Sci.Rev.Metall.,
      1974, 71, 525.
[17] Bruggeman,G.A.,Bishop,G.H.,and Hartt,W.H.,Nature and
      Behavior of Grain Boundaries, edited by Hsun Hu, Plenum
      Press, 1972, p.83.
[18] Watanabe,T.,Kokawa,H.,and Karashima,S.,The 14th Symposium
      on High Temperature Strength of Materials, The Society of
      Materials Science of Japan, 1976, p.1.
[19] Kokawa,H.,Watanabe,T.,and Karashima,S.,To be published.
[20] McLean,D., Phil.Mag.,1971, 23, 467.
[21] Watanabe,T.,Suzuki,M.,and Karashima,S.,Proceedings of the
      Fifth International Conference on the Strength of Metals
      and Alloys, Aachen, Pergamon Press, 1979, p.445.
[22] Fujino,M, and Taira,S., Mechanical Behaviour of Materials,
      Pergamon Press, 1979, Vol.2, p.49.
[23] Watanabe,T.,and Davies,P.W.,Phi.Mag.,A37, 649.

CAVITATION AND FRACTURE OF MICRODUPLEX
CuZnCo AND CuZnCr DURING SUPERPLASTIC DEFORMATION

T. CHANDRA* and I. UEBEL**
*Department of Metallurgy, The University of Wollongong,
Wollongong, N.S.W., 2500, Australia

**Systems Technology Division, Australian Iron &
Steel Pty. Ltd., Port Kembla, N.S.W., 2505, Australia

ABSTRACT

A study has been made of cavitation behaviour during superplastic tensile flow in two commercial α/β brasses one modified with 2wt% cobalt (CuZnCo) and the other with 2wt% chromium (CuZnCr). The chromium and cobalt modified brasses attained maximum neck-free elongations of ~ 200% and ~ 170% respectively at a strain rate of 10^{-3} S^{-1} at 785ºC. At this temperature, both alloys contain equivolume proportions of α and β phases. Cavitation occurred during deformation and has been studied using quantitative metallographic (Q.T.M.) technique. Cavities nucleated at α/β boundaries, triple points and especially at second phase particles.

The extent of cavitation increased with strain in CuZnCo and CuZnCr alloys but a marked difference in the degree of cavitation was observed in these alloys. The CuZnCo cavitated extensively whilst a very few, small cavities were found in CuZnCr alloy under identical testing conditions. CuZnCo attained maximum level of cavitation at an intermediate strain rate of 10^{-3} S^{-1} at 785ºC whereas the extent of cavitation was minimum in CuZnCr under these conditions. It was also found that the alloy with cobalt exhibited a higher flow stress than the one with chromium over a wide range of strain rates at 785ºC. The influence of these variables on cavitation behaviour in CuZnCo and CuZnCr alloys can be interpreted in terms of their effects on cavity nucleation and cavity growth rates and a criterion for failure is suggested.

1.   INTRODUCTION

It is now well documented that during superplastic deformation in tension a number of microduplex alloys cavitate, and the factors which influence the degree of cavitation in Cu-alloys (1-6), Al-alloys (7-9), Fe-alloys (10-12) and most recently in Zn-22%Al alloys (13, 14) have been studied. At least for some alloys, the existence of cavitation would impose a limitation on the mechanical properties of components produced by superplastic blow-forming (15) which would in turn severely inhibit their commercial exploitation. Previous work in the author's department has been concerned with the super-plastic deformation behaviour of Cerium modified α/β brass and it was found that cavitation was markedly reduced by minor additions of cerium (16). Sagat and Taplin (5) reported that a fine grained α/β brass with coarse iron particles in the matrix exhibited superplasticity at $600^{\circ}C$ but extensive cavitation occurred during tensile deformation. The work of Schelleng and Reynolds (3) and more recently in Cu-Zn-Ni alloys (6) have also indicated a significant cavitation in these alloys under conditions of superplastic deformation.

The present work is concerned with identifying other systems which exhibit cavitation during superplastic deform-ation. The aim was to investigate effects of strain, strain rate and temperature on cavitation behaviour in two micro-duplex α/β brasses one modified with cobalt and the other with chromium. Both these alloying elements form second phase particles which are distributed uniformly in the matrix. Cavi-tation in these alloys have been studied employing metallo-graphic and Q.T.M. techniques, and the influence of cavitation on fracture behaviour has been examined.

2.   EXPERIMENTAL

The alloys investigated were two alpha/beta brasses with additions of approximately 2%wt. chromium and 2%wt. cobalt, respectively, which were prepared by Olin Corporation, New Haven, U.S.A. The compositions of these alloys are given in Table 1. The as-received hot extruded alloys had different grain sizes but a stable grain/phase size of ~ 32μm in both brasses was obtained by appropriate heat treatments. The ratio of alpha and beta phase in CuZnCo and CuZnCr alloys was found to be 1:1 at $785^{\circ}C$.

TABLE 1. Composition (wt.%) of CuZnCr and CuZnCo Alloys

Alloy	Cu	Zn	Cr	Co	Bi	Pb	Sn	Cd
CuZnCr	57.6	Rem	1.82	-	.0002	.01	.03	.05
CuZnCo	58.2	Rem	-	1.8	.0002	.01	.03	.05

Cylindrical tensile specimens of 16mm gauge length and 4.5mm diameter were machined and tensile tests were carried out on a floor mounted screw driven, constant crosshead velocity tensile machine fitted with a push button speed selector. A steady temperature with a tolerance of $\pm\ 2^{\circ}C$ over a 100mm length was maintained by means of a three-zone, verticle furnace with independent proportional temperature

control on each zone. Tests were carried out in the temperature range 600-900°C over an initial strain-rate range from
10^{-5} to 10^{-1} S^{-1}. All tests were conducted in air and prior
to testing, specimens were allowed to equilibrate at the test
temperature for 45 minutes. On completion of the test, the
specimens were cooled rapidly and sectioned on the longitudinal
axes for optical metallography. Polishing was carried out by
the method of attack skid technique (17) to preserve the
edges of cavities.

Superplastic behaviour of CuZnCo and CuZnCr was characterized both by the determination of strain rate sensitivity
(m) values and by straining specimens to failure at various
strain rates for a range of temperatures. A Quantimet (Q.T.M.)
image analysing computer was used to determine the volume
fraction of cavities after deformation to different strains
and strain rates. The volume fraction of cavities was measured
in 20 separate areas along the gauge length of specimen.
From these measurements the average volume fraction was
calculated.

3. RESULTS

To determine the optimum temperature for superplastic
deformation, CuZnCo and CuZnCr alloys with an average grain/
phase size of ~ 32μm were tested at various temperatures
(600-900°C) at a constant cross head velocity corresponding
to an initial strain rate of 10^{-3} S^{-1}. The results are
recorded in Figure 1 where it can be seen that maximum tensile
elongations for CuZnCo and CuZnCr alloys occurred at 785°C. It
should be noted that this is the temperature at which these
alloys attain equivolume proportions of alpha and beta phases.

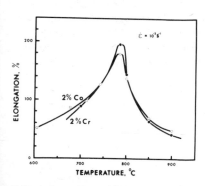

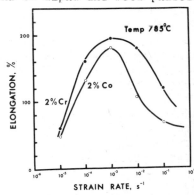

Fig. 1. Tensile elongations to
fracture vs temperature for
CuZnCo and CuZnCr deformed at
an initial strain rate 10^{-3} S^{-1}.

Fig. 2. Tensile elongations
to fracture vs initial
strain rate for CuZnCo and
and CuZnCr deformed at 785°C.

To obtain the optimum strain-rate for this temperature, a
series of tensile tests were conducted at various constant
cross head velocities. Figure 2 shows the plots of percent

elongation versus strain rate for CuZnCo and CuZnCr alloys.
The deformation was uniform at a strain-rate of 10^{-3} S^{-1} and
also at this strain rate the ductility was maximum. The effect
of strain rate on the maximum stress at 785°C in a log stress
vs log strain rate plot is shown in Figure 3. From the slope
of this plot, the strain-rate sensitivity index, m, may be
determined, and it can be seen that a peak in m values for Co
and Cr alloys occurs at intermediate strain rates. The strain
rate sensitivity index, m, as a function of strain rate was
determined from the tangential slope of Figure 3 and these
results are recorded in Figure 4.

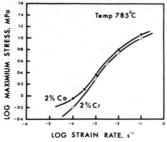

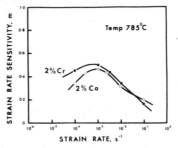

Fig. 3. Effect of initial
strain rate on maximum
engineering stress for
CuZnCo and CuZnCr tested
at 785°C.

Fig. 4. Strain rate sensitivity
vs initial strain rate for CuZnCo
and CuZnCr at 785°C.

A series of photomicrographs showing a typical distribution of cavities along the gauge length of CuZnCo and CuZnCr
specimens deformed under superplastic conditions are presented
in Figures 5 and 6. Cavities are nucleated at various grain
and phase boundaries, triple points and particularly second
phase particules. At low strains, small and few cavities are
observed in chromium modified specimen (Fig.5a), but both
number and volume of cavities increase with strain. Individual
cavities grow to some extent but the larger cavities are formed
by the interlinking of several smaller cavities by internal
necking between them (Fig.5b). It can be seen (Figs.6a and6b)
that cobalt modified brass showed much higher cavitation level
under identical testing conditions. To examine the effect of
strain rate on the level of cavitation, both alloys were
strained to 60% elongation at 785°C at different initial
strain rates. The results are recorded in Figure 7. The
results show that there was a marked difference in the level of
cavitation in CuZnCo and CuZnCr specimens under identical
testing conditions. The CuZnCo showed a maximum in void volume
at an intermediate strain rate at 785°C whilst a minimum in
void volume was attained by CuZnCr alloy under these conditions.
On the other hand, the extent of cavitation was reduced at low
and high strain rate in CuZnCo but reverse was true in CuZnCr
specimens at equivalent strains at 785°C.

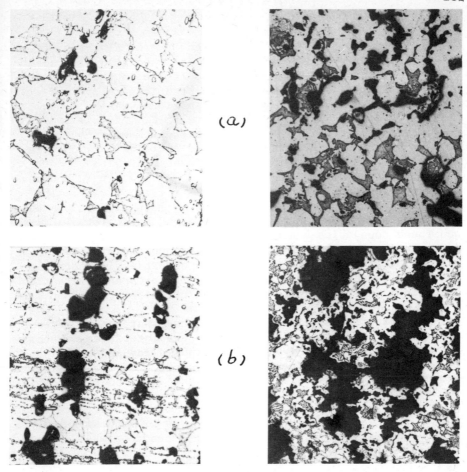

Fig. 5. Microstructure and cavitation in CuZnCr at 785°C and ἑ = 10⁻³ S⁻¹. Tensile axis is horizontal. X 350.
(a) 60% elongation
(b) 140% elongation

Fig. 6. Microstructure and cavitation in CuZnCo at 785°C and ἑ = 10⁻³ S⁻¹. Tensile axis is horizontal. X 350.
(a) 60% elongation
(b) 140% elongation

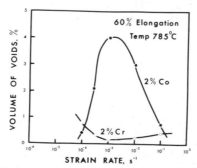

Fig. 7. Effect of strain rate on volume of voids in CuZnCo and CuZnCr deformed to 60% elongation at 785°C.

4. DISCUSSION

Both CuZnCo and CuZnCr alloys show the usual three stage $\log\sigma$ - $\log\dot{\varepsilon}$ curves in which state II is associated with superplastic behaviour and the maximum slopes lie in the range 0.4-0.5. Ductility is very sensitive to both temperature and strain rate, and these alloys show a maximum tensile elongation at a strain rate of 10^{-3} S^{-1} at $785^{\circ}C$. At this temperature, Co and Cr modified brasses have equivolume proportions of α and β phases and such a structure contains maximum area of α and β interfaces. It has been found that with the proportion of area of α/β boundaries at a maximum, grain boundary migration and thus grain growth is restricted (8, 18). The alpha/beta boundaries slide more readily than alpha/alpha and beta/beta boundaries as recently shown by Chandra et al. (19) and these boundaries exhibit the highest strain-rate sensitivity index. Since both grain size stability and grain boundary sliding are important requirements in superplastic materials these are therefore likely sources of the present observation of optimum ductility in CuZnCo and CuZnCr alloys at $785^{\circ}C$.

Quantitative metallography indicated that the preferential sites for cavity nucleation were the interfaces of coarse Cr-rich particles (5µm av.dia) in CuZnCr alloy and the fine Co-rich (0.3µm av.dia) particles in CuZnCo. At low strains, around 50% of the cavities are associated with co-rich particles while at high strains cavities formed preferentially at alpha/beta interfaces (Table 2). In CuZnCr alloy, however, cavities were associated with cr-rich particles at low and high strains under various strain rates. Electron microprobe analysis revealed these particles to be almost pure cobalt and chromium. As F.C.C. cobalt and B.C.C. chromium are quite soft at $785^{\circ}C$ compared with alpha/beta brass, the mechanical properties of these particles, therefore, cannot be responsible for cavity formation, rather the weak interface must be responsible. Besides cavity formation at second phase particles, the preponderance of cavities at alpha/beta interfaces in CuZnCo alloy could be due to several reasons such as : greater sliding at alpha/beta boundary or lower critical energy for cavity nucleation and/or of a decreased ability to accommodate sliding adjacent to this type of boundary.

TABLE 2. Relative Number of Cavities at Different Nucleation sites in CuZnCo and CuZnCr Alloys Strained at $10^{-3}sec^{-1}$

Strain	CuZnCo				CuZnCr			
	α/α	α/β	β/β	Fine Cobalt Particles	α/α	α/β	β/β	Chromium Particles
0.10	12%	39%	0%	49%	0%	5%	0%	95%
0.25	3%	59%	0%	38%	0%	8%	0%	92%
0.50	5%	58%	0%	37%	0%	16%	0%	84%

In a particular alloy system the formation of cavities during high temperature deformation will be dependent on strain, strain rate, temperature, grain size and stress. Many of these parameters are interdependent, and it is very difficult to determine the effect of one parameter without influencing others.

In the present work quantitative measurements have shown that the overall volume of voids produced during tensile deformation in CuZnCo and CuZnCr alloys is influenced by strain rate and strain. The sensitivity of void volume to strain rate and strain is consistent with the idea that all cavity nucleation does not occur early in the test, and the subsequent growth is controlled by sliding and the degree of which determined by strain and strain rate.

It is envisaged that a significant difference in the level of cavitation in CuZnCo and CuZnCr alloys at various strain rates observed in the present investigation could be the result of various combinations of nucleation and growth stages and the latter is determined by sliding and vacancy condensation. Since grain boundary sliding is a pre-requisite for the nucleation of grain boundary cavities and whenever the rate of grain boundary sliding exceeds the rate of accommodation by diffusion or dislocation motion at triple points, ledges and second phase particles, cavity nucleation occurs. In CuZnCr alloy the volume of voids vs strain rate relation exhibits a minimum at intermediate strain rates, where maximum ductility is achieved. The extent of cavitation at high and low strain rate was found to be lower. A high strain rate would lead to high stress concentrations and so a significant amount of cavity nucleation but the growth would be restricted due to the shorter time available for vacancy diffusion to the voids. For a low strain rate, the converse would be true. However, at intermediate strain rates the contribution from the grain boundary sliding to the overall strain is greatest resulting in ductility maximum but both nucleation of cavities and their interlinkage are severely restricted due to high strain rate sensitivity of material which inhibits the interlinkage of adjacent cavities analogous the way external necking is prevented (20).

On the other hand, in CuZnCo alloy at higher and lower strain rates the extent of cavitation is reduced but the maximum in cavitation level is observed at intermediate strain rates. This cavitation behaviour seems to result from the balance between stress and sliding both of which are necessary for cavity nucleation and growth (21). The amount of sliding may be different for different strain rates within stage II of

$\log\sigma - \log\dot{\varepsilon}$ curve. At low strain rates the stress is low and
at high strain rates the contribution due to sliding is reduced.
It is reasonable to suggest that sliding occurs to a maximum
extent at an intermediate strain rate during superplastic
deformation and so the maximum amount of cavitation is associated with maximum amount of sliding. These results are in
accordance with a recent model presented (21) and experimental
findings of Chandra et al. (19) and Fleck et al. (21). The
sensitivity of void volume to strain rate and strain in these
alloys therefore suggests that accommodation is very important
in the development of void once it has nucleated.

A series of photomicrographs (Fig. 5b & 6b) of CuZnCo and
CuZnCr specimens show the formation, growth and interlinkage
of voids in the final stages of deformation. As deformation
proceeds, the frequency of voids increases and they continue
to grow. The main feature of the rupture of these alloys via
cavitation is the extended stage of linkage. The high value of
m inhibits the internal necking between adjacent voids thus
restricting their growth. It has been found in this work that
m decreases gradually with strain during tensile deformation
and this is attributable to grain coarsening and also to the
formation of internal voids. Any microstructural coarsening
destroys the superplastic properties of material by reducing
the strain-rate sensitivity and so higher flow stress is
needed for further deformation.

Material between cavities can be thought of as internal
necks which develop after a strain in an analogous way to the
development of external necks. The low strain rate sensitivity
in the neck between adjacent voids arises primarily due to high
local strain rate between voids (5, 22). When the local strain
rate increases the level of flow stress increases locally
thus further superplastic deformation ceases in these necked
regions and deformation is continued in other parts of the
specimen. A point is reached when the deformation can not be
transferred to another region and rapid void linkage occurs
leading to failure of material. The instability criterion used
for failure in superplastic alloys in that of Brown and Embury
(23). Here the growth strain to a situation where the void
size is equal to the spacing is considered to be critical.
This approach can be adapted to the current situation but with
certain assumptions outlined by Raj (24).

5. CONCLUSIONS

5.1. Cavitation occurs during superplastic tensile straining
of α/β CuZnCo and α/β CuZnCr alloys. Cavities nucleat at α/β
boundaries, triple junctions but <u>primarily</u> at co-rich and cr-
rich second phase particles.

5.2. The extent of cavitation increases continuously with
strain in CuZnCo and CuZnCr alloys but a marked difference in

the degree of cavitation observed in these alloys. The cobalt modified α/β brass cavitates extensively whilst a very few and small cavities were observed at second phase particles in Cr-modified alloy under identical conditions.

5.3. In α/β CuZnCo alloy, the level of cavitation is maximum at intermediate strain rates at 785°C where maximum tensile elongation is attained. On the other hand, a minimum cavitation level is observed in α/β CuZnCr alloy under these conditions of superplastic deformation.

5.4. Crack formation as a result of cavity interlinkage occurs by internal necking between adjacent cavities in these alloys. Final fracture occurs by a multiple growth and interlinkage process.

ACKNOWLEDGEMENTS

The authors would like to express their gratitude to Professor D.M.R. Taplin, University of Waterloo, Canada, and Olin Corporation, New Haven, U.S.A. for providing the experimental materials. Financial support by The University of Wollongong is also gratefully acknowledged.

1. SAGAT, S. BLENKINSOP, P. and TAPLIN, D.M.R. - 'A Metallographic Study of Superplasticity and Cavitation in Microduplex Cu-40% Zn'. J. Inst. Metals, 1972, 100, 268.

2. FLECK, R.G. and TAPLIN, D.M.R. - 'Superplasticity in a Commercial Copper Dispersion Alloy'. Can. Met. Quart. 1972, 11, 299.

3. SCHELLENG, R.D. and REYNOLDS, G.H. - 'Superplasticity and Residual Tensile Properties of a Microduplex Copper-Nickel-Zinc Alloy'. Met. Trans., 1973, 4, 2199.

4. CHANDRA, T., JONAS, J.J. and TAPLIN, D.M.R. - 'A note on the Relationship Between Cavitation and Ductility in Microduplex Brasses' J Aust. Inst. Metals, 1975, 4, 220.

5. SAGAT, S. and TAPLIN, D.M.R. - 'Fracture of a Superplastic Ternary Brass'. Acta Met., 1976, 24, 309.

6. LIVESEY, D.W. and RIDLEY, N. - 'Superplastic Deformation, Cavitation and Fracture of Microduplex CU-NI-Zn Alloys'. Met. Trans., 1978, 9A, 519.

7. MATUKI, K. and YAMADA, M. - 'Superplastic Behaviour of AL-Zn-Mg Alloys'. J. Jap. Inst. Metals, 1973, 37, 448.

8. DUNLOP, G.L., SHAPIRO, E., TAPLIN, D.M.R. and CRANE, J. - 'Cavitation at grain and Phase Boundaries During Superplastic Flow of an Aluminium Bronze'. Met. Trans., 1973, 4, 2039.

9. TAPLIN, D.M.R. and SMITH, R.F. - 'Fracture During Superplastic Flow of Industrial Al-Mg Alloys'. 4th Int. Conf. on Fracture, Waterloo Canada, 1977, Vol. 2, 541, Ed. D.M.R. Taplin, Uni of Waterloo Press, 1977.

10. MORRISON, W.B. - 'Superplasticity in Low Alloy Steels'. Trans. ASM, 1968, 61, 523.

11. HUMPHRIES, C.W. and RIDLEY, N. - 'Cavitation in Alloy Steels During Superplastic Deformation'. J. Mat. Sci. 1974, 9, 1422.

12. SMITH, C.I., NORGATE, B. and RIDLEY, N. - 'Superplastic Deformation and Cavitation in a Microduplex Stainless Steel'. Metal. Sci., 1976, 10, 182.

13. ISHIKAWA, H., BHAT, D.G., MOHAMED, F.A. and LANGDON, T.G. - 'Evidence for Cavitation in the Superplastic Zn-22% Al Eutectoid'. Met. Trans., 1977, 8A, 523.

14. TAPLIN, D.M.R., DUNLOP, G.L. and LANGDON, T.G. - 'Flow and Fracture of Superplastic Materials'. Annual Review of Mat. Sci., 1979.

15. EDINGTON, J.W., MELTON, K.M. and CUTLER, C.P. - 'Superplasticity'. Prog, Mat. Sci., 1976, 21, 61.

16. CHANDRA, T., JONAS, J.J. and TAPLIN, D.M.R. - 'The Mechanical Behaviour of Cerium modified Alpha-beta Brass at High Temperatures'. J. Mat. Sci., 1976, 17, 1843.

17. COCKS, G.J. and TAPLIN, D.M.R. - 'An Appraisal of Certain Metallographic Techniques for Studying Cavities'. Metallurgia 1967, 75, 229.

18. BRIGHT, M.W.A. and TAPLIN, D.M.R. - 'The Effect of Composition on the Superplastic Behaviour of a Cu-Al-Fe Alloy'. CDC-ASM Conf. Paper No. 061/2, Copper Development Assoc. Inc., 405 Lexington Av., N.Y., N.Y., 1972.

19. CHANDRA, T., JONAS, J.J. and TAPLIN, D.M.R. - 'Grain Boundary Sliding and Intergranular Cavitation During Superplastic Deformation of Alpha/beta Brass'. J. Mat. Sci., 1978, 13, 2482.

20. THOMPSON, P.F. - 'A Theory of the Effects of Strain-Rate Sensitivity on Ductile Fracture'. J. Mat. Sci. 1969, 3, 139.

21.   FLECK, R.G., BEEVERS, C.J. and TAPLIN, D.M.R. - 'Hot Fracture of an Industrial Copper Base Alloy'. Met. Sci., 1975, 9, 49.

22.   CHANDRA, T. - 'Ph.D Thesis'. Uni of Waterloo, 1975.

23.   BROWN, L.M. and EMBURY, J.D. - 'The Initiation and Growth of Voids at Second Phase Particles'. Proc. Third Int. Conf. on the Strength of Metals and Alloys, Vol. 1, Cambridge, England, Aus, 1973, 164.

24.   RAJ, R. - 'Nucleation of Cavities at Second Phase Particles in Grain Boundaries'. Acta Met., 1978, 26, 995.

NUCLEATION AND GROWTH OF INTERGRANULAR CAVITIES
DURING CREEP OF TYPE 304 STAINLESS STEEL

I-W. Chen and A. S. Argon

Massachusetts Institute of Technology, Cambridge, Mass.
02139 U.S.A.

SUMMARY

Experimental measurements on intergranular cavity nucleation and growth in Type 304 Stainless Steel at 600°C and
700°C over a wide stress range are discussed in the light of
theoretical considerations for cavity nucleation on stressed
interfaces and models of cavity growth by combined surface
diffusion, grain-boundary diffusion, and matrix power-law
creep. This comparison indicates that when detected at a
size of c.a. 0.1 μm intergranular cavities are well in their
growth range and reflect the kinetics of growth rather than
that of nucleation. Furthermore, it is concluded that the
evolution of the observed cavity size distribution with time
can be accounted for only if cavity nucleation at small sizes
saturates and large cavities undergo accelerated growth.

1.   INTRODUCTION

Polycrystalline materials undergoing creep deformation
at elevated temperatures (T > 0.45 T_m) often fail by intergranular separation following extensive cavitation along
grain-boundaries which are approximately normal to the stress
axis. This significant observation of Greenwood and his co-
workers [1] almost 30 years ago has been followed by intensive
research ever since, motivated both by scientific interest
and by the consequence of failure of structural components in
high temperature applications. Conceptually the cavitation
process can be conveniently divided into two stages i.e.
nucleation and growth. Accordingly, a physical understanding
of either cavity nucleation or cavity growth can be more
readily obtained from separate observations of the two stages.
In most experiments, however, this separation of nucleation
and growth turns out to be very difficult if not impossible.
Furthermore, in most engineering alloys intended for high
temperature applications direct observation of cavities

is hampered by the complex multiphase microstructure typically found along grain-boundaries. On the other hand, indirect measurements such as the density change due to cavitation and the time to fracture are even more difficult to interpret unambiguously. Indeed, the density change method for example will experience considerable difficulty when applied to engineering alloys, at least at the early stage of cavitation, due to possible accompanying phase changes which often take place in these alloys. For the above reasons our understanding of cavitation is still incomplete today.

We have developed two useful techniques [2] recently to aid direct observation of cavities with minimal artifacts. With these techniques we have studied the cavitation process in Type 304 Stainless Steel in some detail. These results and other related developments in modeling are discussed in the present paper.

2. EXPERIMENTS

Commercial Type 304 Stainless Steel was used in this study. The as-received material was solution treated at 1050°C for 0.5 hr., water quenched, and aged at 775°C for 40 hours to precipitate a well defined family of grain-boundary carbides that were stable during subsequent creep cavitation experiments. The grain size of the aged material was 40-50 μm. The size p of grain boundary carbides follows roughly a Poisson distribution ranging from 0.05 μm to 1.8 μm with an average size of 0.4 μm at an average spacing of L = 1.4 μm. Specimens machined from this material were metallographically polished by hand to an optically shiny smooth surface finish. Specimens were crept in a specially constructed creep testing machine described in detail elsewhere [3]. All tests were performed in a vacuum of better than 5×10^{-6} torr pressure at operating temperatures of 600°C and 700°C.

We have found that cryogenic intergranular separation technique can be used conveniently to observe the development of creep cavities. This technique was used for magnesium alloys by Hyam [4], and for b.c.c. metals [5-7] – most notably α-iron by Taplin and Wingrove [5] and more recently by Cane and Greenwood [6]. The austenitic 304 stainless steel subject to the aging treatment prescribed above exhibits severe grain-boundary embrittlement when tested at cryogenic temperature at an impact strain rate. Crept specimens separated by this technique reveal numerous grain-boundary cavities with considerable detail.

A second technique was developed to permit detection of small cavities on a metallographic section without any use of etching. The detail of this technique was described by us

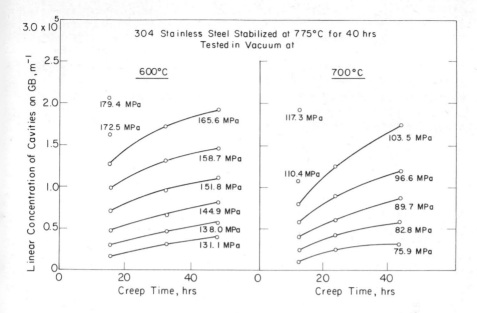

Fig. 1 Increase in linear concentration of cavities on grain-boundaries with creep time, at several stress levels, at 600°C and 700°C. (from Chen and Argon [2])

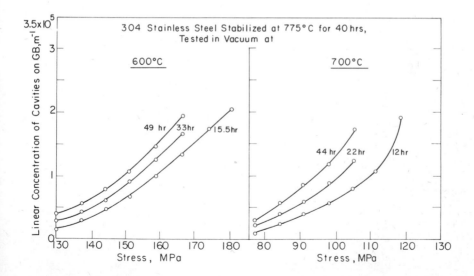

Fig. 2 Increase in linear concentration of cavities on grain boundaries with tensile stress in 304 stainless steel, at several stages in the creep life. (from Chen and Argon [2])

elsewhere [2]. Essentially a brief additional creep straining was employed to reveal cavities on a re-polished section of a crept specimen. By this procedure, intergranular cavities of a size larger than 0.1 μm could be detected readily on the slightly displaced boundaries against the still smooth background without any etching. More recently, prolonged vibratory polishing has been used in this laboratory for heavily crept samples, also with good success in the detection of creep cavities without any etching.

The increase in the linear concentration of cavities measured by the two-stage creep technique, averaged over all grain-boundary orientations, is shown in Fig. 1 for tests at 600°C and 700°C and at a variety of nominal stress levels of the specimen [2]. The same information replotted as a function of nominal stress level after different test times is shown in Fig. 2 [2]. A point of special interest is the dependence of the cavity concentration on the inclination of the surface trace of the boundary to the tensile direction. When the normalized linear cavity concentration at both 600°C and 700°C is plotted against this angle of inclination, all data at all nominal stress levels can be superimposed on a basic distribution as shown in Fig. 3 [2].

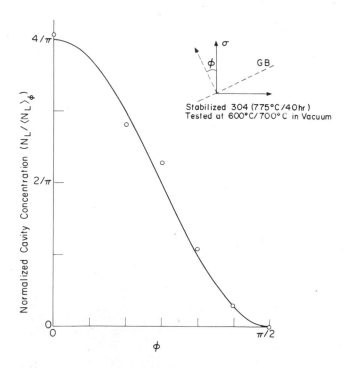

Fig. 3 Dependence of normalized cavity along boundaries on the inclination φ of the boundary with the tensile stress. (from Chen and Argon [2])

The stages of cavity growth measured by the cryogenic fracture technique after the test is shown in Fig. 4 for tests at 700°C and 63.1 MPa [2]. The size distribution of cavities follows roughly a Poisson distribution. The peak of this distribution gradually shifts toward larger sizes as creep strain increases but this is to some extent counteracted by the appearance of new cavities throughout the creep history. Thus Fig. 4 shows that a significant increase of numbers of smaller sizes at a later stage has shifted the peak backward for the largest strain. It is to be pointed out here that there are substantial variations of the cavity density and size distribution among different grain-boundaries and that the distributions in Fig. 4 represent only the averages over the entire specimen. For example, Fig. 5 shows the distributions of cavities in three specific grain-boundaries [2]. Each of the three distributions are distinctly different from the overall distribution. Nevertheless, the number densities, for each size group and as a whole, increase as creep time increases.

TYPE 304 STAINLESS STEEL TESTED AT
700°C AT 63 MPa IN VACUUM

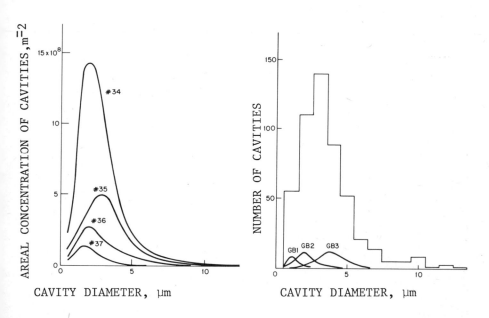

Fig. 4 Variation of cavity concentration with cavity diameter and creep time in Type 304 stainless steel. (#34, 139 hrs; #35, 72 hrs; #36, 36 hrs; #37, 19 hrs) (from Chen and Argon [2]).

Fig. 5 Overall size distribution of cavities averaged over many grain-boundaries. Three curves give size distribution of cavities on three individual grain-boundaries (from Chen and Argon[2])

3.   THEORY OF NUCLEATION AND GROWTH

Argon, Chen and Lau [8] have discussed in detail the
conditions of cavity nucleation. At elevated temperatures
where vacancy concentrations and mobility are high, it is
shown that a high level of interface stress is still required
for decohesion at particles by clustering of vacancies at
such interfaces in the context of classical nucleation theory.
At customary creep temperatures of 0.45 - 0.55 T_m, the required interface traction for decohesion approaches 5 x 10^{-3}E
representing a threshold stress for cavitation. Argon et al.
[8] have shown that, at around this threshold range, the
critical size cavity as a lenticular entity has a major axis
dimension of less than 20 Å and is made up of a cluster of
less than 150 vacancies. From these considerations, the
measurable dilation in the creeping alloys due to cavitation
at the time of nucleation was estimated to be as small as
4 x 10^{-9}. It is thus concluded that performing meaningful
measurements on nucleation rates of critical size cavities
is considerably beyond what is possible by any of the
currently available techniques of sampling. Therefore, any
detectable cavitation is usually well in the stage of growth.

In view of this finding one can alternatively adopt an
operational definition of the nucleation rate as the increase
in the number of cavities that become just detectable (e.g.
0.1 μm in size) per unit time interval during observation.
This concedes that smaller cavities could have gone through
certain uncharacterizable stages of evolution which makes
a direct interpretation based on nucleation theory not
possible since the cavities of a detectable size are already
two orders of magnitude larger than the critical size cavities
at nucleation. Hence the cavities that become just observable
are already fully in the stage of growth, reflecting the
kinetics of the latter process rather than of nucleation.
Indeed, inasmuch as a more refined separation of cavitation
into two distinct stages of nucleation and growth is not
practical, the information such as that shown in Figs. 4 and
5 is probably the most complete kind which can be gathered
experimentally. With the best resolution in size of cavity
and with sufficiently frequent sampling this data should
contain adequate information about both the overall nucleation rate, as operationally defined above, as well as for
kinetics of cavity growth.

Chen and Argon [9] have discussed a model for cavity
growth in which grain-boundary diffusion, surface diffusion
and matrix deformation by power-law creep are coupled
together in determining the shape of a cavity and its growth
rate. When only the quasi-equilibrium shape of a cavity is
considered, as is appropriate for small cavities under small
stress, their result in Eqns. (1) and (2) below is in

excellent agreement with the numerical solutions of Needleman and Rice [10], i.e.

$$\left(\frac{4\pi h(\psi)}{\dot{\epsilon}_\infty a}\frac{da}{dt}\right)_e$$
$$= 2\pi \left(\frac{\Lambda}{a}\right)^3 \left[\ln\left(\frac{a+\Lambda}{a}\right) + \left(\frac{a}{a+\Lambda}\right)^2 \left(1 - \frac{1}{4}\left(\frac{a}{a+\Lambda}\right)^2\right) - \frac{3}{4}\right]^{-1} \quad (1)$$

where $\Lambda = (D_b \delta_b \Omega \sigma_\alpha / kT\dot{\epsilon}_\infty)^{1/3}$ \quad (2)

is a critical diffusion distance within which diffusional
transport is the dominant mechanism of stress relief [9].
For a cavity of a crack-like shape, the growth rate becomes

$$\left(\frac{4\pi h(\psi)}{\dot{\epsilon}_\infty a}\frac{da}{dt}\right)_c$$
$$= \alpha \left(\frac{\Lambda}{a}\right)^{5/2} \left[\ln\left(\frac{a+\Lambda}{a}\right) + \left(\frac{a}{a+\Lambda}\right)^2 \left(1 - \frac{1}{4}\left(\frac{a}{a+\Lambda}\right)^2\right) - \frac{3}{4}\right]^{-3/2} \quad (3)$$

where $\alpha \left(\psi, \frac{D_b \delta_b}{D_s \delta_s}, \frac{\sigma_\alpha \Lambda}{\chi_s}\right) = \frac{4\pi h(\psi)}{[4\sin(\psi/2)]^{3/2}} \left[\left(\frac{D_b \delta_b}{D_s \delta_s}\right)\left(\frac{\sigma_\alpha \Lambda}{\chi_s}\right)\right]^{1/2}$ \quad (4)

is a coupling parameter giving a measure of the ratio of
boundary conductance to surface conductance [9]. In the
above equations, a is the major radius of a lenticular cavity
with an apex angle 2ψ, $h(\psi)$ is a geometric constant depending
on ψ, typically of the order of 0.6, $\dot{\epsilon}_\infty$ and σ_∞ are the
strain rate and stress normal to the grain-boundary, D_b and
δ_b are grain-boundary diffusivity and grain boundary thickness, D_s and δ_s are surface diffusivity and surface "thickness," χ_s is surface energy, kT has its usual meaning,
and subscripts "e" and "c" stand for "quasi-equilibrium" and
"crack-like" respectively. These results are shown in Fig. 6
in which the L-shape curve to the left of the figure corresponds to the growth rate for a quasi-equilibrium cavity
growing by combined diffusional matter transport and power-
law creep while the V-shape curves to the right are for
crack-like cavity growth with different coupling parameters
α which, as given by Eqn. (4), is a function of materials
constants and testing conditions. A transition from a quasi-
equilibrium mode of growth to a crack-like mode is shown in
Fig. 6. The transition occurs when the two modes give the
same growth rate. Physically, small cavities grow most
rapidly by the quasi-equilibrium mode described by Eqn. (1),
while the larger cavities grow preferentially by the crack-
like mode described by Eqn. (3) with an appropriate coupling

parameter α defined by Eqn. (4).

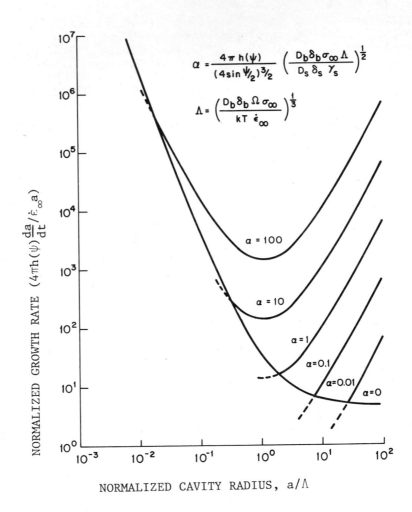

Fig. 6 Growth rate for quasi-equilibrium shape and crack-like cavities by combined diffusional transport and power-law creep. (from Chen and Argon [9]).

4. COMPARISON OF THEORY WITH EXPERIMENTS

When our experimental results on cavity densities are extropolated to short times, they appear to originate roughly at zero time. This indicates that some cavitation starts almost immediately upon the onset of creep deformation. We also note that there is a continued increase of cavities with time, at all stress levels above an apparent threshold value. These observations have been explained by Argon et al. [8] in terms of intermittent grain-boundary sliding [11] that sets up substantial interface tractions at grain-boundary particles over a time period of the order of 10 - 100 seconds following each grain-boundary sliding transient. Furthermore, since observable cavities are well in the stages of growth, it follows that the dependence of the areal density of cavities on the applied stress and the grain-boundary inclination both reflect more the rate of growth of such cavities rather than their rate of nucleation as already remarked above. Since cavities in the sub-micron size range grow primarily by the diffusional transport mechanism of Hull and Rimmer [12], the observed dependence on the normal stress acting across the grain-boundary is to be expected.

As we mentioned earlier the subsequent kinetics of cavity growth after they become observable should be compatible with a proper rendering of the growth theory in conjunction with the data on cavity nucleation. Our results as shown in Figs. 1 and 4 indicate, in agreement with several recent studies of kinetics of cavitation [6,13], that the rate of increase of observable cavities often varies rather slowly with time. For simplicity, we shall first assume a model of constant rate of nucleation of cavities at a single size, which is just detectable, i.e., c.a. 0.1 μm (actually the as-nucleated size of cavities will be very much smaller, but for our computation this is irrelevant). Since cavities of such small size most likely will grow by grain-boundary diffusion [12] at a rate $\frac{da}{dt} \propto 1/a^2$, it follows that any slight variation in cavity sizes at this stage is probably unimportant since the smaller cavities will catch up in a short time with larger cavities by growing much faster than the latter. Thus for our purpose, an initial size a_o shall be taken for each of the newly emerging cavity. The resultant size distribution of cavities can then be estimated by the elementary method to yield

$$n(a) = I \frac{dt}{da}$$

$$= I \frac{2}{\dot{\varepsilon}_\infty} \frac{h(\psi)}{a} \left(\frac{a}{\Lambda}\right)^3 \left[\ln \left(\frac{a+\Lambda}{a}\right) + \left(\frac{a}{a+\Lambda}\right)^2 \left(1 - \frac{1}{4}\left(\frac{a}{a+\Lambda}\right)^2\right) - \frac{3}{4}\right] \quad (5)$$

for cavities growing at quasi-equilibrium shapes and

$$n(a) = I \frac{4\pi h(\psi)}{\varepsilon_\infty a} \frac{1}{\alpha} \left(\frac{a}{\Lambda}\right)^{\frac{5}{2}} [\ln \left(\frac{a+\Lambda}{a}\right) + \left(\frac{a}{a+\Lambda}\right)^2 (1 - \frac{1}{4} \left(\frac{a}{a+\Lambda}\right)^2) - \frac{3}{4}]^{\frac{3}{2}}$$

(6)

for cavities with crack-like shape. In the above equations, n(a) is the size distribution of cavities, and I is the nucleation rate at a single small size a_o. It should be noted that the distribution n(a) has an upper cut-off at the radius a_c which is the radius of the biggest cavity, namely the one nucleated the earliest. These results are shown in Fig. 7 for a number of testing parameters.

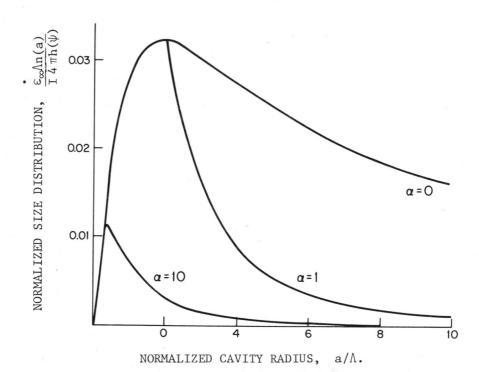

Fig. 7 Cavities size distributions   predicted by Eqn. (5-6). A constant nucleation rate I is assumed at a single small cavity size a_o << Λ. It predicts an upper cut-off for the distributions at a radius a_c corresponding to the largest cavity size. Also the radius at the peak of the distributions decreases with α.

The distribution predicted by this model bears some resemblance to the measured distribution in Fig. 4. Essentially, the initial rise in number density at small cavity sizes is due to the decelerated growth of cavities when the growth rate is mainly controlled by grain-boundary diffusional flow in a manner first described by Hull and Rimmer [12]. In this regime, the growth rate is inversely proportional, roughly to the second power, of the cavity radius a. Consequently the population of cavities will become "jammed" in a narrow range of sizes after they have passed the earlier stage of fast growth. This jam will be relieved only later at a sufficiently large size when accelerated growth becomes possible either due to the power-law creep controlled cavity growth or by the crack-like growth mode. Then, a decrease in the number density at a large cavity size occurs and is accompanied by a wide spread in the cavity size. The integrated area under the distribution curve corresponds to the total number of cavities. It should be noted, that the distribution will always reach a steady state at smaller sizes, due to the assumption of the constant nucleation rate at a single small size of cavity. The only change that takes place as time proceeds is the gradual advance of the distribution front with an increasing cut-off radius a_c corresponding to the radius of the largest cavity. The peak of the distribution at a_p, once reached when $a_c \gtrsim a_p$, will stay at the same position regardless of further growth.

The above development can be extended further to include the possibility of non-steady state growth during a brief period at the beginning of the test. Essentially, we can assume that the steady state in nucleation and growth is reached at $t = t_o$, and that the initial distribution $n(a)$ at t_o can be an arbitrary function. Since at any later time $t > t_o$, the cavity nucleated at t_o will be of a radius a_c and the pre-existing cavities at t_o will have passed a_c, it follows that the distribution up to $a < a_c$ is given by Eqns. (5,6) and Fig. 7, while the only modification which is due to the non-steady state comes at $a < a_c$. If the duration of the non-steady portion of nucleation is short, as indeed it seems to be, judging from most data on the nucleation rate, then the modification at $a > a_c$ is insignificant when a_c is sufficiently large, i.e. when a steady state has prevailed for a long enough time.

We now compare these results with the measured distribution of Fig. 4 of Type 304 Stainless Steel and two other measurements, Fig. 8 of Type 347 Stainless Steel due to Needham and Gladman [13] and Fig. 9 of α-iron due to Cane and Greenwood [6]. It is clear that the simple model described above is not capable to account fully for all of the

300

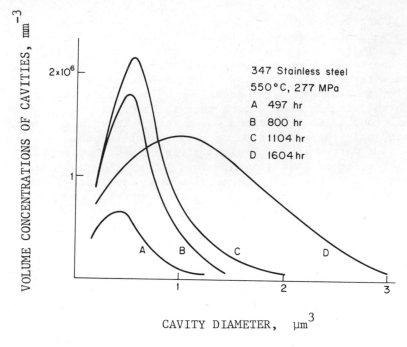

Fig. 8 Size distribution of cavities in Type 347 Stainless
Steel. (from Needham and Gladman [13]).

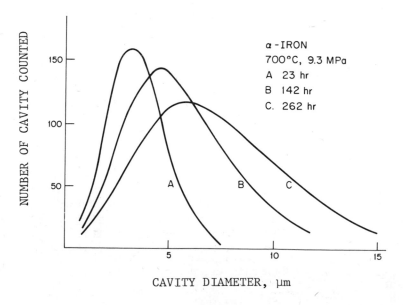

Fig. 9 Size distribution of cavities in α-iron. (from Cane
and Greenwood [6])

experimental observations. Most significantly, the experimental results indicate no steady state distribution for either small cavity sizes or the position of the peak of the distribution. In fact, it should be clear that this discrepancy can not be remedied by the modification of the growth theory given by Eqns. (1-4), since any growth theory will give essentially the same predictions as those in the above two paragraphs, only differing in the detail of the distribution profile.

A closer examination following the argument outlined above indicates that either accelerated nucleation or decreasing growth is necessary to yield an increasing n(a) with time at smaller cavity sizes as is found in 304 and 347 stainless steels, and shown in Figs. 4 and 8 respectively. This implies at a given radius a decrease in local stress or a decrease in diffusivity. Conversely either decelerated nucleation or increasing growth will yield a decreasing n(a) at smaller cavity sizes with time as observed in α-iron and shown in Fig. 9. Since accelerated nucleation was not reported in these studies, this possibility can be dismissed for Types 304 and 347 stainless steels. For α-iron, however, the results of Cane and Greenwood [6] (their Fig. 2 and our Fig. 9) show rather pronounced deceleration of nucleation. Thus it seems that the measured distribution can be understood at least qualitatively; based on a modified growth theory in the case of 304 and 347 stainless steels by the decrease in growth rate for newly observed cavities and in the case of α-iron by a gradual exhaustion of nucleation sites.

The case of decreasing growth deserves more discussion. In our opinion, this pheonomenon which suggests that, at a given size, the cavities which form earlier grow faster than those which form later is closely linked to the heterogeneous nature of cavitation. Since factors affecting cavitation e.g. stresses, diffusivity and particle distribution can all vary from one boundary to the other or even within a specific one, it is very likely that the sites for more favorable nucleation also are ones where growth is expedited, although further speculation for the exact cause of this at this time does not seem to be warranted. Also neglected in this discussion is the possibility of coalescence which, as discussed by Chen and Argon [2], leads to broadening of the distribution as well and should occur more frequently for larger size cavities.

ACKNOWLEDGEMENT: This research was supported by the U.S. Department of Energy under Contract No. EG-77-S-02-4461.

302

REFERENCES

1. GREENWOOD, J. N., MILLER, D.R. and SUITER, J. W. -
    'Intergranular Cavitation in Stressed Metals,' Acta
    Met., 1954, 2, 250.

2. CHEN, I-W. and ARGON, A. S. - 'Creep Cavitation in 304
    Stainless Steel,' Submitted to Acta. Met., 1980.

3. CHEN, I-W. - Creep Cavitation in 304 Stainless Steel,
    Ph.D. Thesis, Dept. of Mater. Sci. and Eng., M.I.T.,
    February 1980.

4. HYAM, E.D. - 'Discussion on Creep Voids in Mg-1% Aℓ
    and Beryllium,' Structural Processes in Creep, Iron
    Steel Inst., London, 1961. p.76.

5. TAPLIN, D.M.R. and WINGROVE, A.L., - 'Study of Inter-
    granular Cavities in Iron by Electron Microscopy of
    Fracture Surfaces,' Acta. Met., 1967, 15, 1231.

6. CANE, B.J. and GREENWOOD, J.N. - 'The Nucleation and
    Growth of Cavities in Iron during Deformation at
    Elevated Temperatures,' Met. Sci., 1975, 9, 55.

7. STIEGLER, J. O., FARRELL, K., LOH, B.T.M. and McCOY,
    H. E. - 'Nature of Creep Cavities in Tungsten,' Trans.
    Am. Soc. Met. 1967, 60, 494.

8. ARGON, A. S., CHEN, I-W. and LAU, C. W. - 'Intergranular
    Cavitation in Creep: Theory and Experiments,' Creep-
    Fatigue-Environment Interactions, Ed. Pelloux, R. M.
    and Stoloff, N. S., Met. Soc. AIME., 1980, P. 46.

9. CHEN, I-W. and ARGON, A. S. - 'Diffusive Growth of
    Grain-Boundary Cavities,' submitted to Acta. Met., 1980.

10. NEEDLEMAN, A. and RICE, J.R. - 'Plastic Creep Flow
    Effects in the Diffusive Cavitation of Grain-Boundaries,'
    Acta. Met., 1980, 28, 1315.

11. CHANG, H. C. and GRANT, N. J. - 'Inhomogeneity in Creep
    Deformation of Coarse Grained High Purity Aluminum,'
    Trans. AIME, 1953, 197, 1175.

12. HULL, D. and RIMMER, D.E. - 'The Growth of Grain-
    Boundary Voids Under Stress,' Phil. Mag., 1959, 4, 673.

13. NEEDHAM, N. G. and GLADMAN, T. - 'Nucleation and Growth
    of Creep Cavities in a Type 347 Steel,' Met. Sci., 1980,
    14, 64.

THE ROLE OF GRAIN BOUNDARY CAVITIES DURING TERTIARY CREEP

R.W. Evans and B. Wilshire

Department of Metallurgy and Materials Technology,
University College, Singleton Park, Swansea, U.K.

SUMMARY

When grain boundary cavity development causes the
acceleration in creep rate during the tertiary stage which
precedes fracture under high temperature creep conditions, the
tertiary creep rate ($\dot{\varepsilon}_t$) increases linearly with the tertiary
creep strain (ε_t) as

$$\dot{\varepsilon}_t = p\, \varepsilon_t$$

where p is a parameter related to the secondary creep
rate. Deviations from this equation are found only just prior
to fracture when the creep rate increases more rapidly than
expected from this linear relationship. This type of behaviour
can be interpreted on the basis that when creep occurs by a
vacancy emission process, irrespective of the detailed
mechanism involved, cavities can act as efficient vacancy
sinks leading to an acceleration in creep rate in zones adjacent to the cavitated boundaries. The deviation from the
linear $\dot{\varepsilon}_t/\varepsilon_t$ relationship can then be accounted for in terms
of the enhanced deformation rates obtained as a result of
stress concentration effects associated with the high incidence of cavities present immediately before fracture.

1.   INTRODUCTION

Although considerable attention has been devoted to the
examination of the processes controlling primary and secondary
creep behaviour and also to the study of the mechanisms of
intergranular creep failure, relatively few investigations
have considered the factors causing the acceleration in creep
rate during the tertiary stage which preceeds fracture. In
creep tests carried out at constant load, an acceleration in
creep rate can originate from the gradual decrease in cross
sectional area as deformation continues. However under constant stress conditions which compensate for the changes in

L

cross sectional area, the commencement of the tertiary stage during tensile creep tests is generally attributable to

    (a)   mechanical instability such as necking,
    (b)   microstructural instability such as overageing or
    (c)   the gradual formation of grain boundary cavities and cracks.

In the present investigation, detailed analyses have been made of the processes whereby an acceleration in creep rate can result from grain boundary cavity development in order to account for the tertiary creep behaviour exhibited by microstructurally stable metals and alloys.

2. INTERRELATION OF CAVITY DEVELOPMENT AND DEFORMATION BEHAVIOUR DURING TERTIARY CREEP

For a wide range of metals and alloys for which the tertiary stage begins as a result of grain boundary cavity development, the variation of the true creep strain (ε) with time (t) over most of the creep life has been shown [1] to be described accurately by the equation

$$\varepsilon = \varepsilon_o + \varepsilon_p (1-e^{-mt}) + \dot{\varepsilon}_s t + \varepsilon_z e^{p(t-t_t)} \tag{1}$$

where ε_o is the initial strain on loading, ε_p is the total primary creep strain, $\dot{\varepsilon}_s$ is the secondary creep rate, ε_z is a constant (~ 0.0001), m and p are parameters related to the rates of deceleration and acceleration in creep rate during primary and tertiary creep respectively and t_t is the time to the onset of tertiary creep. When the parameters in equation (1) were derived from a least squares analysis of the strain/time curves obtained for pure metals, single phase and two phase alloys [1, 2] which are microstructurally stable under creep conditions leading to failure by cavitation, the average difference between the measured creep strains (ε) and those calculated using the derived values of the parameters in equation (1) has been found to be less than 0.0001. However, late in the tertiary stage, deviations from this equation were reported (Fig. 1) which were attributed to opening of the intergranular cracks present immediately prior to fracture contributing to the overall creep strain [2]. For subsequent analysis, it will be convenient to express equation (1) in terms of strain rate and time. Defining ε_t as $\varepsilon - (\varepsilon_o + \varepsilon_p + \dot{\varepsilon}_s t)$ allows equation (1) to be transformed to either

$$\dot{\varepsilon}_t = \dot{\varepsilon} - \dot{\varepsilon}_s = p\varepsilon_t \tag{2a}$$

$$\text{or} \qquad \frac{\dot{\varepsilon}}{\dot{\varepsilon}_s} = 1 + p\,\frac{\varepsilon_t}{\dot{\varepsilon}_s} \tag{2b}$$

where p is proportional to $\dot{\varepsilon}_s$ [1, 2].

2.1. Cavity Development

Density measurements and metallographic studies (3, 4) have established that the fractional void volume varies with the creep strain (Fig. 2) as

$$\Delta v_{/v} \propto (\varepsilon - \varepsilon^*) \qquad (3)$$

where ε^* represents the creep strain at which the void volume/strain relationship becomes linear. The void volume therefore increases linearly with strain throughout the tertiary stage. Comparison of equations (2) and (3) then suggests that the creep rate at any fraction of the tertiary creep life is proportional to the void volume. A direct dependence of the tertiary strain on void volume would be expected from a model for tertiary creep (5) based on cavity growth by absorption of vacancies from the surrounding grain boundary. This process of void growth is equivalent to plating out of atoms on boundaries perpendicular to the applied stress, thereby achieving creep strain in a direction parallel to the stress axis. However, this approach cannot account for the fact that the tertiary creep strains are frequently found

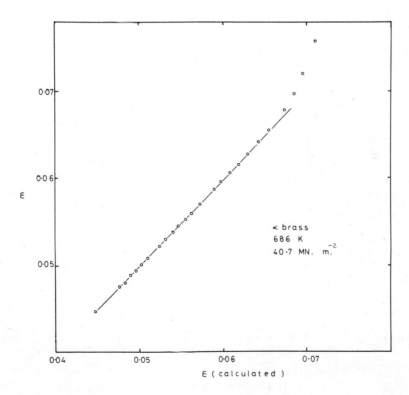

Figure 1   Relationship between the observed creep strain in tertiary and the strain calculated from equation (1).

306

to be considerably greater than the total void volume.
Similarly since the total void volume is usually less than 1%
at fracture (Fig. 2) and is markedly lower than this value at
the end of the secondary creep stage, the initial acceleration in creep rate at the commencement of the tertiary stage
cannot be attributed directly to the reduction in average
cross-sectional area associated with cavity development. This
view is supported by the results of experiments carried out
with pure gold (6). Testpieces crept into the tertiary stage
were annealed to move the grain boundaries away from the
cavities present, without altering the specimen density or the
cavity distribution. On reloading under the original test
conditions the sample did not continue in tertiary. Instead,
the secondary creep stage was restablished. Following this
annealing procedure, tertiary creep began again only when new
cavities were developed on grain boundaries. For materials
which are microstructurally stable during creep, the tertiary
stage therefore occurs only when cavities are developed on
grain boundaries.

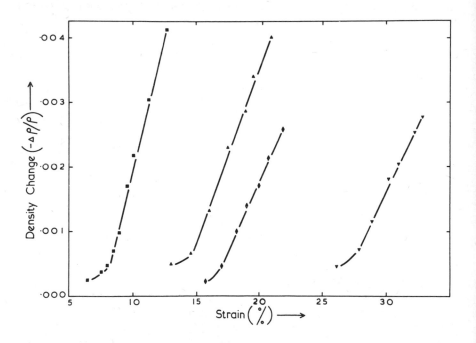

Figure 2   The variation of density change with strain for a
            Ni-0.1% Palladium alloy tested at 776K (4).

$$100.4 \text{ MNm}^{-2} \quad (\blacksquare)$$
$$154.4 \text{ MNm}^{-2} \quad (\blacktriangle)$$
$$177.6 \text{ MNm}^{-2} \quad (\blacklozenge)$$
$$216.2 \text{ MNm}^{-2} \quad (\blacktriangledown)$$

2.2. <u>Deformation Enhancement in Grain Boundary Zones</u>

The importance of cavity enhancement of deformation
processes occurring in the grain boundary regions has been
demonstrated in a recent study of the effects of prestrain on
the creep and fracture behaviour of polycrystalline copper (7).
It can be assumed that the secondary creep rate represents the
deformation rate unaffected by cavity formation. The acceleration in creep rate during the tertiary stage then reflects the
enhancement of the deformation processes due to the presence
of an increasing total volume of cavities. The time dependence
of the tertiary creep strain (ε_t) was then considered by
reference to the 'fractional time' U_t spent in the tertiary
stage which was defined as

$$U_t = \frac{t - t_t}{t_f - t_t} \tag{4}$$

where t_f is the time to fracture. The variation of ε_t
with U_t is illustrated in Fig. 3 for annealed copper and for
samples which had been prestrained by 7.5% at the creep temperature prior to the commencement of the creep test. For

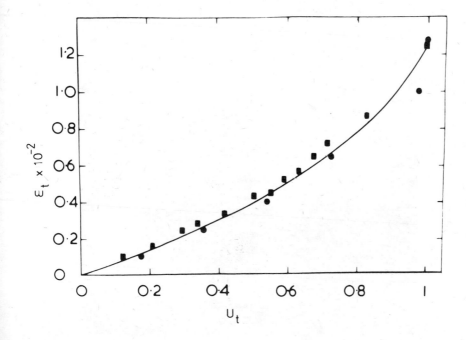

<u>Figure 3</u>   The variation of the tertiary creep strain, ε_t,
with the fractional time, U_t, spent in the tertiary
stage for tests carried out at 686K and 55 MNm^{-2}
for annealed copper and samples prestrained 7.5%
at the creep temperature (7).

the same applied stress level, the tertiary creep behaviour
was unaffected even though the prestrain treatment resulted in
marked reductions in the secondary creep rate and the overall
creep ductility. The cavity distribution at fracture appeared,
however, to be unaffected by prestraining (Fig. 4). Furthermore, measurements of the sliding displacements at grain boundaries indicated that the deformation behaviour in the grain
boundary regions with the prestrained samples was identical to
that for the annealed material. It was therefore concluded (7)
that the prestrain markedly reduced the overall creep rate and
total creep strain by restricting grain deformation without a
corresponding effect on the grain boundary zone behaviour.
The observation that the tertiary creep behaviour was comparable
for the annealed and for the prestrained samples (Fig. 3),
therefore suggests that the acceleration in creep rate is a
consequence of cavitation which enhances the deformation rate
in the grain boundary zones.

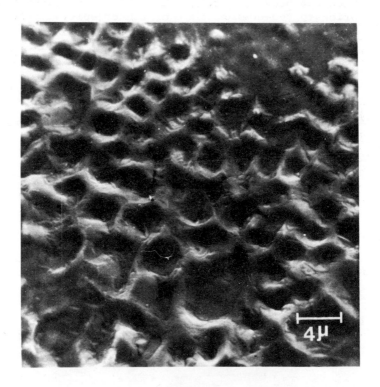

Figure 4   Scanning electron micrograph showing grain
            boundary cavities in annealed copper tested
            to fracture at 686K and 55 MNm^{-2} (7).

3. MODELS FOR TERTIARY CREEP

Consider a stable material deforming in secondary creep. The secondary creep rate matrix $[_s\dot{\varepsilon}_{ij}]$ will be related (8) to the deviatoric stress matrix $[\sigma'_{ij}]$ by the equations

$$[_s\dot{\varepsilon}_{ij}] = \frac{3\bar{\varepsilon}}{2\bar{\sigma}} [\sigma'_{ij}] = A\bar{\sigma}^{n-1} [\sigma'_{ij}] \qquad (5)$$

where $\bar{\varepsilon}$ and $\bar{\sigma}$ are the effective strain rate and stress and A is a constant describing temperature and structure variations in creep rate. The acceleration in creep rate during tertiary may be ascribed either to stress concentration effects which modify $[\sigma'_{ij}]$ in the vicinity of cavities or to a change in the basic deformation process which affects A. Fig. 5 shows a grain boundary whose normal makes an angle θ with the tensile direction. The boundary, at any stage, contains cavities of radius r and spacing 2ℓ and the figure represents a 'cell' containing a single void, the boundary being surrounded by a deformation zone of width 2d.

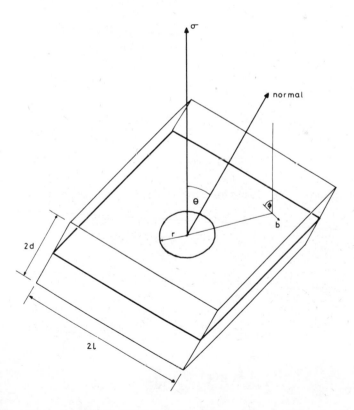

<u>Figure 5</u>   Schematic representation of the deforming zone around a grain boundary cavity.

3.1. Stress Concentration Effects

The lenticular cavities normally developed under high temperature creep conditions approximate more to spheroidal voids than to sharp cracks. The stress and strain concentrations around a spherical cavity in purely elastic materials are considerably smaller than those for elliptical and cylindrical cracks (9). For visco-plastic materials, no analysis for spherical cavities is available, but Nadai (10) has shown that as n increases, the stress concentrations for certain special cases reduce to values considerably below those for elastic behaviour. It does not then appear that there will be any appreciable stress concentrations for the situation depicted in Fig. 5. However, there remains the effect of a reduced stress bearing area which may be estimated as follows.

Erect a set of Cartesian axes with the origin at the cavity centre and one axis normal to the grain boundary plane. For any plane parallel to the grain boundary and at a distance x from it, the stress system when the cavity is present will be $y(x) \left[\sigma'_{ij}\right]$ where

$$y(x) = \frac{4\ell^2}{4\ell^2 - \pi (r^2 - x^2)} \qquad 0 \leqslant x \leqslant r \tag{6}$$

$$y(x) = 1 \qquad\qquad\qquad r < x \leqslant d$$

For this plane, the creep rate $\left[\dot{e}_{ij}\right]$ will be

$$\left[\dot{e}_{ij}\right] = A\, y(x)^{n-1}\, (\bar{\sigma})^{n-1} \cdot y(x) \left[\sigma'_{ij}\right]$$

$$= y(x)^n \left[_s\dot{\varepsilon}_{ij}\right] \tag{7}$$

The creep rate for the whole deforming zone will be $\left[\dot{\varepsilon}_{ij}\right]$ where

$$\left[\dot{\varepsilon}_{ij}\right] = \frac{1}{d} \int_0^d y(x)^n \left[_s\dot{\varepsilon}_{ij}\right] dx$$

i.e. $$\left[\dot{\varepsilon}_{ij}\right] = \frac{1}{d} \left(\int_0^r \left(\frac{4\ell^2}{4\ell^2 - \pi r^2 - \pi x^2}\right)^n dx + \int_r^d dx \right) \left[_s\dot{\varepsilon}_{ij}\right]$$

$$\left[\dot{\varepsilon}_{ij}\right] = g(r) \left[_s\dot{\varepsilon}_{ij}\right] \tag{8}$$

g(r) is the ratio of any strain rate component in tertiary to that in secondary, namely, $\dot{\varepsilon}/\dot{\varepsilon}_s$ (equation 2b). Fig. 6 shows g(r) as a function of void volume per unit grain boundary area

$(\pi r^3/3\ell^2)$ for n = 4 and typical values of ℓ and d. Boundaries
of different θ will have different ℓ and r and a suitable
average of the curves must be taken. However, in all cases,
there is a rapid increase of creep rate with void volume and
hence with ε_t. The shape of the creep curves predicted cannot
therefore be reconciled with the experimental observations contained in equation (2b) except perhaps when deviations from
the linear $\dot\varepsilon_t/\varepsilon_t$ relationship occur just prior to fracture
(Fig. 1) when the total void volume becomes relatively large.

3.2. Enhancement of Deformation Processes

It is here convenient to calculate $\dot\varepsilon_t$. Grain boundaries
with large θ values will not generally act as vacancy sinks
during creep in uniaxial tension but the appearance of cavities
changes the situation. When creep is occurring by a vacancy
emission process, then regardless of the detailed mechanism
involved, cavities may lead to an acceleration in creep rate
by acting as efficient vacancy sinks. Consider a dislocation
at any point in the boundary cell (Fig. 5) whose Burgers
vector makes an angle ϕ with the tensile stress σ. The dislocation will be saturated with vacancies (11) provided that

$$\frac{L}{b} > \exp\left(\frac{\mu b^3}{10kT}\right) \tag{9}$$

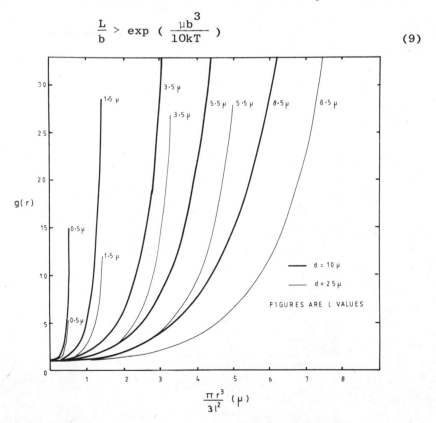

Figure 6   Variation of the function g(r) with void
            volume for various cavity spacings.

312

where L and μ are the average dislocation spacing and shear modulus respectively. Estimation of the terms indicates that this will generally be so for recovery creep and the rate of vacancy emission (and hence creep) will depend on the rate of diffusion between source and sink. The difference in chemical potential (Δf) for vacancies between the dislocation and sink will be (12)

$$\Delta f = \Omega \sigma \cos^2 \phi - \frac{2\gamma \Omega}{r} \tag{10}$$

where Ω and γ are the atomic volume and surface energy. The total number of vacancies entering the void will be $4\pi r^2 \partial f/\partial r$. $\partial f/\partial r$ can be approximated by Δf divided by the mean distance between dislocation and sink. The additional creep rate due to the void can be written as

$$\dot{\varepsilon}_t = \frac{B. \ 8\pi \ r^2 \ \Omega}{\sqrt{2\ell^2 + d^2} - 2r} \left(\sigma \cos^2 \phi - \frac{2\gamma}{r} \right) \tag{11}$$

where B contains the diffusion coefficient and other structural parameters. φ will be symmetrically distributed about π/4 so

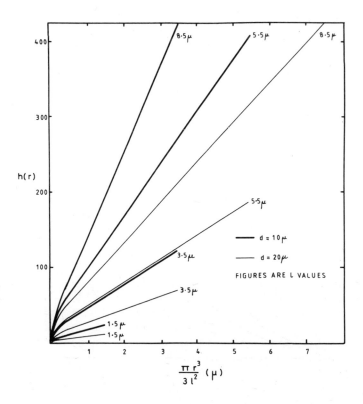

Figure 7   Variation of the function h(r) with void volume for various cavity spacings.

that the mean value of $\cos^2\phi$ will be $\frac{1}{2}$. The mechanism will only operate for $r > 4\gamma/\sigma$ so that if $r_o = 4\gamma/\sigma$,

$$\dot{\varepsilon}_t \propto h(r) = \frac{r^2}{\sqrt{2\ell^2 + d^2} - 2r} \left(\frac{1}{r_o} - \frac{1}{r} \right) \tag{12}$$

Fig. 7 shows h(r) as a function of $\pi r^3/3\ell^2$. Remembering the linear relationship between void volume and strain (equation 3) it is clear that the shapes of the creep curves predicted by this approach are in accordance with equation (2a) except at very small volume fractions of cavities.

4. CONCLUSIONS

1) For a wide range of metals and alloys for which the tertiary stage begins as a result of grain boundary cavity development, the variation of the true creep strain with time over most of the creep life can be described accurately by the equation

$$\varepsilon = \varepsilon_o + \varepsilon_p (1-e^{-mt}) + \dot{\varepsilon}_s t + \varepsilon_z e^{p(t-t_t)}$$

so that

$$\dot{\varepsilon}_t = \dot{\varepsilon} - \dot{\varepsilon}_s = p\varepsilon_t$$

with p related to the secondary creep rate, $\dot{\varepsilon}_s$. Deviations from this equation are found just prior to fracture when the creep rate increases more rapidly than expected from this linear $\dot{\varepsilon}_t/\varepsilon_t$ relationship.

2) Throughout the tertiary stage, the total void volume increases linearly with creep strain. This increasing void volume causes the acceleration in creep rate by the enhancement of the deformation rates in regions of the grains adjacent to the grain boundaries.

3) The exact form of the acceleration in creep rate can be accounted for on the basis that when creep occurs by a vacancy emission process, irrespective of the detailed mechanism involved, small cavities ($r > 4\gamma/\sigma$) can act as efficient vacancy sinks. This form of tertiary curve cannot be reconciled with stress concentration effects at the cavities except during the deviations from the linear $\dot{\varepsilon}_t/\varepsilon_t$ relationship which are found just prior to fracture when a relatively high incidence of cavities is present.

314

5.　REFERENCES

1.　DAVIES, P.W., WILLIAMS, K.R., EVANS, W.J. and WILSHIRE, B. 'An Equation to Represent Strain/Time Relationships during High Temperature Creep'. Scripta Metall., 1969, **3**, 671.

2.　EVANS, W.J. and WILSHIRE, B. - 'The High Temperature Creep and Fracture Behaviour of 70-30 α-Brass'. Met. Trans., 1970, **1**, 2133.

3.　BOETTNER, R.C., and ROBERTSON, W.D. - 'A Study of the Growth of Voids in Copper during the Creep Process by Measurement of the Accompanying Change in Density'. Trans. Met. Soc. A.I.M.E., 1961, **221**, 613.

4.　BOWRING, P., DAVIES, P.W. and WILSHIRE, B. - 'The Strain Dependence of Density Changes during Creep'. Metal Sci. J., 1968, **2**, 168.

5.　HARRIS, J.E., TUCKER, M.O. and GREENWOOD, G.W. - 'The Diffusional Creep Strain due to the Growth of Intergranular Voids'. Metal Sci., 1974, **8**, 311.

6.　DAVIES, P.W., and EVANS, R.W. - 'The Contribution of Voids to the Tertiary Creep of Gold'. Acta Metall., 1965, **13**, 353.

7.　PARKER, J.D. and WILSHIRE, B. - 'The Effects of Prestrain on the Creep and Fracture Behaviour of Polycrystalline Copper'. Mat. Sci. Eng., 1980, **43**, 271.

8.　ODQUIST, F.K.G. - 'Mathematical Theory of Creep and Creep Rupture'. Oxford Mathematical Monographs, Oxford, 1974.

9.　TIMOSHENKO, S. and GOODIER, J.M. - 'Theory of Elasticity', McGraw-Hill, New York, 1951.

10.　NADAI, A. - 'Theory of Flow and Fracture of Solids', McGraw Hill, New York, 1963.

11.　FRIEDEL, J. - 'Dislocations', Addison-Wesley, London, 1964.

12.　BALUFFI, R.W. and SEIGLE, L.L. - 'Growth of Voids in Metals during Diffusion and Creep'. Acta. Metall., 1957, **5**, 449.

MODELS FOR INTERGRANULAR CREEP CRACK GROWTH BY DIFFUSION

D.S. Wilkinson

Dept. of Metallurgy and Materials Science, McMaster University

Hamilton, Ontario, Canada. L8S 4M1

ABSTRACT

The theory of diffusion-controlled crack growth at elevated temperatures has advanced considerably during the past few years. These are reviewed here. We begin with a discussion of models for cavity growth in the absence of a stress concentration. It is shown that two steady-states are possible, depending on whether cavities grow independently or in a linked array. The discussion then moves to models of "homogeneous" crack growth, in which matter diffuses only from the tip of the macroscopic crack. While these models show the right sort of behaviour, they are flawed by their inability to account for microstructural effects, such as the extent of cavitation ahead of the crack. Finally, models for cracks which propagate by the growth of cavities are discussed. It is found that both the shape of the cavities, and the number which grow simultaneously, may vary considerably. Thus a wide range of temperature- and stress-dependencies are possible.

1. INTRODUCTION

One of the most common failure modes for structural materials in use at elevated temperatures (typically between 1/3 and 1/2 of the absolute melting temperature) involves intergranular crack growth. Such fractures exhibit low ductility, and result from the nucleation and growth of cavities along grain boundaries. In addition, service failures are often associated with welds and their heat affected zones, in which large grain size, high residual stress, and high residual impurity concentration can all help to promote intergranular failure.

Concern for this type of failure has spurred interest in developing realistic models for intergranular crack growth at elevated temperatures. To date most work has concentrated on creep brittle materials in which plasticity is limited, and crack growth is controlled by diffusion processes. Although models for plasticity controlled crack growth have recently been produced, this paper will concentrate on diffusion – controlled models, for which the picture is more complete.

2. MODELS OF CREEP CAVITY GROWTH

To understand the development of models for diffusion-controlled crack gowth, one needs to start by considering models for the growth of grain boundary cavities in the absence of stress concentrators. A vacancy diffusion mechanism for cavity growth was originally suggested by Balluffi and Seigle [1], and a model was produced shortly thereafter by Hull and Rimmer [2]. It considers the diffusional growth of cavities on grain boundaries normal to the tensile axis, in specimens of uniform cross section. The model represented a major advance in the understanding of creep fracture, and provided the basis for most subsequent models up to about 1975. Several assumptions (some of which are implicit) are made, among them:

a) that most atom transport is by grain boundary, not lattice diffusion. Since $\delta_b D_b \gg r D_\ell$ at temperatures of interest (r being the cavity radius), this assumption is almost always justified;

b) that surface diffusion is rapid and that cavities are spherical;

c) that the cavity surface and the grain boundary act as perfect sources and sinks for vacancies; and

d) that cavities are uniformly distributed over the grain boundary, and that once steady-state is reached, matter is deposited uniformly over the ligaments between cavities.

The last assumption leads to a particular type of steady-state stress field between cavities as shown in Fig. 1a, and to a linear relationship between cavity growth rate and stress. If cavitation failure is controlled by growth (and not by nucleation, or inter-cavity linkage), the model predicts a failure time for tests on smooth bars, of the form $t_f \sim c^3/\sigma$, where c is the cavity spacing and σ the applied stress.

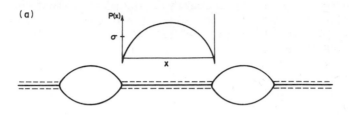

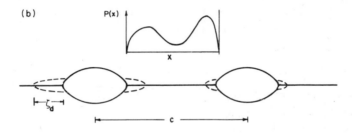

Fig. 1: Two different steady-states are possible in which: (a) matter is deposited uniformly between cavities, and (b) cavities grow independently of one another. Different stress fields result.

A number of authors have tried to extend or improve upon Hull and Rimmer's model while still maintaining assumptions a) through d). Speight and Harris [3] and Weertman [4] developed models using more exact sets of boundary conditions. Dobes and Cadek [5] attempted to include a distribution of cavity sizes to account for effects due to the continuous nucleation of cavities. Vitovec [6] considered the effect of increasing stress on the ligament between cavities as the cavities grow. Finally, Raj and Ashby [7] produced a model which used correct boundary conditions that incorporated many of these improvements. They also modified assumption b), by recognizing that even if surface diffusion is rapid, a variety of equilibrium cavity shapes are possible on grain boundaries and at boundary triple points, and that cavities are rarely spherical.

Despite these attempts at improving Hull and Rimmer's model, they all predict an inverse linear dependence of failure

time on the applied stress, while the weight of experimental
evidence indicates a much more sensitive dependence on stress.
This result follows directly from the assumptions of the models
– especially those of fracture controlled by cavity growth, of
equilibrium shaped cavities, and of uniform deposition of
matter over the grain boundary.

There are several possible explanations which have been
suggested for the discrepancy. It may be that cavity growth
does not control intergranular creep failure; but rather the
nucleation of cavities, or the growth of microcracks during
final failure dominates the fracture process. It may be that
cavity growth is usually plasticity-, rather than diffusion-
controlled. Alternatively, cavities may not be able to
maintain an equilibrium shape as they grow (modified assumption
b)), but rather become narrow and crack-like. Or, matter may
not be deposited uniformly over the grain boundary (assumption
d); but rather the cavities may grow for much of their lives as
independent microcracks. Finally, the growth of cavities may
be constrained by grain boundary sliding or plastic creep in
the material.

Most of these possibilities have been considered in various
models which have been developed during the past few years. A
complete review of these is beyond the scope of the present
paper (a partial review has recently been prepared by Svenson
and Dunlop [8]). Instead, only those approaches which have led
towards a model for crack growth along grain boundaries will be
considered.

3. MODELS OF HOMOGENEOUS CRACK GROWTH

Ashby [9] first suggested a model for crack-like cavities
growing independently of one another. His approximate model
used elliptical shaped cavities. Chuang and Rice [10]
developed a more exact analysis for the shape which a
crack-like cavity assumes, if its growth is controlled by
surface diffusion. They found that the half-width of the
cavity d, depends on the growth velocity v, according to

$$d = \sqrt{2(1-\cos\theta)}\ \left(\frac{\lambda_s}{v}\right)^{1/3} \tag{1}$$

where $\cos\theta = \gamma_b/2\gamma_s$ gives the equilibrium dihedral angle at
the crack tip, and $\lambda_s = \delta_s D_s \Omega \gamma_s/kT$. Thus, as the cavity
grows more quickly, the width of material which diffuses to the
crack tip by surface diffusion decreases.

A different approach was taken by Vitek [11], who assumed
that a microcrack grows with a constant width d, where d is not
determined by the model but is rather an independent variable.
Vitek considered an independent, sharp crack, growing by grain

boundary diffusion. A wedge of matter is deposited along the grain boundary ahead of the crack tip, relaxing the stress concentration there. Far from the crack, the stress is equal to the elastic stress field. A type of steady state results in which the flux of matter and the stress field ahead of the crack are independent of time in the frame of reference which moves at the same velocity as the crack tip (see Fig. 1b). Vitek used a numerical finite difference technique to calculate the stress field and the crack growth velocity for both the transient and steady state. For the latter,

$$\frac{v}{\lambda} = .516 \left(\frac{K}{dG}\right)^4 \tag{2}$$

where $\lambda = (\delta_b D_b \Omega G/kT)$, K is the applied stress intensity factor, and G the shear modulus. At first sight, the assumptions made by Chuang and Rice and by Vitek regarding the effect of velocity on cavity width would appear to be contradictory. However, they merely correspond to different regimes of the same problem in which either grain boundary or surface diffusion is controlling.

These two models both apply to long cracks in which matter diffuses only a short distance ahead of the crack tip (which is why the stress intensity factor K, can be used to characterize the crack). Speight, Beere and Roberts [12] have produced an approximate but analytical model for short cracks. They assume that matter diffuses along the grain boundary over a large distance (ζ_d) as compared to the crack half-length (a). This results in a microcrack growth velocity, $v \sim (\sigma/d)^2$. Not surprisingly, σ rather than K characterizes the stress field for this situation.

A modified version of this analysis can be produced for long cracks in which $\zeta_d << a$ [13]. This parrallels the Vitek [11] model, except that the shape of the wedge of material deposited ahead of the crack is assumed rather than calculated. The solutions differ from one another by only a small numerical factor, i.e. $v \simeq \lambda (K/Gd)^4$. Furthermore, it can be shown using this analysis [13], that irrespective of the ratio (ζ_d/a) being large or small, the distance over which matter diffuses ahead of the crack is given approximately by

$$\zeta_d \simeq 2\sqrt{\frac{\lambda}{v}} \tag{3}$$

This distance corresponds almost exactly to the position of maximum stress ahead of the crack as determined by Vitek's analysis. His results show that both the first derivate of the stress (proportional to the flux of matter) and the second derivative (proportional to the rate of material deposition) are much greater between the crack tip and the maximum in stress, than further away.

These models for homogeneous crack growth have since been extended in two different directions. They can be used to produce a theory for the growth of an array of cavities, in which both the uniform deposition and independent microcrack types of steady-state (Fig. 1a and b) are possible [14,15]. They have also been used as the basis for a more detailed theory of creep crack growth by grain boundary diffusion, in which the true microstructure of the process is analysed. This is what we will consider next.

4. MODELS OF CRACK GROWTH BY DIFFUSION-CONTROLLED CAVITATION

The models for crack growth just discussed are flawed, in that real cracks do not grow by the diffusion of matter directly away from the crack tip. There is generally a considerable amount of cavitation ahead of the crack tip, and cracks propagate by the growth and eventual linkage of these cavities with the main crack. Cavities grow in the stress field of the main crack, which is itself altered by the presence and growth of the cavities. The crack growth rate must therefore depend not only on parameters such as those in eq. 2, but also on microstructural variables such as the spacing between cavities, the shape of the cavities, and their ease of nucleation.

As is the case with diffusion-controlled cavity growth on uniformly stressed grain boundaries, cavity growth ahead of cracks can occur by each of the two types of steady-state already discussed. If cavities are widely separated and/or grow rapidly, then diffusion is limited to a region ζ_d close to the cavities and they behave like independent microcracks. The growth velocity of cavities is then very sensitive to their separation from the crack tip; i.e., only a few cavities are expected to grow at a significant rate ahead of the crack. If the cavities are close together and/or grow slowly, then ζ_d is comparable to or greater than the cavity separation. The diffusion fields of adjacent cavities merge, and they grow by depositing matter evenly over the ligament between adjacent cavities. Cavitation is expected to extend considerably further from the crack tip in this case. Each of these two types of behaviour are discussed below. But first, a discussion of the shapes which cavities can attain is appropriate.

4.1 Shape of Cavities

The shape of the cavities is determined by the relative value of surface and grain boundary diffusivities, as well as cavity spacing and other material parameters. If surface diffusion is rapid enough, cavities with constant curvature will be maintained and the width of each cavity is related to its length by d = $\ell g(\theta)$, where $g(\theta) = (\theta - \sin \theta \cos \theta)/\sin^2\theta$. Once surface diffusion to the cavity tip can no longer keep pace with the flux of atoms into the grain boundary, a change in shape occurs. The detailed shape of the cavity becomes complex and can only be determined by numerically computing the time dependent profile as the cavity grows. (This has been done by Pharr and Nix [16] for the surface diffusion-controlled growth of an isolated cavity.) Alternatively, we can treat the different types of steady-state which eventually result. In this event, the first step is to obtain an estimate of the cavity growth velocity for which an equilibrium shape is not possible. To do that we equate the time required for boundary diffusion to remove an atom from the cavity tip, $(J_b)^{-1}$ (where J_b is the grain boundary flux at the cavity tip) with the time required for surface diffusion to move an atom from the middle of the cavity to the cavity tip. Thus the equilibrium shape is possible only if [14,17]

$$\ell < \ell_c = \frac{\sin \theta}{2 \theta} \sqrt{\frac{2D_s}{J_b} \frac{\gamma_s \delta_s}{kT}} \qquad (4)$$

In general, $J_B = (v\bar{d})/\Omega$, where $\bar{d}$ is related to the geometry of the cavities, such that $\bar{d} = 2 \ell g(\theta)$ for equilibrium cavities, and $\bar{d} = 2d$ from crack-like cavities. The term $\gamma_s \delta_s/kT$ is the drag force for surface diffusion.

For $\ell > \ell_c$, cavities are expected to become crack-like in shape. For long crack-like cavities, Chuang and Rice [10] have shown that if surface diffusion is much slower than grain boundary diffusion, the cavity width is velocity dependent, according to eq. 1. However, at $\ell = \ell_c$, this equation invariably requires a larger cavity width than is actually present. Thus it would appear that for $\ell > \ell_c$, cavities continue to widen as their growth continues (and/or accelerates) with a shape determined by combined surface and grain boundary diffusivities. This is consistent with the numerical analysis of Pharr and Nix [16], who showed that a long transient may be expected before a true crack-like cavity is obtained. In modelling the crack growth process, this transient may be treated by assuming that cavities grow with a constant width. Using $2d = 2\ell_c g(\theta)$, equal to their width at the point where surface diffusion first starts to intervene, gives a lower bound on d. This width is maintained until the Chuang and Rice condition (eq. 1) is reached, after which the cavity width decreases as it accelerates.

4.2 Growth of a Single Cavity Ahead of a Crack

Models have been developed for the growth of cavities ahead of a macroscopic crack. As for homogeneous cracks, this is done using a finite difference technique in which the elastic stress distribution is relaxed by diffusion [11,18]. Thus, the

322

starting point must be an analysis of the elastic stress field surrounding a cavity ahead of a macroscopic crack. This has been done [19,20] for the case of a single, sharp cavity directly ahead of a crack, using plane strain and uniaxial (mode I) loading (Fig. 2).

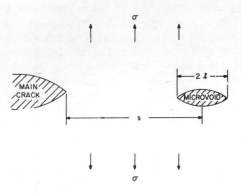

Fig. 2: The model for a plane strain, mode I crack, with a sharp cavity directly ahead of the crack tip.

The complete solution is given elsewhere [19,20]. However, for $\ell/s < 0.5$, the tensile stress surrounding the cavity is well approximated by

$$\sigma_o(x) = \sqrt{\frac{\ell}{2s}} \frac{K}{\sqrt{2\pi|(x-\ell)|}} + \frac{K}{\sqrt{2\pi(s+x)}} - \frac{K}{\sqrt{2\pi s}} \qquad (5)$$

(where the origin of x is taken to be the centre of the cavity). When the cavity is small, it behaves as a microcrack with stress intensity factor $K^* = \sqrt{\ell/2s}\, K$, while $K^* \to K$ as $\ell \to s$.

When cavity growth ahead of the crack is rapid, the cavities grow independently of one another. The stress singularity near the crack tip must be rather sharp in this case. Furthermore, results indicate that the cavity grows with a velocity that changes as $(K^*)^2$ or greater. We therefore expect only one or a few cavities to grow ahead of the crack. Indeed, experimental work [21] indicates this is often the case. A model is therefore constructed in which a single cavity grows ahead of the crack. The model [14,20] results in an integral equation for σ_n, the tensile stress distribution ahead of the crack under steady state conditions (i.e. σ_n is independent of time in the frame of reference which moves with the cavity tip). Typical stress distributions which result from the solution to this equation are shown in Fig. 3 for three values of $v\ell^2/\lambda$, and for $s/\ell = 6$.

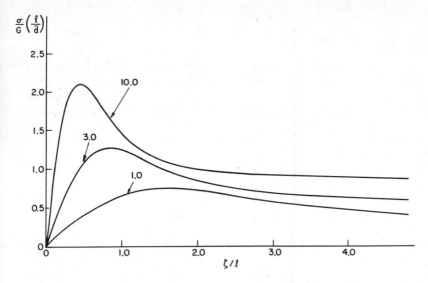

Fig. 3: Steady-state stress distributions for three values of $v\ell^2/\lambda$, ahead of a cavity growing near the tip of a crack.

Consistent with previous work, the position of maximum σ_n, at a distance ζ_d ahead of the cavity tip is given to a good approximation by $\zeta_d = 2\sqrt{\lambda/v}$. The dependence of cavity growth rate v, on the loading conditions, although numerically determined, can be fit within $\pm$ 15% by a set of equations as follows (using $\alpha = \sqrt{\ell/2s}$):

$$\frac{v}{\lambda} = 0.456 \left(\frac{\alpha K^*}{dG}\right)^4 \quad \text{for} \quad \frac{K}{G} > \frac{1.21}{\alpha} \frac{\overline{d}}{\sqrt{\ell}} \tag{6}$$

$$\frac{v}{\lambda} = \frac{0.675}{\ell} \left(\frac{\alpha K^*}{dG}\right)^2 \quad \text{for} \quad 1.21 \, \alpha \frac{\overline{d}}{\sqrt{\ell}} < \frac{K}{G} < \frac{1.21}{\alpha} \frac{\overline{d}}{\sqrt{\ell}} \tag{7}$$

$$\frac{v}{\lambda} = 0.456 \left(\frac{K}{G\overline{d}}\right)^4 \quad \text{for} \quad \frac{K}{G} < 1.21 \, \alpha \frac{\overline{d}}{\sqrt{\ell}} \tag{8}$$

Eq. 8 is identical to eq. 2 (apart from a small numerical constant), derived using the homogoneous crack model. It thus applies to slow cavity growth when the crack and cavity are indistinguishable. These equations really apply only for crack-like cavities, i.e. when the elastic stress analysis is valid. However, a calculation for equilibrium cavities, in which eq. 5 was replaced by a stress field more appropriate to a cavity with a finite root radius, showed these results to be valid, except for very large stress intensity factors [14,20].

The apparent dependence of v on K in eqs. 6-8, varies between K^2 and K^4. However, if the cavity width becomes velocity dependent, such that cavity narrows as its velocity increases, then this dependence is increased towards K^6.

4.3 Multiple Cavity Growth Ahead of a Crack

When cavity growth is slow, diffusional stress relaxation can occur along the grain boundary, over a distance which is comparable to or greater than the spacing between cavities. To a first approximation, this occurs if $\zeta_d > c/2$. From eq. 3, this implies a minimum cavity growth velocity for single cavity growth of

$$\frac{v}{\lambda} = \frac{16}{c^2} \qquad (9)$$

Below this velocity, the diffusion fields of adjacent cavities (or of the main crack and its nearest cavity) overlap. This results in a steady state, in which matter is deposited uniformly over the ligament between cavities, as in the Hull and Rimmer model discussed earlier. There, the average load per cavity is constant as the cavity grows, since the cavities are assumed to grow uniformly in a periodic array. In the presence of a macroscopic crack however, this is not the case, at least for the cavity nearest the crack. As it grows and eventually links up with the crack, the load is transferred to the material ahead of the cavity. Since it is the load per cavity W, which determines the rate of material deposition, and thus the cavity growth rate for the Hull and Rimmer type of steady state, it is important to know how the load changes with cavity size. This can be done using the elastic stress analysis already described. It is found [19,22] that the load decreases very slowly until the cavity has almost joined the main crack. Thus, within an error of less than 15% (for $\ell/s < 0.8$) we can assume that the load on the ligament between cavities is constant, and equal to that on the ligament when no cavities are present; i.e.

$$W_n = \sigma \sqrt{2ac} \; (\sqrt{n} - \sqrt{n-1}) \qquad (10)$$

gives the load on the ligament n cavities removed from the main crack.

The growth rate of individual cavities can be determined analytically using this equation as one of the boundary conditions. The result [22] is

$$\frac{v_n}{\lambda} = \frac{6}{(c-2\ell_n)^2}\left(\frac{W_n}{\overline{d}_n G} - \frac{\gamma_s(c-2\ell_n)}{\overline{d}_n G r_n}\right) \tag{11}$$

where r_n is the radius of curvature of the tip of the nth cavity. For equilibrium cavities r_n is generally large, and the surface energy term can be neglected. For sharp crack-like cavities this is not the case. Both $\overline{d}_n$ and r_n become velocity-dependent with $\overline{d}_n = 2 d_n$ given by eq. 7, and $r_n = d_n/2(1-\cos\theta)$ [10]. By sustitution in eq. 11, $\overline{d}_n$ and r_n can be eliminated as explicit variables for crack-like cavities. This gives [22]

$$\frac{v_n}{\lambda_s} = \left[\frac{3\lambda\gamma_s}{8\lambda_s G(c-2\ell_n)}\right]^3 \left\{\left[\frac{16}{3}\sqrt{\frac{2}{1-\cos\theta}}\,\frac{W_n\lambda_s}{G\lambda}\left(\frac{\gamma_s}{G}\right)^{-2}+1\right]^{1/2}-1\right\}^3 \tag{12}$$

This equation has two limits. If surface curvature does not affect cavity growth, then $v_n \sim W_n^{3/2}$. As the cavity gets sharper, surface curvature begins to affect the growth rate, and the dependence on stress increases towards an upper limit of $v_n \sim W_n^3$. Thus, the uniform deposition steady-state can lead to a dependence on loading which varies between K^1 and K^3.

In the case of multiple cavity growth, an additional parameter which needs to be considered is the number of cavities N, which grow ahead of the main crack. A steady state is reached in which N is constant, and the size of a cavity depends only on its separation from the crack tip. It is found for both crack-like and equilibrium-shaped cavities [22], that the average crack growth rate $\dot{a}$, depends on the parameter $(K/N)^\beta$, where the power β varies between 1 and 3, exactly as the power of W_n in eqs. 11 and 12. N is not determined by this model, but rather by the nucleation of cavities. Possible criteria for N are discussed elsewhere [22]. However, since nucleation can be stress dependent, it is clear that N itself may be a function of K, and the K-dependence of crack growth rate may be even greater than power 3.

Raj and Baik [24] have recently produced a model similar to this, in which a large number of cavities are assumed to grow simultaneously by grain boundary diffusion ahead of a crack. Their model differs from the one just discussed essentially in the assumed stress distribution. The Wilkinson [22] model assumes that the load per ligament is given by the initial elastic load in the absence of cavities. Raj and Baik allow a longer range redistribution of the elastic stress by diffusion. This requires that they use a numerical method to calculate the crack velocity. The effect should be to increase the number of cavities which grow ahead of a crack, and so slow the rate of cavitation near the crack tip. However, their model is unable to include the effect of surface diffusion on cavity shape and growth rate.

5. CONCLUSIONS

The theory of cavitation has been developed to the point where models for creep crack growth, by diffusion-controlled cavitation in the stress intensity field of a macroscopic crack, can be developed. The crack growth behaviour of a material depends not only on applied stress and temperature, but also the shape of cavities and their ease of nucleation. At present, the development of cavity shape, and the growth rate of individual cavities is reasonably well understood. However, a complete analysis of the average crack growth rate requires an integration of the growth rate of each cavity from its nucleation to eventual linkage with the main crack. This has yet to be done for the general case, in which a cavity may grow in several different modes during its life.

Despite the fact that the models are restricted to diffusional cavity growth, they predict a surprising range of behaviour. Since the process can be controlled by grain boundary diffusion, surface diffusion, or a combination of the two, a range of activation energies may be expected. The dependence of growth rate on K may vary between linear and about K^6. There is experimental evidence to support this trend. For example, creep crack growth rates in creep brittle, untempered heats of CrMoV steels [23] show a good correlation with K, and a power of between 3 and 4. For more ductile materials, crack growth rates do not correlate with the stress intensity factor, and a different approach is required. This is one area in which considerable effort will probably be spent in the next few years.

6. ACKNOWLEDGMENTS

Several discussions with Professor V. Vitek are gratefully acknowledged. This work was supported by research grants from Imperial Oil Ltd. and the Natural Sciences and Engineering Research Council, Canada.

REFERENCES

1. BALLUFFI, R.W. and SEIGLE, L.L. - Acta Met., 1957, 5, 449.

2. HULL, D. and RIMMER, D.E. - Phil. Mag., 1959, 4, 673.

3. SPEIGHT, M.V. and HARRIS, J.E. - Met. Sci. J., 1967, 1, 83.

4. WEERTMAN, J. - Scripta Met., 1973, 7, 1129.

5. DOBES, F. and CADEK, J. - Mat. Sci. Engg., 1972, 9, 355.

6. VITOVEC, F.M. - J. Matl. Sci., 1972, 7, 615.

7. RAJ, R. and ASHBY, M.F. - Acta Met., 1975, 23, 653.

8. SVENSON, L.E. and DUNLOP, G. - Intl. Met. Rev., in press.

9. ASHBY, M.F. 1972, unpublished research.

10. CHUANG, T.-J. and RICE, J.R. - Acta Met., 1973, 23, 1625.

11. VITEK, V. - Acta Met., 1978, 26, 1345.

12. SPEIGHT, M.V. BEERE, W.B. and ROBERTS, G. - Mat. Sci. Engg., 1978, 36, 155.

13. WILKINSON, D.S. unpublished research.

14. VITEK, V. - Met. Sci. J., 1980.

15. CHUANG, T.-J. KAGAWA, K.I. RICE, J.R. and SILLS, L.B. - Acta Met., 1979, 27, 265.

16. PHARR, G.M. and NIX, W.D. - Acta Met., 1980, 28, 557.

17. VITEK, V. and WILKINSON, D.S. - Proc. 5th. Intl. Conf. on Strength of Metals and Alloys, Aachen, 1979, p. 321.

18. VITEK, V. WILKINSON, D.S. and TASASUGI, T. to be published.

19. WILKINSON, D.S. and VITEK, V. - Res. Mechanica, 1980, 1, 101.

20. WILKINSON D.S. and VITEK, V. - Res Mechanica Monograph, 1981, in press

21. WILKINSON, D.S. ABIKO, K., THYAGARAJAN, N. and POPE, D.P. - Met. Trans., 1980, 11A, 1827.

22. WILKINSON, D.S. "A Model for Creep Cracking by Diffusion-Controlled Void Growth", to be published.

23. GOOCH, D.J. - Mat. Sci. Engg., 1977, 29, 227.

24. RAJ, R. and BAIK, S. - Met. Sci. J., 1980, 14, 383.

CREEP CRACK EXTENSION BY GRAIN-BOUNDARY CAVITATION

J. L. Bassani

Department of Mechanical Engineering and Applied Mechanics
University of Pennsylvania
Philadelphia, Pennsylvania 19104, USA

SUMMARY

   Recent work by Riedel and coworkers has led to various
descriptions of stationary and moving crack tip fields under
creep conditions. For stationary and growing cracks, several
flow mechanisms (e.g., elastic, time-independent plastic, primary creep, and secondary creep) can dictate the analytical
form of the crack tip field. In this paper, relationship between overall loading and crack velocities are modelled based
upon grain-boundary cavity growth and coalescense within the
zone of concentrated strain in the crack tip field. Coupled
diffusion and creep growth of the cavities is considered. Overall crack extension is taken to be intermittent on a size
scale equivalent to the size of a grain. Numerical results
are presented for a center-cracked panel of 304 stainless
steel.

1. INTRODUCTION

   At elevated temperatures macroscopic cracks in creeping
metals can propagate along grain-boundaries that have been
weakened, often due to environmental attack such as oxidation
or by the formation and growth of grain boundary cavities and
microcracks. A model for environmentally-caused intermittent
crack growth based on local weakening of the grain boundary by
a diffusing species that is followed by sudden cracking just
ahead of the crack tip has been proposed by McClintock and
Bassani [1]. In this paper a model is proposed for intermittent creep crack growth by the formation of large grain-
boundary voids or microcracks followed by sudden separation of
the grain boundary just ahead of the crack tip. These microcracks are assumed to be formed from the coalescense of grain-
boundary cavities that grow due to combined effects of grain-
boundary diffusion and matrix creep [2,3]. Unlike the environmental damaging of the grain boundary, the damaging due to

cavitation is not confined to the immediate vicinity of the crack tip [4].

Several investigators, for example Dimelfi and Nix [5] and Riedel [6] have modeled creep crack growth by a continuous coalescence of grain-boundary cavities with the crack tip. Only relatively simple, time-independent descriptions of the stress field around the crack have been considered so that closed-form relations are obtained between the crack velocity à and a characteristic loading parameter, such as the elastic stress intensity factor K or the steady-state creep, path-independent integral C_s^*. Actual fields around a creep crack after, for example, sudden loading changes or reversals include effects of elastic, plastic, primary creep, and secondary creep deformations [1,7,8] as well as the growth of the crack itself [9]. Surrounding the crack tip there may be distinct regions that are dominated by one or more of these deformation mechanisms. Even under constant load conditions, as the stresses relax due to creep, the extent of each region varies with time. The present model includes an approximate description of the complete stress field and accounts for cavity growth in the nominal region if stresses there are high enough.

We begin with a discussion of the model followed by a review of cavity growth laws by combined effects of grain-boundary diffusion and matrix creep. Then an approximate description is given for the time-dependent state of stress ahead of a growing creep crack. Numerical results for crack growth after a suddenly applied tensile stress are presented in the form of a case study of a center-cracked panel of 304 stainless steel. The trends predicted do not suggest any simple correlation between overall loading and crack velocity, although greater stable crack growth is predicted at applied stress levels that are typical of service conditions as compared with those that are typical of laboratory conditions.

2. CRACK GROWTH MODEL

Two-dimensional problems with planar cracks under Mode I tensile loading conditions are considered. Let r and θ be the polar coordinates centered at the current crack tip, with θ = 0 directly ahead of the crack in its plane. Surrounding the crack, we imagine a collection of grains of mean size d. On grain boundaries perpendicular to the applied tensile axis, there are cavities of current radius ρ and average half-spacing b (see Fig. 1) that can grow under a local, time-dependent tensile stress σ(t) acting across the grain boundary. Two cavities on a grain boundary just ahead of the crack (r ≃ d/2) grow until they coalesce and form a larger cavity or microcrack at which time the grain boundary is imagined to be sufficiently weakened so that its separation

follows instantaneously. The crack then extends by a length
d and the process continues. The effect of zig-zagging on a
size scale d as the crack grows along a grain boundary path
will be neglected. On a scale that is larger than d the
crack is assumed to be straight.

The stress field around a crack is highly non-uniform
and decreases in magnitude with distance r ahead of the
crack so that at any time radii of cavities generally decrease
with increasing distance r. Far from the current crack tip
σ may fall below a threshold stress σ_T [2], and there cavities will cease to grow. We will see that nominal stresses
typical of service conditions may be below σ_T while those
typical of laboratory conditions may be well above. If
considerable cavity growth away from the crack occurs then
with time a rapidly accelerating crack is to be expected.

3. CREEP AND DIFFUSIVE GROWTH OF GRAIN-BOUNDARY CAVITIES

The diffusional growth of grain-boundary cavities was
first investigated by Hull and Rimmer [10] and more recently
by Chuang, Kagawa, Rice, and Sills [11]. It is now recognized
that models of cavity growth by diffusion alone lead to rupture or failure times t_F that underestimate the influence
of the applied stress levels. The growth of cavities by creep
deformation alone has been considered by Hancock [12] and more
recently by Budiansky, Hutchinson, and Slutsky [13]. The
latter investigation showed the importance of moderate triaxial stresses, that will be neglected in this model, in
keeping a more or less spherically shaped cavity under a predominantly tensile load.

A model of cavity growth by combined effects of grain-
boundary diffusion and matrix creep has been proposed by Beere
and Speight [14]. Recently Chen and Argon [3] derived growth
laws along the lines of [14] and demonstrated close agreement
with the detailed finite element results for the coupled problem of Needleman and Rice [15]. The results of Chen and
Argon [3] will be incorporated here.

Let ρ be the current radius of a spherically capped
cavity that makes an angle $\pi/2 \leq \psi < \pi$ with the grain
boundary (see Fig. 1). If a planar array of cavities spaced
2b apart are aligned perpendicular to an applied tensile
stress σ then the rate of increase of the cavity radius due
to diffusion alone is [3]

$$\dot{\rho} = \frac{D'}{2h(\psi)} \frac{\sigma}{\rho^2 f(\rho/b)} \tag{1}$$

where $D' = D_b \delta_b \Omega/KT$ ($D_b \delta_b$ = grain boundary coefficient,
Ω = atomic volume, K = Boltzmann's constant, and T = tempera-

ture in ^{o}K) and

$$h(\psi) = (\frac{1}{1+\cos\psi} - \frac{\cos\psi}{2}) \sin\psi \qquad (2)$$

$$f(\eta) = -\ln\eta + \eta^2(1-\eta^2/4) - 3/4 \qquad (3)$$

The influence of the normal stress acting on the grain boundary at the cavity tip has been neglected in eq. (1) (see [15]
for a discussion of this effect).

The rigid-grain model from which eq. (1) is derived assumes that overall deformation is accommodated by a diffusional plating of atoms along the entire grain boundary between
the cavities. On the other hand, if diffusion only has to
account for plating up to a distance $\rho + \ell \leq b$ from the cavity
center while deformation in the annular region extending from
$\rho + \ell$ to b is primarily due to creep, then cavity growth will
be enhanced by creep. Needleman and Rice [15] have shown
that this diffusion distance ℓ is given in terms of D', σ,
and the corresponding creep strain rate $\dot{\varepsilon}$ as

$$\ell = (D' \sigma/\dot{\varepsilon})^{1/3} \qquad (4)$$

For the cavity growth calculations $\dot{\varepsilon}(\sigma)$ will be chosen from
the power law relation for steady state creep deformation

$$\dot{\varepsilon} = \dot{\varepsilon}_{so} (\sigma/\sigma_o)^{n_s} \qquad (5)$$

where $\dot{\varepsilon}_{so}$ is the steady state creep strain rate at the unit
stress σ_o and n_s is the creep exponent. As σ increases
the distance ℓ over which diffusion must accommodate the
total deformation decreases and creep deformation plays an
increasingly important role.

Chen and Argon [3] suggest modifying eq. (1) by replac-
ing the factor $f(\rho/b)$ with $f[\rho/(\rho+\ell)]$ whenever $\rho + \ell < b$.
Then the cavity growth-rate equation becomes

$$\dot{\rho} = \frac{D'}{2h(\psi)} \frac{\sigma}{\rho^2 f[\rho/(\rho+\ell)]} \qquad (6)$$

where

$$\ell = \min \{ D'\sigma/\dot{\varepsilon}, b-\rho \} \qquad (7)$$

The stress $\sigma(t)$ that enters into eqs. (6) and (7) will be
chosen from the continuum crack tip fields outlined in the
next section.

4. ASYMPTOTIC STRESS AND FLOW FIELDS FOR CREEP CRACKS

4.1 Stationary Cracks. Riedel and Rice [7] and Riedel [8]

have derived various singular stress fields that exist
around a stationary crack tip under creep conditions when one
deformation mode dominates in the far field and perhaps another at the crack tip. In the spirit of small-scale-yielding
approximations, Riedel's [8] results can be pieced together
to demonstrate the evolution and disappearance of regions
surrounding the crack tip that are dominated each by either
elastic, time-independent plastic, primary creep, or secondary
creep deformations [1]. The uniaxial tension, stress-strain-
rate $(\sigma-\varepsilon)$ relation is modeled in terms of Young's modulus E,
plasticity exponent N , primary creep exponent n_p and hardening exponent p > o , secondary creep exponent n_s , and
several normalizing quantities $(\ )_o$ as

$$\dot{\varepsilon} = \dot{\sigma}/E + \frac{1}{N}\ \dot{\varepsilon}_o\ (\sigma/\sigma_o)^{\frac{1}{N}\ -\ 1}\ (\dot{\sigma}/\sigma_o)$$

$$+ \dot{\varepsilon}_{po}\ (\sigma/\sigma_o)^{n_p}\ (\varepsilon/\varepsilon_{po})^{-p} + \dot{\varepsilon}_{so}(\sigma/\sigma_o)^{n_s} \tag{8}$$

Initial loading of a crack in a material that obeys eq.
(8) may give rise to a nominal stress region surrounding an
elastic singularity, a plastic singularity, and a fracture
process zone within which the actual cracking takes place.
The two-dimensional elastic crack tip stresses are given in
terms of the stress intensity factor K and known nondimensional functions $f_{ij}(\theta)$ as

$$\sigma_{ij} = \frac{K}{\sqrt{2\pi r}}\ f_{ij}(\theta) \tag{9}$$

The singular fields associated with time-independent
"plastic" and creep deformations are each of the HRR type
[1,8]. A singular plastic field, that may be embedded within
or engulf the singular elastic field, is given in terms of a
J integral value that sets its amplitude, the crack length
a , and the nondimensional functions $I_{1/N}$ and $\tilde{\sigma}_{ij}(\theta; 1/N)$ as

$$\sigma_{ij}/\sigma_o = \left(\frac{J}{\sigma_o\ \varepsilon_o\ a\ I_{1/N}}\ \frac{a}{r}\right)^{\frac{N}{N+1}} \tilde{\sigma}_{ij}(\theta; 1/N) \tag{10}$$

To within 2% accuracy, for plane strain, McClintock [16] gives
the approximation

$$I_n \approx 10.3\ \sqrt{0.13 + 1/n}\ - 4.8/n \tag{11}$$

As time progresses, creep deformation can lead to crack
tip regions governed by other singular stress fields. At a
crack tip where the primary creep rates dominate the total
strain rates, the stresses are given in terms of a time-
dependent amplitude $C_p(t)$ as

$$\sigma_{ij}/\sigma_o = \left\{ \frac{C_p(t)}{[(p+1)\dot\varepsilon_{po}/\varepsilon_{po}]^{\frac{1}{p+1}} \sigma_o \varepsilon_{po} a I_m} \right\} \left(\frac{a}{r}\right)^{\frac{1}{m+1}} \tilde\sigma_{ij}(\theta;m) \tag{12}$$

where

$$m = n_p/(p+1) \tag{13}$$

If secondary creep rates dominate at the crack tip, then the stresses are given in terms of a time dependent amplitude $C_s(t)$ as

$$\sigma_{ij}/\sigma_o = \left[\frac{C_s(t)}{\sigma_o \dot\varepsilon_{so} a I_{n_s}} \frac{a}{r} \right]^{\frac{1}{n_s+1}} \tilde\sigma_{ij}(\theta; n_s) \tag{14}$$

The approximate field matching of Riedel and Rice [7] and Riedel [8] can lead to simultaneous existence, albeit in separate regions, of the stress fields in eqs. (9), (10), (12), and (14). This matching assumes approximate path independence of the J integral. Proceeding from this innermost crack tip field out to the nominal field we will calculate the amplitude of each field in terms of the amplitude of its surrounding field. Certain restrictions will arise, and the reader is referred to [1 and 8] for a more complete discussion. The plausibility of each restriction will be demonstrated in terms of a case study on 304 stainless steel.

If $n_s > n_p/(p+1)$ the secondary creep region may be embedded in a primary creep region. The amplitude $C_s(t)$ of the secondary creep field in eq. (14) is given in terms of the amplitude $C_p(t)$ of the surrounding primary creep field in eq. (12) as

$$C_s(t) = \frac{n_s+p+1}{(p+1)(n_s+1)} \frac{C_p(t)}{t^{p/(p+1)}} \tag{15}$$

If $n_p/(p+1) > 1/N$ the primary creep region may be embedded in a plastic region and

$$C_p(t) = J/(\frac{n_p+p+1}{p+1} t)^{\frac{1}{p+1}} \tag{16}$$

where J is the amplitude of the plastic field given in eq. (10). For $1/N > 1$ the plastic region may be embedded in an elastic region and

$$J = (1 - \nu^2) K^2/E \tag{17}$$

where ν is Poisson's ratio. With σ_∞ denoting the nominal far field stress and F_K a number depending on geometry and loading that is catalogued for several cases in the handbook by Tada et al. [17],

$$K = F_K \sigma_\infty \sqrt{\pi a} \qquad (18)$$

Any one of the fields described above may, at some time, disappear due to large primary or secondary creep deformations. Furthermore, for high initial loading, the plastic field will dominate over the elastic one. If the secondary creep region wipes out the primary creep region but is within a plastic region, then

$$C_s(t) = \frac{J}{(n_s + 1) \, t} \qquad (19)$$

If this secondary creep region also wipes out the plastic region but is within the elastic region, then eq. (19) still holds with eq. (17) substituted for J. This also applies if the primary region of eqs. (12) and (16) engulfs the plastic but is within the elastic.

For high initial loading or long times, in which case the elastic region given by eq. (9) does not exists, the amplitude of the outer most HRR field can be given in terms of the fully plastic solutions of Hutchinson, Needleman, and Shih [18] and others, for various geometries and loadings.

When the plastic field is embedded in the nominal field,

$$J = F_J \, \sigma_o \, \varepsilon_o \, a (\sigma_\infty/\sigma_o)^{\frac{N+1}{N}} \qquad (20)$$

where F_J depends on geometry and far field loading conditions. When the primary creep field is embedded in the nominal field, $C_p(t)$ approaches its steady state value C_p^* given by

$$C_p^* = F_p \, \sigma_o \, [(p+1) \, \dot{\varepsilon}_{po}/\varepsilon_{po}]^{\frac{1}{p+1}} \, \varepsilon_{po} \, a \, (\sigma_\infty/\sigma_o)^{\frac{n_p+p+1}{p+1}} \qquad (21)$$

Finally, when the secondary creep field is embedded in the nominal field, $C_s(t)$ approaches its steady state value C_s^* given by

$$C_s^* = F_s \, \sigma_o \, \dot{\varepsilon}_{so} \, a \, (\sigma_\infty/\sigma_o)^{n_s+1} \qquad (22)$$

The order suggested above for embedding the various crack tip solutions depends on the restrictions stated for the exponents N, n_p, p, and n_s. If, for instance,

M

$1/N > n_p/(p+1)$ or n_s , then time-independent plasticity may persist at the crack tip within the creep zones. The limiting case of nonhardening plasticity $N = 0$ has been considered by Leckie and McMeeking [19]. In that case, the non-singular stresses at the crack tip are limited by the Prandtl solution.

4.2 <u>Growing Cracks.</u> For the steady growth of cracks in a material that deforms both elastically and in secondary creep, Hui and Riedel [9] have found an asymptotic solution that surprisingly depends only on crack velocity and not on any far-field boundary condition. With $\dot{a}$ denoting the crack velocity, in terms of the nondimensional function $\beta(n_s)$ (which is of order unity for plane stress or strain) and $\hat{\sigma}_{ij}(\theta; n_s)$,

$$\sigma_{ij}/\sigma_o = \beta(n_s)\left(\frac{\sigma_o \dot{a}}{E \dot{\varepsilon}_{so} r}\right)^{\frac{1}{n_s-1}} \hat{\sigma}_{ij} (\theta; n_s) \qquad (23)$$

This field dominates closest to the crack tip and its amplitude is set solely by $\dot{a}$.

In the next section, we will consider the growth of a crack under a suddenly applied remote stress at $t = 0$ that is held constant. The stress fields in eqs. (9), (10), (12) and (14) not only change with time due to stress relaxation effects, but they also change due to crack extension. This latter effect will only be accounted for through the change in crack length a , while time-independent stress changes due to intermittent and sudden growth of the crack are neglected. For the crack velocities calculated this can be shown to be a reasonable approximation.

5. A NUMERICAL EXAMPLE

Crack growth is now calculated for a plane strain center-cracked panel of 304 stainless steel with an initial crack length to width ratio equal to ½. Applied stress levels are taken to model both laboratory and service conditions at an operating temperature of 873°K. The material parameters that enter into eqs. (6) and (8) have been compiled from the data on 304 stainless steel reported in references [1,2 and 15]. A summary of the data is given in Table 1. Although the creep properties tend to vary with stress levels, as noted in [1], only average values have been listed in Table 1.

5.1 <u>Times for cavities to coalesce.</u> We begin with a calculation of the time t_F required for two cavities of initial radius $\rho_o = 2 \times 10^{-7}$m and spacing $b = 4 \times 10^{-6}$m to coalesce under a constant tensile stress σ . The growth of the cavities is assumed to be self-similar so that the spherically capped shape is retained. From eqs. (6) and (7) this time is given by

$$t_F = \int_{\rho_o}^{b} (-\frac{1}{\dot{\rho}}) \; d\rho$$

$$= \frac{2h(\psi)}{D' \; \sigma} \int_{\rho_o}^{b} \rho^2 \; f[\rho/(\rho+\ell)] \; d\rho \tag{24}$$

A plot of t_F and ℓ versus σ is shown in Fig. 1. At low stress levels $\ell+\rho_o > b$ and the cavities grow primarily by diffusion. At somewhat higher stress levels, where $\ell+\rho < b$ throughout part of the growth period, coupling occurs between the diffusive and creep mechanisms. At very high stresses (that are typical of those around cracks) where $\ell+\rho < b$ through most of the growth period, the cavities grow primarily by creep and in that case $t_F \simeq$ constant x $\sigma^{-n}s$.

5.2 Crack growth calculations. Imagine at time $t = 0$ that a remote tensile stress is suddenly applied to the center-cracked panel shown in Fig. 2. Crack growth will initiate along the x_1 axis ($\theta = 0$) when two cavities closest to the crack tip (at $r = d/2 = 2.5 \times 10^{-5}$m) coalesce under a tensile stress $\sigma = \sigma_{22}$ ($r = d/2$, θ, t). Since the factors $f_{22}(\theta) \simeq \tilde{\sigma}_{22} (\theta; n) \simeq 2$ in eqs. (9), (10), (12) and (14) in

TABLE 1. Some properties of 304 stainless steel at 873°K referenced in eqs. (6), (7), (8) and (17).

$$\sigma_o = 50 \text{ MN/m}^2$$

$$E = 145,000 \text{ MN/m}^2 , \quad \nu = 0.3$$
$$\varepsilon_o = 5 \times 10^{-5}, \quad N = 0.3$$
$$\dot{\varepsilon}_{po} = 10^{-13} \text{ sec}^{-1} , \quad n_p = 13$$
$$\varepsilon_{po} = 0.01 , \quad p = 2$$
$$\dot{\varepsilon}_{so} = 10^{-11} \text{ sec}^{-1} , \quad n_s = 7$$

$$D' = 2.34 \times 10^{-32} \text{ m}^5/\text{N} - \text{sec}$$
$$\psi = 70°$$
$$\rho_o = 2 \times 10^{-7} \text{ m}$$
$$b = 4 \times 10^{-6} \text{ m}$$
$$d = 5 \times 10^{-5} \text{ m}$$
$$\sigma_T = 100 \text{ MN/m}^2$$

338

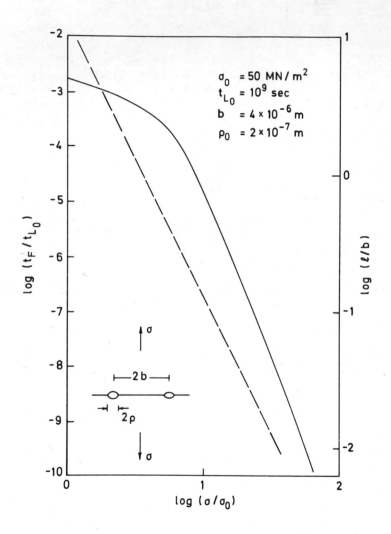

Fig. 1. Time to coalescense t_F (solid line) from eq. (24) and diffusion length $\ell = (D' \sigma / \dot{\varepsilon})^{1/3}$ (dashed line) for cavities under tensile stress σ.

the range $0 \leq \theta \leq \pi/3$ for all n considered, any value of
θ chosen in this range will lead to roughly the same results.
The stress σ acting on each cavity ahead of the crack tip
is given as a function of r and t by one of eqs. (9), (10),
(12) or (14) in their regions of dominance while far from the
crack tip σ is taken to be the remote tensile stress σ_∞.

The initiation time t_1, for crack growth can be calculated from the first of eq. (24) for the cavities at r = d/2.
At this time, the crack suddenly extends a distance d.
Other cavities have also grown during $0 \leq t \leq t_1$ if the
local tensile stress acting on any one exceeds σ_T at some
time within this period. The process then repeats itself so
that the time t_i at which the i th increment of growth
occurs is

$$t_i = t_{i-1} + \int_{\rho_i}^{b} (\frac{1}{\dot{\rho}}) \, d\rho \qquad (25)$$

where $\rho_i = \rho \ (r = d/2, t_{i-1})$ just after the $(i-1)$ th increment of growth. A 40 point Gauss-Legendre quadrature formula
was used to evaluate eq. (25) and explicit Euler time integration was used to calculate $\rho(r > d/2 , t)$.

Two remote stress levels are considered; one that is
typical of power plant service conditions where $\sigma_\infty = 50$ MN/m^2
so that a 0.01 steady-state creep strain is attained in a life
time $t_L = 10^9$ sec $\approx$ 32 years, and the other typical of laboratory conditions where $\sigma_\infty = 150$ MN/m^2 so that a 0.01 steady-
state creep strain is attained in $t_L = 4.6 \times 10^5$ sec $\approx$ 5.3 days.
The crack growth is calculated up to a = 1.5 a_0.

Figures 2 and 3 are plots of crack extension $\Delta a = a - a_0$
and crack velocity $\dot{a}$ versus time at $\sigma_\infty = 50$ MN/m^2. Also
plotted in Fig. 3 is the local tensile stress σ (r = d/2, t)
just ahead of the current crack tip. At this overall stress
crack initiation begins at t = 1.3x10^{-4} t_L = 1.3x10^{-5}
sec $\approx$ 1.5 days. Initially the stresses at the crack tip relax, but after about 10% growth the stresses increased with
$\dot{a}$ according to eq. (23). After 50% growth the crack tip
velocity was $\dot{a}$ = 3x10^{-8} m/sec or about 2x10^{-3} m/day.

Figures 4 and 5 are corresponding results for σ_∞ =
150 MN/m^2. In this case crack growth begins at
t = 5.6x10^{-5} t_L = 26 sec. The stress then dramatically relaxes
and for some time the crack velocity decreases. After
t $\approx$ 10^{-3} t_L = 460 sec the local stresses increase due to the
increasing crack length as given by eq. (14) and the crack
approaches instability at a growth rate approaching 0.1 m per
hour. Just prior to this very rapid growth period
$\dot{a} \approx 2 \times 10^{-6}$ m/sec = 7.2x10^{-3} m/h.

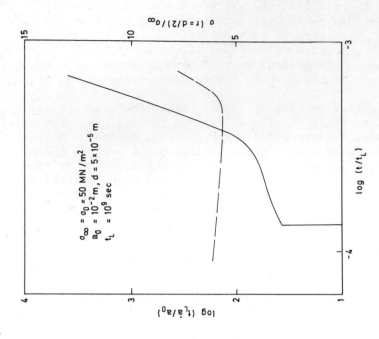

Fig. 3. Crack growth rate $\dot{a}$ (solid line) and stress on cavities closest to crack tip $\sigma(r = d/2)$ (dashed line) vs time at $\sigma_\infty = 50$ MN/m^2.

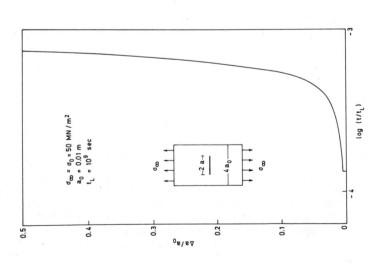

Fig 2. Change in crack length $\Delta a = a - a_0$ vs time at $\sigma_\infty = 50$ MN/m^2.

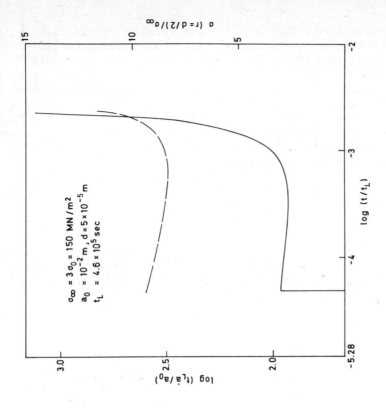

Fig. 5. Crack growth rate å (solid line) and stress on cavities closest to crack tip $\sigma(r = d/2)$ (dashed line) vs time at $\sigma_\infty = 150$ MN/m².

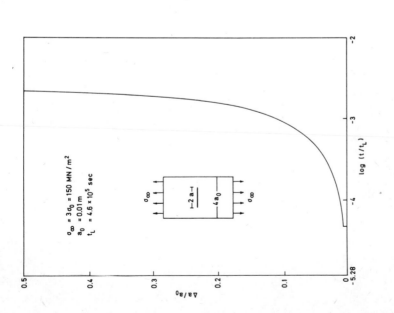

Fig. 4. Change in crack length $\Delta a = a - a_0$ vs time at $\sigma_\infty = 150$ MN/m².

The predictions in Figs. 2-5 are roughly in agreement with the experimental results of Koterazawa and Iwata [20] on a double edge-cracked plate of 304 stainless steel. At $923^{\circ}K$ in an air environment, they measured crack growth rates in the range of 10^{-9} - 3×10^{-7} m/sec for net section stresses in the range of 120 to 220 MN/m^2. In the two cases considered here, that are at slightly lower temperatures, the crack length to width ratio varied from 1/2 to 3/4 during the growth period so that the net section stresses varied from 100 to 200 MN/m^2 and 300 to 600 MN/m^2. The growth rates predicted are also in qualitative agreement with the experiments of Sadananda and Shahinian [21] on a compact tension specimen at $866^{\circ}K$ in both air and vacuum. Their correlations are between $\dot{a}$ and K, for K between 20 and 60 $MN/m^{3/2}$.

In retrospect, the initial crack length chosen was too large to be of practical interest at the stress levels considered. Further numerical studies will be undertaken for other crack configurations and stress levels.

ACKNOWLEDGEMENTS:

This work was supported by the Department of Energy under Contract EG-77-S-02-4461 and the University of Pennsylvania. Helpful discussions with Professors A.S. Argon, I.W. Chen and F.A. McClintock are gratefully acknowledged.

REFERENCES

1. McClintock, F.A. and Bassani, J.L., "Problems in Environ-
   mentally-Affected Creep Crack Growth", to appear in the
   proceedings of the IUTAM Symposium on Three-Dimensional
   Constituitive Relationships and Ductile Fracture, Eds.
   J. Zarka and S. Nemat-Nasser, Dourdan, 1981.

2. Argon, A.S., Chen, I-W. and Lau, C.W., "Intergranular Cavi-
   tation in Creep: Theory and Experiments", Creep-Fatigue-
   Environmental Interactions, Eds. R.M.N. Pelloux and N.
   Stoloff, AIME New York, 1980, p.46.

3. Chen I-W. and Argon, A.S., "Diffusive Growth of Grain-
   Boundary Cavities", submitted to Acta Met., 1980.

4. Gooch, D.J., Haigh, J.R., and King, D.L., "Relationship
   Between Engineering and Metallurigical Factors in Creep
   Crack Growth, Metal Sci., 1977, p. 545.

5. Dimelfi, R.J. and Nix, W.D., "The Stress Dependence of the
   Crack Growth Rate During Creep", Int . J. Fract., 1977, $\underline{13}$,341.

6. Riedel, H., "The Extension of a Macroscopic Crack at Elevated Temperatures by the Growth and Coalescence of Micro-
   voids", presented at the IUTAM Symposium on Creep in
   Structures, Leicester, 1980.

7. Riedel, H. and Rice, J.R., "Tensile Cracks in Creeping
   Solids", Brown University Report E(11-1), 3084-64, submitted for publication to ASTM, 1979.

8. Riedel, H., "Creep Deformation at Crack Tips in Elastic-
   Visco-plastic Solids", Brown University Report MRL E-114,
   1979.

9. Hui, H., and Riedel, H., "The Asymptotic Stress and Strain
   Field Near the Tip of a Growing Crack under Creep Conditions", Int. J. Fract., 1980, $\underline{16}$, no. 6.

10. Hull, D. and Rimmer, D.E., "The Growth of Grain-Boundary
    Voids under Stress", Phil. Mag., 1959, $\underline{4}$, 673.

11. Chuang, T.-J., Kagawa, K.I., Rice, J.R. and Sills, L.B.,
    "Non-Equilibrium Models for Diffusive Cavitation of Grain
    Interfaces", Acta Met., 1979, $\underline{27}$, 265.

12. Hancock, J.W., "Creep Cavitation without a Vacancy Flux",
    Metal Sci., 1976, $\underline{10}$, 319.

13. Budiansky, B., Hutchinson, J.W. and Slutsky, S., "Void
    Growth in Viscous Solids", Mechanics of Solids, Eds. H.G.
    Hopkins and M.J. Sewell, Pergamon Press, Oxford, 1981.

14. Beere, W. and Speight, M.V., "Creep Cavitation by Vacancy Diffusion in Plastically Deforming Solid", Metal Sci., 1978, 12, 172.

15. Needleman, A., and Rice, J.R., "Plastic Creep Flow Effects in Diffusive Cavitation of Grain-Boundaries," Division of Engineering Report, Brown University, Providence, R.I., 1980.

16. McClintock, F.A., "Mechanics in Alloy Design", Fundamental Aspects of Structural Alloy Design, Ed. by R.I. Jaffee and B.A. Wilcox, Plenum, 1977, p.147.

17. Tada, H., Paris, P.C., Irwin, G.R., The Stress Analysis of Cracks, Dell Research Corp., Hellertown, PA, 1973.

18. Hutchinson, J.W., Needleman, A., and Shih, C.F., "Fully Plastic Crack Problems in Bending and Tension", Fracture Mechanics, Eds. N. Perrone et. al., University Press of Virginia, Charlottesville, N.C., 1978, p. 515.

19. Leckie, F.A., and McMeeking, R.M., "Stress and Strain Fields at the Tip of a Stationary Tensile Crack in a Creeping Material", to appear in Int. J. Fract., 1981.

20. Koterazawa, R. and Iwata, Y., "Fracture Mechanics and Fractography of Creep and Fatigue Crack Propogation at Elevated Temperature", J. Engng. Mat. Tech., 1976, 98, 296.

21. Sadananda, K., Shahinian, P., "Effect of Environment on Crack Growth Behavior in Austenitic Stainless Steels Under Creep and Fatigue Conditions", Met. Trans. A., 1980, 11A, 267.

DEVELOPMENT OF A MICROFRACTURE MODEL FOR HIGH RATE TENSILE
DAMAGE

L. Seaman, D. R. Curran, and D. A. Shockey

Poulter Laboratory, SRI International, Menlo Park, California

94025, U.S.A.

SUMMARY

A nucleation-and-growth (NAG) model representing brittle
microfracture has been constructed and applied to high-rate
fracture of several metals, rocks, plastics and other
materials. This paper relates the model processes of nucleation, growth, coalescence and fragmentation to the observed
processes and outlines a method for fitting the model to data.
An experimental program suitable for obtaining the fracture
parameters is described. Recent simulations of impact fracture tests in a polycarbonate are displayed to illustrate the
modelling approach.

1. INTRODUCTION

A model (NAG) describing the Nucleation And Growth of brittle microfractures arising from dynamic tensile stresses has been developed and incorporated into a computer sub-routine. This paper presents the basic fracture processes--nucleation, growth, coalescence, fragmentation--and the stress-strain relations. The experimental data required for quantifying the fracture processes are outlined. Following synthesis of the model, simulations of fracture experiments are made and compared to data on a polycarbonate. We expect that a similar microfracture model can be constructed to represent creep, fatigue, and other lower rate phenomena.

2. ELEMENTS OF THE MODEL

The major features of a brittle microfracture model have been deduced from impact data on Armco iron,[1-4] polycarbonate,[5] oil shale,[6] novaculite quartz,[7,8] several steels,[9,10] and propellant.[11] In spite of the great varity in these materials, the fracture processes are qualitatively similar.

Our observations indicate that the initiation of micro-fracture occurs by the production of cracks in a plane normal to the direction of maximum tension and at a time when this tensile stress exceeds a critical stress level. (Some evidence[12] is now suggesting that rock fracture may occur under a compressive stress if there is significant shear strain.) The nucleation rate for microcracks appears to be a strong function of the stress normal to the plane of fracture. Thus the nucleation process in the model has been made to depend on the tensile stress amplitude and its orientation.

Under continued tensile loading the microcracks grow. Our data indicate that the cracks always occur in a range of sizes, as shown in Figure 1 for Armco iron[2-4]. This figure contains cumulative size distributions obtained by counting the cracks on a cross section of a target impacted on the upper plane face by a shorter cylinder. The stress duration is greatest near the center of the target, where the damage is greatest, and diminish in planes on either side. Clearly the number and size of cracks increase with load duration; hence, a nucleation and growth process is indicated. The rate of growth appears to be approximately proportional to the crack size; large cracks grow faster than small ones. Therefore, the model growth processes require a crack-size distribution, not a single crack size, and a growth rate which is crack-size dependent.

In impact experiments coalescence and fragmentation appear to occur by the intersection of cracks.[7,8,9] If one

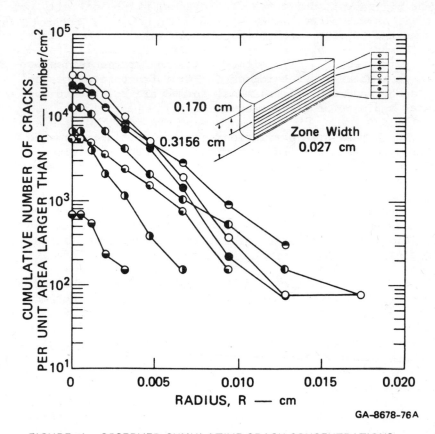

FIGURE 1   OBSERVED CUMULATIVE CRACK CONCENTRATIONS
ON A CROSS SECTION OF AN IMPACTED CYLINDRICAL
TARGET OF ARMCO IRON

plane of fracture dominates, the cracks, which initially do not lie in one plane, grow out of their plane to intersect other cracks and to form a rough fracture surface[5] When cracks are in many planes[8] they intersect more orthogonally to form chunky fragments. In either case, the fragmentation process is a natural extension of the microfracture processes. The number and size of fragments formed are related to the number and size of microcracks.

The stress-strain relations for the partially damaged material, unlike the fracture processes, cannot be directly deduced from the impact experiments. However, it is essential that the stress-strain relations be modified by damage to provide the observed focusing of damage at spall planes and to reproduce the measured "pull-back" signals of particle velocity measurements.[13] Stress-strain relations are obtianed from elastic or inelastic theoretical analyses of material containing cracks.

Finally, a solution procedure is required in the model. The model or subroutine is provided with a strain increment tensor and from that must calculate the stress tensor and the current microcrack size distribution. Because the cracking processes depend on the stresses in a complex way, an iterative numerical technique is required.

3.   DATA REQUIREMENTS

The microfracture processes can be outlined in considerable detail qualitatively, but quantitative determination of the parameters in these processes requires a series of well planned and conducted impact experiments. The series of impacts should span the stress range from just below the fracture threshold to beyond the level of damage of interest. Especially important are three to five impacts producing small to moderate damage. Such impacts cause little damage-induced stress relaxation; hence, the fracture rates can be found fairly independently of the stress relaxation processes. These impacts must provide quantifiable damage (countable cracks with measurable lengths and orientations) with gradual variations through the target. A range of damage stress levels and stress durations are important for reliable determination of the microfracture parameters.

A description of the steps taken in reducing a set of data to obtain fracture parameters illustrates the foregoing data requirements. A hypothetical case is shown in Figure 2. Figure 2(a) shows a section of a target disk that was impacted simultaneously over one face by a similar disk about half as thick. The section shows a multitude of cracks, with longer and more numerous cracks in a central strip. The impact damage is essentially uniform at a given axial position, and

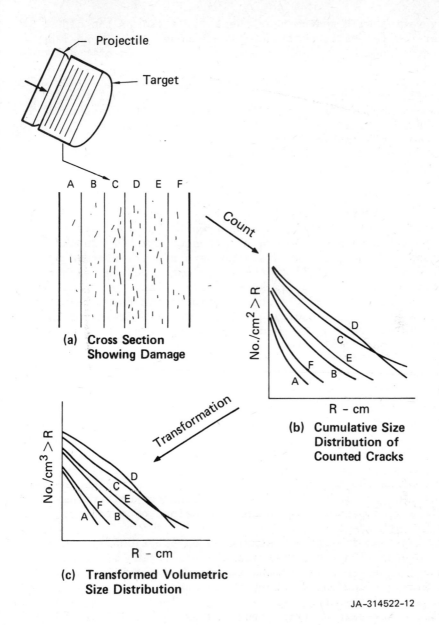

(a) **Cross Section Showing Damage**

Count

No./cm² > R

D
C
E
F
B
A

R – cm

(b) **Cumulative Size Distribution of Counted Cracks**

Transformation

No./cm³ > R

D
C
E
F
B
A

R – cm

(c) **Transformed Volumetric Size Distribution**

JA–314522-12

FIGURE 2   ACQUISITION AND TRANSFORMATION OF CRACK
COUNT DATA FROM CROSS SECTIONS OF IMPACTED
CYLINDRICAL TARGETS

only varies with radius near the outer edges of the disk.
After the target is sectioned, the exposed face is polished
to reveal the microcracks clearly. From photos of the section,
the cracks are counted and measured and then displayed in a
cumulative size distribution, as in Figure 2(b). This graph
(and Figure 1) shows the approximately exponential form
commonly observed in these distributions. It is essential for
the analyses that there be a large enough number of cracks to
provide a reasonable statistical sample, that they display a
range of sizes, and that a gradual variation of damage with
distance through the specimen be evident.

Following collection of the surface count data, the distribution is transformed by a statistical method to the
volume size distribution, a bulk property of the material.
For targets in which the cracks lie approximately in one plane,
the apparent crack lengths and numbers are transformed by the
method of Scheil[14,15] to the true length and the number per
unit volume. When the cracks occur in a range of orientations,
then a transformation[16] which accounts for the orientations
must be used. The resulting volumetric size distribution is
shown in Figure 2(c). Again, the distributions have an approximately exponential form.

In the next steps, these size distributions are used to
obtain quantitatively the parameters for the nucleation,
growth, and coalescence equations. These data are prepared by
fitting each of the distributions to an exponential form

$$N_g = N_o \exp(-R/R_1) \tag{1}$$

where N_g is the number per unit volume with a radius greater
than R, N_o is the total number of cracks per unit volume, and
R_1 is a size parameter with the dimensions of a radius.

Next the impacts are simulated computationally with a
wave propagation code, treating the fracturing material with
a model that accounts for the elastic and plastic behavior,
but does not include fracture. From these simulations the
stress histories are determined at the locations corresponding
to the crack-size distributions. These histories are then
approximated as square waves of amplitude σ and duration Δt.

With the parameters obtained from the fracture distributions and the stress histories, the nucleation and growth laws
can be quantified. An exponential nucleation relation and
viscous growth law are used here for illustration, but these
need not be used if the data suggest other forms.

$$\left(\frac{\partial N}{\partial t}\right)_{R=o} = \dot{N}_o \exp\left(\frac{\sigma - \sigma_{no}}{\sigma_1}\right) \qquad \text{(nucleation)} \qquad (2)$$

$$\left(\frac{\partial R}{\partial t}\right)_n = T_1(\sigma - \sigma_{go}) \, R \qquad \text{(growth)} \qquad (3)$$

The nucleation and growth parameters to be determined in these relations are $\dot{N}_o$, σ_{no}, σ_1, T_1, and σ_{go}. The quantities N and σ may be recognized as tensors here. But for this first estimate of the parameters it is sufficient to consider only the direction of maximum tensile stress. The threshold stress, σ_{no} or σ_{go}, is determined by taking a stress value between the largest tensile stress for which no damage occurred and the smallest for which there was damage.

The nucleation parameters are obtained from a plot (as in Figure 3) of $\partial N/\partial t \doteq N_o/\Delta t$ versus $\sigma - \sigma_{no}$ from a series of tests. Values from the plane of maximum damage in each test should be used, but parameters from other planes can also be used, provided there is little damage-caused stress relaxation. The slope and intercept of the trend line give σ_1 and $\dot{N}_o$.

To fit the data to the growth law, that law is first integrated to the form

$$\ell n\left(\frac{R_1}{R_n}\right) = T_1(\sigma_m - \sigma_{go})\Delta t \qquad (4)$$

Then the R_1 values are plotted versus the tensile impulse $(\sigma_m - \sigma_{go})\Delta t$ as in Figure 4. The intercept and slope provide R_n, the value of R_1 at nucleation, and T_1, the growth rate parameter. T_1 is the reciprocal of a viscosity.

With a set of fracture parameters in hand, the process used in determining the parameters can be examined. The plots made in Figures 3 and 4 are rigorous tests of the reliability or scatter of the data and of the suitability of the nucleation and growth model to represent the data. But many approximations and simplifications have been made in reducing the data to obtain the needed fracture parameters. To overcome these limitations, a fracture model containing the preceding nucleation and growth laws is constructed, and then the experiments are simulated with a wave propagation code containing the model. By repeated simulations and adjustments of the fracture parameters, a set of parameters is obtained which represents the series of tests.

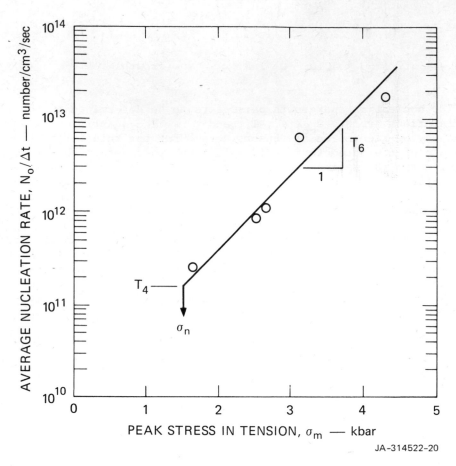

JA-314522-20

FIGURE 3   PLOT OF OBSERVED NUMBER OF CRACKS WITH COMPUTED STRESS σ_m AND
DURATION Δt FOR OBTAINING NUCLEATION RATE PARAMETERS T_4 AND T

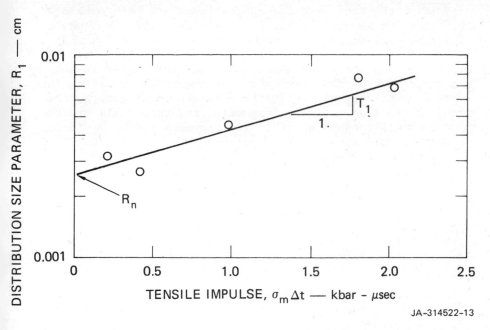

JA-314522-13

FIGURE 4   PLOT OF OBSERVED CRACK SIZES WITH COMPUTED STRESS σ_m AND
DURATION Δt FOR OBTAINING GROWTH RATE PARAMETERS R_n AND T_1

After the basic fracture parameters have been obtained from experiments with low levels of damage, the model may be fitted to coalescence and fragmentation data. The gradual transition from fracture to fragmentation is illustrated in Figure 5. The model fragmentation parameters refer to the shape and size of the fragments. In the model we assume that there is a natural transition from cracks to fragments, that large cracks lead to large fragments, and that small cracks lead to small fragments. Then the fragment radius R_f (a representative dimension of the particle) is $R_f = \gamma R$, where γ is a factor near 1.0 and R is the crack radius. Similarly, the number of fragments N_f is related to the number of cracks required to form the faces of the fragments: $N_f = \beta N$, where β is a factor of 1/3 to 1/5. The volume of each fragment is

$$v_f = T_f R_f^{\,3} \tag{5}$$

where T_f is a numerical constant, about 4. These values of β, γ, and T_f, plus the assumption that fragments of all sizes are geometrically similar, came from our observations in fine-grained quartz[7,8] as illustrated in Figure 6.

To determine the fragmentation parameters β, γ, and T_f, an experiment must be conducted to complete fragmentation. The fragments are then counted by size and arranged in a cumulative size distribution. The test is then simulated, using a trial set of fragmentation parameters. The calculations are repeated with revised values of β, γ and T_f until the breadth of the fragmentation zone and the cumulative size distribution match the experimental values with acceptable accuracy. Of course, more reliable fragmentation parameters are obtained by matching results to a series of tests in which different levels of fragmentation have been attained.

4. MODEL FEATURES

The microfracture model is constructed to represent the observed microfracture processes in a statistical manner, and to compute the macroscopic or average stresses. Thus the model does not represent individual cracks, but an entire size distribution. The crack counts are in units of number per unit volume, so that the computed crack counts are indifferent to the finite-difference cell sizes used in simulations.

The model features are of two types: those based on data and those from analyses. The nucleation, growth, and coalescence processes are obtained more or less directly from the fracture observations, as outlined above. Thus we can be sure that the form of the functions and the parameters are

Creep and Fracture of Engineering Materials and Structures

Creep and Fracture of Engineering Materials and Structures

Edited by:

B. Wilshire
Department of Metallurgy and Materials Technology,
University College, Swansea.

D. R. J. Owen
Department of Civil Engineering, University College, Swansea.

Proceedings of the International Conference
held at University College, Swansea,
24th - 27th March, 1981.

PINERIDGE PRESS

Swansea, U.K.

05729546

First published, 1981 by
Pineridge Press Limited
91, West Cross Lane, West Cross, Swansea, U.K.

ISBN 0-906674-10-7

Copyright © 1981 by Pineridge Press Limited.

British Library Cataloguing in Publication Data

Creep and fracture of engineering materials and structures
    1. Materials – Creep – Congresses
    2. Materials at high temperatures – Congresses
    3. Fracture mechanics – Congresses
    I. Wilshire, B.   II.Owen, D. R. J.
    ISBN 0-906674-10-7

Printed and bound in Great Britain by
Robert MacLehose and Co. Ltd.
Printers to the University of Glasgow

D
620. 1123'3 (AE
+ D620.1126)
CRE

PREFACE

The response to the Call-for-Papers from authors
throughout the world has indicated the widespread interest in
the subject of 'Creep and Fracture of Engineering Materials
and Structures'. The papers accepted for publication in the
Proceedings have been separated into six sections covering
the major areas of interest, namely, creep mechanisms,
deformation processes in particle-strengthened alloys, creep
fracture processes, creep and fracture of ceramics,
materials behaviour at elevated temperatures and the design
and performance of components and structures.

The Proceedings are printed from direct lithographs of
authors manuscripts and the editors cannot accept
responsibility for any inaccuracies, comments or opinions
expressed in the papers. However, the organisers wish to
thank the authors for presenting their work and ideas in the
context of the overall position currently reached in the area
relevant to the theme considered. In this way, the
Conference provides both an overview of the different
approaches being developed in various centres active in the
field of creep and fracture and an indication of the
principal avenues along which future activities should be
directed.

The organisers wish to acknowledge the generous
sponsorship of the Conference provided by the United States
Air Force, European Office of Aerospace Research and
Development and the Department of the Navy, Office of Naval
Research. The sponsorship received from Mand Testing
Machines Ltd., Eurotherm Ltd. and Automatic System
Laboratories Ltd. towards the social programme of the
Conference is also gratefully acknowledged.

B. WILSHIRE   D.R.J. OWEN

Swansea, March 1981.

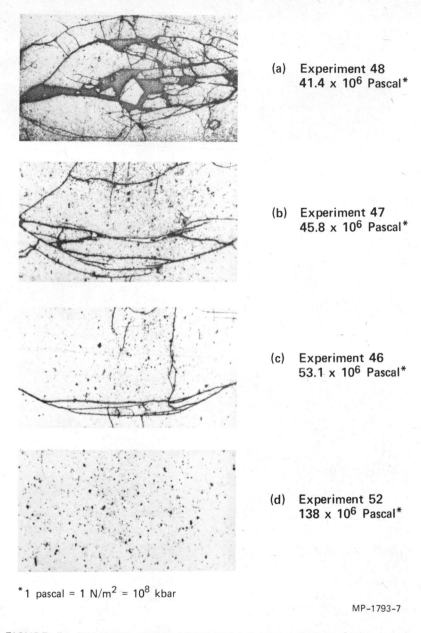

(a)   Experiment 48
      41.4×10^6 Pascal*

(b)   Experiment 47
      45.8×10^6 Pascal*

(c)   Experiment 46
      53.1×10^6 Pascal*

(d)   Experiment 52
      138×10^6 Pascal*

*1 pascal = 1 N/m^2 = 10^8 kbar

MP-1793-7

FIGURE 5   POLISHED CROSS SECTIONS OF ARKANSAS NOVACULITE
           SPECIMENS SHOWING THE EXTENT OF FRACTURE DAMAGE
           PRODUCED BY INCREASING LEVELS OF IMPACT STRESS

           Impact direction was from top to bottom.

UNFRAGMENTED PORTION

RADII 149 TO 210 microns

RADII GREATER THAN 1000 microns

RADII 74 TO 149 microns

RADII 420 TO 1000 microns

RADII 37 TO 74 microns

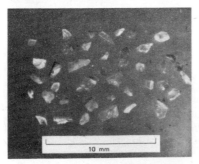

RADII 210 TO 420 microns

RADII 18 TO 37 microns

MP–1793–8A

FIGURE 6   PHOTOMICROGRAPHS OF FRAGMENTS OF ARKANSAS
NOVACULITE OBTAINED FROM AN IMPACT EXPERIMENT

appropriate for the material and the strain rates of the
tests.

The stress-strain relations, the stress-relaxation process, the crack-opening relation, and the solution procedure
are based on analyses. The crack-opening relation is the
elastic expression from Sneddon;[17] hence, neither plastic
opening, permanent set, nor the rate dependence of opening
are provided. The stress-strain relations and stress-relaxation processes are based on our conjecture that the imposed
strains $\Delta\varepsilon^t$ can be separated into portions $\Delta\varepsilon^s$ taken by the
solid, intact material and portions $\Delta\varepsilon^f$, which represent the
crack opening. The solution procedure for this highly non-
linear problem gives results which are strongly dependent on
the strain increment size; therefore, the imposed strains are
cut into acceptable sizes and the calculation is performed by
a series of subcycles. In this way the solution is made
fairly independent of the imposed strain increment so that the
solution procedure is not a source of uncertainty.

Most of the model is described in Reference 5, but the
solution procedure has been outlined only in unpublished
reports.[5,9,10] The solution begins with an estimate of the
solid strain tensor $\Delta\varepsilon^s$, based on the imposed strain tensor
and the current effective moduli. With this estimate,
stresses are first computed and then crack nucleation, growth,
and opening strain are computed. The sum of the estimated
solid strain and fracture strain is compared with the imposed
strain. A regula falsi interpolation technique is used to
obtain revised estimates of $\Delta\varepsilon^s$ for later iterations.

5.   COMPARISONS OF MODEL PREDICTIONS WITH EXPERIMENTAL DATA

ON POLYCARBONATE

A series of plate impact experiments was performed on
polycarbonate[5] to study high-rate fracture in a transparent
material. In each case 0.65-cm thick disks of polycarbonate
were impacted by 0.31-cm thick flyer plates of polycarbonate
as indicated in Table 1. Damage with a range of intensities
was observed near the midplane in each target, as shown in
Figure 7. A family of cumulative crack-size distributions for
one of the targets is shown in Figure 8. These fracture
results were of interest in the current study because
all the cracks formed with essentially the same orientation.

Shot No.	Flyer Thickness (cm)	Target Thickness (cm)	Velocity (m/sec)	Peak Tensile Stress (GPa)
5	0.305	0.658	152.2	0.183
6	0.317	0.645	142.8	0.171
7	0.317	0.653	137.2	0.164

Table 1
Polycarbonate Impact Configurations

358

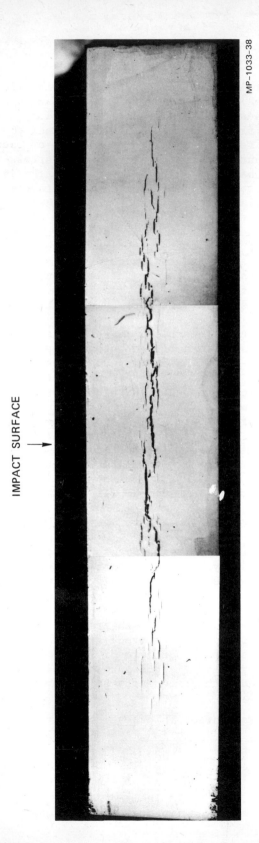

IMPACT SURFACE →

MP-1033-38

FIGURE 7   POLISHED SECTION THROUGH POLYCARBONATE TARGET DISK NO. 5 SHOWING THE DISTRIBUTION OF CRACK TRACES INTERSECTING THE PLANE OF POLISH

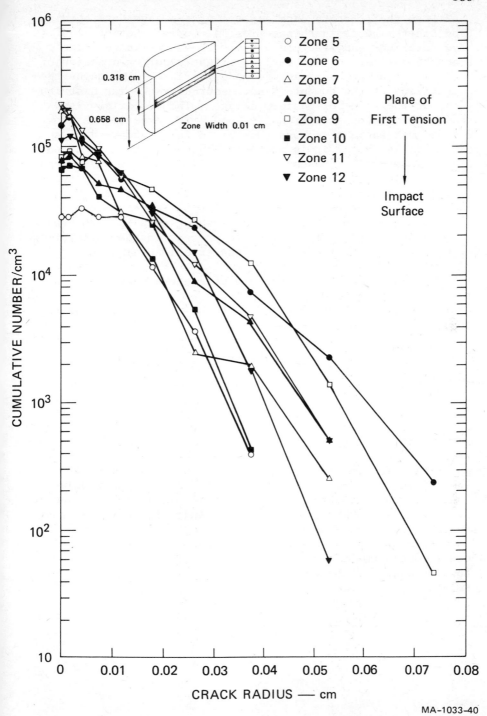

FIGURE 8   VOLUMETRIC CRACK SIZE DISTRIBUTIONS IN POLYCARBONATE
TARGET NO. 5 AT SEVERAL DISTANCES THROUGH THE TARGET

From the fracture data and the stress gage data obtained from other impacts in this series, a complete set of constitutive relations and fracture parameters was computed. The fracture parameters were revised slightly during a series of simulations using the fracture model. Figure 9 contains the comparisons of the computed and measured crack-size distributions on the plane of maximum damage. The comparison is reasonable if not exceptionally good. The experiments show damage just above threshold (No. 7) to fairly heavy damage (No. 5). In Shot 7 it is premised that nucleation occurred in several subsequent phases. This combination of a small amount of nucleation and considerable growth produces the observed distribution with a shallow shape. On the other hand, in Shots 5 and 6 there is more damage, so the stresses are reduced significantly, causing a lesser growth in later tensile phases. These effects, while present in the calculations, do not appear as strongly as in the experimental data.

The large number of fracture parameters determined in these simulations are listed in Table 2. We cannot claim that a unique set has been obtained, but the results are sensitive to all the fracture parameters. The fragmentation parameters β, γ, and T_f are not fully defined in these low-damage results, but a value for the combination $\beta\gamma^3 T_f$ is determined.

6. DISCUSSION

A computational model which closely follows the microfracturing process in many materials has been developed and applied to fracture in a polycarbonate. The model predictions represent reasonably well the fracture processes in this material. In addition, the modelling technique aids in identifying the features of nucleation, growth, and coalescence of cracks and the role of stress relaxation. By varying these features of the model, we can study, for example, the suitability of alternate nucleation functions. The main advantage of the micromechanics approach to modelling over the usual continuum or macro approaches may be that it leads towards a deeper appreciation of the physical processes themselves.

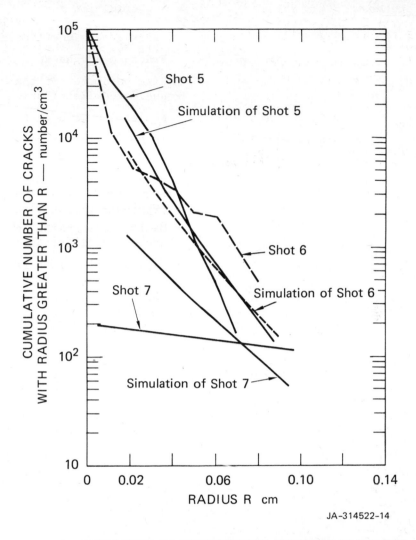

JA–314522–14

FIGURE 9   COMPARISON OF COMPUTED AND OBSERVED CRACK-SIZE
DISTRIBUTIONS ON PLANES OF MAXIMUM DAMAGE
IN THREE POLYCARBONATE TARGET DISKS

Parameter	Value	Definition
T_1	5×10^{-4} cm^2/dyn/sec	Crack growth coefficient.
K_{IC}	3.6×10^7 dyn/cm$^{3/2}$	Fracture toughness.
R_n	2×10^{-3} cm	Size parameter of the nucleated distribution.
$\dot{N}_o$	7×10^{10}/cm^3/sec	Nucleation rate coefficient.
σ_{no}	1.60×10^9 dyn/cm^2	Threshold stress for nucleation.
σ_1	1.41×10^8 dyn/cm^2	Denominator of argument in exponential nucleation function.
R_{nb}	1×10^{-2} cm	Maximum crack size in the nucleated distribution.
β	0.33	Ratio of number of fragments to number of cracks.
γ	1.0	Ratio of fragment radius to crack radius.
T_f	4.0	Coefficient of fragment volume.

Table 2 Fracture and Fragmentation Parameters

for Polycarbonate

REFERENCES

1. Barbee, Jr., T. W., Seaman, L., Crewdson, R., and
    Curran, D., - 'Dynamic Fracture Criteria for Ductile and
    Brittle Metals', J. of Materials, 1972, $\underline{7}$, 393.

2. Barbee, Jr., T. W., Seaman, L., and Crewdson, R. C.,
    - 'Dynamic Fracture Criteria of Homogeneous Materials',
    Technical Report No. AFWL-TR-70-99, to Air Force Weapons
    Laboratory, Kirtland Air Force Base, N. Mex., Nov. 1970.

3. Seaman, L., Barbee, Jr., T. W., and Curran, D. R.,
    - 'Dynamic Fracture Criteria of Homogeneous Materials',
    Technical Report No. AFWL-TR-71-156 to Air Force Weapons
    Laboratory, Kirtland Air Force Base, N. Mex., Dec. 1971.

4. Shockey, D. A., Seaman, L., and Curran, D. R., - 'Dynamic
    Fracture of Beryllium Under Plate Impact and Correlation
    with Electron Beam and Underground Test Results', Final
    Report No. AFWL-TR-73-12 to Kirtland Air Force Base,
    N. Mex., Jan. 1973.

5. Curran, D. R., Shockey, D. A., and Seaman, L., - 'Dynamic
    Fracture Criteria for a Polycarbonate', J. Appl. Phys.,
    1973, $\underline{44}$, 4025.

6. Murri, W. J., et al. - 'Determination of Dynamic Fracture
    Parameters for Oil Shale', Final Report to Sandia Labora-
    tories, Albuquerque, N. Mex., Feb. 1977.

7. Shockey, D. A., Curran, D. R., Seaman, L., Rosenberg, J. T.,
    and Petersen, C. F., - 'Fracture of Rock Under Dynamic
    Loads', Int. J. Rock Mech. Sci. Geomech. Abstr., 1974, $\underline{11}$,
    303.

8. Shockey, D. A., Curran, D. R., Austin, M., and Seaman, L.,
    - 'Development of a Capability for Predicting Cratering
    and Fragmentation Behavior in Rock', Final Report, to
    Defense Nuclear Agency, No. DNA 3730F, May 1975.

9. Shockey, D. A., Seaman, L., Curran, D. R., DeCarli, P. S.,
    Austin, M., and Wilhelm, J. P., - 'A Computational Model
    for Fragmentation of Armor Under Ballistic Impact',
    Final Report No. DAAD05-73-0025, Phase 1, to U.S. Army
    Ballistic Research Laboratories, Aberdeen Proving Ground,
    Dec. 1973.

10. Seaman, L., and Shockey, D. A., - 'Models for Ductile
    and Brittle Fracture for Two-Dimensional Wave Propaga-
    tion Calculations', Final Report No. DAAG46-72-C-0182, to
    Army Materials and Mechanics Research Center, Watertown,
    MA., Feb. 1975.

364

11. Murri, W. J., and Curran, D. R., - 'Fracture Model for High Energy Propellant', Quarterly Progress Report No. 2, Contract P. O. 7250109 under W-7405-ENG 48 to Lawrence Livermore Laboratory, Livermore, CA., Sept. 1980.

12. McHugh. S. L., - 'Effect of Confining Stress on Crack Nucleation', Internal SRI Report, Menlo Park, CA., Jan. 1981.

13. Cochran, S., and Banner, D. J., - 'Spall Studies in Uranium', J. Appl. Phys., 48, No. 7, July 1977, 2729.

14. Scheil, E., - 'Die Berechnung der Anzahl und Grossenverteilung kugelformiger Kristalle in undurchsichtigen Körpern mit Hilfe durch einen ebenen Schnitt erhältensen Schnittkreise', Z. Anorg. Allgem. Chem., 1931, 201, 259.

15. Scheil, E., - 'Statistiche Gefügeuntersuchungen I', Z. Metallk., 1935, 27, 199.

16. Seaman, L., Curran, D. R., and Crewdson, R. C., - 'Transformation of Observed Crack Traces on a Section to True Crack Density for Fracture Calculations', J. Appl. Phys., 49, (10), Oct. 1978, 5221.

17. Sneddon, I. N., and Lowengrub, M., - 'Crack Problems in the Classical Theory of Elasticity', John Wiley and Sons, New York, 1969.

SECTION 4

CREEP AND FRACTURE OF CERAMICS

MECHANISMS OF CREEP DEFORMATION AND FRACTURE IN SINGLE AND
TWO-PHASE Si-Al-O-N CERAMICS

M.H. Lewis, B.S.B. Karunaratne, J. Meredith and C. Pickering.

Department of Physics, University of Warwick, Coventry, U.K.

ABSTRACT

A programme of research has been conducted on the microstructure and high-temperature creep and fracture of single
and two-phase Si-Al-O-N ceramic 'alloys' based on the β Si_3N_4
structure. Single-phase ceramics undergo a grain-boundary
diffusional creep with a long-term variation in activation-
energy due to de-segregation of grain boundary impurities
into a surface SiO_2 oxidation layer. These oxidation-heat-
treated ceramics exhibit zero primary creep and an activation-
energy for steady-state creep of 820 kJ mol^{-1} which may represent an intrinsic grain-boundary diffusion mechanism
($D_b \sim 5 \times 10^{-17}$ m^2 s^{-1}). The presence of triple-junction
silicate residues of the sintering mechanism is shown to be
responsible for profuse internal cavitation with resultant
anomalies in measurement of creep parameters and relatively
short times to failure.

Two-phase (β + yttrogarnet) ceramics, which may be
fabricated via pressureless sintering, exhibit power-law creep
dominated by plasticity of the matrix (garnet) phase.

1. INTRODUCTION

1.1 High Temperature Creep of Ceramics. Experimental data
for high-temperature creep of engineering solids may be interpreted via the general equation for strain rate ($\dot{\epsilon}$) as a
junction of stress (σ) and grain size (d) :

$$\dot{\epsilon} = A \left(\frac{\sigma}{G}\right)^n \left(\frac{b}{d}\right)^m D\frac{Gb}{kT} \tag{1}$$

N

where
(A = constant
(G = shear modulus
(b = Burgers vector
(D = $D_o \exp \frac{-Q}{RT}$ is the diffusion coefficient and Q the activation energy.

Micromechanisms for creep are normally identified via the exponents n and m by comparison with theoretical values together with activation energy Q experimentally determined for a particular mechanism via alternative experiments.

The determination of creep mechanisms for engineering ceramics has been inhibited by the difficulty of performing experiments on semi-brittle solids with uniaxial stresses at comparatively high temperatures, the precise measurement of comparatively low strain-rates at such temperatures, the complexity of possible rate-controlling mechanism with at least two diffusing species and greater difficulty in control of microstructure and its direct examination, by comparison with metallic alloy systems. However, a number of analyses have been made for ceramic oxides (MgO [1-5], Al_2O_3 [6-12], UO_2 [13]), which demonstrate a variation in rate-controlling creep mechanism with stress and temperature with trends predicted via exponents in the constitutive equation (1). Hence 'power-law' (n ~ 3, m ~ 0) creep, involving lattice dislocation motion, is favoured at high stresses whilst diffusional creep (n ~ 1, m = 2 or 3 for 'Nabarro-Herring' creep or 'Coble' creep respectively) dominates at lower stresses. The importance of diffusional mechanisms in ceramics is anticipated from their relatively high Peierls stresses for dislocation motion and generally small grain size. An extreme example of this trend should be that of ceramics based on the β Si_3N_4 crystal structure which has high covalency and, usually, sub-micron grain size. The various studies of creep in commercial Si_3N_4 ceramics of near theoretical density are, however, ambiguous in this respect with non-integral (n > 1) stress exponents (Table 1) but have deformed microstructures which show little evidence for dislocation motion. Explanations of 'non-Newtonian' grain-boundary sliding or the influence of observed cavitation on internal stresses have been used in qualitative interpretation of the data. There is increasing evidence for the effect of cavitation in oxide as well as β Si_3N_4 ceramics and recent modelling [14] of the influence of purely elastic multiple crack-opening on creep strain rate and its stress dependence. In practice there is undoubtedly a superposition of creep strain from the various mechanisms, including diffusional flow.

One of the important features of the research on β Si_3N_4 ceramic 'alloys' reviewed in this paper is the demonstration of a transformation to purely diffusional creep behaviour which accompanies the removal of grain-boundary cavitation. The research is based on the use of the 'alloying' concept to generate a fully-dense ceramic without

detectable grain—boundary phases which are normally residues
of the liquid silicate—phase sintering aid.

Si_3N_4 Type	Temperature (°C)	Test Mode & Stress (MNm^{-2})	Stress Exponent (n)	Activation Energy $(kJ\ mol^{-1})$	Ref.
HS 130	1149—1260	Tension (30—100)	2.2		15
	1175—1275	70		535	
HS 130	1300—1400	Bend (60—120)	1.7	588	16
Lucas HPSN	1350	Compression (222—350)	2.1—2.4	650	17
HS 130	1250—1400	Compression (80—210)	1.8—2.0		18

Table 1.

1.2 β—Si_3N_4 Ceramic Alloys.   The principle of Si_3N_4 ceramic
alloying utilises the crystallographic requirement for
simultaneous solid—solution of Al^{3+} and O^{2-} in a Si^{4+}/N^{3+}
compound without creation of constitutional vacancies. Hence
the progressive crystallisation of the liquid—phase components
Al,Si,O,N during fabrication by hot—pressing results in a
single—phase ceramic of general composition $Si_{6-x}Al_xO_xN_{8-x}$
[19]. Substitution levels up to x∼4 are possible but a
restriction to x∼1 has thus—far been dictated by the greater
homogeneity and smaller grain size obtained with large amounts
of α—Si_3N_4 in the initial particle mixture. The origin of
this effect is believed to be in the homogeneity in distribution of surface SiO_2 on α particles which is an important
source of liquid sintering aid. The sintering kinetics and
hence final density are improved by minor additions of
metallic ions (usually added as oxides) such as Mg^{2+} and
Mn^{3+} which reduce the liquid silicate viscosity [19].

A precise analysis of grain—boundary residue content
requires an application of high—resolution transmission
electron microscopy, in which adjacent crystals are 'lattice'
imaged; a technique first applied to commercial Si_3N_4
ceramics [20]. Examples of such images for the β'Si—Al—O—N
ceramic alloys used in the creep programme are reproduced in
Fig.1 together with conventional diffraction—contrast images
at lower magnification. The lattice images enable one to
define the boundary structure within ∼6Å and conclude [21]
that two—grain interfaces may contain liquid phase residues
of only a few atom layers in thickness i.e. similar to grain—
boundary impurity segregation in metals. Auger electron

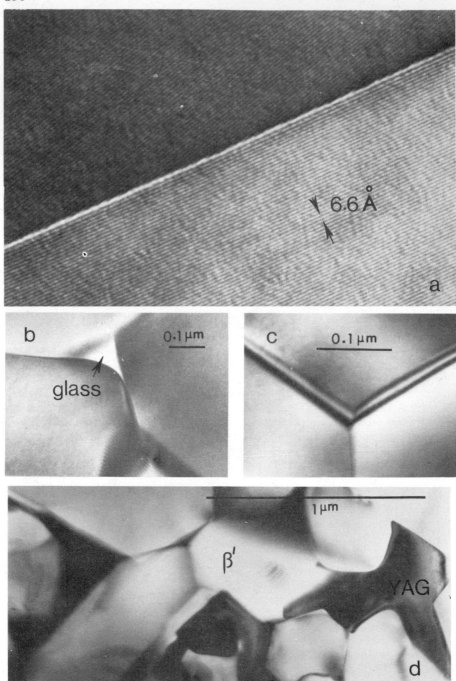

Fig. 1. 6.6Å lattice image of a 2-grain interface (a) and triple-junction structure in cavitating and non-cavitating single-phase ceramics (b and c respectively). Fig. 1d is the microstructure of a 2 phase (β' + YAG) ceramic.

spectroscopy of fracture surfaces [19] provides direct qualitative evidence for segregation of Mg, accidental impurity Ca
and oxygen. The most important microstructural feature which
dictates creep and fracture behaviour is the presence of
silicate residues within grain triple-junctions which provide
incipient cavity nuclei [21]. These residues are only visible
(Fig. 1) in β' materials which deviate from the 'balanced' O/N
ratios of the solid-solution formula or in which the silicate
components are wrongly partitioned between liquid and β'.
This is promoted by transition-metal oxide additives in place
of MgO [21].

In Section 2 of this paper we review our previous
research [21, 22] in which the significance of grain-boundary
microstructure on 4-point bend creep and fracture behaviour
was identified and present new confirmatory evidence from
compressive creep tests. The compressive creep rates enable
a measurement of grain-boundary diffusion coefficient and a
study of variation in diffusion rate with impurity segregation.
In Section 3 we compare the creep behaviour of a two-phase
(β' + yttrogarnet) ceramic alloy with that of the single phase
ceramics. The two-phase ceramics may be prepared by pressureless sintering, a feature which is important in the economical
fabrication of complex engineering components. Their microstructural evolution has previously been described [23] and
an example of the final microstructure of β' crystals (substitution level $x \sim \frac{1}{2}$) in a semi-continuous matrix of the garnet
phase (approx. composition $Y_6Al_7Si_3O_{21}N_3$) is shown in Fig. 1c.
The function of the Al 'alloying' element in these materials
is that of increasing the flexibility in crystallography of
second phase and permitting some   variation in O/N ratio of
both β' and matrix phases [23].

2. SINGLE PHASE CERAMIC ALLOYS

2.1 <u>Creep Data</u>. Fig. 2 summarises the compressive creep
data obtained from incremental stress tests and makes comparison with similar data obtained previously using 4-point
bend tests [21]. The disadvantages of compressive loading for
ceramics are (i) the difficulty in precise measurement of the
very small axial strain and (ii) the problem of ensuring
axial loading without introducing frictional 'end-effects'
associated with short specimens. In this work the 3x3x5mm
specimens were loaded between 'Refel' SiC platens. The
advantages of compressive loading are (i) it provides a
nearly homogeneous stress system avoiding the uncertainties
in determining creep stress exponents and absolute creep rates
and (ii) it enables a greater range of stress application in
exploring changes in creep mechanism via determination of
stress-exponent.

Fig. 2a shows that the bend-tests overemphasise true
creep rates, measured as an outer-fibre tensile strain rate,

370

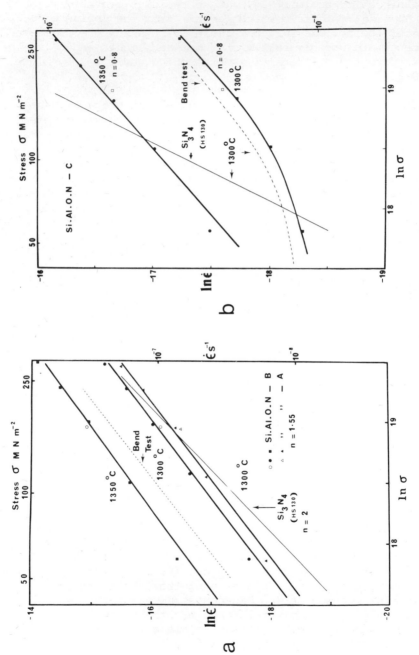

Fig. 2. Stress–exponent data for (a) cavitating and (b) non–cavitating single phase ceramics determined from compressive creep tests.

by a factor of ~2 but reproduce the non-integral exponents for ceramics A and B. The letters denote two different hot-pressings of a ceramic with a mixed (1 wt.% Mn_3O_4/MgO) sintering aid with microstructures characterised by minor triple-junction residues (Fig. 1). The creep rate and n>1 exponent is similar to commercial Si_3N_4 ceramics and has previously been shown to be characteristic of the onset of grain-boundary cavitation [21] which produces a variable value for n and a lack of meaningful comparison with creep models.

Fig. 2b, for a ceramic (denoted C as in previous publications [21, 22]) without observable grain junction residues, exhibits a much lower creep rate and a good correlation between bend and compressive data. The anomalous behaviour in incremental stress tests, during the first ~200 hours of bend-creep testing, is replicated for the compressive tests hence confirming its origin as a non-steady-state creep resulting from a change in grain-boundary structure [21]. The 1300°C data in Fig. 2b has been obtained with increasing stress increments; the reverse trend (n>1 tending to n~1 with increasing time) occurs with decreasing increments. The 1350°C data has been obtained with increasing increments following 72 hours of the transient creep effect and hence is nearer to the steady-state behaviour. For longer times of test or if specimens are pre-annealed in an oxidising environment for ~1000 hours the exponents for bend and compressive tests approach the n=1 value characteristic of diffusional creep. During this time there is a progressive reduction in creep rate.

The Arrhenius plots used to estimate the activation energies for creep show a consistent value for ceramics A and B (Fig. 3a) in agreement with the bend data [21]. The value shown for ceramic C (an average for the plotted 'curve') is not meaningful in view of the significant change in creep rate for a given temperature which occurs within the time scale of the temperature changes. However, a meaningful linear plot is obtained from the heat-treated ceramic of 818 ± 22 kJ mol^{-1}. Fig. 3b is an example of the change in activation energy induced by oxidising heat-treatment. The influence of surface oxidation on this change is evident from the intermediate behaviour of specimens cut from a position further removed from this surface (labelled 'bulk – heat treated' in Fig. 3b).

A similar long-term oxidising heat-treatment of the 'cavitating' ceramics A and B produces a dramatic reduction in creep rate (Fig. 5), raising the activation energy to ~820 kJ mol^{-1}, with an n=1 stress-exponent accompanying the suppression of cavitation [22]. However, this change is comparatively slow such that it does not result in variations in n and Q within the time-scale of stress and temperature cycling experiments.

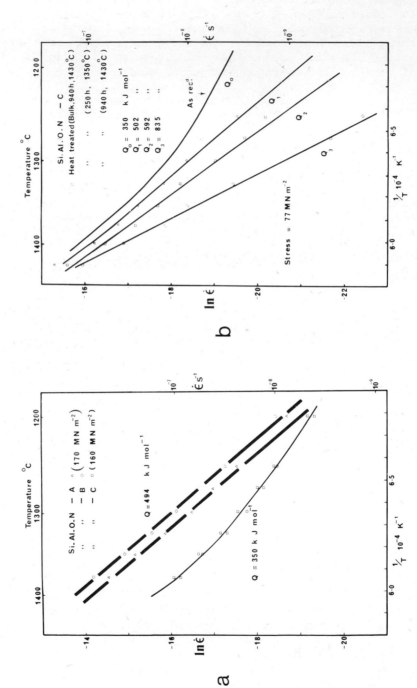

Fig. 3. (a) Activation energy analysis for ceramics A, B & C (compressive creep).
(b) Activation energy analysis for ceramic C (bend creep) with varying heat–treatment.

2.2 <u>Diffusional Creep Mechanisms</u>.   The origin of both
transient phenomena lies in the desegregation of metallic ion
grain-boundary residues via their extraction into a surface
'reservoir' of SiO_2.   This induces a crystallisation of the
residual, nitrogen-containing, silicate phase as β' hence
removing the easy cavity-nucleation mechanism within the
liquid-containing triple junctions.   Two-grain interfaces
increasingly approach the structural state of a 'pure' grain
boundary with increased activation-energy for diffusion and
reduction in creep-rate.   A comparison of the microstructure
of crept specimens before and after heat-treatment (Fig. 4)
confirms the susceptibility to cavitation of ceramics A and B
and its suppression with heat-treatment.

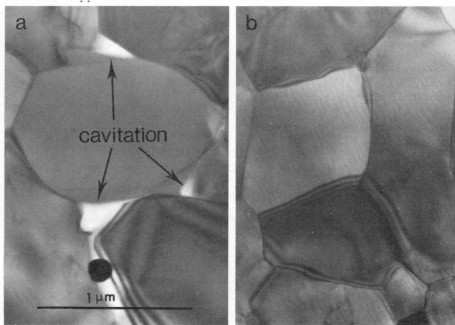

Fig. 4.   Electron micrographs typical of bend creep specimens
          of (a) ceramic B and (b) ceramic C or heat-treated
          ceramic B.

The stress-exponent and sensitivity of creep-rate to
grain-boundary composition provide sufficient evidence for a
'Coble' grain-boundary diffusion controlled creep mechanism.
The compressive creep data and good correlation with bend
data enable quantitative determination of grain-boundary
diffusion coefficients.   The steady-state creep rate $\dot{\epsilon}$ is
given by [24],

$$\dot{\epsilon} = \frac{14\pi \; \sigma \; \Omega \; \delta D_b}{kT \; d^3}$$

where Ω, the atomic volume, $\sim 1.06 \times 10^{29}$ m^3.

Grain sizes for the heat-treated ceramics B and C were
measured from TEM sections to be 1.38 and 1.44 μm, respectively, compared to 1.18 and 1.25 μm for the unheat-treated
ceramics. This limited grain growth (in 1000 hours at 1400°C)
is responsible for a small part (~ 15%) of the reduction in
creep rate. Hence, the grain boundary diffusion parameters
δD_b are:

$$\text{Ceramic B, } \sim 6 \times 10^{-26} \text{ m}^3 \text{ s}^{-1})$$
$$\text{Ceramic C, } \sim 3 \times 10^{-26} \text{ m}^3 \text{ s}^{-1}) \text{ at } 1375°C$$

For the unheat-treated ceramic C(δD_b) is more than one order of
magnitude higher and, in view of the negligible change in a
δ (the boundary width which ~ 7Å), must result from the
sensitivity of diffusion coefficient to impurity 'segregation'.
This trend is the reverse of that observed in metallic
systems. Oxide ceramics have comparatively high boundary
diffusivities; e.g. $\delta D_b \sim 4 \times 10^{23}$ m^3 S^{-1} for Al$_2$O$_3$ at 1375°C
[25], a difference which must result from the high covalency
of Si-N bonds adjacent to the grain boundary. Hence for
Al$_2$O$_3$ activation energies are ~ 420 kJ mol^{-1} [25] compared to
~ 820 kJ mol^{-1} for the desegregated Si-Al-O-N grain boundaries.

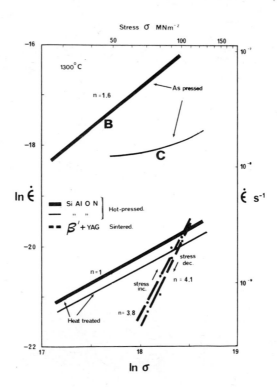

Fig. 5. Stress exponent analysis for the β' + YAG ceramic
with a comparison of single phase ceramics B and C
before and after heat-treatment.

3.   TWO–PHASE CERAMIC ALLOYS

The pressureless sintering of ceramic alloys to near-theoretical density is dependent on the presence of a relatively large volume fraction of liquid-sintering medium. For β' – YAG ceramics this is an yttrium alumino-silicate (with dissolved nitrogen) which, on rapid cooling, solidifies as a glassy matrix surrounding β' crystals. The creep test specimens were either subjected to slow cooling from the sintering temperature (~1800°C) or subsequently heat-treated at 1400°C to crystallise the matrix phase as a Si and N-substituted YAG phase [23].

Incremental stress tests exhibit a variation in creep stress-exponent (n) with time; the result of a long-term reduction in creep rate similar to that observed in single-phase ceramics during grain-boundary desegregation. In this case, however, values of n are $\gg 1$, independent of increasing or decreasing stress increments, and for very long times approach a value ~4. The stress-exponent plot (Fig. 5) shows that the final creep rates at low stresses (bend creep) are similar to the heat-treated single-phase ceramics at 1300°C. The activation-energy for creep is ~900 kJ mol^{-1}.

A tentative explanation of this 'power-law' creep behaviour is that of matrix plasticity involving a glide or climb of YAG crystal dislocations. The specific rate-controlling mechanism is not clear in view of the variability in stress-exponent not only with time but method of load-application; recent compressive tests covering a larger stress range exhibit values of $n \sim 2$. There is no direct evidence, from electron-microscopy of deformed specimens, for dislocation activity in the YAG phase but this presents the difficult problem of being in low volume fraction and highly electron absorbing. Also, by comparison with superplastic metallic alloys which have similar stress-exponents and mechanisms involving accommodation of boundary-sliding by an (infrequently observed) crystal plasticity, the strain levels in ceramic bend specimens are limited to $\sim 4\%$. The progressive reduction in creep rate may result from an oxidation-induced change in matrix composition which increases the activation energy for dislocation climb. This composition-change, which may involve an increase in O/N ratio as well as a reduction in Y and Al content, involves long-range diffusion within a much larger volume than for grain-boundary desegregation in single phase ceramics. Hence the change in creep rate occurs over a longer period of time in specimens of similar dimensions.

The experimental evidence (n$\gg$1 and unusually large Q) does not favour alternative mechanisms such as viscous flow of incompletely crystallised matrix/β' interface layers or of diffusional creep involving a progressive change in diffusion path length via impingement of β' crystals.

From an engineering viewpoint the pressureless-sintered ceramics exhibit surprisingly low creep-rates and an absence of creep-cavitation or failure of bend specimens within the geometrical limits of strain for bend tests up to ~300 hours at 1300°C.

4. CREEP FRACTURE

A qualitative comparison of the susceptibility to creep fracture of the range of ceramic alloys is illustrated in Fig. 6a-e. Fig. 6a is typical of the nominally single-phase ceramics A and B in which cavity interlinkage produces multiple surface cracking within 1% strain. Failure occurs via the dominant surface crack and the regions of sub-critical growth (cavity interlinkage) and fast fracture are visible on the fracture surface (Fig. 6). Intergranular fracture dominates in both regions but the rough surface topography of the sub-critical region (Fig. 6f, g) results from a large 'Process-zone' of cavitation at the primary crack tip. A reduced susceptibility to cavitation in heat-treated specimens results from the lower density and smaller size of triple-junction residual phase within which cavities nucleate during grain-boundary sliding. Fig. 6b is a heat-treated ceramic B in which the incidence of surface sub-critical crack growth is reduced. Fig. 6c is typical of either ceramic C or a long-heat-treated ceramic B in which cavitation is absent for the maximum measured strain ($\sim$ 4%) to a maximum test temperature of 1500°C.

The two-phase ceramic is also resistant to cavitation for similar levels of creep stress and strain. This supports the microscopic evidence for the absence of a residual interfacial glassy phase following YAG crystallisation. However, there is a critical temperature for tests in an oxidising environment above which rapid creep failure occurs. Figs. 6d and e represent bend specimens creep tested in air at 1300°C and 1330°C respectively. A marked increase in oxidation rate at $\sim$ 1330°C is accompanied by complete crystallisation of the, previously liquid, silicate oxidation layer and catastrophic failure. The origin of this behaviour is believed to be a reversion of 'near-surface' matrix (YAG) phase to the liquid state following SiO_2 enrichment. These liquid 'intrusions' catalyse the oxidation rate due to enhanced outward diffusion of metallic ions, thus reducing the SiO_2 oxidation-layer viscosity and promoting rapid crystallisation. The intrusions represent instabilities on the oxidation front which are fracture nuclei.

A quantitative study of sub-critical crack growth has previously been made [21,22,25] for the single-phase ceramics from measurements of crack velocity (V) as a function of

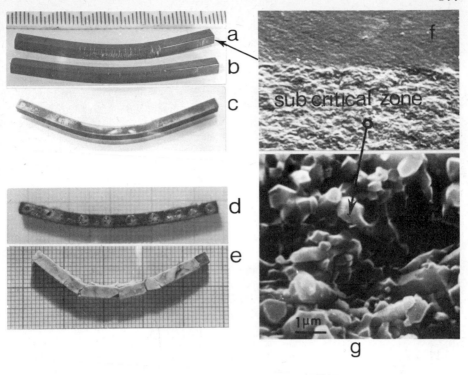

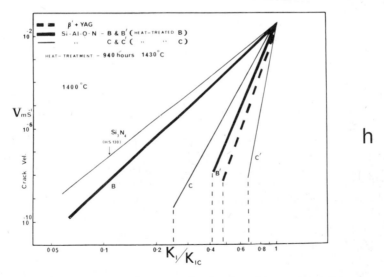

Fig. 6. Bend-creep specimens of single-phase ceramics in different stages of heat-treatment (a, b and c) and of the β' + YAG ceramic above and below 1320°C (e and d respectively).
(f and g) Fracture surfaces of sub-critical and fast fracture zones typical of specimen 6a.
(h) Comparison of crack velocity (v) – normalised stress-intensity (K₁) for various ceramics.

normalised stress intensity factor (K_1/K_{1c}). The data gives
an excellent fit to a 'power-law' relation of the form $V =$
constant K_1^m with the exponent m being a measure of
susceptibility to sub-critical crack growth. The plot is
reproduced in Fig. 6h, without data points, for a temperature
of $1400°C$. The plot for the two-phase (β' + YAG) ceramic is
also included and is shown to exhibit the high-exponent
behaviour typical of non-cavitation ceramics together with the
'threshold' K_1 (dotted lines) below which cracks do not
advance. This data was obtained on double-torsion specimens
in vacuum, which supports the mechanism for oxidation-induced
degradation at a lower temperature in the two-phase ceramic.
The high-exponent ($m \sim 10$-40) behaviour of growth from a
single crack front may be fitted either to a diffusive growth
model [25] or to a thermally-activated interfacial 'bond-
breaking' model [26].

The greatest significance of the fracture data, from the
engineering viewpoint, is the demonstration that both in
single phase and in completely-crystallised two-phase
Si-Al-O-N ceramic alloys it is possible to suppress creep
cavitation and hence the incidence of sub-critical crack
growth at low stresses. Hence apart from impressive creep-
failure data there is the possibility of an approach to
'superplastic' deformation. However it is unlikely that the
achievable strain rates (dictated, for example, by low
diffusion coefficients) will enable simple 'post-sintering'
shaping operations to be economically feasible.

ACKNOWLEDGMENTS

The research forms part of a collaborative programme
with the Lucas laboratories. We wish to thank Dr. Roger Hunt,
Dr. John Lumby and the Directors of Lucas Industries for this
continued support.

REFERENCES

[1]   Vasilos, T., Mitchell, J.B. and Spriggs, R.M., J.Amer.
      Ceram. Soc., 1964, 47, 203.
[2]   Passmore, E.M., Duff, R.H. and Vasilos, T., ibid., 1966,
      49, 594.
[3]   Tagai, H. and Zisner, T., ibid., 1968, 51, 303.
[4]   Zisner, T. and Tagai, H., ibid., 1968, 51, 310.
[5]   Terwilliger, G.R., Bowen, H.K. and Gordon, R.S., ibid.,
      1970, 53, 241.
[6]   Folweiler, R.C., J.Appl. Phys., 1961, 32, 773.
[7]   Warshaw, S.I. and Norton, F.H., J.Amer. Ceram. Soc.,
      1962, 45, 479.

[8] Heuer, A.H., Cannon, R.M. and Tighe, N.J., 'Ultrafine Grain Ceramics', Ed. Burke, J.J. et al., Syracuse Univ. Press, 1970, 339.

[9] Mocellin, A. and Kingery, W.D., J.Amer. Ceram. Soc., 1971, 54, 339.

[10] Hollenberg, G.W. and Gordon, R.S., ibid, 1973, 56, 140.

[11] Cannon, W.R. and Sherby, O.D., ibid., 1973, 56, 157.

[12] Lessing, P.A. and Gordon, R.S., J. Mat. Sci., 1977, 12, 2291.

[13] Chung, T.E. and Davies, D.T.J., Acta. Met., 1979, 27, 627.

[14] Evans, A.G. and Rana, A., ibid., 1980, 28, 129.

[15] Kossowsky, R., Miller, D.G. and Diaz, E.S., J.Mat. Sci., 1975, 10, 983.

[16] Din, S.U. and Nicholson, P.S., ibid., 1975, 10, 1375.

[17] Birch, J.M. and Wilshire, B., ibid., 1978, 13, 2627.

[18] Seltzer, M.S., Amer. Ceram. Soc. Bulletin, 1977, 56, 418.

[19] Lewis, M.H., Powell, B.D., Drew, P., Lumby, R.J., North, B. and Tayor, A.J., J.Mat. Sci., 1977, 12, 61.

[20] Clarke, D.R. and Thomas, G., J.Amer. Ceram. Soc., 1977, 60, 491.

[21] Karunaratne, B.S.B. and Lewis, M.H., J.Mat. Sci., 1980, 15, 449.

[22] Karunaratne, B.S.B. and Lewis, M.H., ibid., 1980, 15, 1781.

[23] Lewis, M.H., Bhatti, A.R., Lumby, R.J. and North, B., ibid., 1980, 15, 103.

[24] Coble, R.L., J. Appl. Phys., 1963, 34, 1679.

[25] Cannon, R.M. and Coble, R.L., 'Deformation of Ceramic Materials', Eds. Bradt, R.C. and Tressler, R.E., Pergamon 1974, 61.

[26] Lewis, M.H. and Karunaratne, B.S.B., ASM Symposium on Fracture Mechanics Techniques in Ceramics, Chicago, 1980 (in press).

[27] Lewis, M.H. and Karunaratne, B.S.B., 'International Conf. on Fracture, 5', Cannes 1981 (in press).

DEFORMATION OF POLYCRISTALLINE α-SiC[x]

A. DJEMEL, J. CADOZ, J. PHILIBERT

Laboratoire de Physique des Matériaux, C.N.R.S.
1, Place Aristide Briand - 92190 Meudon Bellevue. FRANCE.

Abstract : Hot pressed and sintered α-Sic are tested by compressive creep in argon between 1300°C and 1500°C at stresses between 500 MPa and 1700 MPa.

Two stages are observed in the ln $\dot{\varepsilon}$-ln σ curves :

- Below a critical stress σ_o, the deformation is controlled by diffusional creep (certainly Coble creep)

- and above σ_o, diffusion still controls the deformation but through cavitation until the fracture. The cavities lie in grain boundaries oriented parallel to the stress.

1. INTRODUCTION

It has been generally accepted that creep in non-ductile ceramics like SiC, is a diffusion controlled phenomenon at low stress σ. In this kind of material, in which the number of glide systems is insufficient, the creep mechanism may be viewed as that of grain-boundary sliding accommodated by lattice or grain-boundary diffusion, identified by a stress exponent n = 1 in the generalized creep rate $\dot{\varepsilon}$

$$\dot{\varepsilon} = A \, \sigma^n \, \exp \, (- \Delta H/kT) \tag{1}$$

Where A is a constant, σ is the applied stress, ΔH is activation enthalpy for diffusion through the crystal or via grain boundaries, and kT has its usual meaning.

When σ is higher than a critical value σ_o , localized cracks or cavities may occur which provide an important stress relieving mechanism. This would allow increasing proportions of strain by crack propagation to occur with increasing stress.

---

[x] Based on part of a thesis submitted by A. DJEMEL for a "Thèse de 3è cycle", Université d'Orsay - FRANCE.

This last mechanism is considered as a transition between diffusional creep and instantaneous fracture.

The first studies of polycristalline SiC creep have been done by Glenny et al. [1] but no mechanism were proposed Farnsworth and Coble [2] determined n $\sim$ 1 and $\Delta H \sim$ 300 kJ/mole, using bending tests with $\sigma \sim$ 200 MPa between 1900°C and 2200°C in argon. Later Francis and Coble [3] studied the influence of the porosity and grain size under the same conditions as [2]. They suggested that creep could be controlled by the diffusion of carbon along grain-boundaries.

No creep studies on SiC have been done at stress above the critical value σ_o for the appearance of cavities and delayed fracture. The purpose of this paper is to present our results obtained by creep below and above σ_o at temperatures between 1300°C and 1500°C in argon, a temperature range important for applications and never studied using compressive tests.

2. EXPERIMENTAL PROCEDURE

2.1 Materials characterization

Three kinds of material have been used : Table 1.

	CGE hp (hot pressed)	CGE s (sintered)	GE hp
Densification aids	1 % B_4C	1 % B + 1 % C	B and C
Porosity	4 %	6 %	4 %
Microstructure	equiaxed grains	large elongated grains	Equiaxed grains
Size (mean intercept)	3.5 µm		5 µm
Observations	pores at the triple points	Pore at the triple points and inside grains	pores at the triple points

CGE : Compagnie Generale d'Electricité - 91460 Marcoussis - FRANCE
GE : General Electric Company - Schenectady N.Y. - U.S.A.

Table 1 : Microstructure and porosity for the materials studied.

The densities of the samples were calculated from the sample weights and dimensions. The polytypes of SiC were identified by Xray diffraction. They show that the three materials are α type. A scanning electron microscope was used to study grain size, porosity, cavities and the type of fracture.

2.2 Sample preparation

The samples tested have been diamond cut to $1.5 \times 1.5 \times 5 \text{ mm}^3$ Samples study for metallography were polished and grain boundarires were revealed by etching for 30 mn with boiling NaOH.

2.3 Experimental apparatus

Figure 1 is a schematic of the loading system used in the creep machine described elsewhere [4].

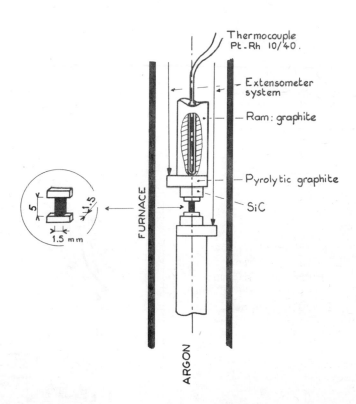

Figure 1 : Schematic of the loading system

Compressive tests were performed between 1300°C and 1500°C in argon, at stresses between 500 MPa and 1700 MPa, with a P_{O_2} between 5 to 10 ppm which correspond to the active SiC oxydation field [5]. The rams were made of graphite with pyrolytic graphite ends. To minimize indentation of the pyrolytic graphite plates, two additional SiC samples, identical to the specimen, were placed at the ends of the specimen as whown in the detail of Figure 1. These were replaced for each experiment. An extensometer system allowed the deformation of the sample to be measured continuously with a resolution less than 0.1 μm .

2.4 Mechanical tests

Typical constant load creep curves ln $\dot{\varepsilon}.\varepsilon$ are shown Figure 2 for a sample deformed sequentially at different stresses. A steady-state flow stress is apparent. The corresponding strain rate has been plotted as ln $\dot{\varepsilon}$ versus ln σ for different temperature. Figure 3. These curves present two stages :

Figure 2 : Typical creep curves ln $\dot{\varepsilon}$-ε for different loads - at 1490°C, in argon.

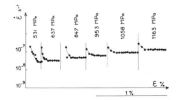

Figure 3 : Strain-rate/stress relation ships for different SiC materials at 1490°C and 1330°C expected two curves CGE hp at 1430°C and 1385°C.

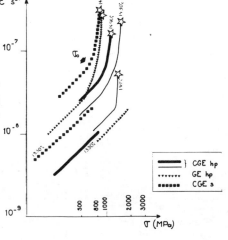

i) The first has a slope equal to 1 at the lowest stresses.

ii) The second stage has a much higher slope and leads to fracture of the sample.

Temperature changes were also performed during creep at constant stress. The curves showing logarith of the steady-state strain rate versus 1/T for different stresses, also present two regions. Figure 4.

Figure 4 : Temperature dependance of strain-rate for different loads on SiC hp material.

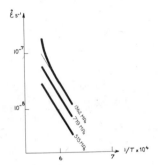

The activation energy for the three kinds of material has been determined for the low stresses-low temperature region and are given in the Table 2.

Temperature : °C	σ_o MPa		
	CGE hp	CGE s	Ge hp
1490	740	500	475
1430	940		
1385	1200		
1330	> 1650	> 620	> 740
$\Delta H \quad \sigma < \sigma_o$	300 kJ/mole	170 kJ/mole	210 kJ/mole

Table 2 : Different values for ΔH, and evolution of σ_o with the temperature.

386

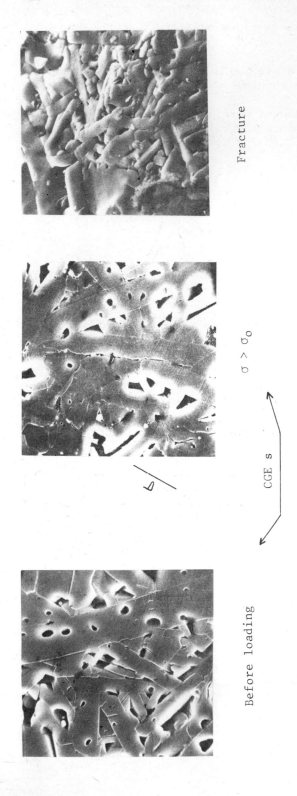

Before loading

CGE s

$\sigma > \sigma_o$

Fracture

Figure 5 : Scanning electron micrographs of the microstructure of SiC for different materials and at different levels of deformation.

Scale : 3μm   except when notified

σ is the direction of the applied stress.

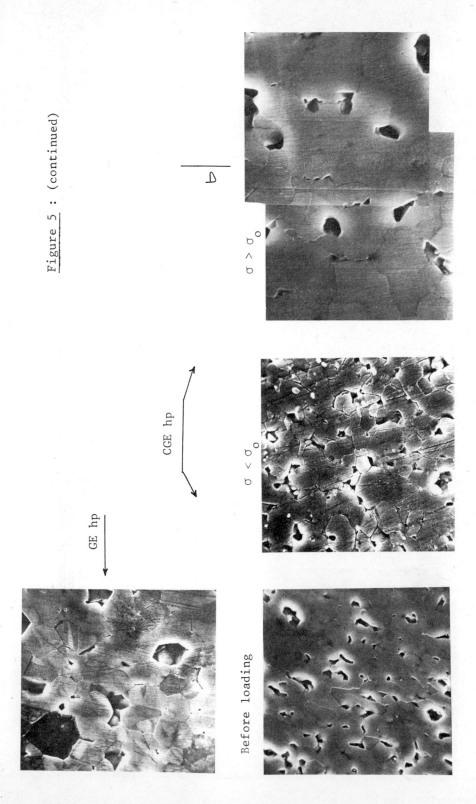

Figure 5 : (continued)

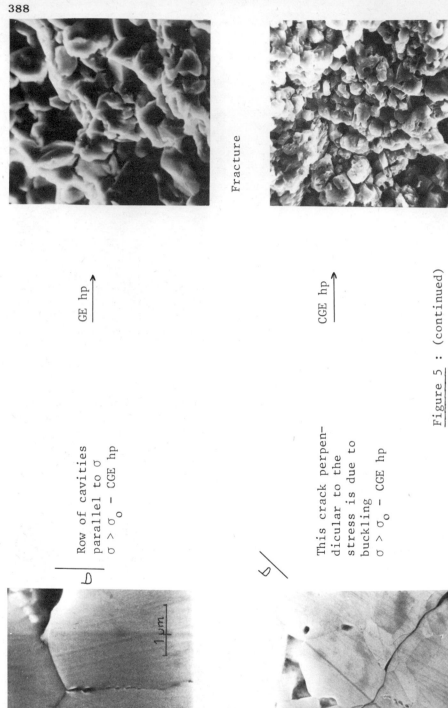

Fracture

GE hp

CGE hp

a) Row of cavities parallel to σ
$\sigma > \sigma_o$ – CGE hp

b) This crack perpendicular to the stress is due to buckling
$\sigma > \sigma_o$ – CGE hp

Figure 5 : (continued)

The results obtained from ln $\overset{\cdot}{\varepsilon}$ - ln σ curves or ln $\overset{\cdot}{\varepsilon}$ - ε curves are quite comparable.

- In the first stage where n = 1 for the three materials, sintered SiC creeps at higher rate than both hot pressed materials. Figure 3. The ΔH values are also not the same for the different materials.

- In the high stress region, ln $\overset{\cdot}{\varepsilon}$ - σ curves of the sintered SiC and G.E. material go to a nearly identical value for the fracture stress, where as fracture occurs at a higher stress for the CGE hp at 1490°C. The evolution of the fracture stress σ_R is identical to σ_o : σ_R increases as T decrease.

The fracture surfaces examined by S.E.M., Figure 5 show no apparent second phases. The only pores observed were located at triple grain junctions exept in the sintered material where some pores were detected within the grains. No apparent difference was observed between the microstructure before and after deformation in the first regime of the ln ε - ln σ curve. But micrographes obtained from the second regime showed rows of cavities along the boundaries approximatively parallel to the direction of the applied stress. The fracture surfaces obtained at the end of the ln $\overset{\cdot}{\varepsilon}$ - ln σ of the three materials indicated that failure is intergranular. The fracture plane was within 30° parallel to the stress. Some cracks have also been observed perpendicular to the applied stress. In this case, the failure is inter and transgranular.

3. DISCUSSION

3.1 Stress below σ_o

Below the critical value σ_o, n = 1. No evidence for differentiating Nabarro-Herring or Coble mechanism can be shown from these results. But bending tests made on α-SiC at about 2000°C with different porosities [3] showed that diffusion along grain boundaries is more effective than bulk diffusion.

In the other hand, activation energy values given by Farnsworth [2] and Francis [3] are of the same order as our values i.e. : $\sim$ 300 kJ/mole, but smaller than bulk self diffusion values of either Si or C : the lowest are more than 600 kJ/mole. [6], [7], [8]. These low values : $\Delta H \sim$ 300 kJ/mole for CGE hp, 170 kJ/mole for CGE s and 210 kJ/mole for GE hp might be explained by diffusion along grain boundaries containing different impurities. Studies of diffusion of Si and C along SiC grain-boundaries would be very useful to understand this behaviour.

In this stress range, porosity plays the major role, as compared to grain size. It is well known that steady state creep rate of polycristalline ceramics increases with increasing porosity [9]. It is easy to understand that measured creep rate differs from a theoritical dense material because :

  - The effective stress is larger than the applied stress.

  - Intergranular porosity affects the grain boundary sliding processes.

  -Intergranular pores may act as sources or sinks for dislocations-mechanism possible for some orientations of grains of α-SiC [10] [11]. Effectively the CGE s material has the highest porosity : 6 % (instead of 4 % for the others) and intragranular pores are observed. In this material, which consistantly deforms at a higher rate, all these mechanisms could be enhanced by the elongated grains and the higher specific boundary area of CGE s compared to equiaxed grains in CGE hp and GE hp materials. Otherwise, grain size difference for CGE hp and GE hp seems not to be important for the level of $\ln \dot{\varepsilon}$-$\ln \sigma$ curves.

3.2 Stress above σ_o

  Under σ_o, stress concentrations will not be sufficiently high to promote crack initiation and stress relief occurs by diffusional mechanisms. When the stress reaches a critical value σ_o, crack initiation and growth will occur and become an important deformation mechanism. This deformation process is more and more important as far as stress increases until to the fracture stress of the material [12]. Figure 3.

  Our observations show that cavities Figure 5 are observed at the beginning of the second stage. They are responsable for the weakening of the boundaries between grains along the compression axis. They are the stress relieving process. Using different results : [15] [16] [17], we propose the following mechanism:

  The tests are performed in compression. Below σ_o, grains flatten and broaden by Coble creep. Figure 6a). At the same time, the forces acting on the grain boundaries, tend to separate grains along facets parallel to the applied stress. Figure 6b). The deformation is accomodated by glide along grain boundary but the latter is not the controlling mechanism : it is easier to glide than to separate two grains. The deformation is due to a) + b). When the applied stress exceeds σ_o , mechanism a) is no longer competitive, i.e. Coble creep is not sufficient to maintain the cohesion of the grains and the boundaries tend to separate. But as it is difficult to shear a single crystal with out using local defects such as dislocations, the separation : mechanism b), is produced by cavities. Figure 6c). Balluffi and Seigle [18] have shown

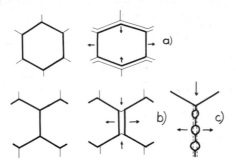

Figure 6 : A schematic indicating the sequence of cavity
formation :
a) by Coble creep
b) mechanical separation of the grains, due to
the applied stress
c) formation of cavities

⟶   Shows the principal forces acting.

that sufficient vacancies can be produced at transverse grain
boundaries because of the tension acting across the boundary
to produce cavities. These vacancies collapse along the grain
boundaries in local zones by nucleation on impurities or pre-
existing voids where they form cavities. The short-range diffusion of vacancies along grain boundaries may provide an important mechanism for the short range diffusion needed for
cavity growth. The cavities may then grow along the grain
boundaries until unstable fracture occurs. As observed in our
experiments, the decrease of fracture stress with grain size
would be expected if the flaws grow rapidly to facet size
which allow instability to occur earlier. In the $\ln \dot{\varepsilon} - \ln \sigma$
curves, one can see that GE hp and CGE hp exhibit similar be-
haviour below σ_o, but different behaviour above σ_o : σ_o and
σ_F are lower for the larger grain size material (GE hp versus
CGE hp whose porosities are comparable). In our experiments,
this sub-critical crack growth via interlinkage of intergranular cavities is sufficiently slow to induce entirely intergranular fracture and tertiery creep. Figure 5. On the other
hand, buckling of the sample can generate high tensile stresses and cracks moving at high velocities -Figure 5 - giving
rise to intra and inter granular cracks. These observations
suggest a similarity between the relation (1) and the relation
$v \propto K_I^m$ where v is the velocity of the crack growth and K_I is
the stress intensity factor, assuming $K_I \propto \sigma$ the applied
stress. Plotting $\ln \dot{\varepsilon} - \ln \sigma$, we see Figure 7 that the slope
n = m tends to values between 10 and 20. These values of the

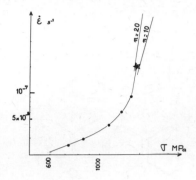

Figure 7 : Strain-rate/stress relation ship for SiC hp at 1430°C.

exponent in a "K_I-v" curve correspond to the stage I or II where the delayed fracture is generally thought to be controlled by a corrosive phase in the flaws – for SiC [20]. But in our case, it is certainly controlled by diffusion of C or Si both above as well as below σ_o. At a higher crack velocity resulting from buckling on some samples, the crack fractures the grain and has an elastic behavior corresponding in our case to the beginning of the 3rd stage [21].

Saisse et al. [22] have tested samples of CGE hp in three point bending. The strain rate before the fracture was about $\sim 10^{-4}s^{-1}$ and their σ_F-T curve increased up to 1500°C then decreased quickly. In our experiments, σ_F decreased in all experiments from 1300°C to 1500°C. This difference can be related to the difference strain rates, as we were using $\dot{\varepsilon} \sim 10^{-7}s^{-1}$ instead of $10^{-4}s^{-1}$, from which a smaller K_{IC} is expected in our case. A similar behaviour has been observed in an other SiC material deformed in air [23] where K_{IC} decreases with decreasing the deformation speed for a given temperature. They interprete it by a moisture effect, but in such a mechanism is unlikely in our experiment where the tests were carried out in argon with less than 10 ppm of oxygen. This thermal behaviour we observed – the decreasing of σ_F with increasing T – is consistant with a diffusion mechanism so that we were expecting a decrease of K_{IC} where temperature increases.

Acknowledgements

The authors are indebted to B. Pellissier for the technical part of this work, to Professor A. Solomon for reading the manuscript and to Mme M. Miloche for the SEM Studies. The Materials were provided by Dr. Broussaud (CGE) and Dr. Prochazka (GE).

This work was supported by a DRET contract n° 79/640.

REFERENCES :

1. GLENNY E. and TAYLOR T.A. - 'High-temperature properties
    of ceramics and cermets'.
    Powder Met, 1958, 1-2, 189-226

2. FARNSWORTH P. and COBLE R.L. - 'Deformation behaviour of
    dense polycristalline SiC'.
    J. Amer. Ceram. Soc. 1966, 49, 5, 264-268.

3. FRANCIS T.L. and COBLE R.L. - 'Creep of polycristalline
    silicon carbide'
    J. Amer. Ceram. Soc. 1968, 51, 2, 115-116.

4. GERVAIS H., PELLISSIER B., CASTAING J. - 'Fluage pour es-
    sais en compression à hautes temperetaures de matériaux
    ceramiques'
    Rev. Int. Htes Temp. et Refract., 1978, 15, 43-47

5. SINGHAL S.C. - 'Thermodynamic analysis of the high-tempe-
    rature stability of silicon nitride and silicon carbide'
    Ceram. Intern. 1976, 2, 3, 123-129

6. GHOSHTAGORE R.N. and COBLE R.L. - 'Self diffusion de sili-
    con carbide'
    Phys. Review, 1966, 143, 2, 623-626

7. HONG J.D., Ph. D. Thesis, 1978, North Carolina State
    University

8. HON M.H. and DAVIES R.F. - 'Self diffusion of 14C in poly-
    cristalline β-SiC'
    J. Mater. Sci., 1979, 14, 2411-2421
    and - 'Self diffusion of 30Si in polycristalline β'SiC'
    ibid, 1980, 15, 2073-2080

9. LANGDON T.G. - 'Dependence of creep rate on porosity'
    J. Amer. Ceram. Soc., 1972, 55, 12, 630-631

10. EVANS A.G. and LANGDON T.G. - 'Structural ceramics'
    Prog. Mater. Sci., 1976, 21, 171

11. KOICHI NIIHARA - 'Slip systems and Plastic deformation of
    silicon carbide single crystals at high temperatures'.
    J. Less. Common. Metals, 1979, 65, 155-166

12. CROSBY A. and EVANS P.E. - 'Creep in non-ductile ceramics'
    J. Mater. Sci., 1973, 8, 1759-1764

13. EVANS A.G. and RANA A. - 'High temperature failure mecha-
    nisms in Ceramics'
    Acta Met., 1980, 28, 129-141

14. EVANS A.G. - 'Deformation and failure caused by grain boundary sliding and brittle cracking'
    Acta Met. 1980, 28, 1155-1163

15. HULL D. and RIMMER D.E. - 'The growth of grain boundary voids under stress'
    Phil. Mag., 1959, 4, 673

16. NEEDLEMAN A. and RICE J.R. - Plastic creep flow effects in the diffusive cavitation of grain boundaries'.
    Acta Met., 1980, 28, 1315-1332

17. CHUANG T.J., KAGAWA K.J., RICE J.R. and SILLS L.B. - 'Non-equilibrium models for diffusive cavitation of grain interfaces'
    Acta Met., 1979, 27, 265-284

18. BALLUFFI R.W. and SEIGLE L.L. - 'Growth of voids in metals during diffusion and creep'
    Acta Met., 1957, 5, 449

19. BIRCH J.M., WILSHIRE B., - 'The compression creep behaviour of silicon nitride ceramics'
    J. Mater. Sci., 1978, 13, 2627-2636

20. HENSHALL J.L., ROWCLIFFE D.J., EDINGTON J.W. - 'K_{IC} and delayed fracture measurements on hot-pressed SiC'
    J. Amer. Ceram. Soc., 1979, 62, 1-2- 36

21. McHENRY K.D. and TRESSLER R.E. - 'Fracture toughness and high-temperature slow crack growth in SiC'
    J. Amer. Ceram. Soc., 1980, 63, 3-4- 152

22. SAISSE H., BROUSSEAU D., DUMAS J.P., CHERMANT J.L., MOUSSA R., OSTERSTOCK F. - 'Fracture behaviour of silicon carbide : influence of microstructure temperature and environment'
    To be presented at :
    "ASTM Symposium on Fracture Mechanics Methods For Ceramics Rocks, and Concrete", June 23-24 1980 Chicago.

23. SRINIVASAN M. and SRINIVASAGOPALAN S. - 'The application of single edge notched beam and indentation techniques to determine fracture toughness of alpha silicon carbide'
    Ibid.

PORE BEHAVIOUR IN FINE GRAINED UO$_2$ DURING SUPERPLASTIC CREEP

T.E. Chung* and T.J. Davies[†]

* Department of Materials Engineering and Design, University of Technology, Loughborough, U.K.
† Department of Metallurgy, UMIST/University of Manchester, Manchester, U.K.

SUMMARY

Investigations of the compression creep behaviour of uranium dioxide have shown that fine grained UO$_2$ can deform superplastically. Whereas the Ashby-Verral creep model can satisfactorily describe the grain translations, pore shape and creep rates observed experimentally, it has been shown that it is necessary to produce a microstructure consisting of fine grains with small pores at triple-point grain boundary junctions in order to obtain extensive ductility. The relation ship between these pores and the overall deformation process is discussed in this paper.

1. INTRODUCTION

When uranium dioxide is used in a nuclear reactor fuel pile the material, in the form of pins, is subjected to high stresses as a result of the high thermal gradients and the build-up of fission products within the pins. These stresses can cause the brittle material to crack extensively unless they could be relaxed rapidly by some plastic deformation process, such as high rate creep.

Uranium dioxide possesses many of the creep character-istics of ceramic materials as a class. Creep strain rates, for example, are low (typically in the region of $10^{-4}hr^{-1}$ for an applied stress of 20 MN m^{-2} at a temperature of $\sim$1700K), and the maximum creep strains achieved before fracture at all stress levels (whether tested in compression or in bending) are below those achieved in metals (<5% as compared to $\sim$10% for most metals). It is unlikely that these low creep rates will be sufficient to affect the degree of stress relaxation necessary to prevent brittle fracture. It is equally unlikely that the low creep ductility would prevent cracking eventually even if the creep rates were sufficient.

An attempt was made to produce high-ductility, high-creep rate uranium dioxide by getting it to exhibit superplastic properties. It was hoped that this could be achieved by reducing the grain size of the polycrystalline material to a mean value of about 1 μm.

2.   EXPERIMENTAL

Compression creep tests were carried out on upright cylindrical test specimens prepared from a commercial grade of uranium dioxide supplied by the UKAEA (Harwell). By varying the powder milling treatment and the sintering conditions a range of specimens having mean grain sizes of between 2 μm and 10 μm was obtained. Full details of the test rig are given in ref. 1.

On creep testing all of the specimens which had grain sizes between 3 and 10 μm exhibited conventional ductilities, i.e. of <5% strain. Unusually high deformation strains were, however, obtained in the case of some 2 μm specimens.

3.   RESULTS

Parts of the results of this work have been published previously [1,2]. They show that whilst cylindrical specimens of the coarser-grained materials (such as those with 10 μm and 6.3 μm grain size) exhibited low compressive creep ductilities (<5.5% strain to failure), the 2 μm grain size specimens exhibited unusually high ductilities under the right combinations of stress and temperature. Figure 1 shows two specimens which had been reduced in length by 41.7% and 56.9% respectively with no signs of the onset of catastrophic failure. A third specimen (not shown) achieved a compressive strain of 68.6%, again without catastrophic failure. Furthermore, these

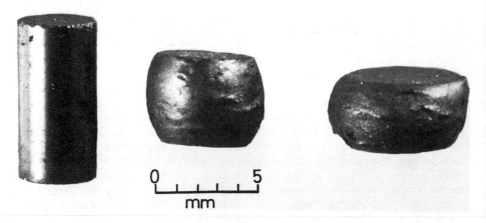

Fig. 1.   Two superplastically deformed UO_2 specimens shown alongside an as-sintered specimen.

high strains were achieved at strain rates which were approximately two to three orders of magnitude faster than those for
10 μm specimens subjected to the same compressive stresses and
temperatures (Figure 2).

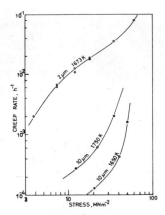

Fig. 2.   Creep rates for 2 μm and 10 μm grain size specimens
at ⌐1700K.

4.   METALLOGRAPHIC OBSERVATIONS

When the microstructures of these superplastically-
deformed specimens were studied it was very apparent that
extensive cavitation and cavity growth had occurred during the
creep deformation process. In view of the recent interest on
the subject of cavitation during superplastic deformation [3-7],
it is proposed in this paper to concentrate on the behaviour
of these cavities and to propose explanations for the resistance of an apparently brittle material to the propagation of
cracks from potential crack sources.

Figure 3 shows the microstructure of a 6.3 μm material of
⌐96% theoretical density. The grains are equiaxed and pores
are found in the interior of grains as well as at grain boundaries. This material is non-superplastic.

Figure 4 shows the pore distribution in a 3 μm specimen at
a point when it was about to fail (ε = 3.5%). It can be seen
that pores on grain boundaries parallel to the compressive
stress direction had linked to such an extent that a zone of
separated grains lying at an angle of ⌐30° to the stress direction is clearly visible. At that point in its creep life
failure by shear was imminent, and would have occurred in a
manner similar to that of a cylinder tested in plane compression mode (Figure 5).

A closer look (Figure 6) at the pores shows them to be
angular in shape and fairly sharp at their tips. It will be
shown later that the presence of pores alone does not prescribe

o

398

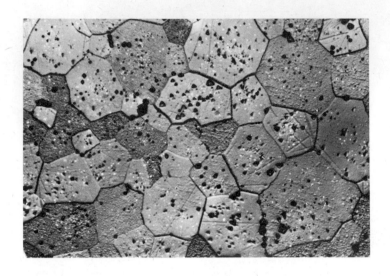

Fig. 3. Microstructure of 6.3 μm grain size material. x1500

early brittle fracture. Instead the moment of fracture is
determined by the stress fields at the extremities of the
pores or linked pores.

Figure 7 shows the structure of the as-sintered 2 μm
material. Note that in this material the majority of the pores
are located at triple point junctions of the fine grains. The
significance of these pores will be shown later.

Figure 8 shows the pore structure and distribution in a
specimen which had had its length reduced by 41.7%. In spite
of this large deformation the grains retained their original
equiaxed shape and the grain boundaries are not cracked in the
manner shown in Figures 4 and 6. Extensive cavitation has,
however, taken place at grain boundaries in the form of
stringers of relatively large, but well-rounded pores.

Figure 9 shows the microstructure of a specimen which had
its length reduced by 68.6%. Here the pore stringers are no
longer present. Fewer, but larger, pores are seen instead. A
feature of these enlarged pores is that they appear to lack
preferred orientation in the compressive stress direction.

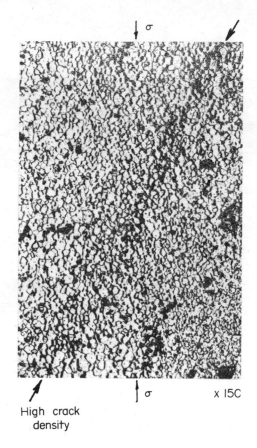

σ

x 15C

High crack
density

Fig. 4. Plane of high crack density seen in 3 μm specimen at
a strain of 3.5%.   x150

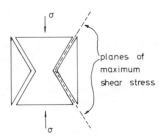

σ

planes of
maximum
shear stress

σ

Fig. 5. Failure mode of non-superplastic UO$_2$ specimens.

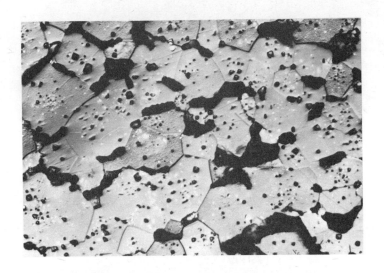

Fig. 6.   Angular cracks in non-superplastic specimen after 3%
          strain.

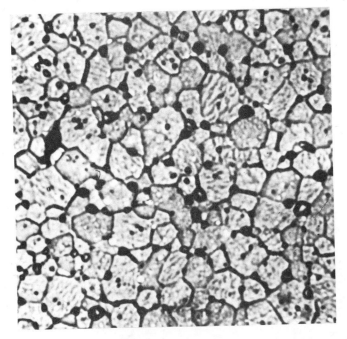

Fig. 7.   Microstructure of as-sintered 2 μm material showing
          pores at triple-point junctions.   x3000

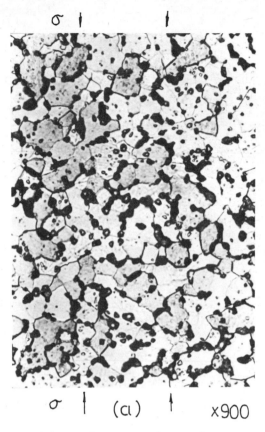

Fig. 8. Pore stringers in a specimen after a strain of 41.7%.
x900

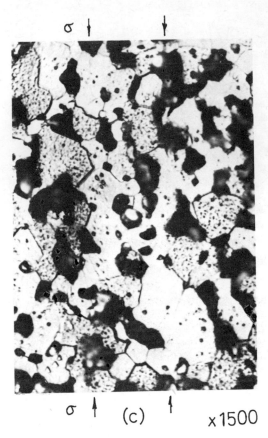

σ ↓ ↓

σ ↑ (c) ↑ x1500

Fig. 9. Large pores in specimen after strain of 68.6%. x1500

5. DISCUSSION

5.1. Mechanism of Deformation

A study of the grain size, stress and temperature dependence of the deformation rates [1] suggests that the deformation process involved at stresses in the range 10-30 MN m^{-2} is that of diffusion creep. The two well-known theories of diffusion creep (Nabarro-Herring [8,9] and Coble [10]), however, do not appear capable of explaining the observations described as they both predict grain elongation with increasing creep deformation.

One deformation mechanism which is capable of accounting for the lack of grain elongation is that proposed by Ashby and Verrall [11]. The deformation is envisaged to take place in units of four grains and the sequence showing the tensile deformation of such a unit is shown in Figure 10. During the deformation process a combination of lattice diffusion, grain boundary diffusion, and grain boundary sliding occurs simultaneously to produce deformation rates which are faster than

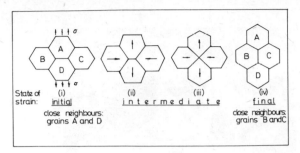

Fig. 10. Grain movements prescribed by Ashby-Verrall model
          for superplastic deformation.

those predicted by the Nabarro-Herring and Coble creep rates
by 1-3 orders of magnitude. Furthermore, this mechanism is
able to account for the retention of the original equiaxed
grain shape during deformation.

If the deformation is now performed in compression and
pores are introduced at the triple point junctions of grains
(as in the as-sintered 2 μm material), we would have the deformation sequence shown in Figure 11. Notice that the pores will
link, distort and reorientate during this deformation sequence.
Some evidence of such grain movements may be seen in Figure 12
(ε = 56.9%).

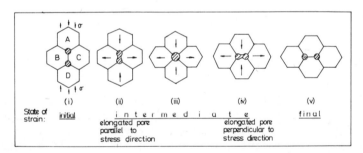

Fig. 11. Modified Ashby-Verrall model containing pores at
          triple-point junctions of grains.

5.2. Cavity growth and resistance to cracking

Whilst the Ashby-Verrall mechanism satisfactorily accounts
for the grain translations, pore shapes and creep rates, it
does not explain the observed cavity growth. Neither does it
explain the high resistance of an apparently brittle ceramic
material to cracking.

It is generally accepted [12] that pores and cavities
occur and grow on two possible locations within a polycrystalline material during creep. They can be located along grain
boundaries which are perpendicular to the stress direction (if

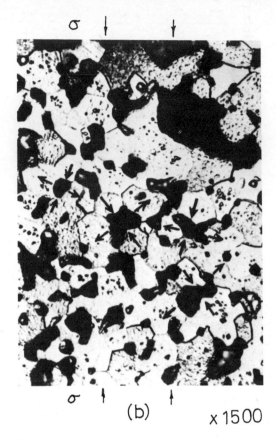

Fig. 12.   Likely grain movements (indicated by arrows) in
           superplastic specimen.   x1500

tensile) or parallel to the stress direction (if compressive).
These cavities tend to be spherical in shape and may be seen
as stringers of pores in Figure 8.   The other cavities are
located at triple-point junctions and are usually wedge-shaped.
Such cavities are due to the translation of the grains and the
inability of the accommodation process (whether diffusional or
dislocational) to keep pace with the translations.   Evidence
of this type of cavity growth is very difficult to obtain as
the wedge-shaped cracks propagate very easily in an unstable
manner in brittle materials.   It is therefore not likely to
account for any significant proportion of the cavity growth in
the coarse-grained uranium dioxide.   It can, however, account
for much of the cavity growth in the fine-grained (2 μm)
material, although the shape of the cavities will not be that
of a wedge.

Consider a spherical pore surrounded by three small
grains.   After a small increment of translation of the grains
according to the Ashby-Verrall model the spherical pore would
be slightly distorted, as shown in Figure 13.   When this dis-

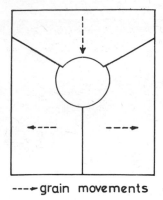

----▶grain movements

Fig. 13. Distortions in an ideal spherical pore as a result of
grain movements.

tortion occurs, surface self-diffusion of the uranium and
oxygen ions would immediately and rapidly cause the two
recesses to be filled in (in a similar way that cavitated creep
fracture surfaces in metals have been observed to be smoothed
out [13,14]) and the cavity rendered harmless as a crack-
initiating defect. As the creep deformation proceeds the
cavities are thus able to increase in size without acting as
stress-raisers, contrary to the way that wedge-shaped cracks do
in coarse-grained materials. Indeed, they could grow to form
large holes in the centre of a specimen (Figure 14) and then
shrink again as these are squeezed out by further deformation
(Figure 15).

Fig. 14. Large voids in the interior of specimen (ε = 41.7%).

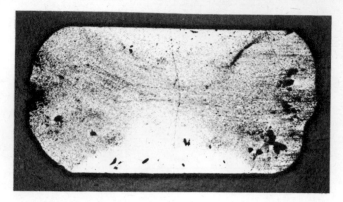

Fig. 15. Absence of large voids in the centre of the specimen after strain of 56.9%.

6.   CONCLUSIONS

It has been shown that,

i)   fine-grain uranium dioxide can deform superplastically when creep-tested in compression,

ii) the Ashby-Verrall model is capable of explaining the grain translations, pore shapes and creep rates,,

iii) the presence of tiny pores at the triple-point junctions of grains is necessary to prevent the formation of damaging wedge-shaped cavities at triple-point junctions of grains,

iv) cavity growth, reorientation and migration took place as the creep deformations proceed.

REFERENCES

1.   CHUNG, T.E. and DAVIES, T.J. – Acta Metall., 1979, 627.

2.   CHUNG, T.E. and DAVIES, T.J.– J.Nucl.Mat., 1979, 79, 143.

3.   SMITH, C.I., NORGATE, B. and RIDLEY, N.– Metal Science, May 1976, 182.

4.   AHMED, M.M.I., MOHAMMED, F.A. and LANGDON, T.G.– J.Mat. Sci., 1979, 14, 2913.

5.   MILLER, D.A. and LANGDON, T.G. – Trans.JIM, 1980, 21, 123.

6.   MILLER, D.A. and LANGDON, T.G. – Met.Trans.A, 1979, 10A, 1869.

7. LANGDON, T.G. - Advances in Materials Technology in the Americas - 1980, vol. 2, 1980, 55.

8. NABARRO, F.R.N.- Proc.Bristol Conf.of Strength of Solids (London Phys.Society, 1948).

9. HERRING, C. - J.Appl.Phys., 1950, $\underline{22}$, 437.

10. COBLE, R.L. - J.Appl.Phys., 1963,$\underline{34}$ (6), 1679.

11. ASHBY, M.F. and VERRALL, R.A. - Acta Metall., 1973, $\underline{21}$, 149.

12. ASHBY, M.F. - Fracture Mechanics: Current Status, Future Prospects (Conference), Cambridge University, 1979, 1.

13. BURNS, D., JAMES, D.W. and JONES, H. - Metal Sc.J., 1973, $\underline{7}$, 204.

14. BAGGERAD, A. and SHUTAS, J. - Scand.J.Met., 1973, $\underline{2}$, 149.

CREEP FRACTURE IN CERAMIC POLYCRYSTALS

A. G. Evans and C. H. Hsueh

Materials and Molecular Research Division, Lawrence Berkeley

Laboratory, and Department of Materials Science and Mineral

Engineering, University of California, Berkeley, California 94720

ABSTRACT

Creep rupture in ceramics occurs by the nucleation, growth and coalescence of cavities in localized, inhomogeneous arrays. The cavities grow by diffusive mechanisms; a process which has previously been analyzed for uniform cavity distributions. However, experimental results indicate that the inhomogeneity exerts substantial perturbations upon the failure sequence and hence, on the failure time. The principal inhomogeneity influence resides in the development of constraints that limit the cavity growth rate. The failure sequence in the presence of inhomogeneity effects is examined in this paper. A model is developed that accounts for the experimentally observed damage accumulation features and predicts the influence of inhomogeneity upon the failure time.

1. INTRODUCTION

High temperature failure in ceramics often occurs through the gradual growth and coalescence of cavities by diffusive processes, until a macrocrack is formed that subsequently propagates to failure [1,2,3]. The process is inhomogeneous, involving preferential cavitation in certain regions of the polycrystalline array [2,3] (Fig. 1). The intent of this paper is to examine the role of cavitation inhomogeneity upon the cavity evolution process and hence, upon the time to failure.

The principal sources of inhomogeneity are assumed to derive from local variations in the grain boundary or surface diffusivity (attributed to grain orientation effects, augmented by impurity distributions) [2], or in the dihedral angle (i.e.,

410

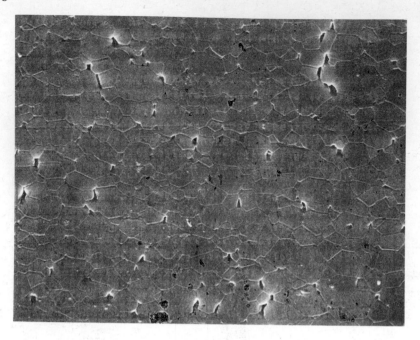

Fig. 1.   A scanning electron micrograph of cavity arrays in
          Al_2O_3, showing preferred regions of cavitation [2].

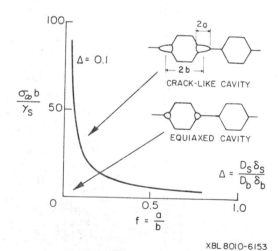

XBL 8010-6153

Fig. 2.   A schematic illustrating the two dominant cavity con-
          figurations, equilibrium triple point cavities, and
          crack-like cavities and the nature of the transition.

the ratio of the boundary to the surface energy). However similar (but less pronounced) trends could result from a grain size distribution. The inhomogeneity is needed to obtain preferred sites for initial cavitation, but appears to be of secondary importance for the zone spreading and coalescence aspects of failure.

Differences in matter transport rates associated with local variations in grain boundary diffusion, and/or inhomogeneous cavitation result in local stresses. These stresses are induced by the constraint of the surrounding material, and tend to suppress the original differences in matter transport [4,5,6]. The constraint thus provides some stability to the inhomogeneous cavity arrays and thereby contributes importantly to the rupture time [5]. An approach for estimating the level of constraint is described in the second section of the paper. Then, the initial formation of cavities, subject to the appropriate constraints, is examined. At this stage, the cavities are located at triple junctions and exhibit an equilibrium morphology [2]. Thereafter, a transition to crack-like morphology ensues (Fig. 2) [2,7]. This transition is of particular significance to creep rupture in polycrystalline ceramics [2], because it signals the onset of more rapid cavity growth, subject to reduced constraint. The characteristics of the transition are examined in the fourth section. Ultimately, the cavitation zone begins to spread laterally and creates the failure initiating macrocrack [2]. The spreading process is discussed in the fifth section of the paper.

The present analysis pertains to situations in which the failure strains are greater than a few times the elastic strain. Small strain failure processes based on statistical accumulations of cavities have been examined in previous studies [3].

2. LOCAL STRESSES

Inhomogeneous diffusion or localized cavitation creates local stresses that can differ substantially from the applied stress. Determination of these local stresses is central to the analysis of creep rupture. The stress distributions in a polycrystalline aggregate are complex. But, an approximate solution to the problem, suggested by Rice [5], can be used to simulate the principal features of the stress field. This solution is derived from the related problem of a crack in a viscoelastic solid. The solution requires matter to be deposited in a zone of length, d (Fig. 3a,b) at a rate which differs from the average mass transport rate in the surrounding material, due to differences in diffusivity and/or to localized cavity growth. This additional matter deposition $\Delta\delta$ exerts normal stresses $\Delta\sigma$ on the region d, similar in magnitude to the stresses created by a crack subject to internal pressure (Fig. 3c). The average magnitude of this stress for a circular zone under steady conditions (see Appendix) is given by;

$$\Delta\sigma = - \frac{3\pi\Delta\dot{\delta}\tau_\infty}{2\dot{\gamma}_\infty d} \tag{1}$$

where τ_∞ and $\dot{\gamma}_\infty$ are the remote shear stress and shear strain-rate, respectively. The local normal stress, σ_ℓ^i, acting over this region is thus;

$$\sigma_\ell^i = \sigma_\infty + \Delta\sigma . \tag{2}$$

Stresses of opposite sign are created outside region d (Fig. 3d). When d encompasses many grains, there are sufficient diffusion paths into the tip region (from the intersecting grain boundaries or from the lattice) that the external stresses can be adequately represented by the equivalent solution for a crack [8];

$$\overset{\circ}{\sigma}_\infty = \sigma_\infty - x\,\Delta\sigma/[x^2-(d/2)^2]^{1/2} + \Delta\sigma \tag{3}$$

where x is the distance from the center of the zone, d. For smaller zones, some localized flow along grain boundaries relaxes the singularity [9], but probably has little influence on the stresses a small distance away from the zone tip. Equation (3) is thus expected to provide a reasonable estimate of the external stresses for most problems of present interest.

The general creep rate of the material is assumed to be dominated by grain boundary diffusion, such that the rate of displacement $\dot{\delta}$ of neighboring grain centers is given by [10];

$$\dot{\delta} = g\dot{\gamma}_\infty = \frac{\beta\Omega D_b \delta_b \tau_\infty}{kTg^2} \tag{4}$$

where $D_b\delta_b$ is the grain boundary diffusion parameter, g is the grain diameter and β is a constant $\approx 14\pi$.[I] The local stress can thus be expressed by;

$$\sigma_\ell^i = \sigma_\infty - \frac{3\Delta\dot{\delta}kTg^2}{28\Omega D_b \delta_b}\left(\frac{g}{d}\right) . \tag{5}$$

[I] The choice of the coefficient $\beta = 14\pi$ is a approximate one. The magnitude of β depends upon the morphology of the grains, and the exact value pertinent to ceramic polycrystals is debatable at this junction. Different choices of β will lead to different absolute constraint levels, but the general trends and the final implications are essentially unaffected by the specific choice.

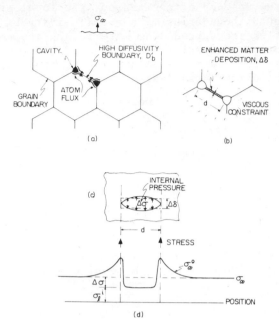

Fig. 3.   Schematics indicating localized diffusive flow from
cavities and the resultant development of constraint
(a) the cavity configuration and atom flux, (b) the
thickening displacement between cavities, (c) a crack
under internal pressure, and (d) the resultant stress
field in a viscous solid.

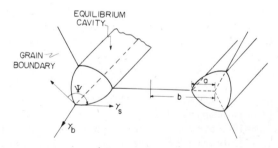

a) EQUILIBRIUM CAVITIES

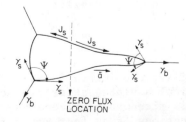

b) CRACK-LIKE CAVITY

Fig. 4.   The geometry of the cavity arrays used for analysis
a) equilibrium cavities, (b) crack-like cavities.

   In the absence of cavitation, the local stress at an array
of contiguous boundaries with a diffusivity D_b' that differs
from that of the surrounding material D_b can be readily deduced.
The deviation in displacement rate can be estimated from the
difference in creep rate in the two regions (by recognizing
the stress differential) as

$$\Delta\dot{\delta} \approx \frac{14\pi\Omega\delta_b}{kTg2} \left[D_b'\sigma_\ell^i - D_b\sigma_\infty\right] . \tag{6}$$

Eliminating $\Delta\dot{\delta}$ from Eqs. (5) and (6) then gives,

$$\frac{\sigma_\ell^i}{\sigma_\infty} = \frac{(D_b/D_b')[1 + (2/3\pi)\,(d/g)]}{[1 + (2/3\pi)(d/g)(D_b/D_b')]} . \tag{7}$$

   The local stress at steady state is thus appreciably diminished on boundaries with high local diffusivities, when the relative zone dimension d/g is small ($\lesssim 10$). However, it is noted
that the boundaries parallel to the applied stress are the source
of matter deposition and hence, the reduction in local stress
given by Eq. (7) only pertains when these boundaries also support
high diffusivities. For one or two contiguous boundaries with
high diffusivities much smaller changes in the local stress
should thus be anticipated.

3.   INITIAL CAVITATION

   The nucleation of cavities at triple junctions is assumed
for present purposes to occur, quite readily, at junctions between boundaries with atypical diffusivities or dihedral angles.
The important issues concerned with this nucleation assumption
have been afforded detailed consideration in a separate publication [11]. It is simply noted here that triple junction nucleation can occur at relatively low stress levels (typical of the
applied stresses employed in creep tests) when the local dihedral
angle ψ is small (e.g., $\psi \sim 80°$; as observed for Al_2O_3[2]).
Nucleation is thus expected to occur at a certain fraction of
grain junctions soon after the application of stress. Some subsequent nucleation at more resistant triple junctions (due to
grain boundary sliding instabilities) may also occur. But, in
the present analysis, the observed differences in cavity size
are considered to derive predominantly from differences in
cavity growth rate, rather than from different nucleation times.

   The cavities given initial consideration are equilibrium-
shaped cylindrical cavities at triple junctions (Fig. 4a). This
configuration is presumably preceded by spheroidal cavities along
three grain junctions or at four grain junctions [2]. However,
equilibrium shaped cavities extend rapidly after initial nucleation, and their growth into a cylindrical shape is assumed to
provide an insignificant contribution to the cavity evolution
time.

The system considered for the present analysis consists of several contiguous boundaries that exhibit atypical diffusivities D_b' or D_s' (relative to the average diffusivities D_b or D_s) or a low dihedral angle, ψ'. Cavities are assumed to nucleate and grow uniformly at each triple junction encompassed by these boundaries. Emphasis is placed on the behavior at boundaries in which $D_b' > D_b$. The additional matter deposition associated with cavity growth can then be considered to occur predominantly along the boundaries (with diffusivity D_b') connecting the cavity array. The matter deposited on these boundaries for uniform deposition between cavities is then (for unit width)

$$\dot{V}_{eqm} = 2b\Delta\dot{\delta} \tag{8}$$

where 2b is the separation between cavity centers (Fig. 4a) and $\dot{V}$ is the rate of cavity volume change. The volume of an equilibrium-shaped, cylindrical, triple junction cavity is (for unit width)

$$V = 3\sqrt{3}\ a^2\ F(\psi)/4 \tag{9}$$

where a is the distance of the cavity tip from the original site of the triple junction (Fig. 4a) and

$$F(\psi) = 1 + \frac{\sqrt{3}\ [\psi - \pi/3 - \sin\ (\psi - \pi/3)]}{2\ \sin^2(\psi/2 - \pi/6)}\ . \tag{10}$$

The rate of volume change is thus;

$$\dot{V}_{eqm} = 3\sqrt{3}\ a\ \dot{a}_{eqm}\ (F(\psi)/2)\ . \tag{11}$$

The cavity velocity is related to the additional matter deposition, from Eqs. (8) and (11) by;

$$\Delta\dot{\delta} = 3\sqrt{3}\ \dot{a}_{eqm}\ f\ F(\psi)/4 \tag{12}$$

where $f = a/b$.

The matter deposition is also related to the level of the local stress over the intervening boundaries and to the local diffusivity, D_b'. The standard result is;

$$\Delta\dot{\delta} = \frac{3\Omega D_b'\delta_b}{kTb^2}\ \frac{[\sigma_\ell^i - (1-f)\sigma_o]}{(1-f)^3} \tag{13}$$

where σ_o, the sintering stress, is given by;

$$\sigma_o = 2\gamma_s h(\psi)/\sqrt{3}a$$

$$h(\psi) = \sin\left[\psi/2 - \pi/6\right] \quad . \tag{14}$$

The cavity velocity is thus,

$$\dot{a}_{eqm} = \left(\frac{4\Omega D_b' \delta_b}{\sqrt{3}kTb^2}\right) \frac{[\sigma_\ell^i - \sigma_o(1-f)]}{F(\psi)\ f(1-f)^3} \quad . \tag{15}$$

The magnitude of the local stress pertinent to Eq. (15) is deduced by noting that the matter deposition given by Eq. (13) must be compatible with the development of the local stress induced by the constraint of the surrounding material (Eqn. 5); then,

$$\sigma_\ell^i = \frac{\sigma_\infty(1-f)^3 + (9/28)\sigma_o(g^3/b^2d)(1-f)(D_b'/D_b)}{(1-f)^3 + (9/28)(g^3/b^2d)(D_b'/D_b)} \quad . \tag{16}$$

Combining Eqs. (15) and (16) the final relation for the cavity velocity, expressed in dimensionless form, becomes;

$$\dot{a}_{eqm}\left(\frac{kTb^3}{\Omega D_b \delta_b \gamma_s}\right) = \frac{4\left[(\sigma_\infty b/\gamma_s)f - (2/\sqrt{3})h(\psi)(1-f)\right]}{\sqrt{3}\ F(\psi)f^2\left[(1-f)^3(D_b/D_b') - (9/28)(g^3/b^2d)\right]} \quad . \tag{17}$$

The variations of the cavity velocity with the dominant variables $(\sigma_\infty b/\gamma_s,\ \psi,\ D_b'/D_b)$ are illustrated in Fig. 5 for the important case, $g \approx 2d \approx 4b$ (i.e., one high diffusivity boundary with a cavity at each triple junction). The equilibrium growth rate exhibits the typical rapid increase after initiation followed by a continuous decrease with increase in cavity size. This decrease causes the growth rate to fall below that for crack-like cavity growth at a certain relative cavity size, f* (Fig. 5). The transition to crack-like growth is thus considered to occur at f*, and the equilibrium growth curves are terminated at this location.

4. CAVITY GROWTH: THE CRACK-LIKE TRANSITION

The transition to crack-like morphology presages the onset of relatively rapid cavity extension and thus initiates the rupture process. The growth of crack-like cavities is thus analyzed as a basis for examining both the transition and the subsequent cavity extension characteristics. This analysis is facilitated by noting that both the cavity profile and the atom flux at the tip of well developed crack-like cavities depend on the instantaneous cavity velocity; viz. the prior, equilibrium morphology of the cavity is of minor significançe [7,11]. The growth process can thus be adequately treated by focusing on the tip region, and neglecting complex morphological changes that may be occurring in the vicinity of the cavity center. Also, for present purposes the meniscus instability is neglected, because the wavelengths needed to permit the growth of perturbations is larger than the grain facet dimension for typical fine grained ceramics [2].

Commencing with the expression for the surface flux at the tip of a crack-like cavity [7].

$$\Omega J_s = 2 \sin(\psi/4) \, \dot{a}_{crack}^{2/3} (D_s \delta_s \Omega \gamma_s / kT)^{1/3} \tag{18}$$

where $D_s \delta_s$ is the surface diffusion parameter, and noting that the surface flux is related to the volume rate of matter removal, up to the zero flux position (Fig. 4b), by;

$$J_s = \dot{V}_o / 2\Omega \tag{19}$$

and that the matter removed from the cavity tip must be deposited on the grain boundary, in order to satisfy matter conservation,

$$\dot{V}_o = \dot{\Delta}\delta b \tag{20}$$

the boundary thickening rate becomes;

$$\dot{\Delta}\delta = 4 \sin(\psi/4) \, \dot{a}_{crack}^{2/3} (D_s \delta_s \Omega \gamma_s / kTb^3)^{1/2} \quad . \tag{21}$$

Combining Eq. (21) with the relations for the boundary transport problem (Eq. 13)[II] and for the local constraint (Eq. 5) permits the cavity velocity and the local stress to be derived. The velocity is given by;

$$v^{2/3}\Delta^{1/3} [(4/3)(1-f)^3(D_b/D_b') + (3/7)(g^3/b^2 d)] + 2v^{1/3}(1-f)\Delta^{-1/3}$$

$$= \sigma_\infty b / \gamma_s \, \sin(\psi/4) \tag{22}$$

where $v = \dot{a}_{crack}(kTb^3 / D_b \delta_b \Omega \gamma_s)$ and $\Delta = D_s \delta_s / D_b \delta_b$. For initial cavitation, the parameters $\sigma_\infty b / \gamma_s$, D_b'/D_b and g^3/db^2 are all > 1, whereupon Eq. (22) reduces to;

[II] The sintering stress σ_o is given by $\sigma_o = 2\gamma_s \sin(\psi/4)$ $(\dot{a}kT/D_s \delta_s \gamma_s)^{1/3}$.

418

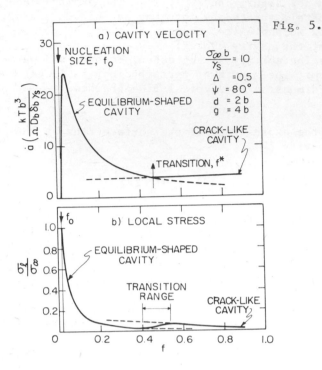

Fig. 5. a) Cavity velocity as a function of the relative cavity length indicating the equilibrium to crack-like transition. b) The local stress in the equilibrium and crack-like regions.

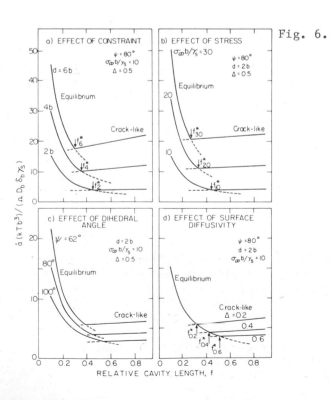

Fig. 6. Plots of cavity velocity for several choice of the important variables: (a) the effect of constraint, (b) the influence of the applied stress, (c) the effect of the dihedral angle, (d) the effect of small changes in the surface diffusivity.

$$v = \frac{[(7/3)\ (\sigma_\infty\, b/\gamma_s \sin(\psi/4))\ (db^2/g^3)]^{3/2}}{\Delta^{1/2}}$$

$$\times \left\{ 1 - \left[\left(\frac{21\gamma_s\ \sin(\psi/4)}{\sigma_\infty\, b} \right) \frac{1}{\Delta} \left(\frac{db^2}{g^3} \right) \right]^{1/2} (1-f) \right\} \quad (23a)$$

which, upon further simplification, becomes

$$v^{2/3} \approx \frac{[\sigma_\infty b/\gamma_s\ \sin(\psi/4)]\ (db^2/g^3)}{\Delta^{1/3}} . \quad (23b)$$

An almost steady velocity is thus anticipated in the crack-like region during initial cavitation. However, when the constraint is reduced in the later stages of cavitation, cavity accelera-tion is to be anticipated and Eq. (23) should not be used. The equivalent expression for the local stress is;

$$\frac{\sigma_\ell^i}{\sigma_\infty} = \frac{2(1-f)[(7/3(\sigma_\infty\ b/\gamma_s\ \sin(\psi/4))\Delta(db^2/g^3)]^{1/2}}{\Delta(\sigma_\infty b/\gamma_s\ \sin(\psi/4))} . \quad (24)$$

The trends in the cavity velocity anticipated by the above analysis are plotted in Figs. 5 and 6 for values of $f > f^*$ (i.e., in the range where the crack-like velocity and the level of the local stress exceed the equivalent velocity and stress pertinent to the equilibrium cavities.) Firstly, the strong influence of the constraint upon initial cavitation ($d \approx 2b$) is noted by com-paring the solution for $d = 2b$ with that for $d \approx 6b$ (Fig. 6a). The effect of the constraint is manifest at the very earliest stages of cavity growth and continues to be amplified as the extension proceeds. The importance of the constraint is also evident from the computed levels of the local stress, shown in Fig. 5b. Note that the stress adjusts to a larger level follow-ing the crack-like transition, because the rate of volume change (and hence, the development of constraint) is smaller for this mode of cavity extension. However, the reduced volume of crack-like cavities with smaller dihedral angles does not lead to a lower constraint. This behavior presumably arises because the rate of cavity extension at small ψ is sufficiently enhanced that the matter deposition cannot be so effectively accommodated by the creep of the surrounding material. Finally, it is re-empha-sized that the development of constraint leads to a relatively invariant cavity velocity in the crack-like regime. The signifi-cance of this result will become apparent later. As expected, the magnitude of the applied stress (Fig. 6b) has a substantial effect on the cavity velocity, over the entire range. However, other notable effects of the stress include the decrease in the transition size, f^*, as observed experimentally [2], and the sub-stantial reduction in the critical nucleation size, f_o, given by;

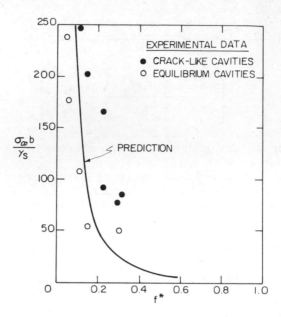

Fig. 7.   A comparison of crack-like and equilibrium cavity for Al$_2$O$_3$ with the predicted transition length, f*: Δ = 0.5, ψ = 80°, d = 2b.

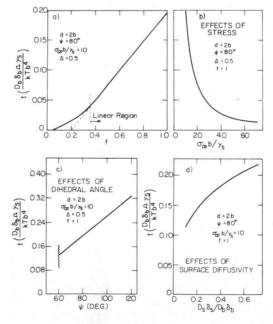

Fig. 8.   The cavity propagation time characteristics: (a) the variation of time with cavity length, (b) the effect of strain on the time taken to reach f = 1.0, (c) the effect of dihedral angle, and (d) the influence of surface diffusivity.

$$f_o = \left[1 + \frac{\sqrt{3}}{2\,\sin(\psi/2 - \pi/6)}\left(\frac{\sigma_\infty b}{\gamma_s}\right)\right]^{-1}. \tag{25}$$

The material parameters with the dominant influence upon inhomogeneous cavitation are the local values of the dihedral angle, ψ (Fig. 6c), and the ratio Δ of the surface to boundary diffusivity (Fig. 6d). Smaller values of these quantities encourage cavitation (a situation which is likely to develop in the presence of atypical impurity concentrations at certain grain boundaries within the material). This may account for an observation that crack-like cavities exhibit relatively small dihedral angles [2]. The local grain boundary diffusivity has a negligible influence on the cavity velocity. However, variations in this diffusivity would exert a much more significant influence on cavity growth at smaller values of g/b (as might prevail for the cavitation along two grain junctions observed in metal systems). Also, it is recalled that high local boundary diffusivities have been invoked in the present analysis in order to allow all of the matter excluded by the cavities to deposit on the cavitating boundaries. More diffuse matter deposition occurring in the absence of variability in the boundary diffusivity would result in reduced constraints (mostly in the equilibrium growth regime) and some enhancement of cavitation in the presence of small local values of ψ or D_s.

The transition f* between the equilibrium and crack-like modes of cavity growth predicted by the above analysis can be compared with experimental data [2] obtained for Al_2O_3. The prediction, illustrated in Fig. 7, appear to adequately separate observations of the two cavity types; hence, some credence in the preceding analysis is established.

5.   THE PROPAGATION TIME

The time taken for cavities to extend across grain facets is of principle importance for the creep rupture process. If cavity nucleation occurs soon after steady state is established, the time t_p needed to create a cavity of relative length f is simply,

$$t_p = b\left[\int_{f_o}^{f^*} \frac{df}{\dot{a}_{eqm}} + \int_{f^*}^{f} \frac{df}{\dot{a}_{crack}}\right]. \tag{26}$$

Some typical propagation times are plotted in Fig. 8. When the dihedral angle or the local surface diffusivity are small and/or the stress is relatively large. Most of the time required to develop a full facet length cavity is dictated by the growth in the crack-like mode (as might be anticipated from the velocity diagrams). The initial cavitation that occurs in local regions of a creeping polycrystal (due to small local values of ψ or D_s) can thus be approximately characterized by the constant velocity relation (Eq. 23); whereupon the propagation time becomes;

$$t_p \approx 0.28f \left(\frac{kT}{D_b \delta_b \gamma_s \Omega}\right) \Delta^{1/2} \left(\frac{g^{9/2}b}{d^{3/2}}\right) \left(\frac{\gamma_s \; \sin(\psi/4)}{\sigma_\infty b}\right)^{3/2} . \quad (27)$$

If the initial cavitation process consumes the major portion of the rupture process, Eq. (27) will also provide an approximate estimate of the failure time. However, the conditions wherein the approximation obtains can only be ascertained by examining the subsequent cavity evolution, as manifest in the zone spreading process.

6. ZONE SPREADING

The incidence of zone spreading is contingent upon the development of enhanced tensions and thus, accelerated cavity growth, around the periphery of the cavitation zone. A detailed solution of this problem involves complex considerations. Hence, a simplified analysis which provides a preliminary estimate of zone spreading effects is developed in this section. The analysis considers that cavitation firstly occurs along a single boundary for which one (or both) of the parameters that dominate the cavitation rate (ψ or D_s) is appreciably different from that for the surrounding material. The local stress outside the cavitation zone, on the contiguous boundaries, is larger than the applied stress and given by Eq. (3). The stress decreases quite rapidly with distance from the cavitation boundary and clearly encourages enhanced cavitation rates on those boundaries at the periphery of the cavitation zone (Fig. 9). The cavitation rates in this peripheral zone are presumably non-uniform, and cavity growth must continuously modify the local stress acting on the original cavitating boundary. A complete solution of this problem is beyond the scope of the present paper. Instead, a simplified intermittent spreading procedure is adopted. The procedure consists of the following steps. Cavity growth in each peripheral zone is assumed to occur uniformly (i.e., two uniformly approaching cavities on each peripheral boundary, Fig. 9) at a stress equal to the average stress over that boundary, determined from Eq. (3); while cavitation on the original boundary continues at the initially deduced local stress. Then, at a time t* when the cavity length in the cavitation zone and in the peripheral zone becomes equal (Fig. 10), the cavitation zone is considered to advance to the boundary of the peripheral zone, i.e., the cavitation zone size increases from b to 3b. The process is then continued by considering the growth in the next peripheral zone, with a new value of the local stress assigned to the cavitation zone (based on the increase in the zone size, d). The next value of the zone spreading time is computed (including the growth behavior in the new peripheral zone up to the time t*).

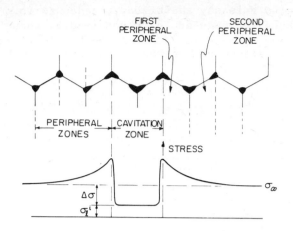

Fig. 9. A schematic indicating the cavitation and peripheral zones considered in the analysis.

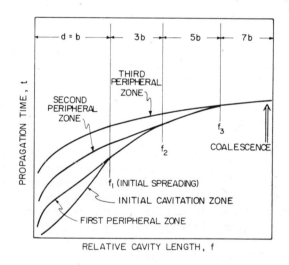

Fig. 10. A schematic indicating the stages of zone spreading through three peripheral zones, prior to coalescence.

Proceeding in this way the time t_i needed to form a discrete macrocrack can be deduced, as schematically shown in Fig. 10.

The average stress in the peripheral zone of length d, deduced from Eqn. (3) is

$$\langle \sigma_\infty^o \rangle = \sigma_\infty - \frac{\Delta\sigma}{8b(2b+d)} \left[2(d+4b) \sqrt{4b^2+2bd+d^2} \ln\left(\frac{d+4b+2\sqrt{4b^2+bd}}{d} \right) \right] + \Delta\sigma$$

(28)

The cavity velocity in this zone is deduced by substituting $\langle \sigma_\infty^o \rangle$ from Eq. (28) for σ_∞ in Eq. (17) (the equilibrium regime) or Eq. (23) (the crack-like regime). The cavity propagation times are then determined, using Eq. (26). These times are compared in Fig. 11 with those on the initial boundary for several choices of ψ', ψ, D_s and D_s'.

Some general cavitation characteristics are established before examining the zone spreading process. Firstly, small local surface diffusivities can not be the source of preferred cavitation, because the equilibrium cavity growth process is independent of D_s (although earlier transitions to crack-like cavitation can certainly be attributed to deviations in D_s). A prerequisite for the appearance of preferred cavitation is thus the existence of a dihedral angle smaller than the average value. Zone spreading considerations are therefore based on the premise that initial cavity development on certain boundaries resides in a small dihedral angle. However, subsequent cavity development on these boundaries can be further enhanced by small surface diffusivities.

The zone spreading process can be conveniently separated into three regimes. Firstly, when the deviations in ψ and D_s are small, and the absolute values are close to the average values for the material, zone spreading occurs very rapidly, while the cavities are still quite small (Fig. 11a). Failure from these regions is expected to occur quite slowly, at a rate similar to that for a homogeneous material. This regime is of the least practical significance, because failure initiates in regions of more substantial inhomogeneity. Conversely, when there are appreciable local deviations in both ψ and D_s, a cavity can extend fully across a grain facet before signifi- cant cavitation can be induced on the contiguous boundaries (Fig. 11b). The cavitation can then be regarded as an essen- tially independent process. This cavitation regime is likely to pertain in isolated regions[III] during the early stages of

---

(III) The number of these regimes would be dictated by the proba- bility of locating a boundary with small values of both ψ and D_s, based upon the appropriate statistical distributions of ψ and D_s.

failure, and explains the observation of premature full-facet sized cavities [2]. Again, however, this mode of cavitation has little influence upon the failure process, because the full-facet cavities have a minor effect on cavitation in the contiguous boundaries and do not, therefore, lead to the generation of macrocracks (as noted experimentally) [2]. An intermediate regime, that consists of appreciable deviations in ψ, but small deviations in D_s, if of principal import with regard to failure (Fig. 11c). Cavity propagation and coalescence under these conditions occurs most rapidly. Such regions are thus considered to be the principal sites for failure initiation.

The trends in cavity propagation during zone spreading suggest that a large proportion of the failure time in the intermediate region should be consumed while the cavity is contained along one or two grain facets. The approximate expression for the failure time (Eq. 27) that pertains during this period should thus provide a first-order estimate of failure. Comparing this relation with that for Coble creep (Eq. 4), the following expression for the failure time, t_f, emerges;

$$ t_f \, \dot{\varepsilon}_\infty \approx 4\pi(g/d)^{3/2} \, (\gamma_s/\sigma_\infty b)^{1/2} \, \sin(\psi/4)^{3/2} \, (D_s\gamma_s/D_b\gamma_b)^{1/2} . $$

$$ (29) $$

The analysis thus anticipates a strong interdependence of the failure time and the steady-state creep rate, as generally observed (the Monkman-Grant relationship); although an additional dependence of the failure time on the stress $(\gamma_s/\sigma_\infty b)^{1/2}$ emerges from the present analysis. Important effects of the local dihedral angle and of the diffusivity ratio are also predicted, in the sense that small values of these parameters encourage failure.

7. DISCUSSION

An analysis of cavitation has been conducted that indicates some important characteristics of cavitation induced failure. The principal impetus for failure is the presence of local variability in material properties, especially in the dihedral angle ψ or in the diffusivity along newly created cavity surfaces, D_s. Such variability in ψ or D_s could arise from crystalline anisotropy, but appreciable effects of impurities may also be involved.

Large local deviations in the dihedral angle and in the diffusivity appear to be relatively innocuous, because the isolated full-facet sized cavities which form in these regions do not enhance the cavitation rate on contiguous boundaries. However, if there are a relatively large proportion of boundaries with a high cavitation susceptibility, premature failure may occur from contiguous accumulations of these boundaries. The probabilistic aspects of failure under similar conditions have been examined

426

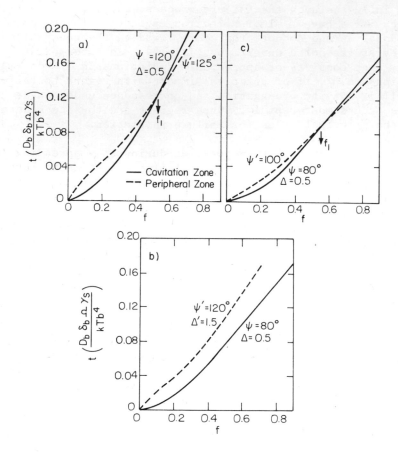

Fig. 11. The zone spreading process characteristics: (a) small
deviations in ψ and D_s, (b) large deviations in ψ and
D_s, c) appreciable deviations in ψ, but small deviations in D_s: $\sigma_\infty b / \gamma_s = 10$.

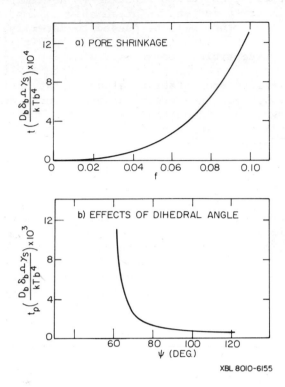

XBL 8010-6155

Fig. 12.   Sintering characteristics: (a) the variation of pore shrinkage time with pore length for $\psi = 80°$, (b) the effect of ψ on the pore shrinkage time for $f = 0.1$.

428

in a previous paper [3].[IV] The material characteristics which lead to these two extremes of creep rupture are not immediately evident, although impurity content is presumably an important consideration. Further study will be needed to determine the material classes that exhibit these two types of cavitation behavior on highly susceptible boundaries.

Exclusion of failure from statistical accumulations of susceptible boundaries results in a creep rupture process dominated by the cavity propagation time on those boundaries with appreciable deviations in dihedral angle but only small deviations in the diffusivity along the cavity surfaces created at the boundary. The zone spreading is then sufficiently limited that appreciable constraint can be maintained on the initial cavitating boundary; but enough stress enhancement is experienced by the contiguous boundaries that zone spreading and eventual rupture can be initiated. An approximate relation for the failure time that obtains under these conditions indicates important influences on rupture of the steady-state creep rate (i.e., Monkman-Grant behavior), on the local dihedral angle ψ and on the ratio of the surface to the boundary diffusivity, D_s/D_b. Small values of ψ or D_s/D_b encourage creep rupture. Such effects should be apparent in creep rupture experiments. It has already been observed, in fact, that cavitating boundaries in Al_2O_3 exhibit smaller dihedral angles than those typically measured during sintering or grain boundary grooving experiments [2] (~80° compared with 100-120°).

In regions of relative uniformity, with properties close to the material average, cavitation is expected to develop homogeneously, by virtue of a rapid zone spreading process. The stress in these regions thus remains at a level essentially similar to the applied stress. However, the time taken to induce extensive cavitation and ultimate cavity coalescence must still exceed the cavitation rate in regions of low ψ and/or D_s; presumably because the more rapid cavitation associated with the smaller ψ or D_s is not sufficiently counteracted by the development of constraint. This trend is evident from a comparison of cavity propagation times in regions with a relatively uniform, average dihedral angle (Fig. 11a) and regions with a particularly low dihedral angle (Fig. 11c).

[IV] The previous study considered creep rupture occuring, without constraint on cavity growth, in a material subject to strains on the order of a few times the elastic strain. However, the result derived on this basis would also be relevant to failure occurring at large creep strains if the constraints were essentially negated by requiring an initial probabilistically determined cavitation zone that encompasses several grain boundaries.

Finally, some correlations between creep rupture and sintering are examined. As low ratio of the surface to boundary diffusivity is a prerequisite for initial stage sintering [14]. Most ceramic polycrystals should thus be susceptible to the cavitation failure processes described in this paper. Also, it is probable that regions of a polycrystalline aggregate that are the last to sinter to full density are also the regions subject to cavitation during creep. Pore removal rates during final stage sintering can be determined by setting σ_∞ to zero in Eq. (17):

$$\dot{a}_{sinter} = - \frac{8 \sin(\psi/2 - \pi/6)}{3F(\psi)f^2 (1-f)^3 \frac{D_b}{D_b'} + \frac{9}{28} \frac{g^3}{b^2 d}} \tag{30}$$

Some typical sintering characteristics are plotted in Fig. 12.

Inspection of Fig. 12 indicates that pores with small dihedral angles, $\psi \to \pi/3$, will be removed very slowly; cavities which, as already noted, extend most rapidly. Such regions are the principal candidates for creep rupture initiation. It may be surmised, therefore, that the addition of solutes that enlarge the dihedral angle should encourage final stage sintering as well as retarding creep rupture. Prospects for identifying solutes with this capability should be explored in future studies.

ACKNOWLEDGMENT

This work was supported by the Division of Material Sciences, Office of Basic Energy Sciences, U.S. Department of Energy under Contract No. W-7405-ENG-48. Some of the original concepts were devised by one of us (AGE) while under contract from the Defense Advanced Research Projects Agency, contract MDA 903-76c-0250, administered by the University of Michigan. Invaluable discussions with J. R. Rice are gratefully acknowledged.

REFERENCES

1. R. C. FOLWEILER, J. Appl. Phys., 1961, 32, 773.

2. J. R. PORTER, W. BLUMENTHAL and A. G. EVANS, to be published.

3. A. G. EVANS and A. S. RANA, Acta. Met., 1980, 28, 129.

4. B. F. DYSON and D. MCLEAN, Metal Sci. J., 1972, 6, 200.

5. J. R. RICE, to be published.

6. G. H. EDWARD and M. F. ASHBY, Acta Met., 1979, 27, 1505.

P

7. T. -J. CHUANG, K. I. KAGAWA, J. R. RICE and L. B. SILLS, Acta Met., 1979, 27, 265.

8. A. E. GREEN and W. ZERNA - Theoretical Elasticity, Clarendon Press, Oxford, 1968.

9. A. G. EVANS, J. R. RICE and J. P. HIRTH, J. Am. Ceram. Soc., 1980, 63, 368.

10. R. RAJ and M. F. ASHBY, Acta Met., 1975, 23, 653.

11. R. M. CANNON and A. G. EVANS, to be published.

12. G. M. PHARR and W. D. NIX, Acta Met., 1979, 27, 1615.

13. Schaum's outline series, Theory and Problems of Continuum Mechanics.

14. M. F. ASHBY, Acta Met., 1974, 22, 275.

APPENDIX

RELATION BETWEEN THE LOCAL STRESS AND THE MATTER DEPOSITION

A relation for the local stress is derived by commencing with the corresponding solution for a crack in an elastic solid. The average opening δ of a disc-shaped crack of diameter d in a homogeneous linear elastic solid subject to a uniform internal pressure σ is given by [5];

$$\delta = 4(1-\nu)\sigma d/3\pi G \qquad (A-1)$$

where ν is the Poisson ratio and G the elastic shear modulus. By the analogy between linear elastic and viscoelastic materials [13], the equivalent solution for a viscoelastic solid can be obtained if G is replaced by $\bar{Q}/\bar{P}$ and ν by $(3\bar{P}K-2\bar{Q})/(6\bar{P}K + 2\bar{Q})$, where K is the bulk modulu, $\bar{P}.\bar{Q}$ is the Laplace transform of the operator P.Q. Equation (A-1) can thus be directly transformed into the equivalent viscoelastic solution

$$\bar{\delta} = \left[\frac{3K + 4\eta s}{6K + 2\eta s}\right]\left(\frac{4\bar{\sigma}d}{3\pi \eta s}\right) \qquad (A-2)$$

where s is given by the Laplace transform, $\bar{F}(s) = \int_{o}^{\infty} e^{-st} F(t)dt$ and η is the viscosity. Inversion of Eq. (A-2) gives;

$$\delta = \frac{2d\sigma}{3\pi\eta}\left[t - \frac{\eta}{K}(e^{-3Kt/\eta} - 1)\right]$$

$$\dot{\delta} = \frac{2d\sigma}{3\pi\eta}(1 + 3e^{-3Kt/\eta}) \qquad (A-3)$$

which for steady state reduces to;

$$\dot{\delta} = \frac{2d\sigma}{3\pi\eta} \qquad (A-4)$$

The viscosity η is also related to the remote values of the shear strain rate, $\dot{\gamma}_{\infty}$, and the shear stress, τ_{∞};

$$\eta = \tau_{\infty}/\dot{\gamma}_{\infty} \qquad (A-5)$$

The final result may thus be rewritten as

$$\dot{\delta} = \frac{2d\dot{\gamma}_{\infty}\sigma}{3\pi\tau_{\infty}} \qquad (A-6)$$

SECTION 5

MATERIALS BEHAVIOUR AT ELEVATED TEMPERATURES

ANELASTIC RELAXATION, CYCLIC CREEP AND STRESS RUPTURE OF γ' AND OXIDE DISPERSION STRENGTHENED SUPERALLOYS

J.K. Tien, D.E. Matejczyk, Y. Zhuang and T.E. Howson
Henry Krumb School of Mines, Columbia University, New York, New York 10027, U.S.A.

The cyclic creep and stress rupture behavior of two representative particle strengthened alloys, a nickel-base superalloy strengthened by γ' precipitates and an oxide dispersion strengthened nickel-base alloy, has been studied at 760°C as a function of frequency, with the load cycled between a fixed maximum load and a value near zero. In the frequency range studied, it was found that the minimum creep rate decreased with increasing frequency and the rupture life generally increased with increasing frequency. (For the superalloy, however, the rupture lives decreased at the highest frequencies used, apparently due to frequency induced increase in degree of planar slip.) An analysis was made of the plastic and anelastic strains that occur during the on-load and off-load periods, and of dislocation structures during cyclic creep. The observed behavior and the apparent material strengthening during cyclic loading was found not to be attributable to a cyclic strain hardening effect but rather to the effects of anelastic strain that is accumulated during the on-load period and recovered during the off-load period.

1. INTRODUCTION

In recent years many investigators have carried out stress change experiments during creep deformation in order to study basic details of the rate controlling creep mechanisms. There exists some controversy over behavior observed after small stress reductions. Some investigators [1-6] report that after the elastic contraction of the specimen a positive creep rate is immediately observed, while other investigators [7-12] report that after the elastic contraction of the specimen there is an incubation period of zero creep rate before positive creep deformation resumes. No such controversy exists, however, over behavior observed after large stress reductions. After large stress drops significant strain relaxation is observed. This anelastic behavior is seen in pure metals and single phase alloys and also recently in highly creep resistant superalloys [13,14]. Indeed, it was shown that particle strengthened systems may exhibit considerably more anelastic strain relaxation than pure metals or simple alloys, possibly because dislocation bowing between strong pinning points may be especially pronounced in particle strengthened alloys, i.e., the back stresses formed during creep in superalloys are high [15-17]. These strain relaxation observations show that during creep there is on loading a component of time dependent strain that is recoverable on unloading.

This fact suggests that in a cyclic creep and stress rupture experiment in which the stress is repeatedly removed and reapplied, anelastic strain should play a role in the cyclic creep and stress rupture properties, particularly when the magnitude of the stress change is large. Furthermore, since the amount of anelastic strain obtained after a stress drop is a function of the period of unloading, the frequency of loading and unloading should also affect the cyclic creep and stress rupture behavior. It is the purpose of this paper to study the cyclic creep and stress rupture behavior of two representative particle strengthened systems which are also heat resisting engineering alloys. The behavior will be studied as a function of frequency, large stress changes will be used and the role of anelastic relaxation will receive special attention.

There may be some confusion arising from the description of these experiments as cyclic creep and stress rupture as opposed to stress controlled low cycle fatigue (LCF). It appears that the label used is a matter of definition and does not necessarily result from the fact that fundamentally different deformation mechanisms are being studied, since the lower end of the usual LCF frequency range and the upper end of the cyclic creep and stress rupture frequency range typically overlap. The cyclic creep test is basically the same test as stress controlled LCF, where R is zero or greater. However, in work under the heading LCF of superalloys, in no instance has the role of anelastic strain been considered.

Low cycle fatigue work on superalloys and on other systems at elevated temperatures have, on the other hand, studied the effect of loading and microstructural variables on crack initiation and crack growth. One problem has been defining when a microstructural defect can be considered to be a crack, i.e., defining when a crack is a crack. In the following analysis of our cyclic creep and stress rupture (or stress controlled fatigue) results, we will discuss a potential means for answering this question.

2. MATERIALS AND EXPERIMENTAL PROCEDURE

The alloys used in this study are two representative particle strengthened high temperature alloys. One is Udimet* 700, a high strength wrought nickel-base superalloy for disc applications with analyzed chemical composition in weight percent of Ni-15.1%Cr-17.5%Co-4.9%Mo-4.15%Al-3.25%Ti-0.08%C-0.029%B-0.14%Fe. A four-stage heat treatment results in equiaxed grains with an average size of 300 μm, discrete γ'-enveloped grain boundary carbides which inhibit grain boundary sliding at the experimental temperature and about 36 vol. % of cuboidal γ' precipitates with an average length on edge of 0.3 μm. The other alloy selected is Inconel** alloy MA 754 which is a mechanically alloyed oxide dispersion strengthened (ODS) nickel-base alloy for vane and burner can applications, with analyzed composition in weight percent of Ni-20.1%Cr-0.25%Al-0.50%Ti-1.4%Fe-0.06%C-0.13%N-0.3%O-0.8%Y_2O_3. Thermomechanical processing has resulted in an elongated grain structure with a grain aspect ratio between 5 and 10. The oxides, ranging in size from 5 to 100 nm with an average size of 14.1 nm are very uniformly dispersed. We have already extensively studied the structures and creep and stress rupture properties of both of these superalloys and our results are published elsewhere [17-19].

Experiments were carried out on modified constant load creep machines which are described in detail in other reports [20,21]. The load is cycled automatically with a mechanism which is critically damped to provide the load changes over a 15 second period with no load overshoot. The stress is cycled between a fixed maximum stress, chosen as the stress to cause rupture in 50 to 100 hours of static load, and a minimum stress near zero but still positive in order to retain tension in the load train. The hold time at the upper stress was in all cases equal to the hold time at the lower stress. Static creep tests were carried out for comparison with the cyclic tests, using as the stress the maximum stress of the cyclic experiment.

The test temperature was 760°C for each alloy, a temperature at which static creep and stress rupture of these alloys has also been studied. Seven hundred sixty degrees centigrade represents a service disc temperature for Udimet 700 and a lower than usual application temperature for MA 754. In addition, at this temperature in these alloys the creep stresses used do not exceed the at-temperature yield stresses so that plastic strain loading increments, which have complicated interpretation of results in other cyclic creep and stress rupture studies, do not occur.

The hold times chosen range from 10 hours down to five

---

*Trademark of the Special Metals Corporation
**Trademark of the INCO family of companies

minutes. Starting from the initial loading of the specimen, the longest of these hold times results in uninterrupted loading into the steady state region, but not in any case into the tertiary region of creep. All tests were taken to stress rupture.

3. EXPERIMENTAL RESULTS

Fatigue data in the literature are usually presented in terms of the number of cycles to failure as a function of the test variable, which in our case is the cycling frequency. The results for Udimet 700 are presented in this way in Fig. 1. In Table I, the elongations to failure of the cyclically crept specimens and also, for comparison, the elongations to failure of the specimens crept under static load are given. Our data in this range of frequencies are consistent with other LCF results reported for Udimet 700 [22,23]. In other work it was found that fatigue lives (cycles to failure) increased by a factor of about 100 as the frequency was increased from two to 600 cpm. With further increases in the frequency to 60,000 cpm, the fatigue life was reduced by a factor of about seven. This behavior was attributed to frequency induced changes in slip character and fracture mode. At low frequencies (and at the correspondingly low strain rates), crack initiation and propagation was intergranular and slip was reasonably diffuse, as in creep and stress rupture in the same temperature range. In the frequency range where the fatigue lives were a maximum, crack propagation was transgranular and the slip character was planar. At the highest frequencies, failure remained transgranular but deformation was concentrated into even fewer slip bands which accelerated crack initiation and propagation and resulted in reductions in the fatigue lives. This explanation of the results did not consider any possible role of anelastic strains in the fatigue behavior.

Much more information in addition to that shown in Fig. 1 can be obtained, starting with a careful bookkeeping of the strains that occur during each on-load and off-load period. Examples of these strains, measured in the steady state region, are shown for both long hold time and a short hold time tests in Figs. 2 and 3. In the low frequency test, Fig. 2, there is an initial loading strain that is entirely elastic and then a region of primary creep consisting of an initially high and then decreasing strain rate which, if the hold time is long enough, reaches a steady state. In fact, this steady state creep rate, measured during a single loading period in the middle of the cyclic test, was found to correspond to the steady state creep rate observed during a static creep test. Upon load removal there is an elastic contraction identical in magnitude to the elastic loading strain followed by a region of anelastic strain recovery. Similar behavior is observed in the higher frequency test, Fig. 3. The main difference is that less strain is accumulated during each loading period in the high

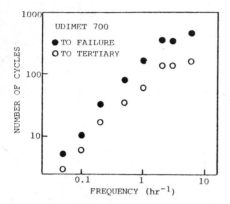

Fig. 1. Number of cycles to failure and number of cycles to tertiary as a function of frequency for Udimet 700 at 760°C with a square wave load cycle between 552 MPa and 41 MPa.

Applied Stress	Period	Elongation at Rupture (%)
static creep 552 MPa	-	9.9
		9.8
cyclic creep 552-41 MPa	10 min	7.0
	20 min	7.6
	30 min	15.0
	1 hr	13.6
	2 hr	14.9
	5 hr	11.5
	10 hr	9.1
	20 hr	8.1

Table I. Udimet 700 rupture elongation at 760°C.

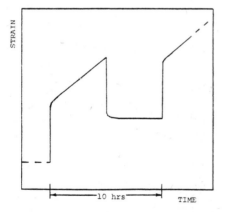

Fig. 2. Schematic strain-time curve at 0.1 cycles per hour for Udimet 700 in the minimum creep rate region.

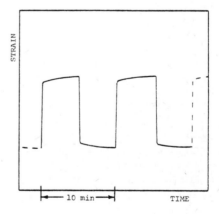

Fig. 3. Schematic strain-time curve at 6 cycles per hour for Udimet 700 in the minimum creep rate region.

frequency test than in the lower frequency test. It is important to note again that there are no plastic loading strains in these experiments. In one other study of cyclic creep and stress rupture in a single phase stainless steel [24], in which the on- and off-loading strains were carefully measured, the stresses required to give significant creep deformation at the test temperature chosen resulted in large plastic loading increments that complicated analysis of the results.

The magnitude of the strains accumulated during the on-load periods can be compared to the magnitude of the strains recovered during the off-load periods by plotting these strains as a function of each on- or off-loading period. This is done for Udimet 700 in Figs. 4 and 5 for tests which span the range of frequencies used. When the hold time is relatively long, as in Fig. 4, the total strain accumulated during the on-load half of the cycle is significantly greater than the strain recovered during the off-load half of the cycle. As the frequency is increased, however, Fig. 5, the strain accumulated during the on-load period decreases and becomes closer in magnitude to the strain recovered in the off-load period. It was also found that, at low frequency, the magnitude of the anelastic strain recovered remained almost unchanged from the beginning to the end of the test, whereas at high frequency, the magnitude of the anelastic strain recovered increased as the test progressed. At all frequencies strain was recovered during the off-load periods, but in the high frequency tests, a considerable fraction of the strain accumulated during the on-load half of the cycle was recoverable.

The effect that this phenomenon has on steady state creep rates and rupture lives in the cyclic creep and stress rupture tests of Udimet 700 is described in the following figures. First, several static and cyclic creep tests are compared by plotting total strain versus time on load, Fig. 6, and by plotting strain rate versus time on load, Fig. 7. Figure 7 shows clearly that the minimum creep rate decreases with increasing frequency. This behavior is further illustrated in Fig. 8, which shows the minimum creep rate versus frequency. The frequency effect already described is entirely consistent with this behavior. At high frequency when much of the forward strain is recoverable and the net strain per cycle is small, the creep rate is lower than at a low frequency, when a smaller fraction of the forward strain is recoverable and the net strain per cycle is relatively large. It is also possible that the cycling frequency may affect creep rates by influencing the slip character, i.e., by causing slip to become less diffuse and to become more planar in character as the frequency is increased. Evidence for this changeover has, as will be shown, been observed in the frequency range used in this study of Udimet 700, and the influence of this change on minimum creep rate is still being evaluated.

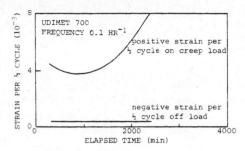

Fig. 4. Strain in each half of the load cycle (while on load, and while off load) throughout the Udimet 700 cyclic test at 0.1 cycles per hour.

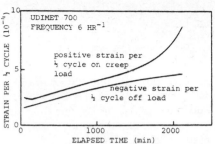

Fig. 5. Strain in each half of the load cycle throughout the Udimet 700 cyclic test at 6 cycles per hour.

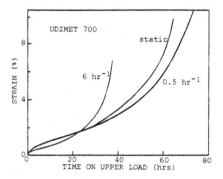

Fig. 6. Udimet 700 strain versus time on load for static load and for two frequencies of cyclic load.

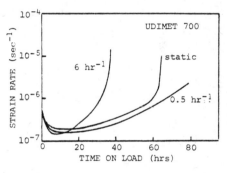

Fig. 7. Envelope strain rate (from the preceeding figure) versus time on load for Udimet 700.

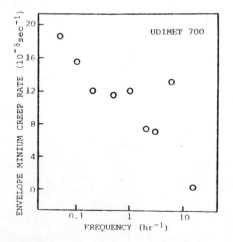

Fig. 8. Frequency dependence of Udimet 700 envelope minimum creep rate.

The effect of frequency on rupture life in Udimet 700 can be seen in Figs. 6 and 7. The cyclic rupture life, defined as the time on load to failure, is at low frequencies nearly the same as the static rupture life. As the frequency is increased the cyclic rupture lives become longer than the static rupture lives, but as the frequency is increased still further, the cyclic rupture life crosses over to become less than the static rupture life, even though the cyclic creep rate at high frequency is less than the static creep rate. Figure 9 illustrates this trend and offers a possible explanation for it. At low frequencies the slip character is apparently diffuse and rupture, as seen in the fractograph, is intergranular. At high frequencies the presence of slip bands as shown in the corresponding fractograph is evidence that slip is more planar in character. While fracture is still primarily intergranular at the higher frequencies, it appears that crack initiation may be enhanced by the increased concentration of deformation into slip bands. It is important to note that the plot of cycles to failure versus frequency may not yield any of this information on the effect of frequency on the cyclic stress rupture life. The details of Fig. 9 are not necessarily apparent from Fig. 1.

It may be possible to estimate, from Fig. 7, the time or the number of cycles to crack initiation. If the end of steady state and the beginning of tertiary creep are taken to be the point at which cracks can first be considered to have formed, then the portions of the total life involved in, respectively, crack initiation and crack growth and the time to crack initiation can be determined. This is especially marked in cases where the onset of tertiary creep is clear. In Fig. 1, the number of cycles to the beginning of tertiary creep determined from creep rate plots, as in Fig. 8, along with the total cycles to failure are plotted against frequency.

The experimental results for MA 754 are similar in many ways to the results already presented for Udimet 700. As mentioned, MA 754 is also an alloy strengthened by second phase particles, but unlike Udimet 700 the strengthening particles are incoherent and non-deformable. Figure 10 presents the results for MA 754 in the standard way, showing the number of cycles to failure as a function of frequency. The behavior is much like that for Udimet 700 shown in Fig. 1. Table II lists data on elongation at rupture for both the static and cyclic creep tests.

The careful analysis of the loading and unloading strains that was carried out for Udimet 700 and shown in Figs. 4 and 5 was much more difficult to accomplish for MA 754. The reason for this was that the anelastic strains as well as the plastic strains (especially at high frequencies) were smaller for the ODS alloy than for Udimet 700, making it difficult to measure strains during every cycle and to construct graphs similar to

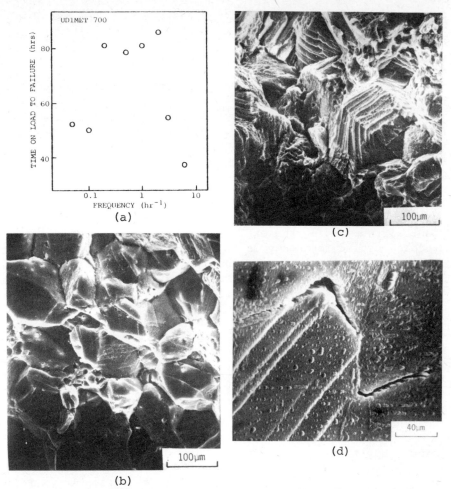

Fig. 9. (a) Frequency dependence of Udimet 700 time-on-load to failure. (b) Fractograph from static load test showing typical intergranular fracture observed for low loading frequencies. (c) Fractograph from a higher frequency (3 hr^{-1}) cyclic test showing an increase in degree of planar slip. (d) Micrograph of the side of the specimen in (c) showing slip bands intersecting a grain boundary crack.

Fig. 10. MA 754 number of cycles to failure versus frequency at 760°C with a square wave load cycle between 221 MPa and 41 MPa.

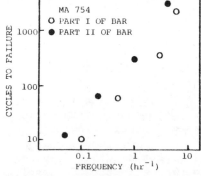

those shown in Figs. 4 and 5. A different approach was taken to shed light on the effect of frequency. For a given test, the value of the anelastic strain measured during an off-load period in the steady-state region was compared to the strain accumulated over the same period during the steady state region of a reference static test. These pairs of values are compared for three different frequencies in Table III. As with Udimet 700, in MA 754 at low frequency the strain accumulated during the on-load half of the cycle is much larger than the strain recovered during the off-load period. At higher frequencies the strain recovered during the off-load period is nearly un-changed, as Table III shows, and a rough estimate of creep strain over the time of the on-load period, using the static creep rate, gives strains smaller than the recovered strain. This shows that the magnitude of the anelastic strain is sub-stantial, supporting the explanation that the effect of in-creasing cyclic frequency is to decrease the ratio of unrecov-erable to recoverable strain that occurs during the on-load period. The effect of cyclic frequency on minimum creep rates and rupture lives are demonstrated in the following figures.

First, the total strain versus the time on load and the strain rate versus the time on load for a static and two cyclic tests are shown in Figs. 11 and 12. Figure 13 is a plot of minimum creep rate versus frequency, and Fig. 14 shows rupture lives, defined as time on load to failure, as a function of frequency. Again it is seen that the minimum creep rate de-creases with increasing frequency, and the effect is even more pronounced in the ODS alloy than in the γ' strengthened super-alloy. Also the rupture lives increase with increasing fre-quency, but there is no reduction in rupture life at the highest frequency, as was seen for Udimet 700 (see Fig. 9). This is not surprising, as there was no transition to extremely planar slip character observed in MA 754. The fracture sur-faces of cyclic and static tests appeared the same. Some dif-ferences were noted between fracture surfaces of specimens from one part of the billet that exhibited relatively high ductility as opposed to fracture surfaces of specimens from a different part of the bar that exhibited relatively low duc-tility (see Fig. 10 and Table II). These differences were not a result of cyclic versus static loading, however.

For MA 754, it is again possible to identify the number of cycles to crack initiation by identifying in Fig. 13 the number of cycles to the onset of tertiary creep. These results are shown in Fig. 15.

The cyclic strengthening in these two materials, as re-flected through decreases in steady state or minimum creep rates with increasing cycling frequency, may not be caused by the effects of recoverable anelastic strain as has been dis-cussed, but alternatively, could be due to a cyclic strain hardening effect. The results of two critical experiments

Applied Stress	Period	Elongation at Rupture (%)	
static creep 221 MPa	–	5.5 4.6 14.0 6.1	
		from part 1 of bar	from part 2 of bar
cyclic creep 221-41 MPa	10 min		
	15 min	6.9	12.1
	20 min	6.6	
	1 hr		11.3
	2 hr	3.8	
	5 hr		14.4
	10 hr	6.0	
	20 hr		9.0

Table II. MA 754 rupture elongation at 760°C.

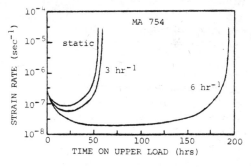

Δtime	Measured Anelastic Strain	(Static 221 MPa Creep Rate)×(Δtime)
5 min	0.8×10^{-4}	4.7×10^{-5}
1 hr	1.0×10^{-4}	5.7×10^{-4}
5 hr	1.8×10^{-4}	2.8×10^{-3}

Table III. Anelastic strain recovered per cycle compared to typical creep strain.

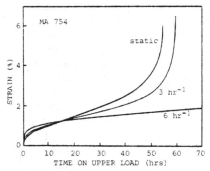

Fig. 11. MA 754 strain versus time on load for static load and for two frequencies of cyclic load.

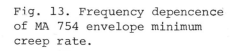

Fig. 12. Envelope strain rate (from the preceeding figure) versus time on load for MA 754.

Fig. 13. Frequency depencence of MA 754 envelope minimum creep rate.

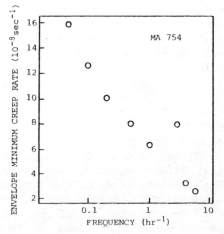

indicate that cyclic strain hardening does not explain the behavior observed. First, the dislocation densities during static
and cyclic creep of MA 754 were determined at identical values
of accumulated strain. If the experimental results were indeed
a result of material hardening due to cyclic loading then the
cyclically crept material would have a higher dislocation density than the statically crept material. In fact, just the
opposite was observed. Similar dislocation structures were
observed for both conditions, and the dislocation density in
the specimen crept under a static load of 221 MPa to a strain of
1.5 percent was $6.4 \times 10^9 \text{cm}^{-2}$, while the dislocation density
of the specimen cyclically crept between 221 MPa and 41 MPa at
a frequency of 6 hr^{-1} to a strain of 1.5 percent was found to
be $3.9 \times 10^9 \text{cm}^{-2}$.

In the second experiment using Udimet 700, load cycling
was interrupted during the steady state region of a cyclic test
(frequency of 15 hr^{-1}) and the specimen was then crept to failure under static load. After load cycling was stopped, the
creep rate was observed to increase to the value characteristic
of the static load. It has already been mentioned that in a
cyclic test with a low frequency the steady state creep rate
reached during the on-load part of the cycle corresponded to
the steady state creep rate of the static test. It appears,
then, that the experimental results cannot be interpreted in
terms of a cycling induced change in the microstructure leading
to cyclic strain hardening.

4. CONCLUDING REMARKS

In the cyclic creep of both Udimet 700 and MA 754, a frequency dependent reduction in creep rate has been observed.
This reduction is attributed to an anelastic mechanism which
retards the accumulation of nonrecoverable creep strain and
which has a more pronounced effect the higher the frequency.
Experimental support for the anelastic explanation of the observed behavior comes from measurement of the magnitude of the
anelastic strains that occur during off-load periods, and comparison of these measurements to measurements of the forward
strains that occur during on-load periods. Results of an
experiment in which the loading mode is changed from cyclic to
static in the middle of a test, plus measurements of dislocation
densities indicate that the experimental results are not attributable to cyclic strain hardening. Also, if the beginning
of tertiary creep is considered to mark the point at which
cracks are formed and crack propagation begins, then it can be
concluded that in the present experiments a majority of the
test life is spent in the formation of cracks and a minority of
the life is spent in crack propagation.

The physical origin of the anelastic strains is thought to
be the unbowing of bowed dislocation segments and also possibly
dislocation motion driven by long range back stresses arising

from dislocation substructures. In the case of MA 754, in which dense dislocation substructures are not observed, [19,25] the back stress giving the anelastic strain is probably primarily due to unbowing of dislocations and is thus on the order of or less than the Orowan stress for that ODS alloy at the test temperature. Identification and measurement of the back stress in the γ' strengthened superalloy appears to be complicated. Work to identify these back stresses in superalloys is in progress at Columbia on two fronts. One is an analysis of low strain rate stress strain curves to identify features which independently relate to the creep resisting stresses in particle strengthened alloys [16]. The second, now being undertaken, is a careful evaluation of fatigue hysteresis loops, an approach which can identify the components of the resisting stress including the friction stresses and back stresses acting on dislocations [25]. In addition to this work, cyclic creep and stress rupture experiments at different temperatures, frequencies and with different loading schemes are being completed.

5. ACKNOWLEDGEMENT

The authors gratefully acknowledge the support of this work by AFOSR under Grant AFOSR-78-3637, and by NSF under Grant DMR77-11281.

6. REFERENCES

1. Ahlquist, C.N. and Nix, W.D., Scr. Metall., 1969, 3, 679.
2. Ahlquist, C.N. and Nix, W.D., Acta Metall., 1971, 19, 373.
3. Solomon, A.A. and Nix, W.D., Acta Metall., 1970, 18, 863.
4. Pahutová, M., Hostinský, T. and Cadek, J., Acta Metall., 1972, 20, 693.
5. Blum, W., Hausselt, J. and Konig, G., Acta Metall., 1976, 24, 293.
6. Thorpe, W.R. and Smith, I.O., Acta Metall., 1978, 26, 835.
7. Mitra, S.K. and McLean, D., Proc. R. Soc. London, Ser. A, 1966, 295, 288.
8. Davies, P.W., Nelmes, G., Williams, K.R. and Wilshire, B., Met. Sci. J., 1973, 7, 87.
9. Williams, K.R. and Wilshire, B., Met. Sci. J., 1973, 7, 176.
10. Parker, J.D. and Wilshire, B., Met. Sci., 1975, 9, 248.
11. Evans, W.J. and Harrison, G.F., Met. Sci., 1976, 10, 307.
12. McLean, M., Scr. Metall., 1979, 13, 339.
13. Gibbons, T.B., Lupinc, V. and McLean, D., Met. Sci., 1975, 9, 437.
14. Lupinc, V. and Gabrielli, F., Mats. Sci. Eng., 1979, 37, 143.
15. Purushothaman, S., Ajaja, O. and Tien, J.K., "On the Concept of Back Stress in Particle Strengthened Alloys," Strength of Metals and Alloys, Vol. I, Eds. Haasen, P., Gerold, V. and Kostorz, G., Pergamon Press, 1979, p. 251.

446

16. Jensen, R.R., Howson, T.E. and Tien, J.K., "Very Slow Strain Rate Stress-Strain Behavior and Resisting Stress for Creep in a Nickel-Base Superalloy," Superalloys 1980, Eds. Tien, J.K. et al., ASM Press, 1980, p. 679.

17. Ajaja, O., Howson, T.E., Purushothaman, S. and Tien, J.K., Mats. Sci. Eng., 1980, 44, 165.

18. Aning, K. and Tien, J.K., Mats. Sci. Eng., 1980, 43, 23.

19. Howson, T.E., Stulga, J.E. and Tien, J.K., Met. Trans. A, 1980, 11A, 1599.

20. Tien, J.K., Zhuang, Y. and Matejczyk, D.E., "Cyclic Creep and Stress Rupture of an Oxide Dispersion Strengthened Nickel-Base Alloy," submitted to Met. Trans. A, 1980.

21. Tien, J.K., Matejczyk, D.E. and Zhuang, Y., "Cyclic Creep and Stress Rupture of a γ' Strengthened Nickel-Base Superalloy," submitted to Met. Trans. A, 1980.

22. Organ, F.E. and Gell, M., Met. Trans., 1971, 2, 943.

23. Gell, M. and Leverant, G.R., "Mechanisms of High-Temperature Fatigue," ASTM STP 520, 1973, p. 37.

24. Morris, D.G. and Harries, D.R., J. Mats. Sci., 1978, 13, 985.

25. Kuhlmann-Wilsdorf, D. and Laird, C., Mats. Sci. Eng., 1979, 37, 111.

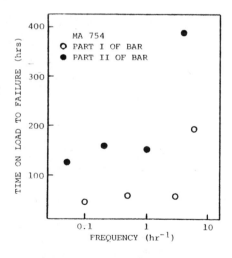

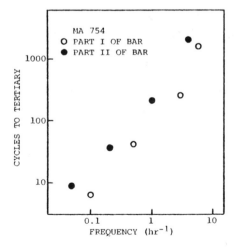

Fig. 14. MA 754 frequency dependence of time-on-load to failure.

Fig. 15. MA 754 number of cycles to tertiary versus frequency at 760°C.

MICROSTRUCTURAL ASPECTS OF THE CREEP OF ALLOYS BASED ON
NIMONIC 80A

Claire Y. Barlow and Brian Ralph

Department of Metallurgy & Materials Science, University of
Cambridge

SUMMARY

A study has been made of relationships between microstructures and creep parameters using bar specimens of Nimonic
80A and a reduced carbon alloy of related composition. The
main technique used for the examination of microstructure was
transmission electron microscopy, at both standard and high
accelerating potentials.

The materials tested were modified from the conventional
Nimonic 80A microstructure by alterations to the distributions
of both the dispersed γ' and the grain boundary phases.
Cellular transformation products were one example of the latter.
Deformation structures and changes in phase distributions
resulting from creep deformation are discussed, and related to
creep properties. Grain boundary cavitation is the cause of
creep failure in these alloys, and examination of polished
sections by light microscopy, using an automatic image analyser,
has allowed the distribution of cracks and cavities to be
compared as a function of microstructure, and correlated with
the creep parameters.

1. INTRODUCTION

1.1 General

The influence of microstructure on the creep of pure materials and simple alloys is reasonably well understood [e.g. 1].
However, commercial creep-resistant alloys have complex microstructures, and the roles of the various components are less
certain. This study is aimed at clarifying some of the significance of the constituent parts of the microstructure of a
commercial alloy, Nimonic 80A. The method followed was to
investigate the creep behaviour of related materials with minor
structural modifications from the commercial alloy.

1.2   The origins of the creep resistance of Nimonic 80A

The nickel-base superalloys rely on a number of micro-structural features for their creep resistance [e.g.2]. The intragranular regions are strengthened by the presence of dispersions of a coherent intermetallic phase, γ', and also by solid solution additions. Both features hinder dislocation movement [e.g.3], and the differential diffusion rates of the solid solution constituents lead to impedance of diffusion creep by generation of concentration gradients [e.g.4]. The grain boundaries are, in Nimonic 80A, heavily decorated with carbide particles. Precipitation on grain boundaries leads to convolution of the boundary plane, which is considered to be of particular significance in reducing or preventing grain boundary sliding [e.g.5]. The microstructure of the matrix adjacent to grain boundaries is also important. During creep, especially dislocation creep [e.g.4], high stresses may build up adjacent to grain boundaries, and the effect is most marked when grain boundary sliding is inhibited [e.g.6]. It is therefore essential that the strength of these regions be properly balanced with those of the matrix and grain boundary. Such modified regions result from the formation of γ' depleted zones at grain boundaries. These more ductile regions are generally considered to be deleterious to creep properties [e.g.7], although they have also been invoked to explain improved creep ductility as a result of enhanced recovery [8]. The presence of cellular grain boundary precipitation [9,10], sometimes accompanied by a γ' depleted zone, is generally considered to produce inferior creep rupture properties [11], so heat-treatments precluding its generation are used for material required for service.

2.   EXPERIMENTAL

2.1 As-heat-treated microstructures. Five distinct micro-structures are discussed here, and their schematic phase distributions together with heat treatments are given in table 1. Structures D and E were designed to have cellular precipitates on the grain boundaries, the two heat treatments giving respectively fine and coarse lamellar spacings. Structure C resulted from a heat treatment to minimise the extent of cellular precipitation in this alloy [9,10]. In practice, C, D and E showed considerable boundary-to-boundary variations, with regions of cellular and discrete precipitation found in all, but in varying amounts.

2.2 Creep tests. The creep tests were performed in air. The temperature was set at 750°C in order for comparisons to be made between specimens; however, a lower temperature was used for material C for reasons of microstructural instability. The stress was increased during the creep tests for some specimens; this was because the low creep rates at the original stresses were expected to give unreasonably long creep lifetimes.

CODE	ALLOY	HEAT TREATMENT	MICROSTRUCTURE 1μm	DESCRIPTION Grain Boundary Phases	DESCRIPTION γ' Distribution	COMMENTS and Mean Linear Intercept Grain size
A	Nimonic 80A	2h 1050/AC/ 18h 700/AC		Discrete $M_{23}C_6$ carbides	Single distribution of discrete particles	As Normally Heat-Treated for Service 25μm
B	"	2h 1050/AC/ 24h 850/AC/ 18h 700/AC		Discrete $M_{23}C_6$ carbides	Bimodal distribution of discrete particles	25μm
C	"	2h 1050/AC/ 2h 850/AC/ 18h 700/AC		Blocky $M_{23}C_6$ particles; some cellular precipitation	As B	40μm
D	Reduced Carbon 80A	2h 1050/AC/ 18h 700/AC		Cellular precipitation; colonies contain laths of $M_{23}C_6$ and γ' in γ.	As A	60μm
E	"	2h 1050/AC/ 24h 850/AC/ 18h 700/AC		Cellular precipitation; some blocky carbide particles: colonies contain γ' and γ.	As B	60μm

Table 1. Heat treatment and microstructures.

2.3 Material preparation and examination. (a) Electron Microscopy. Sections for transmission electron microscopy were cut from the crept specimens and 3mm discs punched from these were ground to ∿120µm thickness. The discs were electropolished to perforation using a 3% solution of perchloric acid in 2-butoxyethanol at 90V and -10°C. Foils were examined both in a JEOL JEM 200A microscope at 200kV, and an AEI EM7 microscope at 1MV. The increased foil thickness penetrable at high voltage allowed cavities in the micron size range, completely contained within the foil thickness, to be observed, and also increased the usable area of each foil. The lower voltage was used where higher resolution was required.

(b) Light Microscopy. Specimens for cavity measurements were cut as longitudinal sections from the crept bars. They were polished to 0.25µm diamond, and then repeatedly alternately etched (in an aqueous solution of 1%HF at 7V) and vibratory polished (using γ alumina and 50% glycerol/water) until a reproducible finish was formed.

3.   RESULTS

3.1 Creep data

Numerical data derived from the creep curves is presented in table 2. Comparison of specimens A and B (commercial alloy) shows that the minimum creep rate (MCR) was significantly greater in B than in A, and the ductility to failure was also increased in B. The reduced carbon alloy specimens (C,D,E) all showed creep rates rather lower than those of the commercial alloy, and very low creep ductilities.

Specimen	°C Temp.	MPa Stress	10^{-6} MCR	h t_f	% ε_f	h $t_{2/3}$	% $\varepsilon_{2/3}$
A [12]	750	154	10	1800	17	350	0.3
B	750	154	24	1956	24	720	1.0
C	700	158 / 220	0.3 / 3.3	930	0.7	432	0.23
D	750	250	24	127	1.8	67	0.26
D1*	750	200 / 280	5 / 38	188	0.6	164	0.24
E1†	750	200 / 280	16 / –	178	1.4	134	0.28

Table 2. Creep Data
* Stress increased after 141h (0.12%ε); † after 169h(0.43%t)
$t_{2/3}$, $\varepsilon_{2/3}$: time , strain on transition secondary to tertiary.

The differences in minimum creep rate between A and B may result from differences in either inter- or intra-granular microstructure. The features of the intergranular structure that are likely to be significant affect the ease of grain boundary sliding. They are the area fraction of boundary occupied by carbide particles, and the inter-particle spacing. It may be assumed [13] that boundary sliding is controlled in part by the degree of roughness of the boundary, which inspection of microstructures indicates increases with carbide particle size. While the inter-particle spacings and areas of carbide-free boundary are greater in B than in A (see table 1), the particles are also considerably larger. Thus the relative amounts of "clean" boundary available to accept matrix dislocations, providing the driving force and mechanism for boundary sliding, are in opposition to the criteria for ease of boundary sliding. While it is impossible to quantify these effects, there is microstructural evidence for increased grain boundary sliding in B, so it would appear that the former factors are dominant in this case.

The intragranular structures refer to the γ' distribution, which is coarser in B than in A. This study has shown (section 3.2) that dislocation looping round particles is a more common deformation mode in B, but is not found in A, where particle cutting is found. This indicates that A is more resistant to dislocation movement than B, so intragranular dislocation processes will be less inhibited in B than in A. This would lead to a higher MCR in B than in A. It is therefore probable that both the inter- and intra-granular microstructures are responsible for the enhanced creep rate.

The significantly lower creep rates of D and E compared with A and B, which have respectively the same γ' distributions, point to marked differences between the alloys in the mechanical properties and behaviour of the intergranular regions. There is microstructural evidence (section 3.2) for severe impedance of boundary sliding in both D and E so the difference in MCR between D and E is best explained solely on the basis of intragranular structures.

The extent of tertiary creep and the strain to failure (ε_f) relate to the ability of the material to accomodate stresses at growing boundary cracks, and on that basis may be expected to increase with the intergranular recovery potential of the material. The reduced carbon samples showed very poor ductility, implying that the intergranular recovery potential is low, and pointing to embrittlement of grain boundaries. The increased tertiary creep strain of B compared with A is in accordance with the above analysis of the mechanical behaviour of the grain boundaries, that where recovery is easier, cavity linkage will be hindered.

3.2   Microstructural observations of specimens crept to failure

Observations were made both of polished and etched longitudinal sections by light microscopy (LM), and of thin longitudinal and transverse sections by transmission electron microscopy
(TEM), at both high voltage (HVEM) and 200kV (CTEM) [14]. The
polished sections were also used for the cavitation analysis
described in section 3.3. The microstructures and heat treatments are described in table 1.

3.2   (a)   3-stage heat-treated commercial material (B)

(i)   TEM observations. Figure 1b shows a typical region
seen by HVEM. The matrix contained only large γ' particles, so
that the original secondary distribution of fine particles had
been lost during creep. This may be compared with the as-heat
treated microstructure shown in figure 1a (CTEM). The γ' particles in the crept specimen are linked by a network of dislocations. The absence of significant numbers of dislocation loops
around γ' particles does, however, imply that the principal deformation mode was particle cutting rather than dislocation
bowing. Moiré contrast was seen at some particles: this is
indicative of a small lattice parameter mismatch or lattice
rotation between particle and matrix, but does not necessarily
require any loss of coherence [15]. Carbide particles (in dark
contrast, labelled C) are seen on the grain boundary in figure
1b. The distributions of neither inter- nor intra-granular

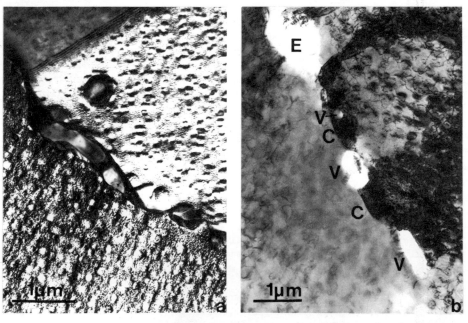

Figure 1. Material B. (a) As heat-treated (CTEM);   (b) Crept
to failure (HVEM). Note the intragranular dislocation network,
and the absence of fine γ' in (b).

carbide particles were significantly affected by the creep test. Grain boundary cavitation was observed, giving a low number density of cavities. Three are visible in figure 1b (labelled V, in light contrast) contained within the foil thickness; the feature designated E is probably a cavity which intersected the foil surface, and the profile of which was modified by etching. It was a feature of the observations of boundary cavitation that cavities were always associated with carbide particles. The cavities observed by TEM were rounded, and tended to be low aspect ratio, although (as in figure 1b) there was a tendency for elongation in the boundary plane. In these specimens, there was generally very little microstructural modification on approach to the grain boundaries, so that observed dislocation structures close to boundaries were comparable to those in the grain interiors. It thus appears that the material strength and properties would not change in these regions.

Figure 2 (CTEM) shows the elongated γ' which was found in some areas of the specimens. It is suggested that such a morphological change could result in regions where predominantly single slip occurred, and probably results from particle cutting, giving displacement parallel to the slip direction. Diffusion would allow the surface area to be reduced, leading to elongated particles. This mechanism requires the modification of particle size or morphology at approximately constant volume fraction of precipitate, implying a potential for recovery which would allow the strength due to the γ' population to remain constant.

(ii) LM observations. A typical region of a polished section of B crept to failure is shown in figure 3. There is a significant amount of grain boundary cavitation, and cavity linkage to give grain boundary facet cracking is seen. The cavities tend to be "blocky", and some w-type cracks were found. Cracked intragranular MC particles are also visible (arrowed) but their presence is not likely to have significantly affected the creep data. A more heavily etched sample is shown in figure 4, and here slip bands are revealed, together with light-etching regions at some grain boundaries. The latter regions may be evidence for Herring-Nabarro creep, where differential element diffusion rates produce inhomogeneities in the matrix solid solution close to boundaries at which accretion or depletion is occurring [e.g.4]. There were no microstructural features from TEM observations corresponding to these LM observations, so that they constitute the only direct evidence for diffusion creep.

3.2 (b) Reduced carbon materials C, D, E

(i) TEM observations. The as-heat-treated structures in these specimens differed principally from those produced in the commercial alloy (A and B) in that the amount of grain boundary

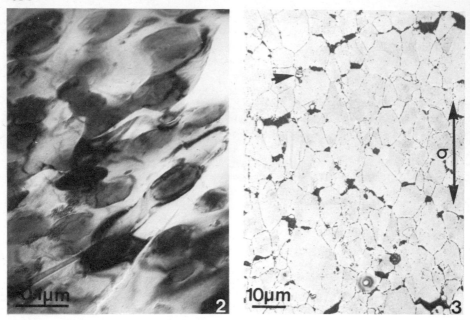

Figure 2. Material B (CTEM). Elongated γ' in single slip region.
Figure 3. Material B (LM). Polished section; stress axis marked.

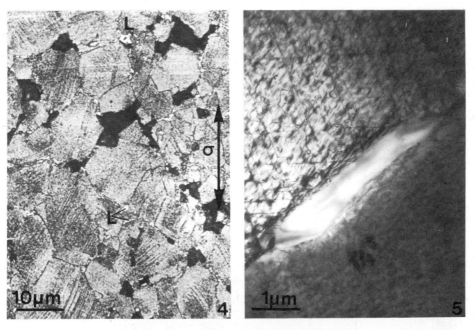

Figure 4. Material B (LM). Etched section; stress axis marked.
Slip bands are seen, and some light-etching boundary regions (L).
Figure 5. Material C (HVEM). Planar arrays and dislocation
loops close to grain boundary carbide.

carbide precipitation was significantly reduced, and the grain boundaries showed cellular precipitation of different constitutions and in varying amounts [9,10]. The appearance of the specimens were qualitatively similar.

The crept specimens showed dislocation sub-structures much closer to those normally found in cold deformed material in that planar dislocation arrays were very obvious. A typical region is shown in figure 5 (HVEM), where two sets of closely spaced planar arrays are seen, together with some dislocations looped around γ' particles. These planar arrays were more noticeable in longitudinal than transverse sections, but were found in both.

The dislocation structures in the grain boundary regions were different from those found in the matrix, and were clearly affected by the phase distributions. Figure 6 (HVEM) demonstrates the dislocation structures typical of γ'-γ cellular colonies. The number density of cavities was very low, with only a small proportion of boundaries showing any cavitation, and cavitated boundaries often contained only a single cavity, as in the example in figure 7 (HVEM). This cavity is seen to be in contact with one of the few grain boundary carbide particles.

(ii) <u>LM observations</u>. The amount of cavitation seen in the polished longitudinal sections, in figure 8, can be seen by comparison with figure 3 to be low. Many of the cavities, particularly in materials D and E, were in the form of grain-facet or multiple-grain-facet cracks, with a tendency towards high aspect ratios. A recurring feature of the spatial distribution of cavities and cracks was that they tended to lie on contiguous boundaries which formed a direct path to the specimen surface. Etched specimens (figure 9) showed clearly the presence of finely spaced slip bands.

3.3 <u>Quantitative light microscopy</u>

The study described here was an attempt to provide a quantitative basis for the comparison of different specimens. A variety of different parameters were derived [16], and interpretations of this data are suggested.

(a) <u>Aspect ratios and orientation</u>. The feret diameters normal, parallel and at 45° to the stress axis were plotted against equivalent circle diameter, thus allowing the mean aspect ratio and orientation for cavities and cracks in each size range to be estimated. The frequency distribution histogram was superposed on this plot. Analysis of this data showed that in all the specimens except B, the small cavities had low aspect ratios, and were not significantly preferentially oriented, while larger cracks tended to be extended normal to the stress axis. B, on the other hand, showed a

Figure 6. Material C (HVEM). Dislocations in cellular region.
Figure 7. Material C (HVEM). Boundary showing cavitation (V).

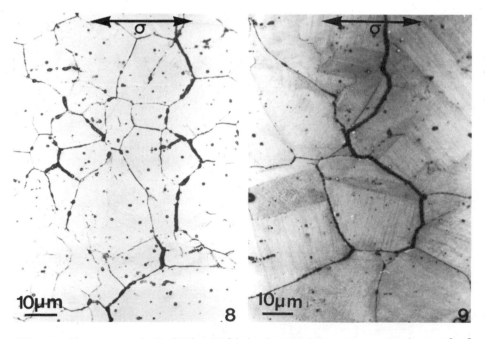

Figure 8. Material C (LM). Polished section; stress axis marked.
The globular intragranular features are carbonitride particles.
Figure 9. Material C(LM). Etched section; stress axis marked.
Fine slip bands are seen.

clear preference for elongation at 45° to the stress axis for
all cavity and crack sizes.

Where grain boundary sliding is inhibited, then the growth
of cavities and cracks is expected to be favoured on transverse
boundaries. However, if significant sliding is possible, those
boundaries experiencing the greatest shear stress (45° boundaries) are expected to show the largest amount of cavitation
and cavity linkage. The data is thus in accordance with the
expectations from the microstructural observations, that grain
boundary sliding occurred fairly readily in B but was inhibited
in the other specimens.

(b) Number densities and cavity sizes. The cavity size
(measured as an equivalent circle diameter) may be described by
either the peak position (derived from the frequency distribution histograms) or by the mean cavity size (d),which is greater
than the peak value. The values are given in table 3, together
with the number densities (N). The peak position indicates the
mean discrete cavity size, while the mean diameter also carries
information on the degree of cavity linkage to form cracks.

Material	Mean d(μm)	Peak position(μm)	Variance(μm)	N (mm^{-2})
A	5.1	2.5	1.4	10.7×10^3
B	4.0	1.5	2.1	3.7×10^3
D	4.6	2.5	3.1	200
D1	4.7	2.5	2.6	170
E	5.7	4.5	3.9	190

Table 3. Cavity number densities and mean sizes.

The number densities in A are very much greater than in B
or in the reduced carbon samples. In particular, those in D,
D1 and E are so low as to preclude statistical analysis of the
data. It is notable that the cavity number densities can be
qualitatively related to the boundary structures. The reduced
carbon materials, with the lowest cavity number densities, have
very low densities of boundary particles. In material B the
particle and cavity number densities are very much lower than
those of material A. There is thus support for the theory that
the cavitation characteristics may be influenced by the boundary
carbide particles.

4.   DISCUSSION

The different materials gave rather dissimilar deformation
characteristics on the scale of both TEM and LM observations.
In this section the differences are summarised, and relationships to property measurements are discussed where possible.

(i) Intragranular deformation. The crept specimen showed
fairly high dislocation densities, with dislocations in planar
arrays or in less regular networks between the γ' particles.

While planar arrays were found in all specimens, only in the large γ' materials (B, C and E) where dislocation bowing was feasible were networks present to any extent. The large γ' materials exhibited higher MCR values (see table 2) than the small γ' materials (A and D), and this is in accordance with the implication of the above observations that the large γ' constitutes more of a hindrance to dislocation movement than the small γ'. It is suggested that the presence of elongated γ' in regions of single slip (figure 2) is evidence for the ability of the γ' to regenerate following deformation, allowing an extended steady-state creep region. The predominance of multiple slip over most of the material would account for most of the γ' being closer to the original spheroidal morphology.

The presence of the very fine slip bands in materials C, D and E may be explained in terms of the behaviour of the grain boundaries. Dislocation nucleation during deformation commonly occurs at grain boundaries [17], and the mechanism of repeated nucleation requires that some local rearrangement of the boundary structure occurs. If, as has been suggested above, the boundaries in these materials have been embrittled, then such rearrangements may be inhibited. Thus, a source would become inoperative after producing only a few dislocations, and a new source would have to be generated in its vicinity.

(ii) Effect of cellular precipitation. The presence of a well-defined transition region between grain interiors and grain boundaries, that is, a cellular colony, gave rise to distinctive dislocation arrays (figure 6). The reduced dislocation density is likely to have arisen from the softer nature of the cellular regions, so that dislocation recovery and annihilation processes could occur more readily. It was notable that cavities were not found in cellular colonies, which is in accordance with this proposal. No definite correlation may, however, be made between the presence of cellular precipitation and changes in creep properties, on account of the alloy embrittlement from other causes.

(iii) Cavitation. The cavity number densities in all the "non-standard" specimens (B,D,D1,E) were lower than in A, and all the grain boundary cavities observed were associated with carbide particles. This observation was particularly significant in the reduced carbon material, where the number density of carbide particles was low, so the probability of a cavity being accidentally so associated was negligible. In all the observed cases, the cavities lay at the carbide particle - grain boundary interface.

It is proposed that the cavity number densities in B as compared with A result both from the increased ease of recovery in B, and also from the reduced number density of boundary carbide particles. The low cavity number densities in the reduced carbon specimens are considered to be in part a reflection on the low strain to failure. However, many of the cavities observed were large, and can better be described as boundary facet cracks, indicating that the boundaries were unable to withstand the growth of cracks. The tendency for cavities and cracks to be linked with surface cracks in these specimens is an indication that diffusion of atmosphere species may be an embrittling mechanism. Carbon is traditionally added to these alloys as a scavenger for embrittling elements, and its reduced concentration here may thus have allowed in-situ embrittlement.

5. CONCLUSIONS

1. The minimum creep rate is dependent on both the inter- and intra-granular microstructures.

2. Grain boundary cavity linkage is inhibited where recovery is easy. Thus an increased potential for recovery in the grain boundary region, and grain boundary sliding, do not always lead to premature rupture, although they will be accompanied by an increase in the MCR.

3. Grain boundary cavities are associated with carbide particles, and a reduced particle number density gives a decreased number of cavities.

4. The effect of cellular precipitation has not been shown to lead to a reduction in creep properties, but the evidence points to accompanying effects being responsible.

ACKNOWLEDGEMENTS

The authors are grateful to Professor R.W.K. Honeycombe for the provision of laboratory facilities. Financial support from the National Physical Laboratory is gratefully acknowledged, as is the interest shown in this study by Dr. B.F. Dyson and Mr. M.S. Loveday of that organisation. Some specialist assistance was given by the European Research and Development Centre of Inco.

REFERENCES

1. GITTUS, J.H. - Creep, Viscoelasticity and Creep Fracture in Solids, Applied Science Publishers, England, 1975.
2. SABOL, G.P. and STICKLER, R. - 'Microstructure of Nickel-base Superalloys'. Phys. Stat. Sol., 1969, 35, 11.
3. KEAR, B.H. - 'Mechanical Properties of γ' Hardened Nickel-base Superalloys', Order-Disorder Transformations, Ed. Warlimont, H., Springer-Verlag, 1974, p.440.
4. BURTON, B. - Diffusional Creep of Polycrystalline Materials, Trans. Tech Publications, 1977.
5. PARKER, J.D. and WILSHIRE, B. - 'Grain Boundary Sliding during Creep of Copper-Cobalt', Mat.Sci.Eng., 1977, 29,219.

Q

460

6. LANGDON, T.G. - 'Grain Boundary Dislocations and Mechanical Behaviour', Mat. Sci. Eng., 1971, 7, 117.

7. TIEN, J.K. and GAMBLE, R.P. - 'Influence of Stress on Grain Boundary Precipitate Morphology during Creep'. Met. Trans., 1971, 2, 1663.

8. FLEETWOOD, M.J. - 'Chromium Distribution around Grain Boundary Carbides in Nimonic 80a'. J. Inst. Met., 1961, 90, 429.

9. BARLOW, C.Y. and RALPH, B. - 'Cellular Transformation Products in Nickel-base Superalloys', J. Mat. Sci., 1979, 14, 2500.

10. BARLOW, C.Y. and RALPH, B. - 'Grain Boundary Migration in Nimonic 80A', Recrystallisation, Ed. Hansen, N., Jones, A.R. and Leffers, T., Risø National Laboratory, Denmark, 1980, p.165.

11. RAYMOND, E.L. - 'Effect of Grain Boundary Denudation of γ' on Notch-Rupture Ductility'. Trans. Met. Soc. AIME, 1967, 239, 1415.

12. DYSON, B.F., LOVEDAY, M.S. and RODGERS, M.J. - 'Grain Boundary Cavitation under Various States of Applied Stress'. Proc. Roy. Soc. Lond., 1976, A349, 245.

13. ASHBY, M.F. - 'Boundary Sliding and Diffusional Creep'. Surface Sci. 1972, 31, 498.

14. BARLOW, C.Y. and RALPH, B. - 'HVEM and Deformed Microstructures', Electron Microscopy, 4, Ed. Brederoo, P. and van Landuyt, J., 7th European Congress on EM Foundation, Leiden 1980, p.352.

15. ASHBY, M.F. and BROWN, L.M. - 'Diffraction Contrast from Inclusions', Phil. Mag., 1963, 8, 1649.

16. DEHOFF, R.T. and RHINES, F.N. - Quantitative Microscopy, McGraw-Hill Book Company, 1968.

17. DINGLEY, D.J. and POND, R.C. 'The Interaction of Crystal Dislocations with Grain Boundaries'. Acta Met. 1977, 27, 667.

ANALYSIS OF THE CAUSES OF SCATTER IN STRESS RUPTURE PROPERTIES
OF A NICKEL-BASE SUPERALLOY

J. Bressers, O. Van der Biest, P. Tambuyser

Commission of the European Communities
Joint Research Centre, Petten Establishment
Petten, The Netherlands

SUMMARY

Scatter in mechanical properties originates from two
sources: test parameter variability and metallurgical variability
over the sample material. By carefully controlling all testing
parameters in a series of stress rupture tests on the nickel-
base alloy Waspaloy, the contribution of test variables to
overall scatter of data was reduced to a negligible level. The
remaining metallurgical scatterband is quantified and the
distribution of several stress rupture properties is analysed.
On the basis of the constancy of the steady-state creep rate,
of the stress exponent and of the apparent activation energy,
it is concluded that the operative deformation mechanism is
similar in all samples. This conclusion is further substantiated
by the invariance throughout the sample series of the microstructural parameters that determine the deformation mechanism.
The wide scatterband observed in the stress rupture life and
specially in the tertiary strain is correlated with differences
in the level of segregation of tramp elements such as Sulphur
and Oxygen to grain boundaries.

1. INTRODUCTION

Mechanical property data measured on high strength alloys
at elevated temperatures display a rather strong dispersion of
data about their mean value. This scatterband is particularly
strong in case of time dependent mechanical properties such
as creep and stress rupture. For most heat resistant steels
for example the time to creep rupture at a certain stress level
shows a considerable scatter, with lives spreading over an
order of magnitude /1/. In terms of stress, which is the measure
of material variability of interest in design, stress rupture
values for any particular cast of steel are usually within
± 20% of the mean value, provided chemical composition and
heat treatment are within the specified ranges.

Mechanical property scatter originates from two sources: test parameter variability and metallurgical variability. Strengthening mechanisms in high strength alloys are complex and may strongly depend on compositional variations and on changes in thermomechanical pretreatment, which both lead to micrometallurgical variability. Test parameter variability includes variability of such factors as testing stress and temperature, bending stress, sample dimensions, surface finish etc. Because of their high stress sensitivity and high apparent energy for activation, high strength alloys are particularly sensitive to stress and temperature variability during creep deformation.

The most common approach towards the scatter problem is to accept scatter as an inevitable fact of life and to define the scatterband width by testing a large number of samples stemming from a single or from several casts under "similar" testing conditions in different laboratories. Statistics are then applied to analyse the data. Correlation calculations are performed on the data in order to discover which metallurgical and/or testing parameter variable(s) correlates with the observed scatter.

The main drawback of such an approach is that metallurgical and test parameter variability effects cannot be fully separated because limits on test parameters are usually too wide or even unknown in interlaboratory testing programmes. It is therefore difficult to isolate conclusively the effect of relevant variables on scatter. The alternative approach is to eliminate the variability of one set of parameters - by, for example, testing under closely controlled conditions of stress, temperature, bending stress etc. - in order to allow a detailed investigation into the effect of the other set of parameters on scatter. Such an approach was adopted in this investigation which aimed at studying the causes of scatter in stress rupture data observed upon quality assurance testing of Waspaloy forged material. The scatter in stress rupture data observed upon testing a number of Waspaloy samples under similar, closely controlled testing conditions is quantified and the data distribution is analysed. It is shown that the scatter can be correlated with segregation of tramp elements like Sulphur and Oxygen to grain boundaries.

2. EXPERIMENTAL PROCEDURES

The material investigated was Waspaloy of one cast which was received in the form of six centerless ground bars having a diameter of 19 mm. The composition in wt.% is, as certified by the manufacturer:

C	Cr	Co	Mo	Ti	Al	Zr	B	Fe
0.046	19.45	14.10	4.25	3.07	1.34	0.067	0.0054	0.78

Cu	Mn	Si	S	P	Ni
<0.01	0.02	0.03	0.003	0.006	bal.

The material was received in the solution heat treated condition (1350 K, 4 hrs, AC). An ageing heat treatment as usually applied to gas turbine discs i.e. 1123 K, 4 hrs, AC + 1033 K, 16 hrs, AC was given to the bar stock in our own laboratory.

All specimens were tested at a stress σ = 550 MPa and at a temperature T = 1003 K, those being the test conditions applied in the quality assurance stress rupture testing of Waspaloy gas turbine disc material. In order to limit data scatter stemming from test parameter variability all test parameters were controlled within narrow limits. Samples were manufactured to close tolerances and tested under true stress conditions in a universal testing machine. Test temperature and stress were controlled to $\pm$ 1.5 K and to better than 0.5% respectively. Bending stresses on each test sample were kept below $\pm$ 3% of the average stress. Samples were always loaded at a constant load rate of 10 MPa/s. More details relating to testing equipment and testing procedure are given in /2/.

Microstructural analysis was based on transmission electron microscopy (TEM) observations coupled with energy dispersive X-ray spectroscopy (EDS), on optical microscopy, on scanning electron microscopy (SEM) and on Auger electron spectroscopy (AES). Fracturing precrept Waspaloy in situ in the AES at liquid nitrogen temperature did not generally result in an intercrystalline fracture surface. Intergranular fracture was obtained by loading precrept Waspaloy in the AES equipment under the abovementioned stress rupture conditions in a specially constructed device /3/. Fracture occurred after 10 h to 15 h after which the bottom half of the sample dropped immediately into a specimen bucket which then was manipulated in the analysis position.

3. RESULTS
3.1. Stress rupture test results

Table I lists the stress rupture test data which were extracted from strain-time recordings of Waspaloy samples tested at 550 MPa and 1003 K. The first and second figures in the sample identification refer to the bar from which the sample was selected and to its position in the bar respectively. The other columns list the secondary strain ε_s, the steady-state strain rate $\dot{\varepsilon}_s$, the tertiary strain ε_t, the strain at fracture ε_f and the rupture life t_r respectively. Some test results are not reliable in ε_t, ε_f and t_r because of unintended test interruptions or faulty true stress control.

Sample	ε_s $(in\ 10^{-2})$	$\dot{\varepsilon}_s\ (s^{-1})$ $(in\ 10^{-8})$	ε_t $(in\ 10^{-2})$	ε_f $(in\ 10^{-2})$	t_r (h)
1.1	0.62	5.0	1.42	2.17	34.2
1.3	n.m.	n.m.	1.26	1.92	36.5
1.6	1.08	5.8	5.35	6.46	51.9
1.9 (1)	1.20	6.1	6.31	7.57	54.5
1.17	1.07	5.5	5.27	6.40	54.0
2.1	1.32	7.7	6.37	7.69	47.5
2.2	1.44	6.1	7.97	9.42	65.5
2.4	1.32	5.7	7.70	9.03	65.0
2.5	0.94	5.3	3.28	4.29	49.1
2.8 (4)	1.53	5.6	8.72	10.30	76.0
2.9	1.29	7.4	7.73	9.05	48.4
2.10	1.66	5.6	10.06	11.72	82.8
2.13	2.15	7.6	10.13	12.33	78.7
2.16	1.34	5.2	9.37	10.71	71.4
3.9 (2)	1.08	7.9	7.69	8.80	38.0
3.13	1.55	6.3	8.68	10.23	68.0
3.17	1.74	6.2	9.50	11.30	77.3
4.5	1.92	6.2	14.03	16.0	85.2
4.13	1.64	4.9	16.31	18.0	93.7
4.22	1.00	n.m.	4.14	5.14	62.3
5.1	1.56	5.5	8.96	10.60	78.8
5.5	1.42	5.4	7.92	9.37	72.5
5.9 (3)	1.44	5.0	9.07	10.59	80.4
5.13	1.54	7.2	6.91	8.53	59.5
5.17	1.24	4.3	9.05	10.29	80.0
5.22	1.70	6.7	10.26	12.03	70.6
6.5	1.68	6.5	10.61	12.36	72
6.17	1.44	5.5	10.33	11.87	72.1

n.m.　　　: not measured
(1)(2)(3): test interrupted after 47 h, 5 h, 57 h respectively
(4)　　　: inaccurate true stress control

Table I　Listing of stress rupture data measured on Waspaloy, tested at 550 MPa and 1003 K.

Stress rupture curves representative for the behaviour of samples with mean and extreme properties are shown in fig. 1, together with the definition of ε_s and ε_t.

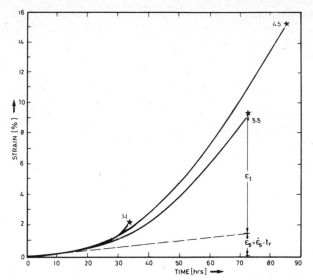

Fig. 1 Stress rupture curves of Waspaloy samples tested
        at 550 MPa and 1003 K, illustrating scatter of
        stress rupture data. Numbers identify samples.

All stress curves are characterised by small anelastic strains
ε_a and small primary strains ε_p (not listed in table I), the
order of magnitude of their sum being 5.10^{-4}. A small secondary
strain is followed by a large tertiary strain range. As an
example of the statistical distribution of the stress rupture
data fig. 2 shows cumulative distributions of ε_t and of t_r. The
solid lines fitted to the data represent a Gaussian distribution
The scatter of the secondary creep strain ε_s, of the steady-
state creep rate $\dot{\varepsilon}_s$ and of the fracture strain ε_f can also be
described by a Gaussian distribution function.

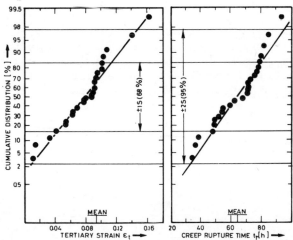

Fig. 2 Cumulative distribution of tertiary strain ε_t
        and of rupture life t_r of Waspaloy samples
        tested at 550 MPa and 1003 K.

Table II summarises the mean value $\bar{X}$, the standard deviation S, and the total scatterband width for the observed stress rupture data distributions.

Property	mean value $\bar{X}$	standard deviation	total scatterband width
$\dot{\varepsilon}_s\,(s^{-1})$	$6,0 \cdot 10^{-8}$	$1,1 \cdot 10^{-8}$	$0,72\ \overline{\dot{\varepsilon}}_s \leqslant \overline{\dot{\varepsilon}}_s \leqslant 1,32\ \overline{\dot{\varepsilon}}_s$
ε_s	$1,40 \cdot 10^{-2}$	$3,36 \cdot 10^{-3}$	$0,44\ \overline{\varepsilon}_s \leqslant \overline{\varepsilon}_s \leqslant 1,54\ \overline{\varepsilon}_s$
ε_t	$7,95 \cdot 10^{-2}$	$3,43 \cdot 10^{-2}$	$0,16\ \overline{\varepsilon}_t \leqslant \overline{\varepsilon}_t \leqslant 2,05\ \overline{\varepsilon}_t$
ε_f	$9,36 \cdot 10^{-2}$	$3,74 \cdot 10^{-2}$	$0,21\ \overline{\varepsilon}_f \leqslant \overline{\varepsilon}_f \leqslant 1,92\ \overline{\varepsilon}_f$
t_r (h)	$64,5$	$16,25$	$0,53\ \overline{t}_r \leqslant \overline{t}_r \leqslant 1,45\ \overline{t}_r$

Table II
Mean values, standard deviations and total scatterband widths for stress-rupture data.

The steady-state creep rate $\dot{\varepsilon}_s$ is subject to a relatively small degree of scatter. Its scatterband width is nearly totally accounted for by the experimental tolerances on the testing parameters σ and T. Moreover $\dot{\varepsilon}_s$ values vary at random when correlated with the strongly scattered tertiary stage data. The secondary strains are more dispersed, but this merely is the result of the scatter occurring in the stress-rupture life t_r ($\varepsilon_s = \dot{\varepsilon}_s \cdot t_r$). It is in the tertiary creep stage that the metallurgical variability causes a strong dispersion of data, as evidenced by the wide scatter in ε_t . Because ε_t contributes a large fraction to ε_f , an almost as wide scatter is observed for ε_f as for ε_t .

Stress rupture properties do not mutually correlate except for the correlation existing between the stress rupture life t_r and the tertiary strain ε_t . The straight line showing this correlation in fig. 3 is a least squares fit to all the experimental data listed in table I which resulted from correctly controlled tests.

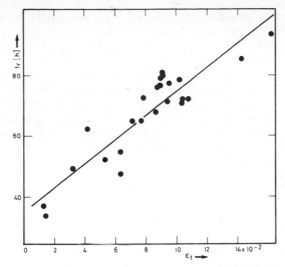

Fig. 3 Correlation between stress rupture lives t_r and tertiary strains ε_t in Waspaloy samples tested at 550 MPa and 1003 K.

Stress rupture tests were supplemented by stress and temperature change tests to determine the stress exponent n and the apparent activation energy Q on samples located adjacent to specimens exhibiting mean and extreme properties. For stresses in excess of approximately 400 MPa and up to 600 MPa n takes values ranging from 10 to 12, typical for nickel base superalloys. The variation in n with respect to the tertiary strains is at random. A similar statement holds for the variation observed in the values of the apparent activation energy Q with tertiary strain. Apparent activation energies are in the range 585 kJ/mole to 620 kJ/mole.

3.2. Microstructural analysis

In order to identify the sources of scatter a number of representative samples having stress rupture properties dispersed over the scatterband was subjected to a detailed microstructural analysis. Microstructure, creep damage, fracture surfaces and grain boundary segregation were investigated.

(a) Microstructure
Average grain sizes in the longitudinal and transverse sections of the bar stock material and in the selected test samples varied between 55 μm and 85 μm. No correlation between grain size and either stress rupture life or fracture strain was apparent.
Grains frequently contain lamellar annealing twins which mostly end within the grain and are bounded by incoherent twin boundaries decorated with elongated $M_{23}C_6$ particles. Occasionally deformation twins are observed in the most deformed specimens.

The matrix strengthening phase is γ'. Precipitates are homogeneously distributed and spherical with diameters ranging from 6 nm to 20 nm. No precipitate growth occurred during the stress rupture testing. Quantitative extraction of the γ' precipitates from five representative samples having properties dispersed over the scatterband pointed to differences of less than 1% in the γ' content. Without any specific preference blocky MC precipitates are observed in the matrix and on the grain boundaries. Both the EDS measurements and the remarkably constant values for the lattice parameter found by X-ray diffraction suggest that the composition of the MC carbides is similar in all samples. M stands mainly for Ti, although Mo, Nb and Zr were also present in order of decreasing concentration. Apart from some faint traces of M_6C observed in Debye-Scherrer recordings, $M_{23}C_6$ is the second carbide precipitate observed. Except for some highly symmetrical boundaries all grain boundaries are decorated with $M_{23}C_6$ carbides. The orientation relationship $<001>_{M_{23}C_6} // <001>_{\gamma}$ is always observed to exist between the carbide and one of the adjacent grains. M stands mainly for Cr and Mo. Again the constancy of the lattice parameter and the EDS observations suggest that no significant compositional differences in $M_{23}C_6$ exist in samples having different stress rupture properties. Both the strongest MC and $M_{23}C_6$ X-ray diffraction peaks have similar intensities, indicating that comparable amounts of both phases are present. The overall carbide content is approximately 0.7 wt.%. The small content and the occurrence of residual γ' particles in the carbide extract prevented a more precise quantitative assessment of the amount of carbides present.

The major conclusion of the phase analysis is that no overall systematic changes exist between specimens having markedly different stress rupture properties. This conclusion holds for the composition and sizes of $M_{23}C_6$, MC and γ' precipitates, for the γ' content and for the content ratio of MC to $M_{23}C_6$. Local differences between samples on the scale of the grain size cannot be excluded. As an example, the morphology and distribution of the $M_{23}C_6$ carbides was seen to vary considerably within single specimens from one grain boundary to another.

(b) Fractography

The morphology of the fracture surface allows a distinction to be made between regions of less ductile crack growth, regions of ductile crack growth and overload fracture areas. The latter are not of interest here. In the less ductile crack growth area the fracture path is fully intergranular. Sometimes shallow dimples can be recognised, whereas other grain boundaries are almost completely void of any features except for slip markings and twin boundaries. In the ductile crack growth area the crack also grows along grain boundaries but rupture in this case is largely plastic as evidenced by the dimples observed on most grain boundaries. The dimple size varies from grain to grain and even from area to area in single grain boundary facets. Grain

boundary facets are connected by transgranular shear surfaces
so that the fracture surface does not display the pronounced
"rock candy" appearance typical for the less ductile crack
growth area.
No obvious differences are however observed between samples
with widely different properties.

(c) Creep damage

The nature and the extent of creep damage in samples with
different stress-rupture properties were determined by microscopic examination of longitudinal sections of fractured
samples, with a limit of detection of approximately 1 μm. Both
the cracks generated at the sample surface and in the bulk were
intercrystalline and of the wedge-type. Inspection of some
partial-life samples revealed no signs of crack formation during
the secondary creep stage. The bulk and the surface crack
densities at fracture were of the order of 0.02 cracks/grain
and 0.01 cracks/grain respectively, the former showing a slight
decrease with increasing ductility to approximately 0.008 cracks/
grain for the longest lived samples. In terms of crack length/
grain an initial increase with increasing ductility followed by
a drop is noted for the bulk cracks. The surface crack length/
grain remains nearly constant over the whole ductility range.

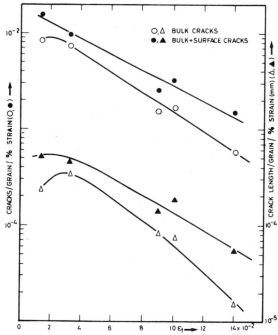

Fig. 4   Bulk and (bulk + surface) crack number densities
and crack lengths per grain and per unit strain
as a function of ε_t for Waspaloy samples tested
at 550 MPa and 1003 K.

Fig. 4 shows crack densities and crack length data per grain, normalised per unit tertiary strain, as a function of tertiary strain (which is directly proportional to t_r). Both the bulk and (bulk + surface) crack generation rates increase with decreasing ε_t if it is assumed that crack nucleation occurs continuously during the tertiary creep stage. The crack length per grain per unit strain also increases with decreasing ductility, pointing to an approximately tenfold increase in crack growth rate over the ductility range investigated.

(d) Grain boundary segregation

Auger peak to peak heights for all the elements detectable in the spectra were measured on the in situ prepared inter-crystalline fracture surfaces. The relevant results, averaged over all intercrystalline areas and normalised with respect to the strongest Ni peak at 848 eV are represented in fig. 5. Experimentally it was verified that the sulphur signal was dependent on the quench rate of the sample, due to the high mobility of sulphur in nickel /3/. Because the quench rate of the samples after fracture was high and the same for all samples, the sulphur signals may be mutually compared although they are probably overestimates of the true grain boundary signals. Phosphorus diffusion to the free surface is substantially slower than diffusion of sulphur /3,4/. So the figures in fig. 5 are thought to reflect the true grain boundary concentration. The measurement of oxygen surface concentration can be influenced by rest gas adsorption. However the pressure in the preparation chamber of the spectrometer was always better than 3 µPa and samples were transferred within minutes after fracture to the analyser chamber. Oxygen diffusion from within the sample may also have contributed but again this contribution

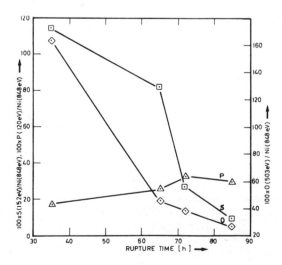

Fig. 5   Change in grain boundary coverage of S, O and P with rupture time for Waspaloy.

will be the same for all samples. Comparison of the results in figure 5 shows that the grain boundary concentrations of oxygen and sulphur are significantly higher in the short lived samples. Phosphorus grain boundary concentration on the other hand is slightly higher in the samples with longer lifetime. The spectra also showed the presence of a small nitrogen peak at 387 eV which was however superimposed on the strong Ti peak at 390 eV and this complicated quantitative assessment. The nitrogen peak however remained small for all fracture surfaces examined.

4. DISCUSSION

The scatter in ε_t and in t_r observed to occur in the Waspaloy material can almost fully be attributed to metallurgical variability over the bar stock. From earlier experimental work /2/ it appeared that out of all test parameters, temperature $\bar{T}$ and stress σ contribute most strongly to scatter. On the basis of the experimental tolerances on T and σ and using the known values for the stress exponent n and the activation energy Q, the maximum contribution of test parameter variability to the scatter of t_r is calculated to be $0.85 \ \bar{t}_r \lesssim \bar{t}_r \lesssim 1.17 \ \bar{t}_r$. Similar values hold for the scatter expected for the secondary creep rate $\dot{\varepsilon}_s$. The experimentally observed dispersion in $\dot{\varepsilon}_s$ data can, therefore, to a large extent be identified with scatter stemming from test parameter variability, whereas the scatter observed for t_r and ε_t is largely due to metallurgical variability.

The relative constancy of the steady state creep properties $\dot{\varepsilon}_s$, n and Q suggests that the mechanism of deformation is similar in all samples. Deformation occurs in planar deformation bands in which dislocation pairs coupled by antiphase boundaries cut through the γ' precipitates. This creates stacking faults which, in a later stage of deformation, spontaneously degenerate into microtwins extending through both the γ and γ' phases. In the tertiary creep stage grain boundary sliding and possibly also the impingement of deformation bands on the grain boundaries create wedge cracks. The dimpled structure observed on the grain boundary facets of the fracture surface suggests that the wedge cracks advance through cavitated grain boundaries. Concurrently with the wedge crack formation in the bulk surface cracks are being generated. Generally, one such surface crack will start to grow and form the principal crack along which the specimen ultimately fails. The similarity of the deformation mechanism in all samples is further substantiated by the invariance throughout the sample series of the microstructural parameters controlling the deformation i.e. grain size and contents, size and composition of the precipitate particles.

The dependence of the crack damage on the tertiary strain (or on its equivalent t_r), combined with the invariance of the steady state properties and of the microstructure, suggest that the scatter in ε_t and t_r is caused by changes in crack-

growth rates and, possibly, also in crack nucleation rates.
The conclusion with respect to the change in nucleation rate
however depends on the assumption made that the generation of
cracks during the tertiary stage occurs continuously. Because
damage measurements on interrupted test specimens were not
consistently made this assumption remains speculative.
The observed change in crack growth rate is however independent
from this assumption and correlates with the differences in
grain boundary chemistry.

This raises the issue of the influence of trace elements
and impurities on the properties of nickel and its alloys. This
has been the subject of a review by Holt and Wallace /5/. From
this review it is apparent that the deleterious influence of
residual sulphur on the high temperature strength and ductility
of unalloyed nickel is well documented. The evidence for the
embrittling effect of sulphur in nickel alloys is not as clear.
It was shown by Schultz /6/ that sulphur additions up to 70 wt
ppm did not embrittle Cr and Ti containing nickel alloys, because
of the formation of Cr and Ti containing primary sulphides or
carbosulphides. Because of this scavenging action of Ti and Cr,
Doherty et al. /7/ felt compelled to study a Cr and Ti free
alloy Ni-Ta-Al with a $\gamma - \gamma$' superalloy microstructure to
study the embrittling action of sulphur. It was shown that as
little as 60 at ppm S was sufficient to cause complete
embrittlement at room temperature. Segregation of sulphur to
grain boundaries was shown by AES and it was demonstrated that
addition of La, Hf and Zr reduced the grain boundary coverage
by S. Segregation of sulphur to grain boundaries in commercially
cast superalloys was shown by Walsh and Anderson /8/. Although
there is strong circumstantial evidence for the embrittling
effect of sulphur at high temperature, there has been no study
to date which has shown a direct relationship between high
temperature creep strength and ductility and the level of grain
boundary coverage by sulphur. Also knowledge on the relationship
between grain boundary segregation of sulphur and other alloying
or tramp elements is limited.
Even less is known about the effect of phosphorus in nickel and
its alloys. Berkowitz and Kane /9/ showed a relationship between
phosphorus segregation to grain boundaries and hydrogen
embrittlement in cold rolled and annealed Hastelloy C-276. This
work involved also AES measurements of P on grain boundaries. On
the other hand, it is known that minute additions of phosphorus
can enhance the malleability of nickel /10/. The effect of gases
such as nitrogen and oxygen on the mechanical properties are not
well documented. Unpublished work by Larson /11/ shows that
stress rupture lives of cast and powder fabricated alloys are
very sensitive to trace amounts of oxygen. Stress rupture live
increases markedly as the oxygen content is decreased below
50 ppm. Actual measurements of oxygen concentration on grain
boundaries have not yet been reported in the literature.

The present results show that segregation of sulphur and

oxygen to grain boundaries reduces drastically the stress rupture strength and ductility of Waspaloy. The present set of analyses do not allow to make more definite statements on the severity of embrittlement induced by either element separately. The results are in agreement with the literature cited above. However it is explicitly shown here that oxygen also segregates to grain boundaries. The results in figure 5 show that phosphorus appears to be less harmful to strength and ductility than oxygen and sulphur; the level of phosphorus segregation is in fact higher in samples with the longer life time.

The variation in properties of the specimens are not completely random. When reviewing the results of table I, it can be seen that some sections of the bar have properties worse than others e.g. part of bar 1 and 2. In view of the correlation which has been found with the presence of oxygen and sulphur, this indicates a variation in trace element content along the bar which should be related to processing. Unfortunately detailed information on processing history was not available and this important link could not be made here.

ACKNOWLEDGEMENT

The authors acknowledge the skilled assistance of Mr. E. Fenske in mechanical testing, and of Messrs. G. Von Birgelen, F. Franck, K. Schuster, P. Helbach and I. Zubani in the microstructural analysis and sample preparation work. They also wish to thank Dr. Joppien from Motoren und Turbinen Union (Munich, D) for the supply of the Waspaloy bar stock.

REFERENCES

1. GOODMAN, A.M. and ORR, J. 'Material Data for High-Temperature Design', Development in High-Temperature Design Methods, Institution of Mechanical Engineers Conference, London, 1979.

2. BRESSERS, J., FENSKE, E. and DE CAT, R. 'Effects of Test Parameter Variability on Scatter of Stress Rupture Data of Waspaloy'. EUR Report, in press.

3. VAN DER BIEST, O., VON BIRGELEN, G. To be published.

4. BURTON, J.J., BERKOWITZ, B.J., KANE, R.D. 'Sulphur Segregation in an Engineering Alloy: Hastelloy C-276'. Met. Trans. A, 1979, 10 A, 677.

5. HOLT, R.T. and WALLACE, W. 'Impurities and Trace Elements in Nickel-base Superalloys'. Int. Met. Rev. 203, March 1976.

6. SCHULTZ, J.W. Ph.D. thesis University of Michigan, 1965 (University Microfilms Inc. Ann Arbor, Michigan No. 65-11030)

7. DOHERTY, J.E., Kear, B.H., GIOMEI, A.F., STEINKE, C.W. 'The Effect of Surface Chemistry on Grain Boundary Strength' in 'Grain Boundaries in Engineering Materials'. Proceedings 4th Bolton Landing Conference (1974) p. 619.

8. WALSH, J.M., ANDERSON, N.P. 'Characterisation of nickel base Superalloy Fracture Surfaces by Auger Electron Spectroscopy' in 'Superalloys: Metallurgy and Manufacture'. Proc. Third. Int. Conf. 1976, p. 127.

9. BERKOWITZ, B.J. and KANE, R.D. 'The Effect of Impurity Segregation on the Hydrogen Embrittlement of a High Strength Nickel Base Alloy in H_2S Environment'. Corrosion NACE, 1980, 36, 24.

10. BIEBER, C.G., DECKER, R.F. 'The Melting of malleable Nickel and Nickel Alloys'. Trans AIME, 1961, 221, p. 629.

11. LARSON, J.M. Quoted in /5/.

THE EFFECT OF SECONDARY PRECIPITATION ON THE CREEP STRENGTH OF 9Cr1Mo STEEL

K.R. Williams, R.S. Fidler* & M.C. Askins

C.E.G.B., Leatherhead, England

*C.E.G.B., Marchwood,   England

SUMMARY

This paper presents the results of an electron optical examination of laboratory creep tested normalised and tempered 9Cr1Mo steel.

Creep takes place by matrix deformation leading to localised areas of a highly recovered dislocation mesh. Within these locally recovered regions, secondary precipitation of an M_2X dispersion takes place. Nucleation of M_2X relies on a dislocation sweeping mechanism, where the local recovery has allowed dislocation movement.

The onset of secondary precipitation leads to a modest increase in creep strength and perturbed stress rupture curves. Evidently, stress rupture extrapolations for design purposes should be re-examined in the light of these findings.

1.   INTRODUCTION

The creep strength of ferritic steels is extremely dependent on the interaction of the inherent dislocation mesh with second phase particles [1,2]. Because ferritic alloys are not heat treated to give the equilibrium volume fraction of the most stable second phase before creep testing, changes in volume fraction, distribution and structure of the second phase inevitably take place with time at high temperature. This occurs by the additional precipitation from solid solution or by transformation of less stable phases to more stable phases.

When normal coarsening (Ostwald ripening with no dislocation enhanced diffusion or concentration of solute) of a dispersion of particles occurs at small engineering creep stresses, it can be shown [3,4] that the effective stress (σ_E) controlling the creep rate is given by the following relationship:

$$\sigma_E(t) = \sigma - \sigma_o(t) = \frac{\alpha Gb(\beta t+1)^{1/3}}{\ell_d} \qquad \dots \quad (1)$$

where $\sigma_E(t)$ is the current effective stress

        $\sigma_o(t)$ "   friction   "

        σ is the applied stress

        ℓ_d is the initial dislocation density

        β is the ratio of the rate of particle coarsening and initial particle size

        t   is the time (hour)

   α, G, b have their usual meaning

Thus normal microstructural degeneration (i.e. growth of particles and the dislocation mesh size) will lead to a continuously increasing effective stress with a commensurately increasing creep rate resulting in pseudo tertiary creep curves [2,5]. Consequently, stress rupture (rupture stress V^s time) curves show a downward changing slope at low stresses and long creep times as a result of this loss of creep strength. However, Sellars [1] has shown that microstructural degeneration can be followed by secondary precipitation, i.e. the coarsening process is reversed. Under these circumstances, the value of β (eqn. 1) changes abruptly, resulting in a substantial modification of the shapes of the creep and stress rupture curves.

In the present work, detailed microstructural examination has been made of laboratory creep tested 9Cr1Mo steel particularly in the range where perturbations occur in the creep and stress rupture curves. This study demonstrates

    (i)   the importance of secondary precipitation on the creep and stress rupture properties of 9Cr1Mo steel

and (ii)   the difficulties in extrapolation of stress rupture data for design purposes.

2. EXPERIMENTAL

The material was received as 25mm bar in the normalised and tempered condition. The composition is given in Table 1. Creep specimen blanks were rough machined from the bar, encapsulated in evacuated silica tubes and heat treated for one hour at 1000°C, AC; 750°C for two hours, AC. The creep blanks were then finally machined to one of two gauge geometries; either 6.4mm diameter with a 63.5mm gauge length or 9.1mm diameter and 79.4mm gauge length. Creep tests were carried out under constant load conditions at temperatures of 475°C, 500°C, 525°C and 550°C. For the present work, several samples suitable for

electron optical examination were taken from the as-received
N+T and the creep tested material up to 30,000 hours duration.

TABLE 1: Composition of Material wt.%

Cr	Mo	C	Si	Mn	S	P	Ni	Cu
8.6	1.04	0.12	0.69	0.48	0.008	0.022	0.22	<0.05

Normal electropolishing techniques developed for $\frac{1}{2}Cr\frac{1}{2}Mo\frac{1}{4}V$
and $2\frac{1}{4}Cr1Mo$ steels [2] were employed, enabling high resolution
HVEM and STEM examination. Furthermore, extraction carbon replication techniques were used in order to examine secondary
precipitation by the micro micro diffraction and EDAX facilities available on the STEM, with a 50Å spot size. (EDAX and
micro micro diffraction analysis was also carried out on both
grain boundary, lathe and interlathe carbides.)

3. RESULTS

The creep testing results have been previously quoted in
Fidler's [6] original work. However, in the interest of com-
pleteness, some of the more relevant results are repeated.

3.1 Stress-rupture Data

The stress-rupture data obtained from the studies of stress
and temperature effects on creep are presented in figure 1, together with the estimated ISO values for the rupture life for
N & T 9Cr1Mo in the temperature range 475→550°C (ISO 1971).
The ISO line appearing in figure 1 for 525°C, was obtained by
interpolating the ISO data which is quoted at ten degree intervals. It can be seen, that the composition and heat treatment
combination for the material used in this study, provides
appreciably greater rupture strength than is predicted from the
ISO data.

The form of the stress rupture curves at the temperature
investigated should be noted. At 475°C the curve is concave
downwards, distinct inflexion points are obtained at 500°C
and 525°C, whereas at 550°C the curve is concave upwards.

3.2 Strain Data

It has been shown previously [6], that the isochronal
strain data fit a relationship of the form

$$\log_{10} \varepsilon_t = \alpha_o + \alpha_1 \log_{10} \sigma \qquad \qquad \cdots \ (2)$$

where the constants α_o and α_1 vary with temperature and time.
The $\log \varepsilon_t$ vs. $\log \sigma$ lines generally diverge at 500, 525 and
550°C with increasing strain and stress, but the opposite behaviour occurs at 475°C up to 1000 hours testing, (see figure
2). Divergence is due to the form of the creep curve often

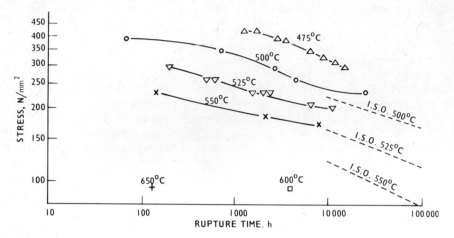

Figure 1: Rupture Durations for normalised and Tempered 9Cr1Mo compared with I.S.O. Data in the Temperature Range 475–550°C

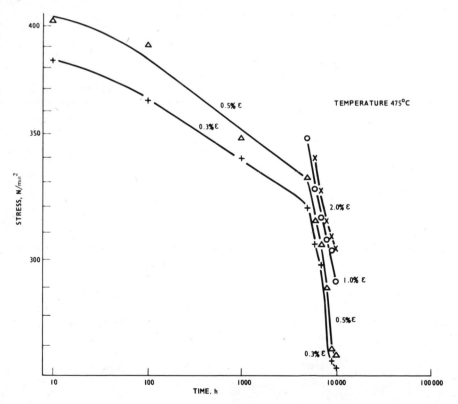

Figure 2: Stress-Time Relationships for strains up to 2% at 475°C

obtained with this material, in which no period of steady state rate of elongation occurs. Instead, after the initial primary extension, the creep rate continuously increases to failure. This can be attributed to recovery processes increasing at a greater rate than the strain hardening, i.e. microstructural degeneration is occurring such that the dislocation structure is not stabilized. At 475°C the divergence is not apparent until after 1000 hours testing. At this temperature the strain-time curves generally show a period of steady state creep, more extensive than at the higher temperatures, prior to rapid elong-ation to failure. Hence it appears that after 1000 hours at 475°C, degeneration of the structure becomes important and the isochronal log σ - log ε lines diverge as at the higher temp-eratures.

However, several creep strain-time curves have been obtained which show perturbations, a marked example of which is shown in figure 3. In this case a steady state creep stage was established at low strain, followed by a period of accelerating creep and then by a further period of steady state creep. Other cases, where the perturbation occurs soon after primary, and at higher temperatures, have been obtained, though no cases at 550°C.

Perturbations are also apparent in the stress rupture curves at 500 and 525°C, where inflexions occur, and at 550°C where the plot curves upwards slightly, with time. It would appear therefore, that the dislocation mesh and particle distribution are interacting in a complex manner over the creep stress and temperature range investigated. Certainly the perturbations appearing in the stress-rupture curves make it difficult to predict long term creep strain and rupture behaviour from short term tests (up to 2×10^4 hours).

3.3 Microstructural Analysis

The microstructural analysis concentrated mainly on the creep tests carried out at 500°C, since the stress-rupture and creep curves show very clear perturbations in this range. Some specimens tested at 475°C and 550°C were also examined.

The as-received N & T structure is shown in figure 4 ob-served by conventional TEM of the three-dimensional structure. Dislocation densities were measured by the methods outlined by Hirsch et al [7] and are presented in Table 2.

Examination of several other creep tested specimens were made above and below the inflexion point B shown on the stress rupture plot, figure 1.

The microstructure of the gauge region of creep specimens above the inflexion (2000 hours at 500°C) shows that normal re-covery creep events have commenced locally in the structure figure 5. The recovery processes have led to a local reduction

480

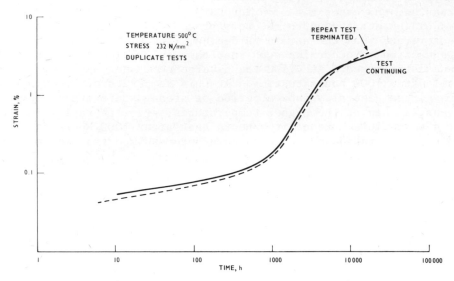

Figure 3: Multistage secondary creep observed in Normalised and Tempered 9Cr1Mo

Figure 4: Transmission Electron Micrograph of a typical As received N&T structure Mag X10,000

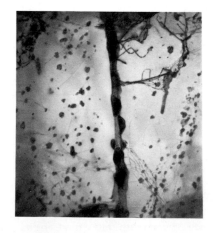

Figure 5: Local Dislocation Recovery (gauge length) after 2,000 hours at 500°C Mag 30,000

in the intrinsic dislocation density, together with the onset of subcell formation and some regions of secondary precipitate. Below the inflexion point, (3×10^4 hours at 500°C) recovery is again inhomogeneous, but additionally in those areas where recovery has occurred, significant secondary precipitation on or near dislocations has taken place, figure 6. The secondary precipitate is up to 500Å diameter and precipitates separately or as a stack-like sequence of thin plates either on or near dislocations (see later discussion).

The analysis of these secondary precipitate plates in thin foils by SAD and EDAX proved impossible because,

    (a)   the conventional large diameter electron beam spot resulted in far too high a matrix X-ray count rate, effectively swamping the precipitate count rate

and (b)   due to the platelike morphology and small volume fraction of secondary precipitate, no precipitate diffraction spots appeared.

Accordingly, extraction carbon replica's were prepared from samples showing secondary precipitation. Micro-diffraction and microanalysis by STEM indicates that the secondary precipitates are of the M_2X type with high chromium content, together with some substitutional iron and molybdenum content.

Figure 7 shows the structure of the grip region of the 3×10^4 hour at 500°C specimen. No secondary precipitate similar to that found in the gauge is visible, and hence the precipitation in the gauge occurs or is enhanced by the presence of creep strain.

Examination of a 1.5×10^4 hour test at 475°C, 302MPa shows a fine general precipitate of a similar, or higher density, than that in the 500°C, 3×10^4 hour specimen's gauge (figure 8). However, the structure of a 550°C, 132MPa, 10^4 hour specimen (figure 9) shows little or no such precipitate, and is very similar to the grip region of the 500°C, 3×10^4 hour specimen.

4.   DISCUSSION

This work confirms the earlier work on 9Cr1Mo steels [6], where it was shown that creep occurs almost exclusively by grain deformation processes, with no evidence of grain boundary sliding or cavitation in the temperature range $475 \rightarrow 550^\circ$C. The present work does however indicate that the deformation mechanism is by no means homogeneous. Only favoured areas in certain grains appear to have undergone recovery creep. This observation is consistent with recent work [3,8] where it was shown that recovery in $\frac{1}{2}$Cr$\frac{1}{2}$Mo$\frac{1}{4}$V steels occurs inhomogeneously during creep, particularly at small engineering stresses. Thus at any time, only small localised volumes of the specimen/component undergo recovery controlled deformation. As recovery

Figure 6: Secondary Precipitation in the Recovered Regions at 30,000 hours duration.
Mag. 10,000

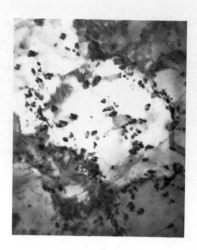

Figure 7: Same as figure 6, but micrograph taken from a sample from the grip region of the specimen.
Mag. 20,000

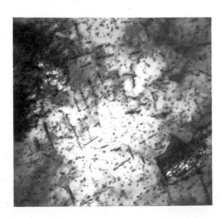

Figure 8: Showing secondary pre-cipitation at 475°C at 302MPa
Mag. 20,000

Figure 9: Typical micro-structure of material creep tested at 550°C and 132MPa
Mag. 20,000

proceeds in these areas, the dislocation density falls from its as-received value (Table 2), to that characteristic of recovery creep at the operating laboratory or service creep stress. Typically for $\frac{1}{2}Cr\frac{1}{2}Mo\frac{1}{4}V$ in the normalised and tempered condition, and operating at 40MPa and 565°C, the time to the onset of homogeneous recovery creep can be substantial ($\simeq$ 100,000 hours). However, 9CrlMo steels have a larger primary precipitate than $\frac{1}{2}Cr\frac{1}{2}Mo\frac{1}{4}V$, with correspondingly much greater interparticle spacings. Accordingly, the primary precipitate dispersion in 9CrlMo, is much less effective in stabilising the inherent dislocation mesh [4]. We therefore expect the inherent dislocation mesh in 9CrlMo steels, to follow the Friedel [9] coarsening kinetics.

Using these ideas, it is possible to explain qualitatively the observed creep behaviour. The dependence of dislocation density (ρ) on applied stress in a homogeneously deforming body is given as

$$\sigma = \sigma_o + \alpha Gb\sqrt{\rho} \qquad \qquad \ldots \quad (3)$$

where the parameters have the same meaning as described in equation 1.

Thus if the applied stress, is capable of generating a dislocation density greater than the inherent normalised and tempered density during the initial plastic extension, then there is every possibility of the material immediately undergoing recovery creep. Under these circumstances the transient creep behaviour represents a period where the material settles down to a condition where $\partial\sigma/\partial t$ (recovery rate) $\propto \partial\sigma/\partial\varepsilon$ (work hardening rate) and secondary creep commences. It is also clear, [10], that given these conditions, transient creep become rather stress sensitive with transient creep strains increasing with increasing stress. However, when the applied creep/service stress is too low for the nucleation of new dislocations, homogeneous recovery creep is only possible following a period of microstructural degeneration [4]. During this period of degeneration, little strain accumulates. Table 2 shows that when localised recovery creep commences, the dislocation density has fallen from its as received value to $1.4\times10^9 cm/cm^3$ at 232MPa and $2.8\times10^9 cm/cm^3$ at 270MPa. Using equation 3, the σ_o contribution would have to be in the region of 150MPa thus enabling us to construct Table 3. The σ_o contribution would be expected to be slightly higher at 475°C.

At the generally high stresses employed in deforming the material at 475°C, recovery creep normally commences immediately on application of the applied stress, leading to stress dependent transient strains above 330MPa. Examination of primary creep strains suggests that below stresses of the order 330MPa, a period of structural degeneration is necessary prior to the onset of recovery creep. This is confirmed by examination of Table 3 where it is clear that applied stresses

Table 2

Material Condition	Creep Duration (hours)	Dislocation Density ρ (cm/cm^3)
As-received	0	7.3 x 10^9
Creep at 270MPa and 500°C	10 000h	2.8 x 10^9
Creep at 232MPa and 500°C	30 000h	1.4 x 10^9

Dislocation density required for recovery
creep of 9Cr1Mo between 450 and 700°C

Table 3

Effective Stress $(\sigma-\sigma_o)$ MPa	Applied Stress MPa	Dislocation Density		Dislocation Mesh Spacing x (cm)
80	230	1.5cm/cm^3	x 10^9	2.6 x 10^{-5}
100	250	2.5	x 10^9	2 x 10^{-5}
150	300	5.7	x 10^9	1.35 x 10^{-5}
200	350	7.6	x 10^9	10^{-5}
250	400	1.7	x 10^{10}	0.8 x 10^{-5}
300	450	3.1	x 10^{10}	0.57 x 10^{-5}

>330MPa are capable of generating dislocation densities greater
than the as received dislocation density, hence immediate recovery creep, while at stresses <330MPa, a period of structural
degeneration is required prior to the onset of equilibrium recovery creep. Under these circumstances, it is to be expected
that the stress rupture plotat 475°C, shows the usual downward
changing slope characteristic of particle hardened materials
undergoing structural degeneration [2]. In the present work,
these normal stress rupture events are clearly perturbed at
temperatures above 475°C, as shown in the stress rupture curves
(figure 1) and creep curves (figure 3).

When secondary precipitation is a possibility, [1] the
normal downward changing stress rupture curves characteristic
of a degenerating particle hardened material are disrupted.
The onset of secondary precipitation alters the normal coarsening kinetics and hence the effective stress described in
equation 1 decreases due to changes in β. Secondary precipi-

tation is clearly evident in many of the microstructures examined, figure 6. It is suggested therefore, that secondary precipitation of M_2X on or near the mobile dislocations created by localised recovery, leads directly to depressed creep rates (figure 3). This creep strengthening effect naturally results in extended creep lives, hence the sigmoidal shaped stress rupture curves, figure 1. Figure 1 additionally indicates the range over which secondary precipitation is effective. At $475^{\circ}C$, the stress rupture behaviour is controlled by recovery processes above 330MPa and by recovery and structural degeneration processes at the lower stresses. At the high stresses used in the $475^{\circ}C$ tests the fine precipitate observed,(probably formed due to the supersaturation of solute at $475^{\circ}C$ after tempering at $750^{\circ}C$,)is presumably not sufficiently finely dispersed to stabilize the dislocation mesh, or noticeably strengthen the specimens.

At intermediate temperatures, we see the change over from recovery creep at high stresses controlled by dislocation degeneration leading to downward changing slopes, to the dislocation locking by secondary precipitation of mobile dislocations created by recovery events, leading to upward changing stress rupture slopes at low stress and long times. At $550^{\circ}C$ the upward changing stress rupture slope suggests some effect of secondary precipitation, although none has been observed (figure 9). Hence it is likely that there is some effect of solute on the dislocation motion, i.e. at $475^{\circ}C$ general precipitation removes solute from the matrix allowing recovery processes to proceed; at $500^{\circ}C$ and $525^{\circ}C$ solute atmospheres gradually build up at the mobile dislocations until precipitation occurs, whilst at $550^{\circ}C$, the supersaturation is too low for precipitation to occur in the atmospheres.

Secondary precipitation in ferritic steels appears to be a rather general phenomenon [1] and depends critically on the stability of the primary carbides and deformation rate. Complex reactions take place during creep between moving dislocations and alloy carbides [1]. It has been suggested [11], that atmosphere formation could cause 'sweeping' of solutes by moving dislocations, thus enabling a marked acceleration in coarsening. Eventually this 'sweeping' process may lead to the direct nucleation of new precipitates on mobile dislocations. Alternatively, dislocation annihilation during recovery creep would result in local regions of high solute concentration, where nucleation and growth of a new precipitate occurs. Evidently the secondary precipitation nucleation sites are ideally positioned for the further blocking of dislocation recovery processes.

The source of substitutional and interstitial elements necessary for secondary precipitation may be either that already in the matrix, or result from the gradual transformation of primary carbides. Obviously, these processes are time and temperature dependent and thus the observation that secondary precipitation gradually becomes more evident as time increases towards the longest test at 500°C (30,000 hours) is reasonable. The secondary carbide distribution is inhomogeneous at 500°C because it can only occur where recovery and hence a source of mobile dislocations enables a sweeping mechanism to operate. Dislocation locking is therefore provided in those areas undergoing deformation. As time increases, (i.e. low creep stress), progressively more of the structure undergoes recovery creep, with the result that dislocation locking becomes more effective, hence the upward changing stress rupture curves.

Studies have been made [12,13] which suggest that improved creep strengths could be achieved by reducing the tempering temperature below 700°C, to provide fine Cr_2C precipitate within the grains. It appears from the present work, that a fine M_2X type precipitate nucleates quite naturally at the lower temperatures ($\simeq 500°C$) and long times during deformation. Of course, to achieve this enhanced creep strength requires some time and deformation, which may be avoided by use of the lower tempering temperatures. M_2X type precipitates are relatively unstable and will transform to more stable, massive carbides at long times and high temperatures. Under these circumstances, the strengthening effect of M_2X formation is expected to be only a transient phenomenon. However, CFR operating temperatures are rather low ($\simeq 520°C$) and the strengthening effect of the M_2X precipitate may persist for a significant part of the service life.

5. CONCLUSIONS

   1. Secondary precipitation of M_2X occurs at long times during creep of the N & T 9Cr1Mo investigated.

   2. The secondary precipitate is confined to localised areas which have undergone recovery during the creep process. The nucleation and subsequent growth of M_2X is dependent on a dislocation 'sweeping' process.

   3. The onset of secondary precipitation leads to modest increases in creep and rupture strengths. Normal structural degeneration processes are perturbed under these circumstances giving rise to sigmoidal shaped creep and stress rupture curves.

6. ACKNOWLEDGEMENTS

This paper is published by permission of the Central Electricity Generating Board.

REFERENCES

1. SELLARS, C.M., - 'Structural Stability During High Temperature Creep'. Creep Strength in Steel and High Temperature Alloys. The Metals Society, University of Sheffield, 1972, p. 20.
2. WILLIAMS, K.R. and WILSHIRE, B. Material Science and Engineering, 1977, 28, 29.
3. WILLIAMS, K.R., CEGB, Report No. SSD/MID/R3/75, 1975.
4. WILLIAMS, K.R. and CANE, B.J. ibid, 1979, 38, 199.
5. SMITH, P. and WILLIAMS, K.R. CEGB Report No. SSD/MID/R75, 1975.
6. FIDLER, R.S. 'Tempering and Stress Relief Heat Treatments in 9Cr1Mo Steel', CERL Note No. RD/L/N9/88, 1977.
7. HIRSCH, P.B., HOWIE, A., PASHLEY, D.N., NICHOLSON, R.B. and WHELAN, M.J. 'Electron Microscopy of Thin Crystals', Butterworths, London, 1965.
8. WILLIAMS, K.R. and WILSHIRE, B. Material Science and Engineering, 1980, to appear.
9. FRIEDEL, J., 'Les Dislocations' Gauthier-Villars, Paris, 1956.
10. EVANS, W.J. and WILSHIRE, B., Trans. Met. Soc., AIME., 1968, 242, 2514.
11. MUKHERJEE, T. and SELLARS, C.M. Met. Trans., 1972, 3, 953.
12. DAY, R.V. and BARFORD, J. 'Structural Examination of a 9Cr1Mo Steel'. CERL Report No. RD/L/R1379, 1966.
13. FIDLER, R.S., 'The Creep of Normalized and Tempered 9Cr1Mo', CERL Report No. RD/L/R1949, 1976.

EXAMINATION OF THE CREEP BEHAVIOUR OF MICROSTRUCTURALLY
UNSTABLE FERRITIC STEELS

K.R. Williams

Central Electricity Research Laboratories, Kelvin Avenue,

Leatherhead, Surrey.

SUMMARY

The inherent microstructural instability of $\frac{1}{2}Cr\frac{1}{2}Mo\frac{1}{4}V; 2\frac{1}{4}Cr1Mo$
and carbon steels creep tested or service exposed at low stresses
is demonstrated. Measurements of important dispersion parameters have been made during creep life and have been found to
follow normal coarsening kinetics. Using the measured time dependent change of the dispersion parameters, a dislocation
source controlled model for recovery creep is used and further
developed.

The model allows the calculation of the Manson-Haferd plot
of log (time to failure) against temperature for unstable steels.
In addition, a classification of material stability is proposed,
based on the ratio of time to fracture, t_f, and time to tertiary
creep, t_t. This classification enables estimates of remaining
creep life to be based either on well established criteria for
stable materials or modifications of these criteria for unstable steels.

1. INTRODUCTION

Currently, several laboratories are engaged in collecting
and collating creep strength and creep rupture data in order to
extend the range of our stress rupture knowledge for design
purposes. This work involves an enormous number of machines
and man hours and will never enable sufficient data to be collected at engineering stresses and temperatures to obviate the
use of extrapolation procedures.

There is therefore, a clear incentive to understand the
mechanisms of creep and creep rupture at long times (>100,000
hours) both as an aid to high temperature design and for remaining creep life estimation on plant nearing its design life
or plant which has operated outside design considerations.

In recent years, microstructural information has become available in $\frac{1}{2}Cr\frac{1}{2}Mo\frac{1}{4}V$ [1,2], Plain carbon steel [3] and $2\frac{1}{4}Cr1Mo$ steels [3,4] both from long term laboratory creep/ageing tests and from plant after extended service. In all cases, these investigations have shown a marked change in the initial particle distribution, leading to an increasing particle spacing with time [1,2,3,4]. Inevitably these microstructural changes lead to loss of creep strength with time giving rise to the characteristic downward changing slope of the stress rupture plot. This loss of creep strength is reflected in the low values of stress dependence (n) of creep, to values well below 4, [1]. The low n values (<4.0) recorded at very low stress levels have, in general, been accounted for in terms of processes such as solute-drag [5], grain boundary sliding [6,7] etc. being rate controlling under these conditions. Alternatively, one can account for these changing, n, values on the basis of a 'friction stress' (σ_0) for creep [1], such that creep does not take place under the full effect of the applied stress (σ), but only under a stress ($\sigma-\sigma_0$) such that

$$\dot{\varepsilon} = A^* \, (\sigma-\sigma_0)^p \, f \, (T) \qquad \qquad \ldots \quad (1)$$

where A^* and p are constant, with $p \simeq 4.0$.

The friction stress (σ_0) has been identified with the microstructural state of the material [1]. However, when the microstructure changes during service, we expect a corresponding change in the current level of σ_0. Recently [8,9] it has been shown how a time dependent σ_0 can be incorporated into the basic recovery creep model described by equation 1.

In this paper, a brief review will be made of the important microstructural changes in low alloy steels, and the dispersion parameters associated with these changes presented. We then make use of this recently acquired microstructural information in deriving a dislocation propagation model for creep. This model is then used to calculate the Manson-Haferd (M-H) parameter for the steels under examination. In addition, a classification of steels into stable or unstable types is suggested over wide stress and temperature ranges.

2.   MICROSTRUCTURAL CHANGES IN $\frac{1}{2}Cr\frac{1}{2}Mo\frac{1}{4}V$ AND $2\frac{1}{4}Cr1Mo$ STEELS DURING LOW STRESS CREEP

The initial creep strength of $\frac{1}{2}Cr\frac{1}{2}Mo\frac{1}{4}V$ steels of ostensibly the same chemical composition is highly dependent on the type of heat treatment. Thus steels austenitised and tempered at similar temperatures for similar times can show markedly different properties, depending on their cooling rate from the austenitising temperature, [1]. Honeycombe,[10] and co-workers have shown clearly the complex nature of the transformation products resulting from changing precipitation kinetics during cooling. Each transformation product is expected to exhibit a different creep response when loaded to low stresses at high temperatures. Normally components are cooled at varying rates, and the transformation products can range from interphase VC precipitation, figure 1, to dislocation VC precipitation figure 2. Creep and hardness tests on specimens with predominantly one or other of these products, indicate the interphase VC with the highest strength, figure 3. This is an important point and will enable the microstructural placement of $\frac{1}{2}Cr\frac{1}{2}Mo\frac{1}{4}V$ steel in the stress rupture band. During long term service, the carbides coarsen gradually in a manner such as to produce a more uniform structure [2,4], figure 4.

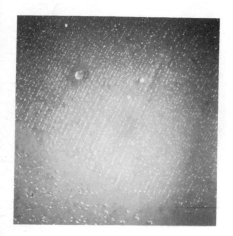

Figure 1. Interphase VC precipitate X15,000

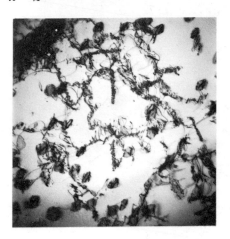

Figure 2. Dislocation VC precipitation X20,000

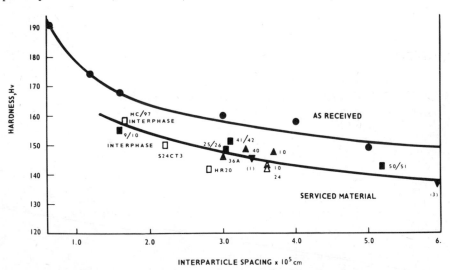

Figure 3. Hardness Vs interparticle spacing (As Received and Service Exposed $\frac{1}{2}Cr\frac{1}{2}Mo\frac{1}{4}V$)

R

In addition to the general increase in size of the
vanadium carbide particles, H-type precipitates became increasingly evident, figure 5. This type of precipitate has
been reported previously [11,12], and its growth is probably
due to the dislocation pipe diffusion of molybdenum to VC
particles attached to the dislocation mesh [10]. All the H-
type precipitates found in this work have indeed been
located on the inherent mesh. Ultimately the central VC core
of the H-type precipitate disappears leaving rods of Mo_2C
[13,14].

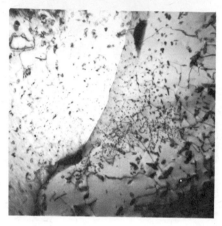 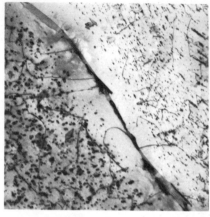

Figure 4. Coarsened carbides
100,000 hours exposure at
40MPa at 571°C Mag X10,000

Figure 5. Formation of PFZ's
and H type precipitate
Mag X15,000

As changes in the size, spacing and form of the carbide
particles occurred, the density of dislocations evident with
the grains changed very gradually, figure 6. The individual
dislocations within the grains were almost invariably found to
be held up at carbides suggesting a gradual loss of creep resistance as particle coarsening takes place. Moreover, growth
of grain boundary carbides leads to the progressive formation
of particle free zones (PFZ) adjacent to boundaries, figure 5.

The high dislocation density generally observed within
these zones demonstrates that creep of $\frac{1}{2}Cr\frac{1}{2}Mo\frac{1}{4}V$ steels at low
stress levels occurs in a highly inhomogeneous manner. This
inhomogeneity of deformation leads to the development of
poorly formed subgrain boundaries in limited regions within
the ferrite grains, after extended periods of service
($\approx$100,000 hours).

Although the transformation kinetics do not generally
allow the formation interphase structures in $2\frac{1}{4}Cr1Mo$, a
general coarsening of the inherent Mo_2C carbide takes place
at long times and low stress in the temperature range
550°C → 680°C, [4].

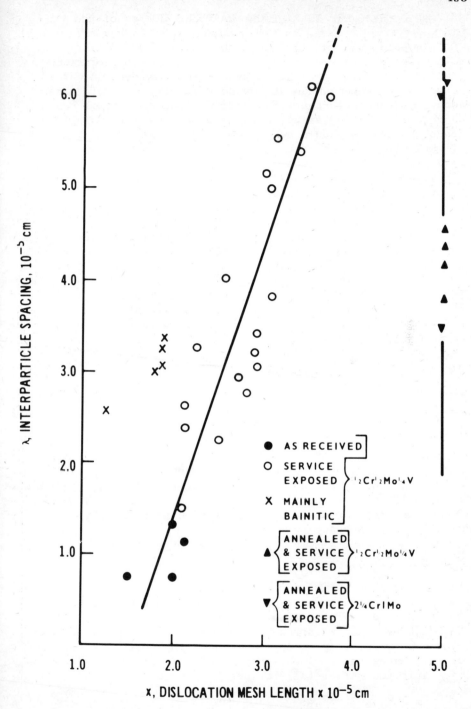

Figure 6. Gradual change of dislocation mesh length with carbide spacing during initial coarsening

3.   RECOVERY CREEP IN PARTICLE HARDENED FERRITIC STEELS

In earlier work [1] it was shown that immediately following small stress reductions during creep of ferritic steels, incubation periods of zero creep rate were recorded before creep recommenced at the reduced stress level. This behaviour indicates that dislocation glide mechanisms cannot be rate controlling since, with these processes, a small decrease in stress should be followed immediately by a new slightly lower creep rate and not by an incubation period of zero creep. Instead, the occurrence of incubation periods suggests that the flow stress of the material must decrease by recovery before creep can begin again after the stress reduction, i.e. creep of ferritic steels is recovery controlled.

The previous observations [1] were interpretable in terms of a recovery model for creep [15], in which diffusion controlled growth of the network leads to the development of link lengths sufficiently long to act as dislocation sources allowing slip to occur (i.e. source controlled creep, SCC,).   In this way a balance between recovery and strain hardening processes leads to the gradual formation of a roughly Gaussion distribution of link lengths in the network during creep [16]. The occurrence of incubation periods would then be anticipated since, immediately after a small stress reduction, even the longest link present would be too small to act as a source, so that creep cannot recommence until the network size grows by recovery.

Essentially, the present results lead to much the same conclusion as our earlier work [1], but with the following new considerations. Due to the high inherent dislocation density in normalised and tempered ferritic steels, application of a small engineering stress (30→50MPa) will not allow the immediate propagation of the dislocation network [9], i.e. SCC cannot occur because

$$\sigma < \frac{\alpha G b}{x} \qquad \qquad \dots \quad (2)$$

where σ is the applied stress
α, G & b have their usual meaning and x is the dislocation mesh spacing.

At relatively higher stresses where $\sigma > \alpha G b / x$ normal SCC can commence immediately on application of the load. Under such circumstances the dislocation-particle interactions give rise to a friction stress (σ_0) for creep such that

$$\sigma_E = \sigma - \sigma_o = \alpha G b \sqrt{\rho} \qquad \qquad \dots \quad (3)$$

where ρ is the dislocation density.

Equation 3 is used directly in rationalising the creep behaviour of commercial steels resulting in our earlier equation 1.

However, as a result of material degeneration with time,

the gradually increasing interparticle spacing allows a commensurate increase in the dislocation mesh length. Eventually, the coarsening can proceed to a stage where $\sigma = \alpha Gb/x_c$ where x_c is the current mesh length, at which point SCC begins. If the dislocation mesh is strongly stabilised by the carbide dispersion, then it is easily possible to show

$$x/\lambda = \text{constant} \qquad \qquad \ldots \ (4)$$

where λ is the interparticle spacing.
Thus, the dislocation mesh length changes at exactly the same rate as the interparticle spacing, (see figure 6), with the growth of the carbides rate controlling [9].

When particles coarsen according to $t^{1/3}$ kinetics [3] it can be shown [8,9],

$$\sigma_E(t) = \frac{\alpha Gb}{x} (\beta t + 1)^{1/3} \qquad \qquad \ldots \ (5)$$

where $\sigma_E(t)$ now represents the time dependence of σ_E

   x is the dislocation mesh length
and β describes the carbide coarsening behaviour.
Equation 5 shows that the effective stress gradually increases with time, and hence the creep rate. We therefore expect the creep curves of unstable materials at low stresses to exhibit almost total tertiary creep. This is indeed the case [1,17] in several commercial steels at engineering stresses.

4.   THE EFFECTS OF INSTABILITY ON TERTIARY BEHAVIOUR

   As the applied stress falls, the proportion of the creep life spent in tertiary increases [1,17]. This observation, together with the reasoning outlined in section 3, provides additional information on the degree of instability of materials For a number of pure metals (i.e. stable materials), both the time to tertiary t_t and the time to fracture t_f, have been shown to be inversely proportional to the secondary creep rate [18] as

$$\dot{\varepsilon}_s = \frac{E'}{t_t} = \frac{E}{t_f} \qquad \qquad \ldots \ (6)$$

where E' and E are constant over a wide range of stresses and temperatures so that $t_f/t_t = k$ (constant). The relationship between t_f and t_t for pure metals has been shown previously [1], giving a k value of 1.5 independent of stress and temperature over the range examined. Similar t_f/t_t ratios can be derived for other materials which are structurally stable during creep. Conversely the results for the ferritic steels included in figure 7 show a marked increase in the t_f/t_t ratio for progressively lower applied stress levels reflecting structural instability during creep. Indeed, the t_f/t_t values offer a means of classifying ferritic steels into stable or unstable categories dependent on the applied stress and temperature.

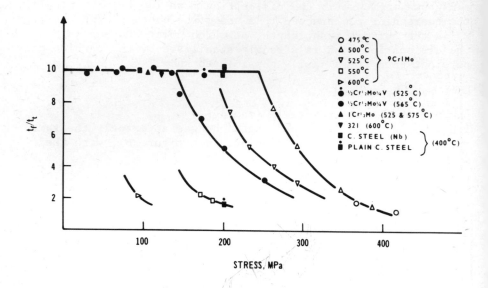

Figure 7. Microstructural stability of several steels with creep stress and temperature

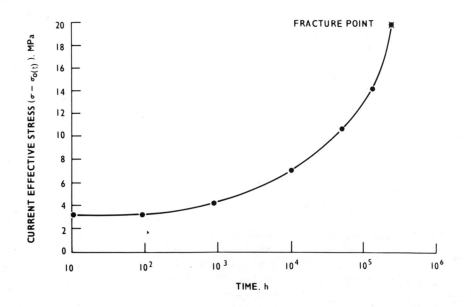

Figure 8. Current effective stress $(\sigma - \sigma_0(t))$ vs time to rupture

In figure 7, an arbitrary maximum value of $t_f/t_t = 10$ has been taken, but this value will depend on the sensitivity and resolution of the strain measuring equipment. Values of t_f/t_f near 10, are therefore classified as 100% unstable materials and are expected to exhibit a creep curve comprising essentially of tertiary behaviour. Under these circumstances equation 5 is expected to apply. At relatively higher stresses/temperatures where $t_f/t_t \rightarrow 1.5$ the strain time behaviour can be described by the well developed Garafalo equation and its modifications [19] such as

$$\varepsilon = \varepsilon_o + \varepsilon_T(1-e^{-mt}) + \dot{\varepsilon}_s t + Ze^{p(t-t_t)} \qquad \cdots \quad (7)$$

where ε_o is the instantaneous strain on loading
ε_T the total transient strain, m a constant relating to the rate of exhaustion of transient creep. Z and p are constants.

Further important indicators of materials behaviour can be obtained from the t_f/t_t relationship. Referring to figure 7, it is evident that plain carbon steels behave as stable materials at $400^{\circ}C$ and 200MPa and above giving rise to classic creep curves with $t_f/t_t = 1.5$. Under these conditions SCC commences immediately on application of the load. However, small additions of niobium (deliberate or otherwise) result in dispersion hardening behaviour with corresponding unstable characteristics at 200MPa and $400^{\circ}C$ (i.e. $t_f/t_t \rightarrow 10$). We therefore have a typical example of the behaviour described in section 3, with SCC inhibited by the dispersion of NbC. It is therefore important that material behaviour should be understood before creep life calculations are made.

5. CALCULATION OF THE MANSON-HAFERD PARAMETER FROM MICROSTRUCTURAL INFORMATION

The basis of predicting rupture data relies on calculation of the rate of change of $\sigma_E(t)$ to a value where SCC can occur almost independently of the remaining carbide dispersion. Physically, the rupture time can be identified with a carbide size (or interparticle spacing) where the dislocation mesh is no longer strongly stabilised. When the carbides have increased to size such that the commensurate growth of the dislocation mesh is x_c where

$$\sigma - \sigma_o = \frac{\alpha Gb}{x_c} \qquad \cdots \quad (8)$$

then significant dislocation propagation will begin resulting in creep rupture.

Because of the wide distribution of dislocation mesh lengths within the matrix, the onset of SCC will be a gradual process as more dislocation meshes exceed the critical value x_c. Equation 5 describes the kinetics of this process, while figure 8 shows $\sigma_E(t)$ plotted against time for a $\frac{1}{2}Cr\frac{1}{2}Mo\frac{1}{4}V$ from measured values of β and x_c at 40MPa and 570°C. This figure shows typical pseudo tertiary form, suggesting the physical description of creep to be reasonable.

The critical carbide size for creep failure can either be calculated using equation 5 and allowing $\sigma_E(t_f) = \frac{1}{2}\sigma \to \sigma$ or alternatively measurements can be made on failed specimens conducted at the appropriate stress and elevated temperatures; always ensuring $t_f/t_t \to 10$ for these tests.

It has already been shown [1,2,3,4] that metastable carbides obey coarsening kinetics of the form

$$\delta^n - \delta_o^{\ n} = Kt \qquad \qquad \ldots \ (9)$$

where n = 2,3 or 5
Metastable carbides of the type M_3C, VC and M_2C coarsening in the range 500→700°C obey the $t^{1/3}$ kinetics [3,4,8]. Assuming the final carbide size is δ_f, then

$$\delta_f^{\ 3} = Kt - \delta_i^{\ 3} \qquad \qquad \ldots \ (10a)$$

where K = K*D
D is the diffusion rate of the diffusing species (i.e. V or Mo in our case) and K* involves geometry and frequency factors, while δ_i is the initial carbide size.

Normally $\delta_i^{\ 3} \ll \delta_f^{\ 3}$; and equation 10 becomes

$$\delta_f^{\ 3} = Kt_f \qquad \qquad \ldots \ (10b)$$

K is the slope of (carbide size)^3 against time plot, and has been measured for VC coarsening in $\frac{1}{2}$Cr$\frac{1}{2}$Mo$\frac{1}{4}$V in the temperature range 565°C→710°C, for M_3C in plain carbon steels [3] and Mo_2C in 2$\frac{1}{4}$Cr1Mo [3,4]. Necessarily, no failed specimens were available at the lower temperatures at the service stresses employed (35→40MPa), however coarsening data to 100,000 hours was available [2,3,4] allowing the temperature dependence of K to be measured as shown in figure 9. Measured values of δ_f have been obtained from several creep tests conducted at 40MPa on $\frac{1}{2}$Cr$\frac{1}{2}$Mo$\frac{1}{4}$V in the temperature range 600°C→710°C. The average value was 3300Å. Similarly, δ_f for Mo_2C in 2$\frac{1}{4}$Cr1Mo between 550°C and 680°C at 35MPa was 3500Å [4]. Substituting these values into equation (10b) with the appropriate K value we arrive at the basic Manson-Haferd lines shown in figure 10. These agree very favourably with the measured lines. The similarity in the slopes of the M-H lines are expected if the mechanism discussed prevails. For example, at a fixed stress and temperature the stability of the dislocation mesh will depend only on the volume fraction, size and spacing and coarsening rate of the dispersed carbides. These values are similar for VC and Mo_2C in $\frac{1}{2}$Cr$\frac{1}{2}$Mo$\frac{1}{4}$V and 2$\frac{1}{4}$Cr1Mo, fig. 10, under low stress creep conditions.

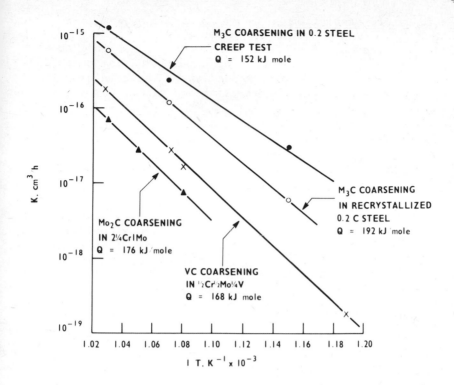

Figure 9. Temperature dependance of the coarsening rates of VC; Mo_2C and M_3C in low alloy steels

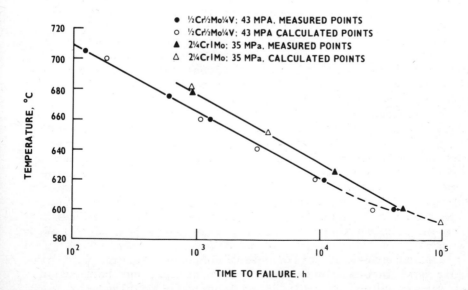

Figure 10. Calculated Manson Haferd curves for $\frac{1}{2}Cr\frac{1}{2}Mo\frac{1}{4}V$ and $2\frac{1}{4}Cr1Mo$ steels

5.1 Creep Life Prediction

The approach to creep failure prediction described above, also enables estimation of remaining creep life from components already exposed at engineering conditions. The remaining creep life (t_{fE}) is readily calculated using equation (10b) as,

$$t_{fE} = \frac{\delta_f^3}{K_T} - \frac{\delta_c^3}{K_T} \qquad \cdots \quad (11)$$

where δ_c is the current particle size, and K_T the appropriate carbide coarsening constant at temperature T.

Because generating plant tends to run at variable temperatures, the measured carbide size (δ_c) will obviously indicate an effective temperature, rather than the design temperature. This effective temperature can be measured by microstructural assessment of the carbide size together with figure 9. The remaining creep life estimate can then be made either
  (a) on the basis of the measured effective temperature
or (b) based on the design temperature, if plant modifications enable subsequent operation to be confined to this value, i.e.

$$t_{fE} = \frac{\delta_f^3}{K_{TD}} - \frac{\delta_c^3}{K_T}$$

where δ_c^3/K_T represents the damage accumulated during variable temperature running, and δ_f^3/K_{TD} represents the damage accumulated under design temperatures to the failure life.

A further important modification of the M-H plot can be made by incorporating ISO-strain lines for constant structures. Such a plot has been produced experimentally for $2\frac{1}{4}$CrlMo steel at 35MPa [4]. By reducing $\frac{1}{2}$Cr$\frac{1}{2}$Mo$\frac{1}{4}$V creep strain data, (i.e. plotting ε against t/t_f) it can be shown that approximately 1% strain accumulates at 80% through the creep life, at low stress ($\approx$40MPa). At this point tertiary creep becomes significant suggesting we are nearing the point where the current interparticle spacing can no longer stabilise the dislocation mesh and the carbide size is approaching the failure carbide size, δ_f. Assuming approximately linear strain accumulation between t=0 and t=0.8t_f, then each iso-strain line will indicate a constant particle size (i.e. constant microstructure) in a similar way to that found experimentally [4].

Measurements of strain during component service will give additional information on creep usage/damage, but because of low strain values accumulated at long times on plant strain monitoring may not be practicable. Accordingly, the microstructural approach suggested here, offers significant benefits. Clearly, to obtain this microstructural information during service, a NDT method of sampling must be available. A technique has been developed in the CEGB whereby small cylinders of material (3mm x 6mm) are extracted from operating plant enabling microscopy and creep life estimation.

6.   CONCLUDING REMARKS

In this paper, an attempt has been made to categorise the creep behaviour of ferritic steels over a wide stress range. It is not necessary to postulate changes in creep mechanisms over this range when the developing creep microstructure is understood. It can be shown that when the stresses are high, normal SCC begins immediately and creep failure can be predicted using the Monkman-Grant rule (equation 6).

$$t_f = \frac{E}{\dot{\varepsilon}_s}$$

When, the stresses are very much lower, material degeneration plays a significant role in determining the rate of SCC. Under these circumstances

$$t_f = \frac{\delta_f^3}{K}$$

These two equations are in fact remarkably similar with the rate processes identified as $\dot{\varepsilon}_s$ and K and the temperature independent constants as E and δ_f^3. For the purposes of creep life prediction, it is important to know when each applies.

The mechanism of failure described does not rely on the nucleation and growth of grain boundary cavities. Cavitation appears to be a relatively rare event in normalised and tempered ferritic steel operating at engineering stresses and temperatures. Examination of creep and service exposed specimens taken to a significant proportion of their failure life fail to reveal substantial numbers of cavities. Always they are well isolated and few in number. Final failure appears to be the result of gross localised internal necking near isolated cavities which have grown in the last 10→20% of life, or local decohesion at massive grain boundary particles by continuum growth. The rupture time is therefore controlled by creep strength criteria rather than cavitation mechanisms. The only creep mechanism available to describe the observed creep rates at fracture is that due to dislocation propagation and recovery. Under these circumstances, an understanding of the dislocation recovery processes must inevitably lead to an understanding of creep failure.

ACKNOWLEDGEMENTS

This paper is published with permission of the C.E.G.B.

REFERENCES

1. WILLIAMS, K.R. and WILSHIRE, B. - Material Sci. Eng., 1977, 28, 289.
2. WILLIAMS, K.R. C.E.G.B., Report Number, RD/L/N192/79, 1979.
3. SELLARS, C.M. - Creep Strength in Steel and High-Temperature Alloys, Sheffield, 1972, 20.
4. HALE, K.F. - Int. Conf. on Physical Metallurgy of Reactor Fuel Elements, Berkeley Nuclear Labs., 1973, 650.
5. RUSSELL, B., HAM, R.K., SILCOCK, J.M. and WILLOUGHBY, G. - Met. Sci. J., 1968, 2, 201.
6. COLLINS, M.J. - Creep Strength in Steel and High-Temperature Alloys, Sheffield, 1972, 217.
7. SILCOCK, J.M. and WILLOUGHBY, G., ibid., 122.
8. WILLIAMS, K.R. - C.E.G.B., Report Number, SSD MID/R3/75, 1975.
9. WILLIAMS, K.R. and CANE, B.J. - Material Sci. Eng., 1979, 38, 199.
10. HONEYCOMBE, R.W.K. - 29th Hatfield Memorial Lecture, University of Sheffield, 1979, Metal Science, 1980, 14, 201.
11. FELIX, W. and GEIGER, T. - Sulzer Tech. Rev., 1961, 3, 37.
12. MURPHY, M.C. and BRANCH, G.D. - J.I.S.I., 1969, 207, 1347.
13. CANE, B.J. and WILLIAMS, K.R. - C.E.G.B., Report Number, RD/L/R1965, 1977.
14. WILLIAMS, K.R. and WILSHIRE, B. - Material Sci. Eng., 1980, to appear.
15. DAVIES, P.W. and WILSHIRE, B. - Scripta Met., 1971, 5, 475.
16. McLEAN, D. - Trans. Metall. Soc. AIME., 1968, 242, 1193.
17. SMITH, P. and WILLIAMS, K.R. - C.E.G.B., Report Number, SSD/MID/R75/75, 1975.
18. DAVIES, P.W. and WILSHIRE, B. - 'Structural Processes in Creep', Iron and Steel Inst., London 1961, 34.
19. DAVIES, P.W., EVANS, W.J., WILLIAMS, K.R. and WILSHIRE, B. - Scripta Met., 1969, 3, 671.

CAVITATION AND CREEP CRACK GROWTH IN LOW ALLOY STEELS

C.D. Hamm[*][+] and R. Pilkington[*]

[*] Department of Metallurgy, Manchester University, Grosvenor
Street, Manchester, U.K.
[+] Now at: C.E.G.B. (S.S.D.), S.W. Region, Bristol.

SUMMARY

   The paper reviews some of the recent models for cavity
growth in the light of new experimental information obtained
from tests carried out on smooth specimens of $1\frac{1}{2}$%Cr $\frac{1}{2}$%V steel.
The implications of these results are discussed and applied to
current models and observations of microscopic aspects of creep
crack growth. It is suggested that the mechanisms of cavity
growth are substantially different in the two different testing
conditions.

1.   INTRODUCTION

   The problem of predicting the long term service lives of
components operating at elevated temperatures is one which is
still a long way from solution. Much effort has been expended
in attempting to extrapolate the results from short term creep
tests, not only to give an accurate prediction of service life,
but also in an attempt to estimate the remanent lives of components which have now reached the ends of their design life,
but which are clearly still in a fit condition for further service. Although such predictions require a detailed knowledge
of the creep deformation behaviour of components it is also
apparent that the correct understanding of cavity nucleation
and growth behaviour in components is required. Since real
components invariably have stress concentrations present of
different magnitudes, it is therefore necessary to understand
cavity nucleation and growth behaviour in test pieces which are
both smooth and pre-cracked in the laboratory context. At the
present time substantial effort has been made in investigating
the microscopic nature of cavity growth with the implicit
assumption that cavity nucleation may not be a major problem.
Although this is not the case the present paper will also concentrate on cavity growth aspects in both the smooth test piece
and cracked test piece situation.

A substantial number of models for cavity growth are now available and, although these are not going to be reviewed in detail (and the present list is by no means exhaustive) some of the important models can be categorized in one of two ways within the context of diffusive cavity growth. The first category can be considered in terms of models of unconstrained diffusion controlled growth, and the second as models of inhibited or constrained cavity growth.

The initial work studying the diffusional growth of cavities was carried out by Hull and Rimmer [1], but the more generally accepted version is that due to Speight and Beere [2] (Eqn. 1). However, with increasing stress the cavity growth rate may be enhanced due to the deformation of the surrounding material and this has been taken into account in the work of Beere and Speight [3], Edward and Ashby [4], Needleman and Rice [5]. This enhanced cavity growth behaviour leads eventually to continuum hole growth as suggested by Hancock [6] and Hellan [7] (Eqn. 2). It is clear however that such uninhibited behaviour may frequently not occur and that it is necessary to take into account other restrictions on cavity growth behaviour. These may be a simple geometrical constraint, as suggested by Dyson [8,9] (Eqn. 3) or alternatively, a restriction on the ease of vacancy creation by the grain boundary dislocations; - this is frequently referred to as the source control of vacancy growth of cavities and has been studied by Ishida and McLean [10] and by Beere [11] (Eqn. 4). It is also possible to inhibit cavity growth by the presence of grain boundary precipitation, this has been studied by Harris [12], Crossland and Harris [13] and also by Dyson [9].

If one then considers the problem of applying this type of knowledge to the growth of macroscopic creep cracks, in the sense of understanding their microscopic growth characteristics, again there are a number of models which are now available which predict creep crack growth rates with varying degrees of success [14-18]. However it has become clear from such studies that there is a need to understand the cavity morphologies in both contexts [19] and also to have fairly accurate knowledge of the different diffusion coefficients [19]. Work by Chuang and Rice [20] and by Chuang et al. [21] has developed criteria for the different types of cavity growth, dependent on the inter-relation between surface and grain boundary diffusion.

It is therefore the purpose of the present paper to examine cavity growth behaviour in a low alloy steel, in uniaxial tension, from both an analytical viewpoint and from a detailed microscopic examination of cavitation morphology. The information thus obtained, by making comparisons with the various theoretical models involved, will then be applied to the available microscopic information from studies of creep crack growth.

2. EXPERIMENTAL

A vacuum melted $1\frac{1}{2}$%Cr $\frac{1}{2}$%V steel of composition given in Table 1 was austenitised for 1h at 1423K before quenching into iced brine to produce a martensitic microstructure with a prior austenite grain size of 80 μm. Specimens were tempered for 24h

C	Si	Mn	S	P	Cr	Mo	Sn	V	Al
0.11	0.02	0.08	<0.01	0.007	1.67	0.008	<0.01	0.64	<0.01

Table 1. Material Composition

at 953K and creep specimens were then tested at 823K and 923K at various stresses. Tests were carried out in vacuum and a number of tests were terminated at various pre-determined strains to permit a quantitative assessment of cavitation damage. The specimens which fractured during the course of creep testing were examined by S.E.M. to determine the nature and quantity of cavitation damage on the fracture surfaces. Specimens from terminated tests were subsequently fractured transversely at 77K to expose the cavitated grain boundary facets. Cavity spacings were measured in terms of an average value calculated as $N^{-\frac{1}{2}}$, where N is the number of cavities per unit area of grain boundary. Approximately 10^3 cavities were measured to give a representative size distribution for each specimen. Grain boundary precipitate sizes and spacings were also monitored.

3. RESULTS

3.1. Cavity Growth Rates

Figure 1 shows the variation of the cavity growth rates as a function of initial stress for specimens tested at 923K. On this figure are plotted the derived growth rates for the three models considered most relevant to the present work, namely that due to uninhibited diffusional growth [2], that due to vacancy source control [10] and that due to geometrical constraint [8]. In addition, the predicted rates for the continuum growth model are shown [6]. For each predictive model, it will be seen that upper and lower curves are presented as a consequence of using the two extremes of cavity radius for each example. This has the effect that a range of cavity growth rates is possible for each particular model dependent upon the applied stress. It will be observed that no mention is made of that model due to particle inhibition [12], since the results of the analysis due to geometrical constraint become almost identical with those predicted from particle inhibition. This latter analysis has been ignored for the sake of clarity. Figure 2 shows a similar plot for the results of tests carried out at 823K.

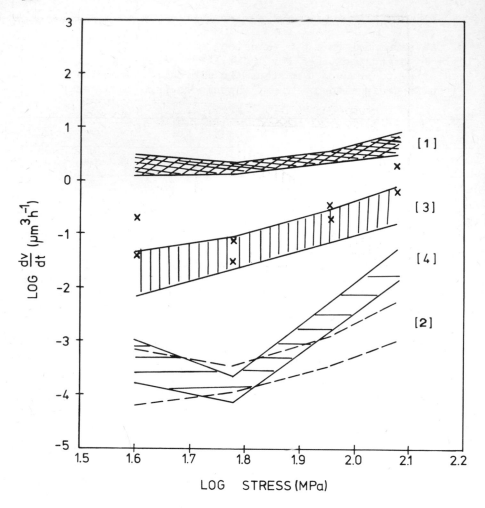

Fig. 1.   Variation of theoretical and experimental cavity
          growth rates at 923K as a function of stress.
          (Nos. [1]-[4] refer to Appendix 1.)

The results shown in Figs. 1 and 2 have been calculated on
the basis of mean cavity data and can be questioned on the
grounds of an apparent inability of the technique to take
account of continuous cavity nucleation. Proposed methods for
measuring either the largest [22] or the nth largest cavity
[23] are reasonable suggestions, except that in the present
situation of inhomogeneous cavity distributions with early and
rapid cavity coalescence, these techniques are found to offer
little if any improvement in the accuracy of such calculations.
The treatment of continuous nucleation by Needham and Gladman
[23] relies heavily on assumptions of cavity distributions
which are found to be inappropriate in the present case. Hence,
in the absence of quantitative cavity nucleation data, it is
not feasible to make any assessment of the effect of continuous
nucleation.

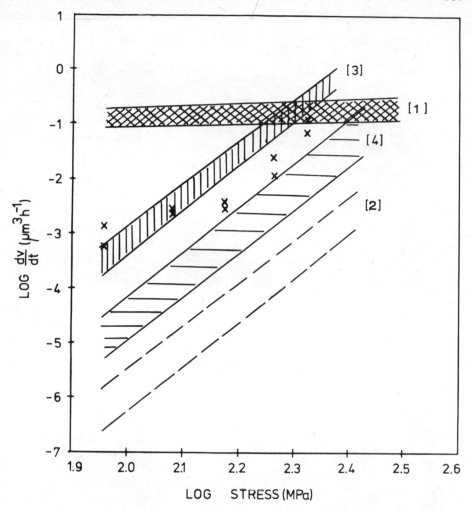

Fig. 2. Variation of theoretical and experimental cavity
growth rates at 823K as a function of stress.
(Nos. [1]–[4] refer to Appendix 1.)

3.2. Cavity Morphologies

Figure 3 is a typical micrograph of the cavitation damage
occurring in a sample tested at 923K. This particular sample
was tested at the relatively high stress of 120 MN m^{-2} and
shows a characteristic dendritic morphology, whereas at low
stress the morphology of cavities becomes much more crystallographic in nature (Fig. 4). However at 823K cavity morphologies tend to be very nearly flat in nature even though some
limited crystallography is observed and Fig. 5 gives a representative indication of the nature of the damage under these test
conditions.

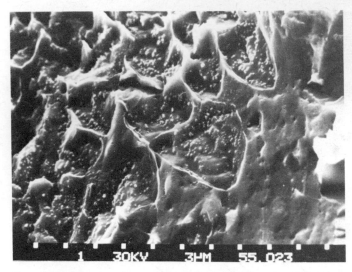

Fig. 3. S.E.M. showing dendritic cavity morphologies from a
specimen tested at 923K and 120 MPa.

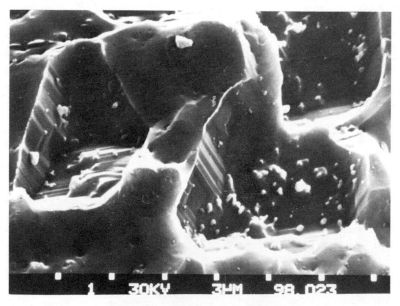

Fig. 4. S.E.M. showing crystallographic cavities from a
specimen tested at 923K and 40 MPa.

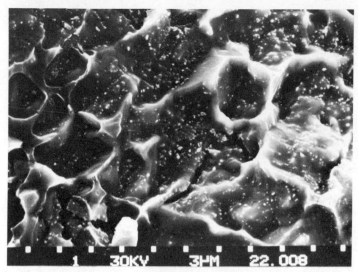

Fig. 5. S.E.M. showing flat coalesced cavities from a specimen tested at 823K and 120 MPa.

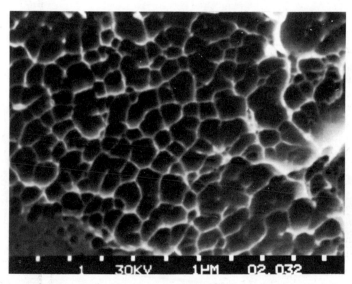

Fig. 6. S.E.M. showing spherical cavities at the tip of a
macroscopic creep crack. (Worswick and Pilkington
[25].)

Similar cavity morphologies have been observed in smooth
test pieces of bainitic $\frac{1}{2}$%Cr $\frac{1}{2}$%Mo $\frac{1}{4}$%V 0.15%C ferritic steel
tested at 823K [24], in the sense that relatively flat cavita-
tion damage occurred in the body of a specimen well away from
the fractured region. If, however, one examines cavitation
damage in the local region at a crack tip it will be seen that
the morphology is completely different in that the cavities

are no longer flat, but have become spherical in nature (Fig. 6).

4. DISCUSSION

4.1. Cavity Growth

The present results illustrate the complex nature of the cavity growth problem and our imperfect understanding of the subject at the present time. It is clear that no single model is capable of assessing the cavity growth behaviour in the present material. At both 923K and at 823K, unconstrained diffusional growth models predict cavity growth rates far in excess of those observed experimentally. It is equally true to say that models related to continuum hole growth [6,7] predict values far smaller than those observed in the smooth test piece results although it will become apparent later that the continuum hole growth models have an important role to play in the context of creep crack growth [15,19]. However it is suggested that cavity growth in the present work is constrained or inhibited from acting in a classical diffusive manner. At 923K the models based on geometrical constraint give a prediction which is slightly less than the experimentally observed values, whereas the prediction based on vacancy source control is substantially slower. It could be inferred from these results that, since both of these predictions are lower bounds, the cavity growth in the present work could be controlled by both geometrical constraint and vacancy source control. At 823K the results are less clear, in that the model based on geometrical constraint appears to predict a faster growth rate except at the lower stresses, whereas the vacancy source control model predicts slower rates throughout. It is clear therefore that few, if any, definitive conclusions can be drawn concerning the merits of any of the particular models that exist at the present time. It is interesting to note, however, that none of the models makes any comment concerning the morphological nature of cavity growth except in the implicit sense of the relationship between grain boundary and surface diffusion coefficients.

4.2. Cavity Morphologies

The evidence of Figures 3 and 5 shows that much of the cavity growth is dendritic, finger-like or crack-like in nature [3,21]. The basis of much recent work studying cavity growth [3,20,21] suggests that this result is to be expected in that at the temperatures under consideration the surface diffusion values could well be expected to be less than those for grain boundary diffusion. Figure 7 is a composite plot of available diffusion data for α-iron, following the initial suggestion made by Beere and Speight [17]. Clearly, on this basis, one would expect to see crack-like growth of cavities at 823K. However when one considers the observations shown in Fig. 6,

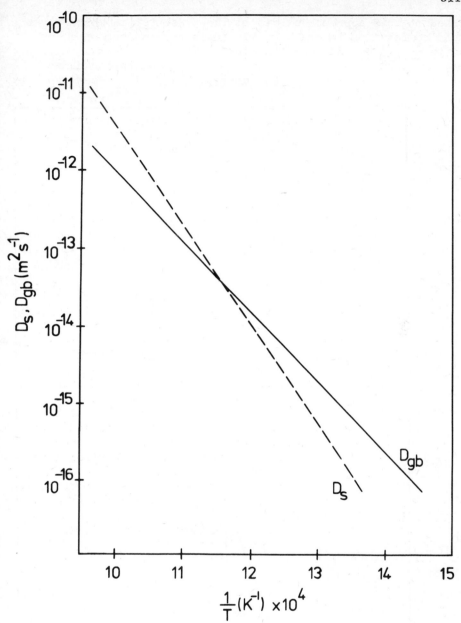

Fig. 7. Mean lines for the variation of grain boundary and surface diffusion coefficients for α-iron with change in
temperature (after Beere and Speight [17]).

that the morphology of cavities at the tips of macroscopic
creep cracks is of a spherical rather than a flat nature, it is
clear that diffusional processes may be of secondary importance
in this latter context [15], and that cavity growth at the tip
of a creep crack may occur in a manner more related to those
predicted by models based on continuum hole growth [6]. A model

based on this approach exists at the present time which gives good correlation in crack growth tests carried out on a low alloy $\frac{1}{2}$%Cr $\frac{1}{2}$%Mo $\frac{1}{4}$%V ferritic steel in both a virgin condition and also after the introduction of severe cavitation damage prior to creep crack propagation [25] (Fig. 8). Recent work by

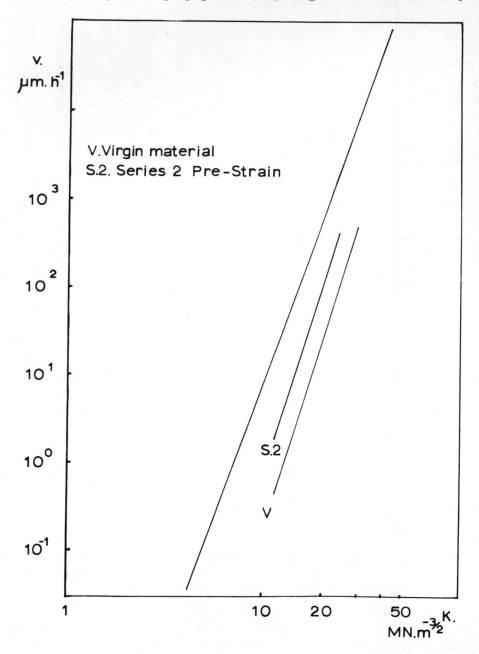

Fig. 8. Variation of theoretical and experimental creep crack growth rates with initial stress factor for $\frac{1}{2}$%Cr $\frac{1}{2}$%Mo $\frac{1}{4}$%V steel at 823K. (Worswick and Pilkington [25].)

Needleman and Rice [5] also suggests that spherical cavities may be expected at the tip of a creep crack, although the explanation in their example is that surface diffusion still plays a role in maintaining the spherical cavity shape, instead of cavity elongation being produced perpendicular to the plane of the grain boundary. Clearly substantial further work is required in this area, particularly with regard to understanding both cavity nucleation and cavity growth under controlled conditions of multi-axial stressing rather than under the more uncertain conditions of creep crack growth.

5. CONCLUSIONS

1. It is clear that most of the models in existence for cavity growth at the present time do not satisfactorily take into account cavity nucleation and there is a clear need to build cavity nucleation characteristics into predictive models.

2. Cavity growth in the present work may be constrained from acting in an uninhibited diffusional manner.

3. A clear distinction must be drawn between the metallography of smooth test piece cavitation and that occurring at the head of a macroscopic creep crack tip. It is clear that the latter favours a continuum mode of cavity growth.

6. ACKNOWLEDGEMENT

One of us (C.D.H.) wishes to acknowledge the support of a Science Research Council studentship during the course of this work.

7. REFERENCES

[1] HULL, D. and RIMMER, D.E. - 'The Growth of Grain-boundary Voids under Stress', Phil.Mag., 1959, 4, 673.

[2] SPEIGHT, M.V. and BEERE, W. - 'Vacancy Potential and Void Growth on Grain Boundaries', Met.Sci., 1975, 9, 190.

[3] BEERE, W. and SPEIGHT, M.V. - 'Creep Cavitation by Vacancy Diffusion in Plastically Deforming Solid', Met.Sci., 1978, 12, 172.

[4] EDWARD, G.H. and ASHBY, M.F. - 'Intergranular Fracture during Power-Law Creep', Acta Met., 1979, 27, 1505.

[5] NEEDLEMAN, A. and RICE, J.R. - 'Plastic Creep Flow Effects in the Diffusive Cavitation of Grain Boundaries', Acta Met., 1980, 28, 1315.

[6] HANCOCK, J.W. - 'Creep Cavitation without a Vacancy Flux', Met.Sci., 1976, 10, 319.

[7] HELLAN, K. - 'An Approximate Study of Void Expansion by Ductility or Creep', Int.J.of Mech.Sci., 1975, 17, 369.

[8] DYSON, B.F. - 'Constraints on Diffusional Cavity Growth Rates', Met.Sci., 1976, 10, 349.

514

[9] DYSON, B.F. - 'Constrained Cavity Growth, its use in quantifying recent creep fracture experiments', Can.Met.Q., 1979, 18, 31.

[10] ISHIDA, Y. and McLEAN, D. ; 'The Formation and Growth of Cavities in Creep', Met.Sci.J., 1967, 1, 171.

[11] BEERE, W. - 'Inhibition of Intergranular Cavity Growth in Precipitate Hardened Materials', J.Mat.Sci., 1980, 15, 657.

[12] HARRIS, J.E. - 'Diffusional Growth of Creep Voids', Proc. Conf. 'Grain Boundaries', Inst.of Met., London, 1976, p.E34.

[13] CROSSLAND, I.G. and HARRIS, J.E. - 'Influence of Inter-granular Precipitates on Diffusional Growth of Creep Voids', Met.Sci., 1979, 13, 55.

[14] DIMELFI, R.J. and NIX, W.D. - 'The Stress Dependence of the Crack Growth Rate during Creep', Int.J.of Fracture, 1977, 13, 341.

[15] MILLER, D.A. and PILKINGTON, R. - 'Diffusion and Deforma-tion Controlled Creep Crack Growth', Met.Trans.A., 1980, 11A, 177.

[16] VITEK, V. - 'A Theory of Diffusion Controlled Inter-granular Creep Crack Growth', Acta Met., 1978, 26, 1345.

[17] BEERE, W and SPEIGHT, M.V. - 'Diffusive Crack Growth in Plastically Deforming Solid', Met.Sci., 1978, 12, 593.

[18] SPEIGHT, M.V., BEERE, W. and ROBERTS, G. - 'Growth of Intergranular Cracks by Diffusion', Mat.Sci.and Eng., 1978, 36, 155.

[19] PILKINGTON, R., MILLER, D.A. and WORSWICK, D. - 'Cavi-tation Damage and Creep Crack Growth', Met.Trans.A., 1981, 12A, in press.

[20] CHUANG, T.J. and RICE, J.R. - 'The Shape of Intergranular Creep Cracks Growing by Surface Diffusion', Acta Met., 1973, 21, 1625.

[21] CHUANG, T.J., KAGAWA, K.I., RICE, J.R., and SILLS, L.B. - 'Non-Equilibrium Models for Diffusive Cavitation of Grain Interfaces', Acta Met., 1979, 27, 265.

[22] CANE, B.J. and GREENWOOD, G.W. - 'The Nucleation and Growth of Cavities in Iron during Deformation at Elevated Temperatures', Met.Sci., 1975, 9, 55.

[23] NEEDHAM, N.G. and GLADMAN, T. - 'Nucleation and Growth of Creep Cavities in Type 347 Steel', Met.Sci., 1980, 14, 64.

[24] JONES, C.L. - 'Creep Fracture in a $\frac{1}{2}$%Cr, $\frac{1}{2}$%Mo, $\frac{1}{4}$%V Ferritic Steel', Ph.D. Thesis, Univ. of Manchester, 1976.

[25] WORSWICK, D. and PILKINGTON, R. - 'The Interrelation between Prior Creep Damage and Creep Crack Propagation in Steels', Proc.5th Int.Conf.on Fracture, 1981, to be published.

APPENDIX 1

Cavity growth equations examined

$$(1) \quad \frac{dV}{dt} = \frac{8K'}{4\ell n \left(\frac{a}{r} \right) - \left(1 - \left(\frac{r^2}{a^2} \right) \right) \cdot \left(3 - \left(\frac{r^2}{a^2} \right) \right)}$$

Ref. 2

$$\text{where } K' = \frac{\pi\Omega.\delta.D_{g.b.} \left(\sigma - \frac{2\gamma_s}{r} \right)}{kT}$$

$$(2) \quad \frac{dV}{dt} = V.\dot{\varepsilon}$$

Ref. 6

$$(3) \quad \frac{dV}{dt} = 4a^2.d.\dot{\varepsilon}$$

Ref. 8

$$(4) \quad \frac{dV}{dt} = \frac{\pi^2}{2}.a^2.S.\dot{\varepsilon} \left[\frac{1 - 2\gamma_s/r\sigma}{1 - r^2/a^2} \right]^n$$

Ref. 11

APPENDIX 2

Symbolism and Data for Cavity Growth Equations

Symbol		Description	Value
V	–	cavity volume	
$2a$	–	cavity spacing	
r	–	cavity radius	
Ω	–	atomic volume	1.525×10^{-29} m^3
δ	–	Grain Boundary width	4.96×10^{-10} m
$D_{g.b.}$	–	G.B. diffusion coefficient	
σ	–	stress	
γ_s	–	surface energy	2.0 J.m^{-2}
k	–	Boltzmann's Constant	1.38×10^{-23} J.K^{-1}
T	–	Temperature	
n	–	creep exponent	4.21 at 923K, 9.12 at 823K
$2S$	–	particle spacing	0.57×10^{-6} m
d	–	grain size	80×10^{-6} m

CREEP-FATIGUE CRACK INITIATION IN ½ Cr-Mo-V STEEL

H.J. Westwood and W.K. Lee

Ontario Hydro Research Division, 800 Kipling Avenue

Toronto M8Z 5S4   Canada

SUMMARY

Under cyclic or "two-shift" loading, critical components of fossil-fuelled electricity generating plant are prone to creep-fatigue cracking in regions of stress concentration. For turbine rotors, Cyclic Life Expenditure (CLE) curves have been developed whereby the fatigue life expended per start up can be related to operational parameters. The fatigue data used in deriving CLE curves are based on crack initiation, making no allowance for growth to critical dimensions. Although the CLE approach is conservative in this sense, the initiation data come from isothermal tests which may overestimate the fatigue resistance under realistic conditions involving temperature cycling. Results of isothermal and in-phase thermal-mechanical low-cycle fatigue tests on ½ Cr-Mo-V steel are presented and demonstrate that isothermal crack initiation data are non-conservative for this material. Comparisons are made with similar work previously reported on 2¼ Cr-1 Mo steel.

INTRODUCTION

Economic factors dictate increasing use of fossil-fuelled electricity generating plant for "peaking" or load-following applications, the extreme case being the so-called "two-shifting" involving a daily start-up, shutdown cycle. Under such cyclic loading conditions, critical components can be subjected to large transient thermal stresses which, together with steady-state creep loading, can cause creep-fatigue cracking in regions of stress concentration and, sometimes, lead to catastropic failures[1].

For two-shift operation, it is desirable to be able to start up and load a generating unit in minimum time. Since the transient thermal stresses increase with start up rate, however, a compromise is needed in terms of the fatigue life expended per cycle. This requirement has led to the generation of cyclic life expenditure (CLE) curves for turbine rotors and, more recently, to the stimulation of a similar approach to performance assessment of critical boiler components.

This paper reviews the CLE approach and attempts to identify some ways in which the method could be refined. In particular,

518

it is suggested that the materials data used for deriving CLE
curves may be inappropriate, and some work on ½ Cr-Mo-V steel is
presented in support of this view.

THE CLE APPROACH

Figure 1 shows schematic CLE curves typical of those supplied by the manufacturers for steam turbine rotors. Basically,
each curve represents a given fraction of fatigue life expended
per operating cycle and shows the maximum heating rates permitted for a given start up to full load metal temperature differential. Such curves are developed for individual rotors,
reflecting their type, diameter, surface geometry, and material
response[2-5], and their availability allows the operator to
calculate the rotor life expended per start, so that ramp rates
can be balanced against damage in the context of the total required service life of the component.

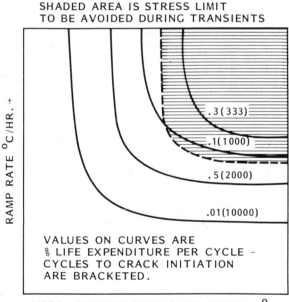

FIGURE 1   Cyclic Life Expenditure Curves (Schematic)

Generation of CLE Curves

To produce CLE curves, three types of input are required
ie,

(i) plant data, specifically transient and steady metal
     temperatures,

(ii) stress analysis,

(iii) materials data.

The method involves heat transfer analysis to determine temperature distribution and gradients, from which thermal stresses can be derived and added to the mechanical stresses, yielding nominal surface elastic stresses and strains. Using the local component geometry, the strain concentration factor (K_ε) is determined from the stress concentration factor (K_t). When stresses are elastic, K_t and K_ε are directly related. After yielding has occurred, K_ε increases and K_t decreases with increasing strain.

For a specific thermal cycle, the local concentrated strain range is given by $K_\varepsilon.\Delta\varepsilon_n$ where the strain range $\Delta\varepsilon_n$ is the maximum difference between the strain levels during the cycle.

With the computed local concentrated strain range value, isothermal low-cycle fatigue data for the component material are used to find the number of cycles to initiate a crack. Finally, the information is used to construct the CLE curves relating surface metal temperature change ΔT and start up ramp rate $\dot{T}$ to the cyclic life expended.

REFINEMENTS OF CLE APPROACH

Fatigue Data Relevance

Low-cycle fatigue data are obtained on smooth bar specimens and relate to crack initiation and early growth in the plastic zone associated with a stress concentrator[6]. Thus the CLE approach is implicitly concerned with crack initiation only, making no allowance for crack growth to critical dimensions. In this sense the approach may be excessively conservative, so that, ideally, the CLE curves should be based on initiation and growth data, the latter being obtained from fracture mechanics type tests.

For this paper, however, attention is focussed on the initiation aspects.

Crack Initiation Data

In power plant components, the fatigue strain cycle is usually associated with changing temperature so that laboratory tests to determine cycles to crack initiation should ideally incorporate both mechanical and thermal cycles. Transient thermal stresses are important in heavy section components and the stress-strain-temperature cycles arising during start up, shutdown conditions are quite complex and difficult to reproduce using conventional fatigue test equipment, though attempts have been made[7]. The practical compromise between isothermal and fully simulated service testing is the thermal-mechanical test, conducted in-phase or out-of-phase, as shown schematically in Figure 2. The in-phase test relates to components

whose steady-state stress is tensile, eg, due to internal
pressure. The out-of-phase test is relevant to compressive
stress situations which can arise on the hotter side of components subjected to steady-state temperature gradients, eg,
boiler tubes.

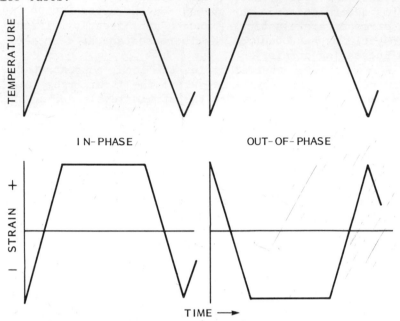

FIGURE 2   Thermal-Mechanical Test Cycles

Use of isothermal data has often been claimed to give
"worst-case" results and thus to be conservative. There is a
growing body of evidence, however, that, for some materials,
the fatigue resistance under thermal-mechanical conditions may
be significantly lower than that predicted by isothermal tests
[8-11]. Such non-conservatism has been demonstrated by comparing isothermal and in-phase thermal-mechanical properties of
$2\frac{1}{4}$ Cr-1 Mo, which is widely used for critical boiler components
and main steam lines[12]. The remainder of this paper reports
a similar study on a second important low-alloy ferritic material, ie, $\frac{1}{2}$ Cr-Mo-V.

EXPERIMENTAL

Material

The material used for this work was from a $\frac{1}{2}$ Cr-Mo-V
closed die forging, the chemical composition and mechanical
properties being given in Table I. The material was in the
normalized and tempered condition and no additional heat treatment was deemed necessary. The microstructure consisted of
ferrite, containing a fine carbide dispersion, plus tempered
bainite.

TABLE I - MATERIAL DATA

Chemical Composition (wt %)

C	Si	Mn	S	P	Ni	Cr	Mo	V
0.14	0.21	0.56	0.018	0.013	0.17	0.35	0.48	0.21

Mechanical Properties

0.2% Proof Strength MPa	Tensile Strength MPa	Elongation %	Reduction of Area %
318	497	30	79

Fatigue Tests

Full details concerning test equipment, specimen design, and procedures have been given previously[12,13]. As in recent work on stainless steel[14], only isothermal and in-phase thermal-mechanical tests were carried out and, for comparative purposes, the same temperature cycle was used as in the work on 2¼ Cr-1 Mo, ie, 300-667°C with a 10 min hold at peak temperature[12]. The isothermal tests were run at 667°C with a 10 min tensile hold. Rationale and jusification for these temperatures have been given[12,15,16,17].

RESULTS

Test Data

Table II presents complete data for all the tests. The maximum tensile stress is the stress at the start of the hold period.

As noted in previous work[12,13], the stress-strain loops for thermally cycled tests were asymmetric on the stress axis, resulting in a mean compressive stress due to the flow stress temperature dependency. Also in these tests, the tensile stress peaked at an intermediate temperature and then reduced significantly before the hold period began. By contrast, the isothermal loops showed essentially zero mean stress and less tendency for the tensile stress to peak, at least in the early stages of a test. In both kinds of test, successive loops were not superimposed, suggesting the ratchetting type of specimen shape change typical of such tests[12].

TABLE II - TEST DATA

In-Phase Thermal-Mechanical

Ramp Time Min	Effective Rate cm/min x10^{-3}	Max Tensile Stress MPa	Max Compressive Stress MPa	Total Strain 2Δεt %	Plastic Strain 2Δεp %	Cycles to Failure Nf
6.25	45.7	136	546	12.4	11.4	4
9.38	17.1	58	507	6.3	6.2	5
9.38	7.1	97	351	1.9	2.02	20
6.25	5.7	97	273	0.80	0.62	45
6.25	15.7	88	390	3.5	3.3	14
6.25	5.7	122	263	0.63	0.60	53
9.38	2.1	88	191	0.20	0.19	257

Isothermal

9.38	10.0	88	146	3.9	3.6	19
9.38	5.0	66	63	1.8	1.3	61
9.38	20.0	57	200	8.4	8.4	10
6.25	5.0	107	107	0.8	0.65	92
6.25	10.0	132	132	2.3	2.2	34
6.25	20.0	166	166	5.3	5.3	18

Fatigue Life vs Strain Ranges

Figure 3 shows the relationship between Nf and plastic strain Δεp for the Isothermal and In-Phase tests. In such plots, the results should extrapolate back to predict the tensile ductility at Nf = 0.25, ie, the case of a tensile test at temperature. Published data on high temperature tensile ductility of ½ Cr-Mo-V were found to be sparse, but indications pointed to a value of around 75% as a reasonable average. This value was thus used to help in selecting the best lines to represent the data. Despite a small amount of scatter, it is readily apparent that the In-Phase thermal-mechanical fatigue life is significantly lower than the Isothermal life.

Figure 4 shows the Nf vs total strain Δεt relationship, revealing the same trend, ie, lower In-Phase than Isothermal fatigue resistance.

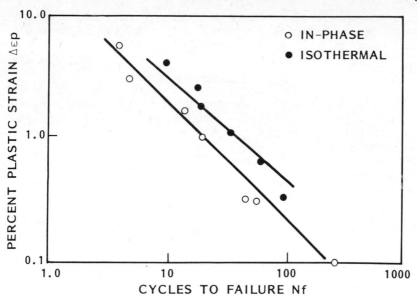

FIGURE 3   Cycles to Failure vs Plastic Strain Range

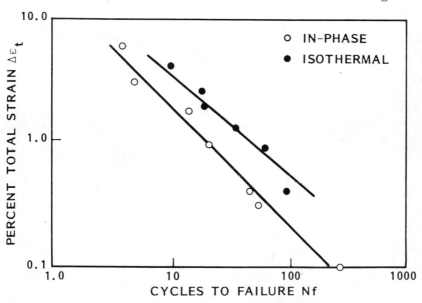

FIGURE 4   Cycles to Failure vs Total Strain Range

Deformation and Fracture Characteristics

Specimens were sectioned longitudinally and examined macro-
scopically in order to characterize deformation behaviour, with
particular regard to shape changes, since these had been associated with fracture in previous work[12,13]. Microscopical examination was also carried out to determine the dominant mode
of fracture. The effects of strain range were investigated for

s

in-phase and isothermal tests and the two test modes were compared at a similar strain range. Note that the macro and micrographs show one side only of the hollow gauge length.

Effect of Strain Range - In-Phase

Figure 5a shows the specimen with total strain range 6.2% which fractured in 4 cycles. Pronounced necking is apparent and a very slight amount of off-centre barreling[12]. Post-fracture mechanical and arcing damage had occurred but the main fracture characteristics were still evident. Necking was estimated at 60% based on the cross-sectional area reduction and, in general, the fracture was essentially ductile in nature.

Figure 5b shows the specimen with total strain range 0.1% which fractured in 257 cycles. Necking is much less pronounced than in the high strain specimen (Figure 5a) and no barreling is evident. Fracture was associated with a blunt surface crack but most of the fracture surface was apparently transgranular cleavage. The secondary cracking was intergranular, however. Some grain growth had occurred adjacent to the fracture.

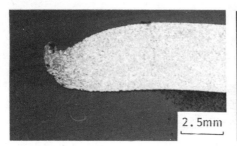 

FIGURE 5   (a) $\Delta\varepsilon t$ = 6.2%   (b) $\Delta\varepsilon t$ = 0.1%

Effect of Strain Range on In-Phase Fracture

Effect of Strain Range - Isothermal

Figure 6a shows the specimen with total strain range 4.2% which fractured in 10 cycles. Pronounced necking is evident and a barely perceptible degree of off-centre barreling. Fracture had involved some growth of an intergranular crack through the thickness, followed by ductile rupture.

Figure 6b shows the specimen with total strain range 1.15% which fractured in 34 cycles. Little or no necking is evident, nor can any barreling be discerned. Fracture involved growth of an intergranular crack about 40% through the thickness followed by fast shear. A secondary crack of similar depth is also evident.

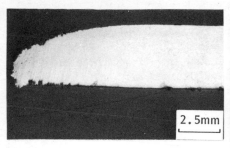 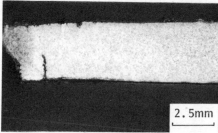

FIGURE 6   (a) Δεt = 4.2%   (b) Δεt = 1.15%

Effect of Strain Range on Isothermal Fracture

Comparison of In-Phase and Isothermal Fractures

Figures 7a and 7b show In-Phase and Isothermal specimens
with total strain ranges of 0.32 and 0.4 respectively which had
failed in 45 and 92 cycles. Differences in deformation and
fracture behaviour are quite apparent. Thus, the In-Phase
specimen was substantially necked and although some surface
crack growth had preceded fracture, the latter was predominantly
ductile in nature. In the Isothermal specimen, no significant
necking was evident and fracture was predominantly due to
cracking from both surfaces, with only around 25% ductile
rupture. In both specimens, some localized but pronounced
ferritic grain growth had occurred in the fracture region.

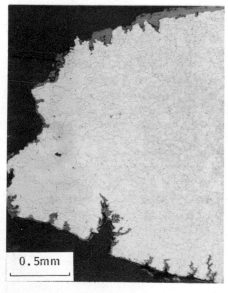 

FIGURE 7   (a) In-Phase   (b) Isothermal

Comparison of In-Phase and Isothermal Fractures

Figures 8a and 8b compare secondary cracking in the long term In-Phase (7a) and Isothermal (7b) specimens which fractured in 257 and 92 cycles respectively. Whilst the crack tip in the In-Phase specimen is clearly proceeding along grain boundaries, it is also apparent that substantial plastic deformation had accompanied earlier growth. This is shown both by the crack width and, indirectly, by the dense sub-structural development in the surrounding grains. By contrast, the Isothermal crack is clearly intergranular and had propagated in a much more brittle manner.

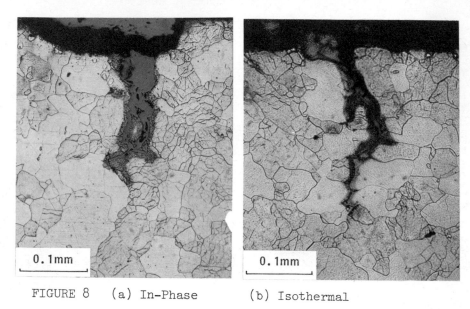

FIGURE 8   (a) In-Phase   (b) Isothermal

Comparison of In-Phase and Isothermal Cracking

DISCUSSION

Fatigue Life vs Plastic Strain Range

As with the previous work[12,13], the linear relation between plastic strain range $\Delta\epsilon p$ and fatigue life Nf can be conveniently represented by the Coffin-Manson equation

$$\Delta\epsilon p = C \ Nf^{-\beta}$$

where C and β are constants[18,19].

For the present results, the following equations fit the data presented in Figure 6:

In-Phase $\Delta\epsilon p = 0.23 \ Nf^{-0.96}$

Isothermal $\Delta\epsilon p = 0.20 \ Nf^{-0.84}$

Comparing the results with those obtained on 2¼ Cr-1 Mo using the same temperature cycles[12], the ½ Cr-Mo-V alloy showed much higher predicted tensile ductility (the extrapolation to Nf = 0.25) ie, around 75% compared with around 25%. The β values were also higher than for 2¼ Cr-1 Mo for which values of 0.76 (In-Phase) and 0.60 (Isothermal) were obtained. For ambient temperature low-cycle fatigue a value of 0.5 is usually observed whilst, when creep becomes significant, the value increases to approach 1. It is surprising to find higher β values for ½ Cr-Mo-V than for 2¼ Cr-1 Mo, since the former material has generally superior creep resistance.

Comparison of the present results with those of other workers is difficult, since substantially lower test temperatures have usually been employed. Coles et al performed reverse-bend low-cycle fatigue tests on ½ Cr-Mo-V at 550°C and found endurances roughly 10X those in the present isothermal tests[20]. The difference can, however, be rationalized in terms of the temperature dependence and failure criteria. An indication of the former may be obtained from work on alpha-iron which showed a reduction in fatigue life of ~2X between 550°C and 667°C for a constant total strain range of 2%[21]. Concerning failure criteria, this was defined as complete fracture in both studies. In the present work, the first appearance of a crack was followed almost immediately by complete fracture of the thin-wall hollow gauge length - in this sense, the results are felt to be an excellent measure of crack initiation. Coles et al, however, compared cycles-to-initiation with cycles-to-fracture and, comparing these with the present results for a common strain range of 0.5%, the Nf values are as follows:

Present Work	Coles et al Initiation	Coles et al Fracture
90	200	800

If the present results are temperature-compensated by the suggested factor of 2X, there is good agreement between the two sets of work in terms of crack initiation. Conversely, without temperature compensation and taking the Coles et al cycles to fracture, the approximate order of magnitude difference can be rationalized.

Aside from the above comparisons, the main point to be made from the present work is that the fatigue resistance of ½ Cr-Mo-V is lower under the realistic In-Phase conditions than under Isothermal conditions. If the Isothermal results in Figure 6 are extrapolated to Δεp = 0.1%, a factor of 2.6 difference in Nf value is predicted. In this respect, the material shows the same behaviour as the 2¼ Cr-1 Mo, though for the latter material, the factor was close to 4 at the same strain range.

Deformation and Fracture Characteristics

In general, the deformation and fracture behaviour varied with type of test and applied strain range. Thus, the In-Phase fractures were more ductile than Isothermal and the influence of crack growth rather than plastic instability on fracture increased with decreasing strain range in both kinds of test. Perhaps the best illustration of the differences in mechanism was provided by the secondary cracking details shown in Figures 8a and 8b.

In the present work, the off-centre barreling characteristic of isothermal and in-phase thermal-mechanical tests[22, 23,24] was only discernible in very high strain specimens, the shape changes being slightly less apparent than in the $2\frac{1}{4}$ Cr-1 Mo specimens[12].

PRACTICAL IMPLICATIONS OF RESULTS

The present results and those on $2\frac{1}{4}$ Cr-1 Mo indicate that, for low-alloy ferritic steels, isothermal data may overestimate the fatigue life under the more realistic thermal-mechanical conditions. Thus, CLE curves based on such data may be non-conservative in terms of crack initiation. For turbine rotors and other smoothly finished components, where the majority of the fatigue life is spent in the initiation stage, this may be important. For other components such as welded pressure vessels, eg, superheater headers, crack initiation may be less important than growth from existing defects. It is not known whether growth, like initiation, may be faster under cyclic temperature rather than isothermal conditions.

CONCLUSIONS

1. Cyclic Life Expenditure curves may be conservative since the creep-fatigue data used relates strictly to crack initiation with no consideration of crack growth to critical dimensions.

2. Tests on $\frac{1}{2}$ Cr-Mo-V steel have indicated that the isothermal test data used to derive CLE curves may overestimate the resistance to crack initiation under the more realistic thermal-mechanical conditions.

3. Fracture of specimens was more ductile under thermal-mechanical conditions.

4. In both types of test, fracture ductility decreased with decreasing fatigue strain range and growth of cracks became increasingly important relative to plastic collapse.

ACKNOWLEDGEMENT

Thanks are extended to Dr. A.D. Batte of NEI Parsons Ltd, Newcastle-upon Tyne, for generous donation of the test material.

REFERENCES

1. KRAMER, L.D. and RANDOLPH, D.D., ASME-MPC Symposium on Creep-Fatigue Interaction 1976, p 1-24.

2. TIMO, D.P. and SARNEY, G.W., ASME Paper 67-WA-PWR-4, 1967.

3. IPSEN, P.G. and TIMO, D.P., Proc. American Power Conference, 1969, p 314.

4. TIMO, D.P., "Thermal Stresses and Thermal Fatigue", Proc. International Conference, CEGB, Berkeley. Littler, D.J. (ed), Butterworths, 1972, p 453.

5. SPENCER, R.C. and TIMO, D.P., Proc. American Power Conference, 1974, p 511.

6. COFFIN, L.F., ASTM STP 520, American Society for Testing and Materials, 1973, p 5-34.

7. SKELTON, R.P., Central Electricity Generating Board Report RD/L/N74/77, 1977.

8. STENTZ, R.H., BERLING, J.T. and CONWAY, J.B., Proc. International Conference on Structural Mechanics in Reactor Technology, Berlin, 1971.

9. LINDHOLM, U.S. and DAVIDSON, D.L., American Society for Testing and Materials, ASTM STP 520, 1973, p 473.

10. SHEFFLER, K.D. and DOBLE, G.S., ASTM STP 520, American Society for Testing and Materials.

11. TAIRA, S., FUJINO, M. and OHTANI, R., Fatigue of Engineering Materials and Structures, Vol 1, 1979, p 295-508.

12. WESTWOOD, H.J. and MOLES, M.D.C., Canadian Metallurgical Quarterly, Vol 18, 1979, p 215-230.

13. WESTWOOD, H.J., Fracture 1977, Vol 2, Waterloo, Canada, 1977, p 755-765.

14. WESTWOOD, H.J., Proceedings, International Conference on Mechanical Behaviour of Materials, Vol 2, Cambridge, England, 1979, p 59-67.

15. ASTM STP 151, American Society for Testing and Materials, p 87.

16. HART, R.V., Metals Technology, January 1976, p 1-7.

17. HALE, K.F., Proceedings, Conference on Physical Metallurgy of Reactor Fuel Elements, CEGB Berkeley Nuclear Laboratory, 1973, p 193-201.

18. COFFIN, L.F., Trans ASME 76, 1954, p 931.

19. MANSON, S.S., NACA Technical Note 2933, NASA Lewis Research Centre, Cleveland, Ohio, 1954.

20. COLES, A., HILL, G.J., DAWSON, R.A.T. and WATSON, S.J., Monograph and Report Series No 32, The Metals and Metallurgy Trust, 1967, p 270-294.

21. WESTWOOD, H.J. and TAPLIN, D.M.R., Metallurgical Transactions, Vol 3, 1972, p 1959-1965.

22. COFFIN, L.F., Trans ASME 82, 1960, p 671.

23. SHEFFLER, K.D., ASTM STP 612, American Society for Testing and Materials, 1976, p 214-226.

24. COFFIN, L.F., ASTM STP 612, American Society for Testing and Materials, 1976, p 227-238.

GRAIN BOUNDARY IMPURITY EFFECTS ON THE CREEP DUCTILITY OF
FERRITIC STEELS
D. P. Pope and D. S. Wilkinson*

Dept. of Mat. Sci. & Eng., Univ. of Pennsylvania, Philadelphia
PA 19104, USA *Dept. of Met. & Mat. Sci., McMaster Univ.,
Hamilton, Ontario L8S 4M1, Canada.

1. SUMMARY

Many investigators have proposed that the low ductility
intergranular failure mode observed in low strain rate (long
failure time) creep tests is facilitated by the segregation
of impurities to grain boundaries. Low alloy steels have received the most attention in this regard, since metalloid impurities from groups IVA to VIA of the periodic table are
known to promote grain boundary failure at low temperatures in
these materials. However, the results of these studies are
far from conclusive. For example, P appears to promote grain
boundary cavitation in some steels but retard it in others.
We review here the rather extensive literature concerned with
the effects of grain boundary impurities on creep ductility
of ferritic materials, with particular emphasis on recent
work (including our own experiments on 2 1/4 Cr - 1 Mo steel).
It will be shown that these impurity effects are at present
poorly understood and that analogies between impurity effects
on high and low temperature grain boundary cracking are mis-
leading.

2. INTRODUCTION

Many of the characteristic features of grain boundary
fracture at elevated temperature were already known by the
early part of this century. For example Figure 1, taken
from a 1912 paper by Huntington [1], shows that the reduction
in area (RA) of Cu samples deformed in tension approaches
a broad minimum between 300 and 500°C, then increases again
at higher temperatures. Huntington showed that several copper
alloys as well as steel and wrought iron behave similarly.
The ductility decrease was known at the time to be accompanied
by a shift in fracture path from transgranular to intergranular, a fact used as strong support for the amorphous cement
theory of grain boundary cohesion. (McLean's Historical intro-

532

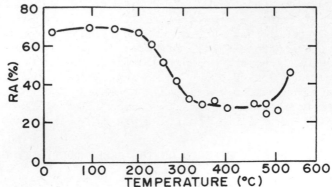

Fig. 1. RA of copper as a function of temperature,
data from ref. 1.

duction [2] gives an excellent description of the debate on
this subject.) The fact that tensile testing at high temperatures and low strain rates produces grain boundary failure
was considered at the time to be a natural consequence of the
viscous nature of the intercrystalline amorphous cement [3];
although it was known then that the grain boundaries of some
metals could be made brittle by small compositional changes,
for example, by the addition of small quantities of bismuth to
gold [4]. In fact, Guertler suggested in 1913 that the tendency for grain boundary failure at elevated temperatures may
be due to the presence of grain boundary impurities [5]. (He
was actually suggesting that a thin eutetic film encases each
grain.) So already by 1913 it was known that metals show a
ductility decrease at elevated temperatures due to increased
amounts of grain boundary failure, that the amount of grain
boundary failure is strain rate and temperature dependent, and
it had been suggested that grain boundary impurities are
involved.

Somewhat later, after it had been observed that some
steels become very notch brittle when operating under load at
450-500°C (e.g. see [6]), an apparent link was made between
the development of high temperature and low temperature grain
boundary brittleness. Since the extent of temper embrittlement was measured by the loss in the room temperature notched
bar impact energy, it was quite natural to measure the extent
of prior creep damage in the same way. (Sachs and Brown [7]
provide an excellent historical perspective of this development.) In some cases, only impact tests could reveal the
damage [8]. The impact tests, performed on samples aged at
elevated temperatures, with and without applied stress, revealed that the combination of stress and elevated temperature
is more damaging than is elevated temperature alone. Since it
was known that low temperature brittleness could be mitigated
by reheat treatement followed by rapid cooling from the tempering temperature, it was again quite natural to apply the same
techniques to samples embrittled under creep conditions.
Thum and Richard [9,10] performed an extensive series of

experiments on various steels in the 1940's using this technique. They found that part of the room temperature impact
energy lost during creep was recovered by heat treatment and
part was not. Sachs and Brown [7] defined the fractional impact energy loss during creep as "creep damage", and that
not recovered by heat treatment as "permanent damage", and
showed that the permanent damage becomes a larger fraction of
the total at long failure times, Figure 2. We know now the

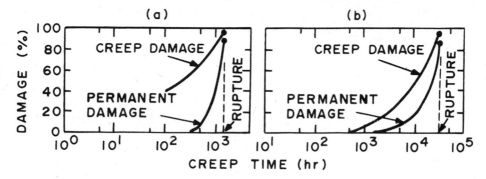

Fig. 2. Creep and permanent damage in a CrMoV steel at 500°C
and a stress of (a) 393 MPa and (b) 159 MPa. Data
from ref. 10 as plotted in ref. 7.

recoverable part of the impact energy loss is due to the
segregation of group IVA - VIA atoms to the prior austenite
grain boundaries, which can be dispersed by reheat treatment,
and the permanent damage is, at least in part, due to grain
grain boundary cavitation. However, since at that time
this distinction had not yet been made, it was logical to
assume that temper embrittlement and creep embrittlement
(a term which appears to have been introduced by Sachs and
Brown [7]) were manifestations of the same physical phenomenon.

The observation by Greenwood et al. [11-13] that grain
boundary failure at elevated temperatures is caused by cavitation demonstrated that the mechanisms of grain boundary
failure at high and low temperatures must be different. Grain
boundary failure at high temperatures involves cavity nucleation, growth and coalescence, whereas low temperature grain
boundary failure involves the nucleation and propagation of
a quasi-brittle crack (we exclude here considerations of
"ductile" grain boundary failure at low temperatures due to
a high density of MnS particles on the boundaries). Since
the work of Greenwood et al., a voluminous literature has
developed on the subject of cavity nucleation and growth at
elevated temperatures, see the reviews by Perry [14] and by
Svensson and Dunlop [15,16]. It appears that the cavities
nucleate at non-wetting grain boundary particles (e.g.
sulfides in steels) or on wetting particles under the influence
of high local stresses caused by grain boundary sliding [17].
Cavity growth is generally thought to be caused by the

movement of vacancies within the grain boundary to the cavities, driven by gradients in the hydrostatic component of the
applied stress [18], although plasticity effects cannot be
ignored. However, even after Greenwood and coworkers [11-13]
showed that the high and low temperature grain boundary failure mechanisms are quite different, it was still widely believed that the same impurities responsible for temper embrittlement must also promote creep embrittlement. In the next
section we will examine the mechanisms by which such impurities might affect grain boundary failure at high temperatures
and show that such an assumption is not justified.

3. PROPOSED EFFECTS OF GRAIN BOUNDARY IMPURITIES ON CAVITY
    NUCLEATION AND GROWTH MECHANISMS

It was proposed in 1956 that impurity segregation could
reduce interfacial energies (both the grain boundary and the
free surface energies) in such a way that grain boundary cohesion is reduced [19,20]. This reduced cohesion can stabilize small cavities and thus increase the density of cavities
available for subsequent growth and coalescence. An example
of this mechanism in ferritic steels was provided by Rellick
and McMahon who showed that grain boundary P can weaken the
normally very tenacious ferrite-carbide interface [21].
These ideas were independently applied to creep cavity nucleation in steels by Abiko et al. [22] and by Seah [23].

Middleton [24] has proposed quite a different mechanism,
one which also might be categorized as a cohesive effect. He
proposed that grain boundary impurities can alter the grain
boundary cavity density by modifying the density of MnS particles in the grain boundary, all of which are non-wetting.
No mechanism is given for this effect, but it presumably is
related to a change in the grain boundary energy which
facilitates the nucleation of MnS. Middleton's work will be
discussed in greater detail later in this paper.

Grain boundary impurities are expected to have a marked
effect on the grain boundary self diffusivity, and therefore
to the extent that cavity growth rates are diffusion controlled, these impurities should affect the extent of cavitation [22,23,25]. Data which demonstrate such an impurity
affect on grain boundary diffusion are very limited; although
Gupta [26] has observed such an effect in a gold-tantalum
alloy. To our knowledge there are no such data on iron-based
alloys. One might expect that impurities would reduce the
grain boundary self diffusivity, and therefore the impurities
would be expected to reduce the cavitation rate.

Cavities are often found to nucleate on sliding grain
boundaries. The sliding process induces stress concentrations
at the grain boundary particles. The more slowly the relaxation of these stresses occurs, by diffusion or plasticity,

the more cavities will be nucleated. Since relaxation very near the particles will be controlled by interfacial diffusion (for sufficiently small particles), a decrease in grain boundary self diffusivity will result in higher void nucleation rates. A model for calculating the interfacial stresses has been developed by Argon et al. [17].

In summary, grain boundary impurities may affect the grain boundary cavity nucleation rate by: (1) changing the cohesion between grain boundary particles and the matrix, (2) altering the grain boundary self diffusion rate and thus modifying the rate of stress relaxation around grain boundary particles, (3) changing the concentration of non-wetting grain boundary particles, e.g. MnS. They may affect the cavity growth rate by changing the rate of grain boundary self diffusion and thereby altering the rate of vacancy accumulation in the cavities.

Based on these considerations, there is little reason to expect a strong link between low and high temperature embrittlement. In fact, it is entirely possible that a steel with a high concentration of, say, P in the prior austenite grain boundaries may be more creep ductile than a P-free material. Several studies which show this to be the case will be cited later in this paper.

4. REVIEW OF EXPERIMENTAL WORK

Grain boundary impurity effects are determined most unambiguously by comparing the creep ductility and/or grain boundary cavitation rate of samples made from high purity material with and without controlled amounts of added impurities. The grain boundary impurity concentration is measured using Auger spectroscopy or, less preferably, is inferred from the shift in the ductile to brittle transsition temperature. Naturally the grain size, strength, etc. of all the samples used in a study must be the same, but these variables are not quite as easily controlled as it might appear. Since impurity additions quite commonly modify the grain growth and austenite decomposition kinetics, the heat treatment must commonly be tailored to the compositions. Many investigators have used commercial material subjected to a standard heat treatment or have used high purity material also subjected to a standard heat treatment, with the result that the compositions and/or the properties were not well controlled.

4.1 2 1/4 Cr-1Mo Steel

Hopkin and Jenkinson, who appear to have been the first to study the effects of deliberate impurity additions on the creep properties of a steel, found that 0.1% Sn has no deleterious effects on creep strength or ductility of controlled

536

purity material, independent of hardness and microstructure
[27]. However, Bruscato found that the RA at 482 and 565°C is
lower for commercial weld metal containing higher concentrations
of As, Sb, Sn and P [28]. Wolstenholme [29] concluded that high
P in commercial weld metal results a deterioration of the
540°C creep properties of notched specimens by enhancing the
cavitation rate and thereby reducing the rupture time (the
rupture strain appears to have been unaffected in this study).
Unfortunately the creep strengths of the high and low P materials were different, making accurate comparisons difficult.
Ferrell and Pense [30] measured the ductile-brittle transition temperature and the creep ductility of commercial material and found that the material with the highest transition
temperature showed the highest creep ductility, i.e., the
creep ductility trends were opposite those of the temper embrittlement trends. This is clearly not in agreement with
[28 and 29] above, but note that all the above studies except for [27] were performed on commercial grade material.
Abiko [22] and Wilkinson et al. [25] found that the notched
bar RA of as-tempered samples was dramatically increased
by additions of Mn, Si and P in various combinations, Figure 3,

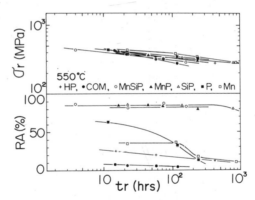

Fig. 3. Strength and ductility of 2 1/4 Cr-1Mo steel tested at
550°C. Note the large ductility differences. From
refs. 22 and 25.

but the ductility of step cooled samples containing Mn and P
was less than that of as-tempered samples. The composition,
grain size and hardness of those materials were carefully
controlled. These results were interpreted in terms of the
expected increase in nucleation rate and decrease in growth
rate due to the grain boundary P. The grain boundary P is
expected to decrease the grain boundary self diffusivity by
reducing the free volume. Auger spectroscopy showed that the
most creep ductile material did, indeed, have the highest
grain bound P content. A parallel study of temper embrittlement of this same material showed that the most creep ductile
materials were the most temper embrittled [31,32,33].
Lonsdale and Flewitt [34] found that P increases the grain
boundary cavitation rate in controlled purity samples.

The strengths of the P-containing and P-free samples were different, but since the P-containing samples were the weaker, the conclusion about the effects of P is probably valid. Needham and Orr [35] found very complex effects of As, Sb, P, S and Si, which were added in various combinations to a pure material. They found that the 550°C creep rupture elongation was reduced by Sb and by Sb+As additions but combinations of As, Sb and Si and As, Sb, P, S and Si increased the elongation. No details were given about microstructural or strength details, so it is not known how they affect the results. Gooch [36], in a study of the effects of residuals on creep crack growth rates, found that the weld material with the highest P and As contents showed the lowest creep crack growth rate at 565°C, but this material also showed the highest secondary creep rate. He concluded that the reduced crack growth rate is only indirectly related to the impurity content through the effects of the impurities on the creep strength. The material with the high P and As levels did show the greatest amount of grain boundary damage during 650 and 700°C stress relaxation tests, however.

In summary, depending upon the study one chooses to quote, it can be claimed that As, Sb, Sn and P improve, impair, or make no change in the creep ductility of 2 1/4 Cr-1Mo steel. If only those results on controlled purity material are quoted we find that: Sn has no effect [27], P is deleterious [34], P is beneficial, in as-tempered samples [22,25], As, Sb, Sn and P have different effects in different combinations [35], and P is not necessarily deleterious [36]. Clearly, no concensus is to be found here.

4.2 Cr-Mo-V Steel

The discovery by Hopkins et al. [37,38] that improved purity could dramatically increase not only the creep ductility but also the creep strength of 1Cr-1Mo-1/4V (1CrMoV) and 1/2Cr-1/2Mo-1/4V (1/2CrMoV) steels led to an increased interest in this problem. The aim of most of the later studies was to determine which impurities cause the deterioration in creep properties. Benes and Skyvor [39] claim that Sn has no effect on the creep ductility of commercial heats of 1/2 CrMoV (as did Hopkin and Jenkinson [27] in their study of 2 1/4 Cr-1Mo), but Sn does reduce the time to rupture. Here again there are problems with uncontrolled strength levels and microstructure. Viswanathan [40] and Viswanathan and Beck [41] found that a commercial heat of 1CrMoV steel with high residual content shows a lower smooth bar creep ductility and is more notch sensitive at 510 and 593 °C than a commercial heat with low residual content. They concluded, however, that differences in Al content caused the disparate results, i.e., As, Sb, Sn and P were not responsible. In their study of the tensile ductility of both commercial and laboratory heats of 1/2 CrMoV steel,

538

Tait and Knott [42] found that As, Sb, Sn, P and Cu reduce the RA and increase the tendency for intergranular failure at elevated temperatures. Tu and Seth [43] and Roan and Seth [44] found a similar result on laboratory heats of 1CrMoV. Roan and Seth found that As, Sb, Sn, P and S reduce the RA, Figure 4, and increase the amount of grain boundary failure

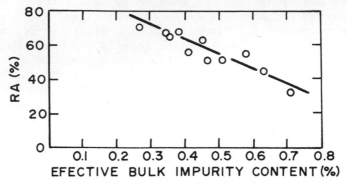

Fig. 4. RA vs total effective impurity content, defined by (16.1Sb + 13.8Sn + 12.6P + 10.5As + 8.8S), of 1CrMoV steel tested at 220MPa and 593°C, from ref. 44.

while P also reduces the rupture time. So if one considers only the results obtained by 1978 on controlled purity labora- tory heats one concludes that impurities reduce the strength and ductility of CrMoV steel [37,38], and the im- purities responsible for this degradation are As, Sb, Sn, P, Cu and S [42-44]. Unfortunately the results of subsequent studies do not support these conclusions. In his synthesis of the results obtained at the National Physical Laboratory on CrMoV steels, Tipler [45] confirmed the earlier results [37,38] on the beneficial effects of improved purity; however when As, Sb, Sn, P, N, S, were individually added to the high purity 1/2 CrMoV material (the material with the high strength and ductility) the 550°C creep properties were affected in a manner very similar to that found by Abiko et al [22] and Wilkinson et al. [25]. It was found that none of the above additions altered the failure time but P and Sb dramatically improved the RA of material austenitized at 1200°C. The ductility enhancement disappeared for failure times longer than 1000 hours, however. As and Sn caused a slight ductility de- crease and N caused no change. The material to which sulfur was added actually showed a light ductility improvement. The results of King [46] on the 550°C tensile ductility and 565°C crack growth rate of commercial and laboratory heats of 1/2 CrMoV steel indicate that As, Sb, Sn and P are all dele- terious. The results of Middleton [24] on the 700°C tensile ductility of 1/2 CrMoV steel, to which we have referred earlier in this paper, indicate that impurity effects are seen through their influence on the grain boundary MnS particle distri- bution. Depending upon whether the impurities increase or decrease the density of MnS particles on the prior austenite grain boundaries, they can either improve or impair the creep ductility. And finally, the recent hot tensile test results of Takasugi and Pope on laboratory heats of 1CrMoV steel indicate that P is beneficial, Figure 5.

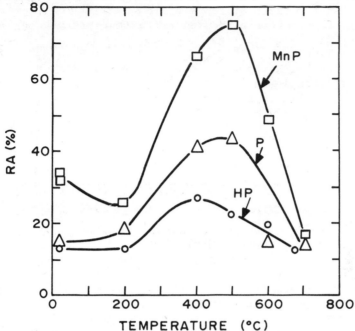

Fig. 5. RA vs. temperature for impurity-free (HP), P-doped (P), and Mn+P-doped (MnP) 1CrMoV samples tensile tested at a strain rate of 4.44×10^{-5} s^{-1}.

We see from the above list of conflicting results that the role of impurities on intergranular cracking of CrMoV steels is also poorly understood. Again, considering only those data on controlled purity material we find that: a low residual content is beneficial [37,38], As, Sb, Sn and P are deleterious [42,44,46], Sn is deleterious [43], Sb and P are beneficial (for failure times less than 1000 hours) while As and Sn are slightly deleterious [45], As, Sb, Sn and P are not necessarily deleterious [24], and P is beneficial [Takasugi and Pope].

Other Steels

Viswanathan has studied the effects of Sb, Sn, P, B, and Ti on the 538°C creep rupture ductility of laboratory heats of 1 1/4 Cr-1/2 Mo steel [40,41,47-50]. He found that Sb, Sn and P have no effect, but B by itself has a deleterious effect. (This latter effect is reminiscent of Middleton's [24] results.) The ductility of the undoped heat for rupture times greater than 1000 hours was found to be less than the ductility of the heats doped with Ti+Sb+Sn, Ti+B+Sb+Sn, Ti+B and Ti+B+P, indicating that B in combination with Ti is not deleterious.

540

The short time creep ductility of commercial and laboratory heats of Ni-Cr-Mo-V steel at 600°C [51,52] appears to be
most adversely affected by B, Cu and Al. As, Sb, Sn, P and S
appear to have little effect. Embrittlement was only observed
in steels austenitized at temperatures above 1100°C, again
suggesting that a mechanism similar to that proposed by
Middleton [24] is operating.

And finally, Kirby and Beevers [53] showed that the high
temperature tensile ductility of pure iron is improved by the
addition of P, Figure 6. Here again the fracture path in the

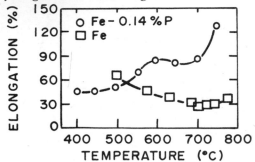

Fig. 6.   Creep rupture strain for pure Fe and Fe-P, ref. 53.

low ductility pure material was intergranular while that in the
high ductility P-doped material was transgranular.

5.   SUMMARY

We conclude from this review of the literature that the
effects of grain boundary impurities on creep ductility of
ferritic steels are, at best, only poorly understood.
Depending upon the relative importance of cavity nucleation
and diffusive cavity growth in determining the ductility,
these impurities, when segregated to the grain boundaries,
may either increase or decrease the ductility. Ample evidence for both effects can be found in the literature, but
there currently is no way to predict what the effect of a
given impurity in a given steel will be. A large part of the
confusion is due, in our opinion, to the assumption by many
investigators that the impurities responsible for temper embrittlement must also promote intergranular creep fracture,
or, at the very best, they may be innocuous-never beneficial.
This point of view has, as we have previously mentioned, a
firm historical but a rather unsound scientific basis. Because of this widespread assumption, results which link
temper and creep embrittlement have been viewed as quite
natural and reasonable, while results which showed them to be
unrelated have received surprisingly little attention. We
believe that the recent ideas about how impurities can separately influence cavity nucleation and cavity growth [22,23,25]
are helpful, but they are not sufficient. We believe that

indirect effects, such as the influence of grain boundary impurities on promoting or hindering the formation of non-wetting grain boundary particles [24], are also important.

ACKNOWLEDGEMENTS

This research was supported by the Electric Power Research Institute and the National Science Foundation MRL Program under Grant No. DMR79-23647.

REFERENCES

1. Huntington, A. K.- 'The Effects of Temperatures Higher than Atmospheric on Tensile Tests of Copper and Its Alloys, and a Comparison with Wrought Iron and Steel', J. Instit. Metals, 1912, 8, pp. 126-148.
2. McLean, D.- Grain Boundaries in Metals, Oxford at the Clarendon Press, Oxford, 1957.
3. Rosenhain, W. and Ewen, D.- 'Intercrystalline Cohesion in Metals', J. Instit. Metals, 1912, 8, pp. 149-185.
4. Arnold, J. O. and Jefferson J.- 'Influence of Small Quantities of Impurities on Gold and Copper', Engineering, February 7, 1896.
5. Guertler, W., Discussion on paper by Rosenhain and Ewen, J. Instit. of Metals, 1913, 10, p. 142.
6. Bailey, R. W.- 'Mechanical Testing of Materials', Proc. Inst. Mechanical Engineers, 1928, 114, pp. 417-452.
7. Sachs, George and Brown, W. F., Jr.- 'A Survey of Embrittlement and Notch Sensitivity of Heat Resisting Steels', Symposium on Strength and Ductility of Metals at Elevated Temperatures, ASTM STP 128, Amer. Soc. for Testing and Materials, 1953, pp. 6-20.
8. Lea, F. C. and Arnold, R. N.- 'Embrittlement of Alloy Steels', Proc. Inst. Mechanical Engineers, 1935, 131, pp. 539-609.
9. Thum, A. and Richard, K.- 'Embrittlement and Damage Ranges of High Temperature Strength Steels Under Creep Stress', Archiv Eisenhuttenwesen, 1941, 15(1), pp. 33-45. (Available in English as Translation No. 1131, Henry Brutcher, ASM, Metals Park, Ohio, USA).
10. Thum, A. and Richard, K.- 'The Damage Line During Creep Stressing', Archiv Eisenhuttenwessen, 1949, 20(7/8), pp. 229-242. (Available in English as Translation No. 2450, Henry Brutcher, ASM, Metals Park, Ohio, USA).
11. Greenwood, J. Neill,- 'Intercrystalline Cracking of Metals', Bull. Inst. Met., 1952, 1, pp. 104-5.
12. Greenwood, J. Neill, 'Intercrystalline Cracking of Brass', Bull. Inst. Met., 1952, 1, pp. 120-21.
13. Greenwood, J. Neill, Miller, D. R. and Suiter, J. W.- 'Intergranular Cavitation in Metals', Acta Met., 1954, 2, pp. 250-58.
14. Perry, A. J.- 'Cavitation in Creep', J. of Mat. Sci., 1974, 9, pp. 1016-39.

542

15. Svenson, L-E. and Dunlop, G. L.- 'The Role of Inter-
    facial Dislocations in the Nucleation of Intergranular
    Creep Cavities', Canadian Met. Quarterly, 1979, 18,
    pp. 39-47.
16. Svenson, L-E. and Dunlop, G. L.- 'The Growth of Inter-
    granular Cavities', Int. Met. Rev., In Press.
17. Argon, A. S., Chen, I. W. and Lau, C. W.- 'Intergranular
    Cavitation in Creep, Theory and Experiments', Creep-
    Fatigue-Environment Interactions, Ed. Pelloux, R. M.
    and Stoloff, N. S., The Metallurgical Soc. of AIME,
    1980, pp. 46-85.
18. Hull, D. and Rimmer, D. E.- 'The Growth of Grain
    Boundary Voids under Stress', Phil. Mag., 1959, 4,
    pp. 673-687.
19. Hopkin, L. M. T.- 'A Note on the Effect of Antimony on
    Hole Formation During the Diffusion of Zinc from Brass
    in Vacuo', J. Instit. Metals, 85, 1956-57, pp. 422-24.
20. McLean, D.- 'A Note on the Metallography of Cracking
    during Creep', J. Instit. Metals, 85, 1956-57,
    pp. 468-72.
21. Rellick, J. R. and McMahon, C. J., Jr.- 'Intergranular
    Embrittlement of Iron-Carbon Alloys by Impurities',
    Met. Trans., 1974, 5, pp. 2439-2450.
22. Abiko, K. A., Bodnar, R. L., and Pope, D. P.- 'Impurity,
    Grain Size, and Hardness Effects on the Notched Bar
    Creep Rupture Ductility of 2 1/4 Cr, 1Mo Steel', Ductil-
    ity and Toughness Considerations in Elevated Temperature
    Service, Ed. Smith, G. V., ASME, 1978, pp. 1-10.
23. Seah, M. P.- 'Impurities, Segregation and Creep
    Embrittlement', Philos. Trans. R. Soc. London A, 1980,
    295, pp. 265-78.
24. Middleton, C. J.- 'Cavitational Control in Steels of
    High Residual Element Content', same as #23, p. 305.
25. Wilkinson, D. S., Abiko, K., Thyagarajan, N. and
    Pope, D. P.,- 'Compositional Effects on the Creep
    Ductility of a Low Alloy Steel', Met. Trans A, 1980,
    In Press.
26. Gupta, D.- 'Influence of Solute Segregation on Grain
    Boundary Energy and Self Diffusion', Met. Trans. A, 1977,
    8A, pp. 1431-1438.
27. Hopkin, L. M. T. and Jenkinson, E. A- 'Creep Resistance
    and Metallographic Structure of 2 1/4% Cr-1%Mo Steel With
    and Without a Tin Addition', J. Iron Steel Inst., 1962,
    200, pp. 356-359.
28. Bruscato, R.- 'Temper Embrittlement and Creep Embrittle-
    ment of 2 1/4 Cr-1Mo Shielded Metal-Arc Weld Deposits',
    Welding Res. Supplement, 1970, pp. 148s-156s.
29. Wolstenholm, D. A.- 'Intergranular Embrittlement of 2 CrMo
    Manual Metal Arc Welds', Grain Boundaries, Institution of
    Metallurgists, 1976, pp. C7-C12.
30. Ferrell, D. E. and Pense, A. W.- 'Creep and Temper Em-
    brittlement of A-542 Steel', Chrome-Moly Steel in 1976,
    Ed. Smith, G. V., ASME, 1976, pp. 29-49.

31. Yu, J. and McMahon, C. J., Jr.- 'The Effects of Composition Carbide Precipitation on Temper Embrittlement of 2 1/4 Cr-1Mo Steel: Part I. Effects of P and Sn', Met. Trans A, 1980, 11A, pp. 277-290.

32. Yu, J. and McMahon, C. J., Jr.- 'The Effects of Composition and Carbide Precipitation on Temper Embrittlement of 2 1/4 Cr-1Mo Steel: Part II. Effects of Mn and Si', Met. Trans A, 1980, 11A, pp. 291-300.

33. Murza, J. C.- 'The Effects of Composition and Microstructure on Temper Embrittlement in 2 1/4 Cr-1Mo Steel', M. S. Thesis, University of Pennsylvania, Philadelphia, PA., 1978.

34. Lonsdale, D. and Flewitt, P. E. J.- 'The Effect of Small Changes in Impurity Element Content on the Creep Life of 2 1/4 Cr-1Mo Steel', Mat. Sci. and Eng., 1979, 41, pp. 127-36.

35. Needham, N. G. and Orr, J.,- 'The Effect of Residuals on the Elevated Temperature Properties of Some Creep Resistant Steels', same as #23, p. 288.

36. Gooch, D. J.- 'Creep Crack Growth in 2 1/4 CrMo Weld Metals: The Suppression of Trace Element Embrittlement by Creep Strength Effects', same as #23, p. 295.

37. Hopkins, B. E., Tipler, H. R., and Branch, G. D.- 'Improvements in the Creep Properties of Cr-Mo-V Steels Through Enhanced Purity', J. Iron and Steel Inst., 1971, 209, pp. 745-746.

38. Tipler, H. R. and Hopkins, B. E.- 'The Creep Cavitation of Commercial and High Purity Cr-Mo-V Steels', Metal Sci., 1976, 10, pp. 47-56.

39. Benes, F. and Skyvor, P.- 'Effect of Copper and Tin on the Creep Resistance Properties of Cr-Mo-V Steel', Hutnicke Listy, 1972, 3, pp.197-200 (Translated from Czech by the British Iron and Steel Institute, paper 10480, Aug. 1973).

40. Viswanathan, R.- 'Effects of Residual Elements on Creep Properties of Ferritic Steels', Metals Engng. Q., 1975, 15 (4), pp. 50-56.

41. Viswanathan, R. and Beck, C. G.- 'Effect of Al on the Stress Rupture Properties of Cr-Mo-V Steels', Met. Trans. A, 1975, 6A, pp. 1997-2003.

42. Tait, R. A. and Knott, J. F.- 'The High Temperature Intergranular Fracture of a Low Alloy Creep Resisting Steel', same as #29, pp. C1-C6.

43. Tu, L. K. and Seth, B. B.- 'Effects of Composition, Strength and Residual Elements on Toughness and Creep Properties of Cr-Mo-V Turbine Rotors", Metals Technology, 1978, 5(3), pp.79-91.

44. Roan, D. F. and Seth, B. B.- 'A Metallographic and Fractographic Study of the Creep Cavitation and Fracture Behavior of 1Cr-1Mo-.25V Rotor Steels with Controlled Residual Impurities', same as #22, pp. 79-97.

45. Tipler, H. R.- 'The Influence of Purity on the Strength and Ductility in Creep of CrMoV Steels of Varied

544

Microstructure', same as #23, pp. 213–233.

46. King, B. L.- 'Intergranular Embrittlement in CrMoV Steels: An Assessment of the Effects of Residual Impurity Elements on High Temperature Ductility and Crack Growth', same as #23, pp. 235–251.

47. Viswanathan, R.- 'Temper Embrittlement and Creep Embrittlement of 1.25Cr-0.5Mo Steels Containing Sb, Sn, P and B as Impurities', Scripta Met., 1974, 8, pp. 1225–29.

48. Viswanathan, R.- 'Effect of Impurities on Creep Properties of 1.25Cr-0.5Mo Steels', Metals Tech., 1975, 2, pp. 245–8.

49. Viswanathan, R.- 'Effect of Sb, P, Sn and B on the Microstructure and Creep Properties of Normalized and Tempered 1.25Cr-0.5Mo Steels', Met. Trans. A, 1975, 6A, pp. 1135–41.

50. Viswanathan, R.- 'Effect of Ti and Ti+B Additions on the Creep Properties of 1.25Cr-0.5Mo Steels', Met. Trans. A, 1977, 8A, pp. 57–61.

51. Presser, R. I. and McPherson, R.- 'Boron Segregation and Elevated Temperature Embrittlement of Ferritic Steel', Scripta Met., 1977, 11, pp. 745–49.

52. Presser, R. I. and McPherson, R.- 'Prior Austenite Grain Boundary Embrittlement of Low Aloy Steel by Boron', same as #23, p. 298.

53. Kirby, B. R. and Beevers, C. J.- 'A Note on the Effect of 0.14 wt% Phosphorus Addition to the Creep Behavior of Alpha-Iron', Scripta Met., 1977, 11, pp. 659–663.

THE ROLE OF STRESS STATE ON THE CREEP RUPTURE OF 1%Cr½%Mo AND 12%Cr1%MoVW TUBE STEELS

R.J.Browne, D. Lonsdale and P.E.J. Flewitt.

Central Electricity Generating Board,
Scientific Services Department,
South Eastern Region,
Gravesend,
Kent. U.K.

SUMMARY

Results of creep tests on uniaxial specimens and tubes of two ferritic steels, 12%Cr1%MoVW and 1%Cr½%Mo, carried out at a temperature of 848K over a range of stresses are described. The tubes were tested under internal pressure both with and without externally applied axial loads. Optical and electron microscopy are used to examine the microstructural changes which lead to creep failure. The results are examined in terms of the contribution of stress state in the strain controlled rupture of the 12%Cr1%MoVW steel and the cavitation controlled rupture of the 1%Cr½%Mo steel. The variation of creep life with changing stress state between the uniaxial creep specimens and the tubes is discussed.

1.   INTRODUCTION

Methods currently employed in the design and residual life estimation of tubes and pipes operating in the creep range are invariably based on data which has been obtained using accelerated isothermal, uniaxial, constant load testing procedures [1-3]. However, the stresses in internally pressurized, thick walled tubes are essentially biaxial and variable through the wall thickness [4]. During creep the initial elastic stress state will redistribute to a stationary state [5]. Throughout the tertiary creep range further redistribution will occur. Thus it would be helpful to identify a single component of stress so that uniaxial creep data could be used to describe the creep life of a tube [5] [6].

Creep life is a combination of the interrelated processes of deformation and fracture. The former is controlled by the motion of dislocations whereas the latter can be governed by either total strain or the formation of cracks or cavities along grain boundaries. At temperatures above $\sim 0.3\ T_m$ the uniaxial creep of metals is frequently approximated to a power law [7]: $\dot{\varepsilon}_s = c\ (\sigma/\mu)^n$ where c and n are materials parameters, μ is the shear modulus and $\dot{\varepsilon}_s$ is the minimum creep rate. As stress is varied the value of n has been observed to change. Although this has been interpreted in terms of the stress dependence of the deformation mechanisms, uncertainty still exists with regard to the controlling mechanism within a given stress range. Moreover, the failure process may also change with stress. At high stresses an obvious mechanism is an extension to the creep regime of low temperature ductile fracture whereas at low stresses grain boundary cavitation occurs [8] [9].

Several investigations have resulted in the general conclusion that two criteria may control creep fracture under complex stress; namely the equivalent stress and the maximum principal stress [1] [3] [10] [11] [12] [13]. In general the Von Mises equivalent stress describes the stress dependence of creep rate and where no or little evidence of intergranular damage is apparent then it is also appropriate to the rupture life. In contrast for materials where grain boundary cracking or cavitation is present then creep life is characterised by a combination of the maximum principal and the equivalent stress.

For tubes, the simple engineering design criterion requires that the representative rupture stress σ_r is calculated using the mean diameter hoop stress, $\sigma_M = \sigma_r = pD/2t$ where p = operating pressure, D = mean diameter and t = tube wall thickness [14]. σ_M is an approximation to the maximum hoop stress in an elastically pressurized thin walled tube. Results of limited duration internal pressure creep tests [15] show an empirical correlation between creep life and uniaxial creep life when $\sigma_r \equiv \sigma_M$. However results of longer term tests on $1\%Cr\frac{1}{2}\%Mo$, $12\%Cr1\%Mo$ and $9\%Cr1\%Mo$ steel tubes indicate that σ_M over-estimates σ_r for conditions typical of service [16] [17]. Thus, residual life estimates of tubes and pipes using σ_M may be unduly conservative.

In this paper creep lives of tubes are compared with those of uniaxial specimens for two ferritic steels, a $12\%Cr1\%MoVW$ and a $1\%Cr\frac{1}{2}\%Mo$, tested at 848K. The tubes have been tested under internal pressure both with and without externally applied axial loads. Microstructural changes leading to creep rupture are described and correlations between experimentally derived tube creep rupture stresses and elastic and steady state creep equivalent and maximum principal stresses are described.

2.   STRESSES WITHIN TUBES

The state of elastic stress in an internally pressurized homogeneous cylinder is given by [17]:

$$\sigma_H = \left\{ p\, R_i^2 / (R_o^2 - R_i^2) \right\} \left\{ 1 + R_o^2 / r^2 \right\} \qquad \ldots (1)$$

$$\sigma_A = p\, R_i^2 / (R_o^2 - R_i^2) \qquad \ldots (2)$$

$$\sigma_R = \left\{ p\, R_i^2 / (R_o^2 - R_i^2) \right\} \left\{ 1 - R_o^2 / r^2 \right\} \qquad \ldots (3)$$

where σ_H, σ_A, and σ_R are the elastic hoop, axial and radial stresses respectively, R_i and R_o are the bore and outer radii and r is the pipe radius.

If the stress dependence of secondary creep is given by $\dot{\varepsilon}_s \propto \sigma^n$ (see section 3), then after redistribution from the initial elastic state of stress, tube stresses [18] will be given by:

$$\sigma_{HS} = \left\{ (2/n - 1)\ (R_o/r)^{2/n} + 1 \right\} \left\{ (R_o/R_i)^{2/n} - 1 \right\}^{-1} p \qquad \ldots (4)$$

$$\sigma_{AS} = \left\{ (1/n - 1)\,(R_o/r)^{2/n} + 1 \right\} \left\{ (R_o/R_i)^{2/n} - 1 \right\}^{-1} p \qquad \ldots (5)$$

$$\sigma_{RS} = \left\{ 1 - (R_o/r)^{2/n} \right\} \left\{ (R_o/R_i)^{2/n} - 1 \right\}^{-1} p \qquad \ldots (6)$$

where σ_{HS}, σ_{AS} and σ_{RS} are the stationary state creep hoop, axial and radial stresses respectively.

The creep deformation of thick walled ferritic steel tubes and pipes has been shown to be controlled by the same form of multiaxial stress as that which governs yielding in ductile materials [19] [20] i.e. the Von Mises equivalent stress:

$$\bar{\sigma} = \frac{1}{\sqrt{2}} \left\{ (\sigma_H - \sigma_A)^2 + (\sigma_A - \sigma_R)^2 + (\sigma_R - \sigma_H)^2 \right\}^{\frac{1}{2}} \qquad \ldots (7)$$

For internally pressurized tubes, the stationary state equivalent creep stress is given by:

$$\bar{\sigma}_s = \frac{\sqrt{3}}{n}\ (R_o/r)^{2/n} \left\{ (R_o/R_i)^{2/n} - 1 \right\}^{-1} p \qquad \ldots (8)$$

In the presence of end-loads, elastic axial and equivalent stresses are increased (equations 7 and 8) and creep stresses are given by the analysis of Finnie [21] using an appropriate numerical solution [22].

Elastic and stationary state creep stress distributions for the present tube tests are given in Figure 1.
From Figure 1, it is apparent that there is a position within the tube wall where the value of $\bar{\sigma}_s$ is insensitive to the value of n and this position is known as the skeletal point. Since it can be shown that $\dot{\varepsilon} \propto r^{-2}$, the deformation rate of any point in the tube wall can be derived from a knowledge of $\bar{\sigma}_s$ at the skeletal point, the skeletal point radius and uniaxial creep

rate. This stress, referred to as the reference stress, σ^*_R, can be derived either approximately using graphical constructions, Figures 1(a) to (d) or it can be calculated using the solution due to Fairbairn and Mackie [23]:

$$\sigma^*_R = \left\{ \sqrt{3}/((R_o/R_i)-1) \right\} \left\{ ((R_o/R_i)^2-1)/(n(R_o/R_i)^{2/n}-1) \right\}^{n/(n-1)}$$

$$\ldots (9)$$

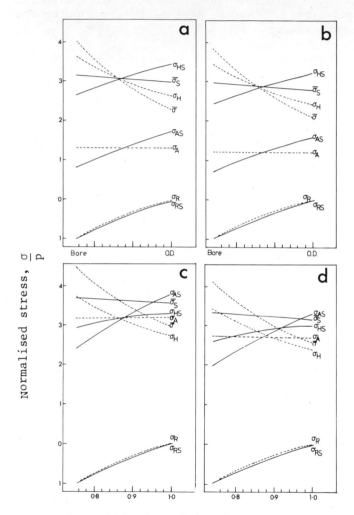

FIGURE 1. Elastic and creep stress distributions in
(a) 12%Cr1%MoVW tubes under internal pressure
(b) 1%Cr$\frac{1}{2}$%Mo tubes under internal pressure
(c) 12%Cr1%MoVW tubes under internal pressure plus end load
(d) 1%Cr$\frac{1}{2}$%Mo tubes under internal pressure plus end load

3. EXPERIMENTAL

The two compositions of stock steel tubing, nominally
12%Cr1%MoVW and 1%Cr½%Mo, used for the experimental programme
were chemically analysed and the compositions are given in
Table 1. The 12CMVW steel tubes were 38mm outside diameter
with a wall thickness of 4.5mm whereas the 1CM steel tubes
were 58mm outside diameter with a wall thickness of 6.5mm.
450mm lengths of tube were fabricated into test specimens by
welding on end-caps and the internal volume of each specimen
was minimised by the inclusion of loose filler bars. The

Code	Composition (%)						
	Cr	Mo	V	W	C	Si	Mn
12CMVW	12.1	1.03	0.30	0.86	0.22	0.33	0.69
1CM	0.84	0.49	-	-	0.09	0.15	0.51

TABLE 1: Chemical Analysis

12 CMVW tubes were heat treated at 1323K and then tempered
at 1013K for 1 hr to produce a microstructure of tempered
martensite. The 1CM tubes were normalised at 1193K and then
tempered at 923K for 1 hr to produce a microstructure
comprising large areas of ferrite with about 10% bainite. The
internal pressure testing procedure has been described
previously [24]. The tubes were tested at a temperature of
848K, controlled to ± 1 deg, and at internal pressures
controlled to ± 1%. End-loads were applied using modified
lever loading creep machines to give a ratio of total
elastic axial stress to mean diameter hoop stress of 0.86.
Uniaxial creep tests were carried out at 848K using either
axially loaded tubes in modified creep machines or
conventional geometry 10.5mm long 3.0mm diameter gauge
specimens, machined from the tubes and tested in a controlled
inert atmosphere. The uniaxial and tube test conditions are
summarised in Table 2.

After testing, samples were cut from both tubes and uniaxial
specimens and were prepared using standard metallographic
techniques. To ensure all grain boundary cavitation or
intragranular voids were revealed a repeated sequence of
polishing on γ-alumina and etching in 2% nital was carried
out [25].

4. RESULTS

The uniaxial creep data in Table 2 gives the stress exponents
of the minimum creep rate, as n = 24 for the 12CMVW steel
and n = 5 for the 1CM steel over the range of stresses

examined. The variation in time to failure for the uniaxial
creep tests is shown in Figure 2 together with the relevant
mean I.S.O. data lines.

TUBES

Material	Specimen No.	Int Press. MPa	Applied End Load MN	σ_M MPa	σ_A MPa	Rupture Life h	Reduct^n of wall thickness %	Cracking Mode
12CMVW	435	38.0	0	140	51	16046	59	Transgranular Axial from OD
12CMVW	452	37.1	0	140	51	19331	41	"
12CMVW	436	36.1	0	130	47	22673	45	"
12CMVW	437	38.2	0.0334	140	124	8601	67	Transgranular circum. from OD
12CMVW	438	35.7	0.0313	130	116	11684	60	"
1CM	415	20.6	0	70	26	5473	14	Intergranular Axial from OD
1CM	425	20.2	0	70	26	5941	22	"
1CM	416	17.7	0	61	22	12546	14	"
1CM	426	17.8	0	61	22	10968	22	"
1CM	427	20.4	0.0287	70	57	4644	24	Axial and circum. from OD
1CM	428	17.8	0.0276	61	53	6591	22	"

UNIAXIAL

Material	Specimen No.	Minimum creep rate h^{-1}	σ_A MPa	Rupture Life h	ε_f %	Cracking Mode
12CMVW	T1	1.4×10^{-6}	145	5108	39	Transgranular
12CMVW	T2	5.9×10^{-6}	155	(3674)	-	—
12CMVW	T3	9.9×10^{-6}	165	12714	43	Transgranular
12CMVW	R1	1.0×10^{-3}	195	73	59	"
12CMVW	R2	-	205	18	51	"
12CMVW	R3	1.1×10^{-3}	185	95	59	"
12CMVW	R4	4.0×10^{-4}	180	211	60	"
1CM	T4	7.8×10^{-7}	61	(9234)	—	-
1CM	T5	1.6×10^{-6}	66	7902	6.0	Intergranular
1CM	T6	1.4×10^{-6}	70	6009	8.8	"
1CM	R5	2.9×10^{-5}	130	802	3.7	"
1CM	R6	7.5×10^{-5}	150	436	5.4	"
1CM	R7	1.1×10^{-4}	160	341	5.7	"
1CM	R8	1.5×10^{-4}	170	262	5.6	"

TABLE 2. Creep tests: conditions and results |All test at 848K; T=tube, R=rod|

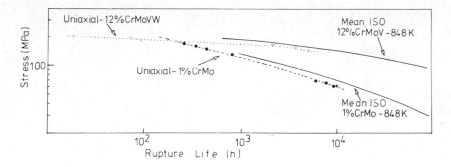

FIGURE 2.   Uniaxial rupture data for the 12%Cr1%MoVW and
1%Cr$\frac{1}{2}$%Mo steels.

The 12CMVW tubes subject to internal pressure contained only
axial cracks, Figure 3(a),whereas those tested with an
additional end-load contained only circumferential cracks
Figure 3(b). By comparison the 1CM steel tubes subject to
internal pressure also contained only axial cracks, Figure 3(c),
whereas those subject to an additional end-load contained a
'crazed' pattern of surface cracks, Figure 3(d).

In the case of the 1CM steel tubes and uniaxial creep specimens
failure resulted from the coalescence of discrete cavities
and small cracks developed on grain boundaries which were
oriented approximately normal to the major stress axis,
Figure 4(a). Within the internally pressurized tube,grain
boundary cavities were observed only in microsections taken
normal to the tube axis and those in the axial-circumferential
plane, whereas they were observed in these sections as well
as the axial-radial plane when end-loads were applied. The
small grain size prevented any assessment of the distribution
of these cavities relative to the stress axis.

The through thickness distribution of cavitation was
determined on samples from tube walls which were
fractured under liquid nitrogen to reveal grain boundaries
oriented normal to the hoop and axial stresses. These
fracture surfaces were examined in the scanning electron
microscope (SEM). Specimens were taken from regions of the
tubes close to failure and at positions where the diametral
strain was about 2%. The amount of cavitation was determined
in each of four equal width zones across the tube wall by
measuring the fractional area, A/A_o, of cavitated grain
boundary from photographs taken on the SEM at 550 X
magnification. Figure 5 (a to d) shows the variation of the
amount of cavitation with position and orientation in the tube
wall. In the internally pressurized tubes cavitation was
predominantly present on boundaries normal to the hoop stress
with little evidence on boundaries normal to the axial stress.

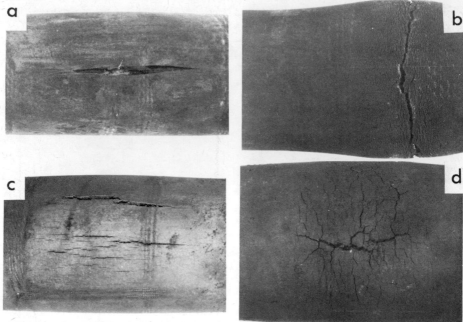

FIGURE 3.   Distribution of cracking on failed tubes of
            12%Cr1%MoVW a)internally pressurized
                         b)plus end load
            1%Cr$\frac{1}{2}$%Mo   c)internally pressurized
                         d)plus end-load

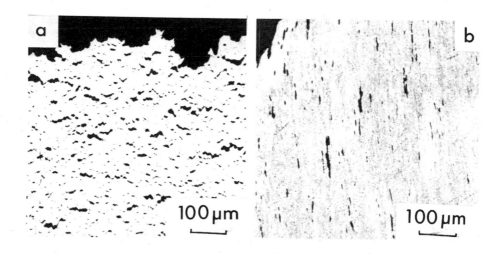

FIGURE 4.   Optical micrographs of uniaxial creep samples
            showing:
            (a) cavitation in 1%Cr$\frac{1}{2}$%Mo steel.
            (b) ductile voids in 12%Cr1%MoVW steel.

In the tubes subject to additional axial loading the degree
of cavitation was similar on boundaries normal to the hoop and
axial stresses. In both cases the amount of cavitation was a
maximum at the outer surface and decreased through the wall
to the bore.

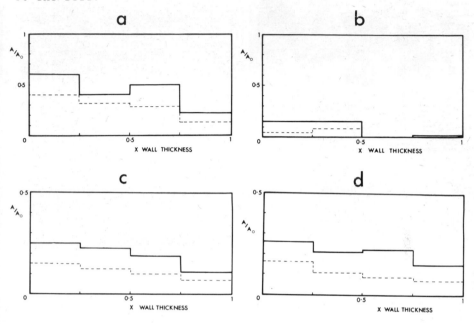

FIGURE 5. Variation of the mean area fraction of cavitation
as a function of position and orientation in the
1%Cr$\frac{1}{2}$%Mo tube wall (a) internally-pressurized tube
normal to hoop stress, (b) internally pressurized
tube normal to axial stress (c) internally
pressurized and end-loaded tube normal to axial
stress (d) internally pressurized tube normal to
the axial stress. | - Corresponds to region of
failure..... corresponds to region of tube
strained to ∿ 2%.|

The cavities were of irregular shape with smooth inner
surfaces, and generally did not exceed 2μm diameter,
Figure 6(a).

In the case of the 12CMVW steel, failure of uniaxial creep
specimens and tubes occurred by local necking following
substantial elongations. In the case of the tubes,necking
within the wall occurred in a direction normal to the maximum
principal stress and the fracture was predominantly
transgranular. Within each tube voids were formed and these
were extensive in the necked region characteristic of the
localised plastic flow, Figure 6(b). Voids were observed
in both the tubes and the uniaxial creep samples even within

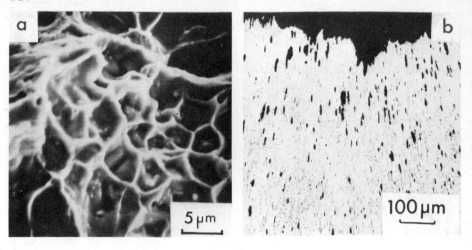

FIGURE 6   a) Scanning electron micrograph showing grain
             boundary cavities in 1%Cr½%Mo tube. (Internally
             pressurized section parallel to axial-radial
             plane)
          b) Optical micrograph of voids in 12%Cr1%MoVW tube
             (Internally pressurized section parallel to plane
             normal to tube axis).

regions of the failed specimen subjected to relatively small
strains ~ 5%. Essentially the voids elongated in the
direction of the maximum principal stress and were thus
distinct from the grain boundary cavities in the 1CM steel.
Thus tubes which failed by axial cracking contained voids
elongated in the circumferential direction, parallel to the
hoop stress and for those containing circumferential cracks
void elongation was in the axial direction. It was difficult
to quantify the size or total volume due to the irregular
void shape. However numbers were measured from optical
micrographs, Figure 7.

5.   DISCUSSION

These results for the two ferritic steels reveal that
different processes control the failure and thus creep life
for the stresses and temperature considered. Since for each
material the failure mechanism identified on the biaxially
stressed tubes was similar to that in the uniaxially stressed
specimens, comparisons of the results are valid and should
provide a measure of the contribution of stress state to the
overall creep life [8].

The major factors which will influence σ_r in internally-
pressurized thick-walled tubes operating in the creep range
are [10] the multiaxial stress rupture criterion and the creep
rate stress exponent (n). For the majority of engineering

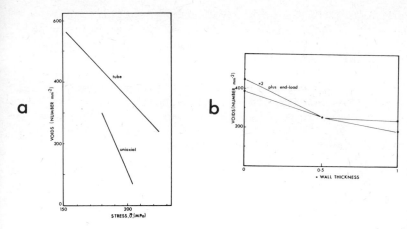

FIGURE 7.   Variation of number of voids in 12%Cr1%MoVW tubes
            a) with stress   b) through wall thickness.

materials failure is controlled by a combination of σ_1 and $\bar{\sigma}$
[1] [8] [10]. The creep rate stress exponent will influence
the rate of stress redistribution and determine the through
wall stress gradients.

Failure of the 12CMVW tubes occurred by a ductile shear mode
with no evidence of intergranular cavitation such that both
deformation and failure should be controlled by $\bar{\sigma}_s$. The voids
which formed during the accumulation of creep strain, Figure
6(b), may well accelerate failure. However, if one invokes a
critical distance criterion for coalescence of voids such that
failure occurs when void height approximates to their
separation [26] [27] it can be shown that they do not
significantly reduce the time to failure in both the uniaxial
or tube specimens.

Figure 1 illustrates that, because of the high value of n
exhibited by the 12CMVW steel, over the stress range of
interest, $\bar{\sigma}_s$ is essentially constant across the tube wall.
This is true for both the internally pressurized and end-
loaded tubes and was supported by the constant dislocation
substructure observed across the tube walls. However, the
hydrostatic stress varies through the wall and this
correlates with the change of number of voids, which is also
maximum at the outer surface, Figure 7(b). Since tertiary
redistributions will be small, the representative rupture
stress, σ_r, should be $\approx \sigma_R^* \approx \bar{\sigma}_s$. The use of σ_M both
overestimates σ_r, (as determined by applying tube lives to the
uniaxial rupture data) in the tube tests and does not reflect
the change in creep life with differing stress state.
However, changes in $\bar{\sigma}_s$ with changing stress state correlate
very well (coeff 0.98) with the observed variations in σ_r.
Moreover the ratio of σ_r to $\bar{\sigma}_s$ is very close to unity.

T

These results show that current design and residual life
estimation methods based on the mean diameter formula may be
unduly conservative particularly in situations where system
stresses are small. Any agreement between σ_r and σ_M in the
end-loaded tubes is fortuitous since failure occurred
circumferentially. Nevertheless, in situations on plant where
high system stresses are sustained throughout service (e.g.
deadweight imbalance in a pipework system), it is evident that
σ_R^* may approach σ_M. Clearly if $\sigma_r = \sigma_R^*$ is used in life
predictions the levels of system stress must be evaluated.
However, it is worth noting that system stresses are not
usually primary stresses and consequently will decay during
early service. End-loads in the present tube tests were
maintained constant throughout such that the typical service
situation was not simulated and system stresses will generally
be less important.

In contrast to the 12CMVW tubes, the 1CM tubes failed by
intergranular cavitation such that creep life should depend on
both $\bar{\sigma}$ and σ_1 [1] [10] [28]. The area fraction of cavitation,
Figure 5, was greatest at the outer surface where the
stationary state hoop stress is maximum, Figure 1. Thus the
cavitation which lead to failure reflects the maximum
principal stress distribution. Recently Lonsdale and Flewitt
[10] have shown that when failure is controlled by the
diffusional growth of continuously nucleating grain boundary
cavities the creep life varies as

$$t_f^{-1} \; \alpha \; \sigma_1^{2/5} \; \bar{\sigma}^{3n/5} \qquad \qquad \ldots (10)$$

This was shown to be valid for a $2\frac{1}{4}\%Cr1\%Mo$ steel. In the
present 1CM steel n has a value of 5 such that the
representative rupture stress, σ_r, will be $[\sigma_1^{2/5} \; \bar{\sigma}^3]^{5/17}$
which will differ only slightly from $\bar{\sigma}$.

For the 1CM tubes the use of σ_M again fails to describe the
variation in creep life for the different stress states.
Changes in the mean value of $\bar{\sigma}_{SM}$ with stress state correlate
(coeff. 0.97) with the observed variations in σ_r obtained
from uniaxial data. However, the correlation is improved
(coeff. 0.98) when $(\sigma_1^{2/5} \; \bar{\sigma}^3)5/17$ is considered.

The present results demonstrate that the mean diameter formula
does not account for changes in creep life with stress state
for two ferritic steels which fail by different mechanisms.
In both cases this can be described by the stationary
state equivalent stress (or reference stress) although for the
1CM tubes an improved description is obtained using a combined
equivalent and principal stress relationship.

REFERENCES

1.   McLEAN, D. DYSON,B.F.and TAPLIN, D.M.R. - Advances in
Research on Strength and Fracture of Materials,
4th Int. Conf. on Fracture, 1977, Waterloo, Canada,
Pergamon, Oxford 1978, Vol. 1, 325.

2.   HENDERSON, J.-Trans. ASME, J. Eng. Mat. & Tech., 1979,
101, 356.

3.   JOHNSON, A.E. - Metall. Rev., 1960, 5, 447.

4.   GOODALL, I.W. LECKIE, F.A. PONTER, A.S and TOWNLEY, C.H.A
- J. Eng. Mat. and Tech. 1979, 101, 349.

5.   KACHANOV, L.M.- Rupture Time under Creep Conditions -
Problems in Continuum Mechanics, 1961, p. 202.

6.   ABO el Ata,M.M. and FINNIE,I. - U.T.A.M. Symp. On Creep
Structures 1970, Springer Berlin.

7.   MUKHERJEE, A.K., BIRD, J.E. and DORN, J.E. - Trans. Am.
Soc. Met., 1969, 62, 155.

8.   ASHBY, M.F., GANDHI, G. and TAPLIN, D.M.R. - Acta. Met.,
1979, 27, 699.

9.   PAVINICH, W. and RAJ, R. - Met. Trans., 1977, 8A, 1917.

10.   LONSDALE, D. and FLEWITT, P.E.J. - Proc. Roy. Soc., A,
(in press).

11   JOHNSON, A.E., HENDERSON, J. and KHAN,B. - Inst. J. Mech.
Sci., 1962, 4, 195.

12.   JOHNSON, A.E., HENDERSON, J. and KHAN, B. - Engineer
(Lond), 1964, 217, 729.

13.   DYSON, B.F. and McLEAN, D. - Met. Sci., 1977, 11, 37.

14.   British Standards Institution, BS 2633: May 1973.

15.   CHITTY, A. and DUVAL, D. - Electrical Research
Association, 1963, No. 5033.

16.   CANE, B.J. and BROWNE, R.J.-CEGB Report, 1979,
RD/L/N/177/78,

17.   ROARK,R.J. - 'Formulas for Stress and Strain,'
McGraw-Hill, London 1965.

18.   BAILEY, R.W. - Proc. Inst. Mech. Eng. 1935, 131,131

19.   CANE, B.J. and BROWNE,R.J. - To be published.

20.   COLEMAN, M.C., PARKER, J.D., WALTERS, J.D. and
WILLIAMS, J.A. - 3rd Int. Conf. on Mechanical
Behaviour of Materials, Cambridge, England, 1979,
Pergamon, Oxford, 1980, p. 193.

21.   FINNIE, I. - Journal of Basic Engineering, 1960, 689.

22.   FIELDING, P.J. - CEGB Report, 1977, SSD/SE/M90/77.

23.   FAIRBAIRN, J. and MACKIE, W.W. - Creep in Structures,
IUTAM Symposium, Gothenburg, 1970.

24.   ELDER, W.J. - CEGB Report, 1971, SSD/SE/R16/71.

25.   LONDSALE, D. and FLEWITT, P.E.J. - Mat. Sci. and Eng.,
1978, 32, 167.

26.   THOMASON, P.F. - J. Inst. Metals, 1968, 96, 360.

27.   BROWN, L.M. and EMBURY, J.D. - Proc. 3rd Int. Conf. on
Strength of Metals and Alloys, Inst. of Metals,
London (1973), p. 164.

28.   CANE, B.J. - 3rd Int. Conf. on Mechanical Behaviour of
      Materials, Cambridge, England, 1979. Pergamon,
      Oxford, 1980, p. 173.

ACKNOWLEDGEMENT.

This paper is published with the permission of the Director-
General, CEGB, South Eastern Region.

EFFECTS OF PLASTIC PRESTRAIN ON THE CREEP
OF ALUMINIUM UNDER BIAXIAL STRESS.
D.W.A. Rees

Engineering School, Trinity College,
Dublin 2, Ireland.

Results are presented from 300-400 hr biaxial stress
creep tests performed under combined tension (σ) and torsion
(τ) on thin walled tubes of EIA aluminium for $T/_{Tm} = 0.32$
in the annealed and prestrained conditions. Plastic strain
values in the equivalent percentage range $-4 < \bar{\varepsilon}_M^P < 7$
were applied by preloading in each of tension-compression
and forward-reversed torsion. Subsequent creep stress levels
were achieved by incremental loading along the radial path
$\tau/_{\sigma} = 1$. Axial (ε_{zz}) and shear ($\gamma_{\theta z}$) strain components were
measured continuously throughout deformation.

A Bauschinger effect, a cross effect and a rotation in
the component strain path were the observed manifestations of
anisotropy in subsequent plasticity and creep due to prior
plastic strain history. Prestrains most beneficial to
biaxial creep resistance were those for which the Bauschinger
effect was absent. It is shown from the normality rule,
that the observed strain paths can be modelled through
translation in a field of yield and creep potentials.

1. INTRODUCTION

Multiaxial creep theory assumes a material that has the same flow properties in all directions irrespective of the direction of straining [1].[+] Its use, therefore, lies in its application to polycrystalline metallic materials where the imposed strains occur over a great number of randomly orient- ated crystal grains that result in approximately isotropic macroscopic behaviour. The assumption of isotropy does, however, present a serious limitation to the theory. It is well known that real materials exhibit a Bauschinger effect where, for example, a history of one directional straining would change the tensile and compressive creep strengths of all other directions. The present investigation further examines anisotropic creep deformation. The objective was to establish experimentally the effects of a known plastic strain history on the subsequent biaxial creep behaviour of a material by comparing the results with those for the material in the near isotropic annealed condition.

An examination of strain vector directions has enabled the application of a model of anisotropic hardening.

Full experimental details of the test rig, instrumentation, material and testpiece preparation are described elsewhere [2].

2. RESULTS AND DISCUSSION

Table 1 gives details of the eight plastic strain values used for the investigation. They are expressed in actual and equivalent strain. Axial and shear prestrains were achieved under approximately the same equivalent strain rate ($\dot{\bar{\varepsilon}}_M^P$). Nominal equivalent strain values $\bar{\varepsilon}_M^P$ are referred to hereafter where annealed material is designated by $\bar{\varepsilon}_M^P = 0$.

[+] References are given under Section 4.

TABLE I   Plastic Strain History (%)

TENSION – COMPRESSION $\dot{\bar{\epsilon}}^P_M = \dot{\epsilon}^P_{zz}$ $= 1.67 \times 10^{-4}\,s^{-1}$		FORWARD-REVERSED TORSION $\dot{\gamma}^P_{\theta z} = 2.75 \times 10^{-4}\,s^{-1}$ $\{\ \dot{\bar{\epsilon}}^P_M = \dot{\gamma}^P_{\theta z}/\sqrt{3} = 1.59 \times 10^{-4}\,s^{-1}\ \}$		
Actual $\bar{\epsilon}^P_M = \epsilon^P_{zz}$	Nominal $\bar{\epsilon}^P_M$	Actual $\gamma^P_{\theta z}$	Equivalent $\bar{\epsilon}^P_M = \gamma^P_{\theta z}/\sqrt{3}$	Nominal $\bar{\epsilon}^P_M$
−4.28	−4	−7.00	−4.04	−4
−3.28	−3	−5.25	−3.03	−3
—	—	−4.68	−2.70	−2.7
−1.87	−2	−3.45	−1.99	−2
0	0	0	0	0
1.98	2	3.52	2.03	2
2.96	3	5.18	2.99	3
3.92	4	6.89	3.98	4
6.80	6.8	—	—	—

2.1. Creep Curves

The tensile (ϵ_{zz}) and shear ($\gamma_{\theta z}$) creep curves under combined axial and shear stresses of $\sigma_{zz} = 16$ N/mm^2 and $\tau_{\theta z} = 15.2$ N/mm^2 are plotted in Figs. 1 and 2 for the prestrains of Table I.

(a) Prior Tension and Compression

Figs 1A and 1B show the effect of prior plastic axial strain on the $\epsilon_{zz}, \gamma_{\theta z}$ component creep curves. For a typical time of 250 hr (Fig. 1A) the tensile creep resistance is increasingly improved by increasing amounts of tensile and compressive prestrain. For the same prestrain value, however, the material exhibited more tensile creep after compressive prestraining than after tensile prestraining. The Bauschinger effect is therefore evident throughout creep. Furthermore, we see from Fig. 1B that the shear creep resistance at 250 hr is increasingly improved by increasing amounts of tensile and compressive prestrain (the cross effect). For the same prestrain value, however, the material exhibited more shear creep after compressive prestraining.

A comparison between the creep curves for annealed and prestrained material in Figs. 1A and 1B shows that a low prestrain has little or no effect in improving creep

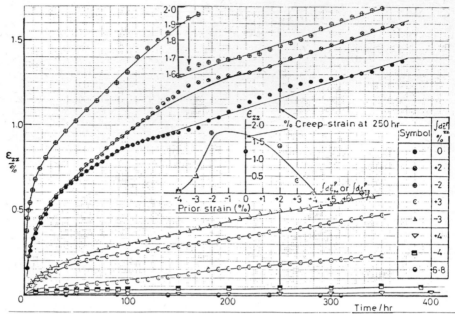

Fig. 1 A ε_{zz} Creep component for prior plastic axial strain $\left(\int d\varepsilon_{zz}^{p}\right)$

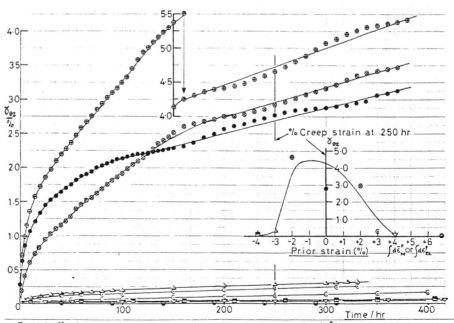

Fig. 1 B $\gamma_{\theta z}$ Creep component for prior plastic axial strain $\left(\int d\varepsilon_{zz}^{p}\right)$

resistance. Indeed a small compressive prestrain yields
more ε_{zz} and $\gamma_{\theta z}$ primary creep strain. There is,however,little
difference in secondary creep rates between the two conditions
of material. Where an improvement in creep resistance is
observed, i.e. for $\bar{\varepsilon}_M^P > \pm 2\%$,both the amount of primary creep
strain and the secondary creep rate are reduced. For
$\bar{\varepsilon}_M^P \geqslant \pm 4\%$ little, if any, primary creep is evident and secondary
creep forms the larger part of subsequent creep strain.

(b) Prior Forward and Reversed torsion

Figs 2A and 2B show similar trends. The Bauschinger
effect is evident in Fig. 2A where for the same prestrain
value the material exhibited more shear creep after reversed
shear prestraining than after forward shear prestraining.
Fig. 2B shows the cross effect of prior plastic shear strain
on the axial creep strain component. Typically after 250hr
increasing amounts of forward and reversed shear prestrain
increasingly improve the tensile creep resistance. A forward
shear prestrain is more effective in this respect.

Little, if any, reduction in both the amount of primary
creep strain and secondary creep rate is apparent for prior
strains of $\pm 2\%$ in Fig. 2A and for -2% in Fig. 2B. However,
greater prestrains are effective in reducing both quantities
irrespective of their sign.

(c) General Comments

Prestrains which are most effective in providing biaxial
creep resistance in a metal are those that avoid a Baushinger
effect and fully exploit a beneficial cross effect. We have
seen that positive prestrains in either tension or forward
torsion do this when the creep stress system is combined
tension-forward torsion.

Originally it was thought that the oscillating secondary
creep region in Figs. 1 and 2 was a further effect of plastic
prestraining. This was rejected when it was also apparent
for the annealed material. In a brief review Garofalo[3]
found that irregular creep curves, though not well understood,
were not uncommon, in polycrystalls. Lawley et al [4],for
example, found oscillating creep curves for an iron-aluminium
solid solution which were attributed, in the absence of the
physical cause,to a periodic structural change in the degree
of order.

(d) Secondary Creep Rates

A correlation between the measured secondary creep rates
and the amount of prestrain is presented in Fig. 3. The mean
secondary shear creep rates ($\dot{\gamma}_{\theta z}$) from Figs 1B and 2A are
plotted logarithmically against $\bar{\varepsilon}_M^P$ in Fig 3(a). A similar
plot for axial creep rates ($\dot{\varepsilon}_{zz}$) from Figs 1A and 2B is given

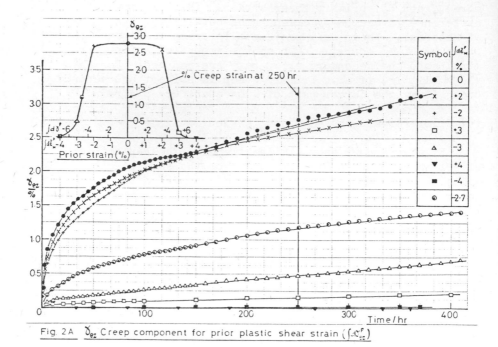

Fig. 2A $\gamma_{\theta z}$ Creep component for prior plastic shear strain $\left(\int d\gamma_{\theta z}^{p}\right)$

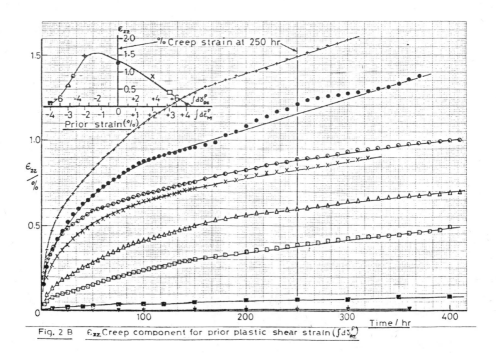

Fig. 2 B ϵ_{zz} Creep component for prior plastic shear strain $\left(\int d\gamma_{\theta z}^{p}\right)$

in Fig. 3(b), while Fig. 3 (c) combines the ordinates of (a) and (b) through an equivalent (von Mises) secondary creep rate. That is,

$$\dot{\bar{\varepsilon}}_M = \sqrt{\dot{\varepsilon}_{zz}^2 + \dot{\gamma}_{\theta z}^2 / 3} \tag{1}$$

which provides a satisfactory correlation. The usefulness of Fig. 3(c) would lie in its ability to also correlate secondary creep rates under any stress system for material with a complex prestrain history. It is interesting to note that the curve in Fig. 3 (c) is approximately symmetrical about $\dot{\bar{\varepsilon}}_M$ when $\bar{\varepsilon}_M^P$ is corrected for subsequent plastic loading strain.

It should be noted that the use of Mises equivalent strain and strain rate definitions is strictly incorrect for prestrained material. Anisotropic definitions are difficult to apply and may not provide an improved correlation.

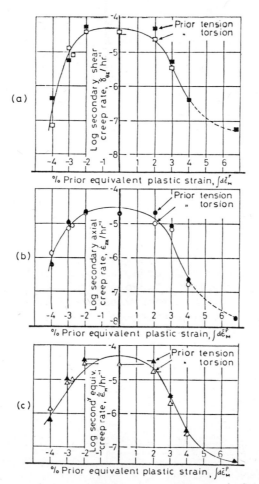

Fig. 3   Secondary creep rate correlation for $\frac{\tau}{\sigma} \approx 1$

2.2. Component Strain Paths

Figs 4 and 5 show the $\gamma_{\theta z}$ vs ε_{zz} paths for loading and ensuing creep in annealed and prestrained material. Loading to the creep test stresses at X was performed incrementally with a time interval of 10 min. allowed for the accumulation of transient creep strain between successive increments. Ensuing 'long time' creep strains are added to X for the hourly times indicated.

(a) Instantaneous and Transient Creep Strain (Fig. 4)

Annealed material ($\varepsilon_M^P = 0$) displays a linear path i.e. the plastic strain increment vector direction ($d\gamma/d\varepsilon$) is constant and independent of stress level. For prestrained material, however, the direction of this vector is seen to be stress dependent with a common feature that with increasing stress the vector rotates towards the direction associated with the annealed material. For any particular stress increment, given in the region between successive symbols in Fig. 4, the resulting instantaneous plastic and transient creep strain components follow the same direction.

The axial strengthening and weakening of the material that is caused by prior tension and compression respectively is reflected in the indicated $d\gamma/d\varepsilon$ values. For prior tension the values are higher than the 2.33 value for annealed material while for prior compressed material the values are lower. For prior forward (+) and reversed (−) torsion the values are generally less than the annealed value. Such behaviour has been explained in terms of a complex interaction between either weakening or strengthening in shear and the associated cross effect in tension [5].

Whenever the instantaneous strain was mostly plastic it was observed that transient creep immediately followed. If on the other hand the instantaneous strain was entirely elastic then no transient creep was observed. For those stress levels where transient creep occurred in the 10 min. interval the component creep curves were found to follow the power law at^m where a and m depended upon current stress and strain history [2].

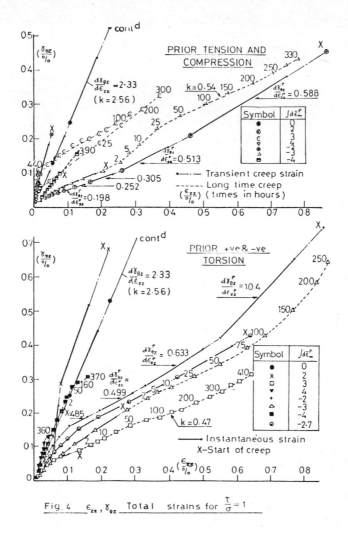

Fig. 4 ϵ_{zz}, $\gamma_{\theta z}$ Total strains for $\frac{\tau}{\sigma} \approx 1$

(b) Creep (Figs 4 and 5)

Fig. 5 shows that after X the constant direction of the
plastic strain increment vector for annealed material is
preserved in the ensuing creep path. An initial rotation
is evident, however, in the creep strain paths for lower
prestrained material which may occur relatively quickly, e.g.
during the first 5 hr for 2% in compression, or it may take
longer, e.g. 250 hr for 3% in reversed torsion. Although
each rotation is a continuation of that observed in $d\gamma/d\epsilon$
under incremental loading there appears to be no relationship
between the time of the rotation and the completion of
primary creep from Figs. 1 and 2.

For larger prestrains, where the subsequent loading
strains were mostly elastic, rotations in ensuing creep were

less evident. It is possible that insufficient time was allowed for enough creep strain to accumulate and begin to nullify the prestrain effects. It is likely to be then that the rotation begins.

Following the rotations in Fig. 5 a linear creep strain path is observed. In comparing the indicated k values from the constitutive relation.

$$\frac{d\gamma}{d\epsilon} = k\left[\frac{\tau}{\sigma}\right] \qquad (2)$$

there is some evidence that they depend upon the magnitude and nature of the prior strain. In this respect a prestrain has a lasting effect on creep.

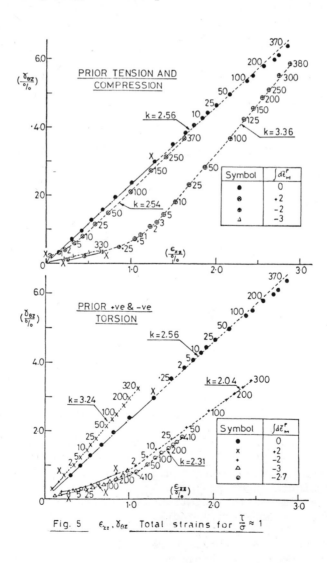

Fig. 5 ϵ_{zz}, $\gamma_{\theta z}$ Total strains for $\frac{\tau}{\sigma} \approx 1$

3. HARDENING MODEL

In Fig. 6 a model of anisotropic workhardening due to Mróz [6] is applied to the present stress paths. The annealed condition of the material is represented by a field of uniform hardening potentials f_o, f_1, ...f_n which are simply inflations of the surface f_O which describes the initial yield condition (Fig. 6(a)). Mróz assumed that this field was given by either the Mises or Tresca function in which case $k = 3$ or 4 respectively in equation (2) and Fig. 6. However, from the normality rule and the linear plastic strain path in Fig. 4 we find from integration of equation (2) that the corresponding field function is

$$f_n = \sigma^2 + k\tau^2 = \lambda_n^2 \qquad (3)$$

where $k = 2.56$ and λ is a monotonically increasing scalar. For simplicity equation (3) is represented by circles of radius λ_n on $\sigma, \sqrt{k}\,\tau$ axes.

Plastic deformation begins when the stress vector makes contact with f_o. Figs 6(b) and (c) show the disturbances in the field which result from prestressing in reversed torsion ($\tau/\sigma = -\infty$) and compression ($\sigma/\tau = -\infty$) respectively. In each case the extent of plastic deformation is modelled by the circles $f_o....f_5$ which translate rigidly with the stress vector without changing their form or orientation. In the process of unloading reverse deformation begins when the stress vector contacts f_o again. Upon completion of unloading the extent of deformation is now represented by the translation of f_o and f_1 to the origin O as shown in figs 6(d) and (e). Note that circles f_6 and f_7 undisturbed by loading, are defined by fig. 6(a) while circles $f_2 ..f_5$, undisturbed by unloading, are defined by Figs. 6(b) and (c).

Now consider the proportionate loading $\tau/\sigma \approx 1$ following prestressing. For simplicity the associated translations in the fields of figs 6(d) and (e) are not shown since the directions of the plastic strain increment vector ($d\varepsilon_{ij}^P$) can be obtained directly from the first quadrant in each figure. At each point of intersection between $\tau/\sigma \approx 1$ and f_n the normal to f_n defines the direction of $d\varepsilon_{ij}^P$ as shown. Encouragingly we see that the model correctly predicts the rotations observed in the strain paths of Fig. 4. Rotations for prior forward torsion and tension, though less obvious in Fig. 4, are supplied by the model corresponding to the respective paths $\tau/\sigma = \infty \rightarrow \tau/\sigma \approx 1$ and $\sigma/\tau = \infty \rightarrow \tau/\sigma \approx 1$ as shown in Fig. 6.

Fig. 5 provides the necessary experimental evidence that the model may be extended to describe anisotropic creep deformation since it is consistent with the restoration of a linear strain path when the tip of the stress vector

$\tau/\sigma \approx 1$ makes contact with circles f_6 and f_7. The extension
to creep is simply made by identifying surfaces $f_0 \ldots f_7$
with creep potentials where f_n increases with strain rate
radius. Fig. 6 is then consistent with the initial rotations
observed in the creep strain paths in Fig. 5 and the subsequent linear paths at a time when the tip of the current
strain rate vector (of magnitude $\dot{\lambda}_n$) lies at a point which
has been left undisturbed by the prestraining operation.

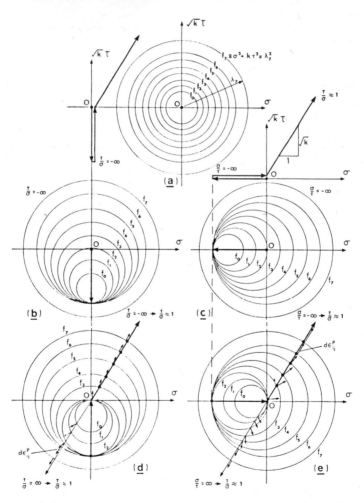

Fig 6 Translations in the fields of uniform hardening flow potentials

The model may be expressed in general mathematical terms. It has been shown [7] that equation (3) can be identified with an anisotropic quadratic form of deviatoric stress function that is a generalisation of the von Mises isotropic hardening function. That is,

$$f(\underset{\sim}{\sigma}') = \frac{1}{2} C_{ijkl} \, \sigma'_{ij} \, \sigma'_{kl} = F(\lambda) \qquad (4)$$

where C_{ijkl} represents 15 independent constants which characterise initial textural anisotropy and $F(\lambda)$ specifies the degree of hardening.

Translated surfaces are defined from the kinematic hardening rule in which the initial yield surface retains its shape and size but translates rigidly with the stress vector. Fig. 6 is consistent with a combination of uniform and kinematic hardening. Shrivastava et al [8] have shown that existing yield functions of this form can all be expressed by invariant theory. That is,

$$f \left\{ I_2(\underset{\sim}{\sigma}'), \, I_2(\underset{\sim}{\varepsilon}^P), \, K_1(\underset{\sim}{\sigma}' \cdot \underset{\sim}{\varepsilon}^P) \right\} = F(\lambda) \qquad (5)$$

which is a function of the second principal invariants of deviatoric stress $I_2 = \frac{1}{2} \sigma'_{ij} \sigma'_{ij}$ and plastic strain $I_2 = \frac{1}{2} \varepsilon^P_{ij} \varepsilon^P_{ij}$ together with the first mixed invariant $K_1 = \sigma'_{ij} \varepsilon^P_{ij}$.

Constitutive relations descriptive of the observed strain paths then follow by employing the flow rule $\dot{\varepsilon}_{ij} = \dot{\lambda} \left(\partial f / \partial \sigma_{ij} \right)$ with equation (4) for the linear region and equation (5) for the non linear region.

The simplest form of equation (5) corresponds to a translation in an initial potential surface that is defined by equation (4). That is,

$$f(\underset{\sim}{\sigma}') = \frac{1}{2} C_{ijkl} \, (\sigma'_{ij} - \alpha'_{ij})(\sigma'_{kl} - \alpha'_{kl}) = F(\lambda) \qquad (6)$$

where α'_{ij} defines the centre co-ordinates of the translated surface in deviatoric stress space. Equation (6) describes the surfaces shown in Fig. 6 for translations which each occur in the direction of a vector joining their centres to the stress point.

4. REFERENCES

1. Soderberg, C.R. 'Plasticity and Creep in Machine Design'.
   Trans. ASME, 1936, 58 (8), 733.

2. Rees, D.W.A. 'Biaxial Creep and Plastic Flow of Anisotropic
   Aluminium.' Ph.D. thesis, C.N.A.A. (U.K), 1976.

3. Garofalo, F. Fundamentals of Creep and Creep-Rupture in
   Metals, Collier-MacMillan, London, 1965.

4. Lawley, A., Coll, J.A. and Cahn, R.W. 'Influence of
   Crystallographic order on Creep of Iron-Aluminium Solid
   Solutions! Trans AIME, Feb. 1960, 218, 166.

5. Rees, D.W.A. and Mathur, S.B. 'Biaxial Plastic Flow and
   Creep of Anisotropic Aluminium and Steel! Inst. Phy.
   Conf. Proc. 'Non linear Problems in Stress Analysis',
   Durham University, Sept. 1977, 185.

6. Mróz, Z. 'On the Description of Anisotropic Workhardening!
   J. Mech. Phys. solids, 1967, 15, 163.

7. Rees, D.W.A. 'A Hardening Model for Anisotropic Materials!
   Accepted for publication in 'Experimental Mechanics',
   Sept. 1980.

8. Shrivastava, H.P., Mróz, Z and Dubey, R.N. 'Yield Criterion
   and the Hardening rule for a plastic solid! Z. Angew
   Math.Mech, 1973, 53, 625.

THE APPLICATION OF THE J-INTEGRAL TO SMALL SPECIMENS OF DUCTILE
MATERIAL TO BE EXPOSED TO HIGH TEMPERATURES AND HIGH LEVELS OF
IRRADIATION

F.R. Montgomery

Department of Civil Engineering, Queen's University,
Belfast, Northern Ireland.

1. SUMMARY

An investigation of the J-integral as a fracture characterisation for ductile metals in small test specimens is discussed.
A stainless steel used in nuclear reactor construction and an
aluminium alloy were used in this investigation to see if there
might be any limiting size below which the J-integral becomes
inaccurate. The results indicate that this lower limit in size
can be defined, particularly for the stainless steel, and that
it is not as low as might be desired for development of a system
of simple, economical, special environment testing. Some comments are made about crack opening displacement measurements.

2. INTRODUCTION

In the new generation of nuclear power reactors of the
liquid metal cooled fast breeder type (LMFBR), austenitic stainless steel is much employed as a structural material. It is a
material which behaves in an elasto-plastic manner, that is, it
fails under load only after the growth of large plastic flow
areas. This makes it very difficult to obtain an analytical
solution to the stress field at the crack tip in a fracture
test, and it is therefore difficult to define the conditions of
fracture. Linear elastic fracture mechanics (L.E.F.M.) tests
are defined as valid only when the test specimen size is of a
certain order of magnitude greater than the crack tip plastic
zone. For the stainless steels in general use in LMFBR construction this would entail a fracture test specimen of very
extensive dimensions, and completely impractical. Any examination of the properties of materials used in reactor construction must take account of the peculiar environment in which they
are called upon to operate, that of high temperature and high
levels of irradiation for components in the immediate vicinity
of the core. This necessitates the irradiation of many specimens before test, but irradiation facilities are at a premium
and generally cannot accept specimens of a size sufficient to
allow of subsequent L.E.F.M. testing. Other methods of test

must be used, that of Rice, the J-integral[1,2] test is thought to be particularly appropriate. With such methods, much smaller test pieces are viable. It was with this in mind that the work described here was initiated, that is, to find just how small could a specimen of a very ductile metal be and still yield a reliable answer of fracture toughness characterisation. Two metals were used, the first an austenitic stainless steel as being most appropriate and the other a ductile aluminium alloy, sufficiently different to act as a check on the method.

3. MATERIALS

3.1 AISI 304 austenitic stainless steel

The material used was of German origin and of composition reported in Table 1. It is seen to be not exactly a true type 304 but is sufficiently close to be referred as such. Its mechanical properties are reported in Table 2.

3.2 Aluminium alloy - Noral D54SO

The composition of this material is reported in Table 3 and its relevant mechanical properties in Table 4.

4. SPECIMENS

4.1 Stainless steel

The specimens of this material range in size from a standard Charpy specimen to a large single edge notched (S.E.N.) plate and are shown in figure 1. All were subjected to low level fatigue loading to produce a short length of fatigue crack at the bottom of their machined notches. The dimensions of all the specimens used, after fatigue crack growth, are reported in Table 5.

4.2 Aluminium alloy

All of the specimens of the aluminium alloy were machined from one 25 mm thick plate of the material. Two different thicknesses of specimen were produced, the first being the full plate thickness of a nominal 25 mm and the second being 10 mm. The narrow specimens were produced by cutting the 25 mm plate in half, retaining one of the original surface unmachined. By retaining an original surface it was possible to ensure all specimens were cut from the same position, relative to the thickness dimension, in the plate. The specimens were manufactured to the porportions shown in figure 3 and fatigue cracks produced to thedimensions reported in table 6.

5. TEST PROCEDURES

5.1 Test configurations

The method of test for the compact tension test (C.T.T.) specimens was standard for this specimen. The compact compression test (C.C.T.) specimens, which are an adaptation of the CTT specimens, were tested as shown in figure 1. The large single edge notched (S.E.N.) plate specimens were loaded in tension

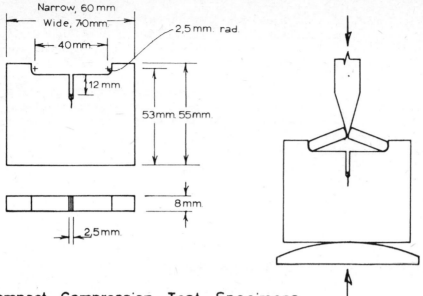

Compact Compression Test Specimens.

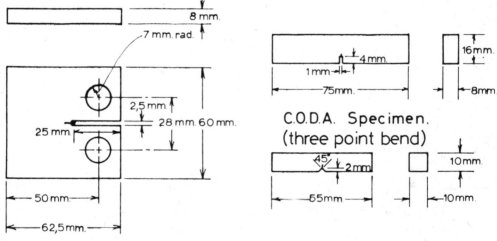

Compact Tension Test Specimen.

C.O.D.A. Specimen.
(three point bend)

Charpy Specimen.

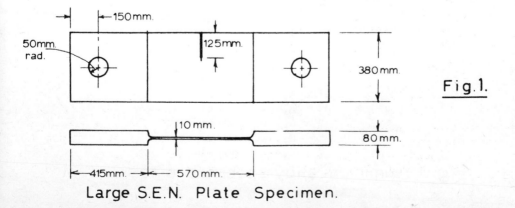

Fig.1.

Large S.E.N. Plate Specimen.

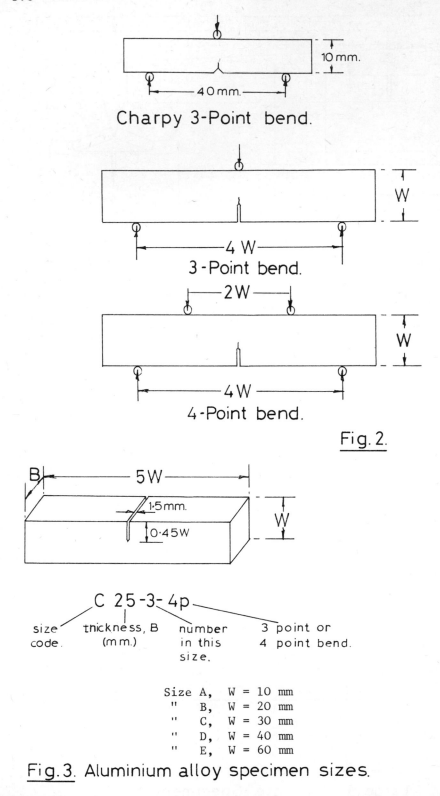

Charpy 3-Point bend.

10 mm.

40mm.

W

4 W

3-Point bend.

2W

W

4 W

4-Point bend.

Fig. 2.

B

5W

1·5mm.

0·45W

W

C 25-3- 4p

size code.

thickness, B (mm.)

number in this size.

3 point or 4 point bend.

Size A, W = 10 mm
"   B, W = 20 mm
"   C, W = 30 mm
"   D, W = 40 mm
"   E, W = 60 mm

Fig. 3. Aluminium alloy specimen sizes.

Material	Chemical Composition by Percentage of Weight						
	C	Si	Mn	P	S	Cr	Ni
German origin	0.04–0.08	≤0.75	≤2.0	≤0.045	≤0.03	17.0–19.0	10.0–12.0
Standard AISI 304	≤0.08	≤1.0	≤2.0	–	–	18.0–20.0	8.0–12.0

Table 1 Stainless Steel Chemical Composition

Temperature	Room temperature		$380^{\circ}C$ $716^{\circ}F$	$540^{\circ}C$ $1004^{\circ}F$
0.2% proof stress	$24kg/mm^2$	33ksi	$14.9kg/mm^2$ 21.2ksi	$14.2kg/mm^2$ 20.2ksi
ultimate strength	$60kg/mm^2$	85ksi	$48.7kg/mm^2$ 69.2ksi	$45.1kg/mm^2$ 64.1ksi

Table 2 Mechanical Properties of AISI 304 Stainless Steel

Material	Chemical Composition by Percentage of Weight				
	Cu	Si	Mn	Mg	Zn
Noral D54SO	0.2	0.08	0.84	0.92	0.04

Table 3   Aluminium Alloy Chemical Composition

0.2% proof stress	ultimate strength	% elongation
$19kg/mm^2$ 27ksi	$22.2kg/mm^2$ 30.9ksi	9%

Table 4 Mechanical properties of aluminium alloy
at room temperature

Specimen	a (mm)	$\frac{a}{w}$	Thickness (mm)
S.E.N. 1	139.0	0.366	9.90
S.E.N. 2	129.0	0.339	9.75
C.T.T. 1	17.7	0.354	8.00
C.T.T. 2	19.2	0.384	8.00
C.T.T. 3	21.4	0.428	8.05
C.T.T. 4	22.85	0.457	8.00
C.T.T. 5	24.6	0.492	8.05
C.T.T. 6	27.35	0.547	8.00
narrow C.C.T. 2	20.75	0.419	7.95
narrow C.C.T. 3	22.5	0.455	7.95
narrow C.C.T. 4	24.2	0.489	8.00
narrow C.C.T. 5	26.8	0.541	8.00
narrow C.C.T. 6	28.6	0.578	7.90
broad C.C.T. 1	18.25	0.359	8.00
broad C.C.T. 2	20.3	0.410	8.00
broad C.C.T. 3	22.55	0.456	8.00
broad C.C.T. 4	25.1	0.507	8.00
broad C.C.T. 5	26.45	0.534	7.90
broad C.C.T. 6	28.8	0.582	7.95
small 3 P bend 1	4.5	0.281	8.00
small 3 P bend 2	7.1	0.444	8.00
small 3 P bend 3	8.0	0.50	8.00
small 3 P bend 4	8.7	0.544	8.00
small 3 P bend 5	9.6	0.60	8.00
Charpy 1	3.6	0.36	10.10
Charpy 2	3.8	0.38	10.15
Charpy 3	4.15	0.415	10.15
Charpy 4	5.1	0.510	10.15
Charpy 5	5.05	0.505	10.15
Charpy 6	4.0	0.40	10.15
Charpy 7	5.35	0.535	10.15

TABLE 5 Dimension of Stainless Steel Specimens

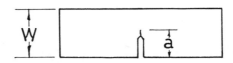

Specimen	a (mm)	$\frac{a}{w}$	Thickness (mm)
A25-1-4p	6.2 invalidated during fatigue loading		
A25-2-4p	5.65	0.565	25.45
A25-3-4p	6.3	0.63	25.45
A25-4-3p	5.6	0.56	25.4
A25-5-3p	5.55	0.555	25.4
B25-1-4p	11.0	0.55	25.45
B25-2-4p	10.5	0.525	25.45
B25-3-4p	10.0	0.50	25.45
B25-4-3p	10.95	0.548	25.45
C25-1-4p	15.5	0.517	25.45
C25-2-4p	16.4	0.547	25.45
C25-3-4p	15.8	0.527	25.45
C25-4-3p	14.0	0.467	25.45
C25-5-4p	15.2	0.507	25.5
D25-1-4p	22.5	0.563	25.45
D25-2-4p	22.0	0.55	25.45
D25-3-4p	20.5	0.512	25.45
D25-4-3p	20.7	0.518	25.5
E25-1-4p	30.0	0.50	25.45
E25-2-4p	30.6	0.51	25.45
E25-3-4p	31.8	0.53	25.45
A10-1-4p	5.4	0.54	11.75
A10-2-3p	5.4	0.54	11.8
B10-1-4p	9.75	0.488	11.7
B10-2-3p	10.3	0.515	11.65
C10-1-4p	14.9	0.497	11.9
C10-2-3p	16.4	0.537	11.65
D10-1-3p	21.1	0.528	11.8
D10-2-4p	19.5	0.488	11.75

Table 6   Dimensions of Aluminium Alloy Specimens

through the holes in their thick end sections. The various other
bend test configurations were as figure 2.

5.2   Crack growth initiation measurement

During testing of all the stainless steel specimens, excepting only the two large S.E.N. plates, crack growth initiation was recorded using an electrical potential field system.
(3,4) It was found that sufficient current could not be generated to enable reliable use to be made of this method during
the S.E.N. plate tests. For these, crack growth initiation was
observed using a system of magnifying lenses focussed on the
crack tip.

The tests on the aluminium alloy were performed in a different laboratory where electrical equipment suited to the potential field method was not available. For these tests it was
found possible to interpret the resulting load displacement
curves to establish the point of initiation of crack growth.
This method had been verified by a process of fatigue marking[4].

5.3   J-integral measurement and calculation

For all the specimens tested a continuous automatic record was made of load versus load point displacement. For all
the stainless steel tests, except the S.E.N. specimens, it was
found possible to apply the compliance method of Begley and
Landes[7] to each group of results to calculate J-integral at
crack growth initiation. The results from only two tests are
not enough for this method, so for the S.E.N. plate tests an
elasto-plastic computer simulation for compliance[6] was used
with the methods of Bucci et al[5] to compute a range of load
versus displacement curves. Those were verified by the limited
experimental results and then used to compute J-integral values
by the method of Begley and Landes, as above.

Rice et al[8] have developed an even simpler technique to
calculate J-integral from the result of only one test of a
limited number of fracture test specimen types. This method
also was used to calculate J-integral for all the stainless
steel specimens.

This last method of Rice et al was the only one used to
compute J-integral for all the aluminium alloy specimens because much confidence was gained in its use during the stainless steel tests.

5.4   Crack opening displacement (C.O.D.) measurement

Three methods were used to measure C.O.D. at crack growth
initiation. On all the stainless steel specimens a magnifying
camera was used to record crack flank movements. C.O.D. was
measured from the photographs produced. To assist this process
a pattern of micro-hardness indentations was imprinted on the
crack tip zone of each specimen.   Further to this, crack opening transducers were used on the S.E.N. plates and the C.T.T.
specimens of stainless steel.

The method of calculation of C.O.D. by calibrated plastic hinge rotation of B.S.-DD19[9], using a hinge point at a position 45% of the ligament length below the crack tip, was used for all the aluminium alloy specimens.

6. RESULTS

From the load versus load point displacement plots obtained autographically for each specimen tested the J-integral at crack growth initiation was calculated using one or more of the methods described. C.O.D. was similarly calculated for each specimen. These results are tabulated in table 7 for stainless steel and table 8 for aluminium alloy.

7. DISCUSSION AND CONCLUSIONS

The results of the calculation of J-integral for the stainless steel specimens, table 7, shows that the single specimen calculation method of Rice et al [8] yields approximately the same values as the compliance method which requires groups of specimens. For the testing of irradiated specimens this has obvious benefits. Because of the confidence gained in the stainless steel tests the single specimen method was the only one used to calculate J-integral values for the aluminium alloy values.

Figure 4, the values of J-integral at crack growth initiation plotted versus remaining ligament length (Q-a) is perhaps the most interesting outcome of this work. It shows a very rapid increase of J-integral at short ligament lengths. It has been proposed by Begley and Landes[7,10] that a limitation on the validity of J-integral tests should be considered and should be of the form; ligament length must be greater than $\frac{\alpha J}{\sigma y}$, where α is a constant determined by experiment. Figure 4 indicates a value for α of between 35 and 55 for the stainless steel, depending on the value one chooses to use for αy, the yield stress, and also on the interpretation of where the line in figure 4 begins to rise. In any case it would seem that, for a ligament length much less than 40 mm, a true value of J-integral cannot be obtained for this material in the thicknesses considered. One could hope, of course, that this might have been much less because it seems to preclude the use of the very small specimens in irradiation and high temperature testing, something to be regretted.

It is seen from table 7 that the C.O.D. values calculated for the stainless steel show quite a variation but that there is no immediately obvious relation between these values and the J-integral values. Both the Charpy and S.E.N. plate specimens at opposite ends of the size spectrum exhibit fairly high C.O.D. at crack growth initiation. One could conclude that C.O.D. is not influenced by the same size limitations as is the J-integral. C.O.D. measurements are, though, more difficult to make on irradiated specimens than are J-integral measurements because of the difficulty of application of suitable gauges and of observation during testing.

Specimen	Ligament length (mm)	C.O.D. initiation (mm)	J.initiation by compliance (kgmm/mm^2)	J.initiation by ref 8 (kgmm/mm^2)
S.E.N. 1	241	0.78	18.25	–
S.E.N. 2	251	0.585	17.80	–
C.T.T. 1	32.3	0.43	26.5	32.2
C.T.T. 2	30.8	0.675	22.0	28.25
C.T.T. 3	28.6	0.54	20.5	26.9
C.T.T. 4	27.15	0.55	18.85	19.99
C.T.T. 5	25.4	0.55	18.85	20.2
C.T.T. 6	22.65	0.51	16.3	19.35
narrow C.C.T. 2	28.75	0.45	–	27.95
narrow C.C.T. 3	27.0	0.65	–	28.35
narrow C.C.T. 4	25.3	0.65	–	27.75
narrow C.C.T. 5	22.7	0.55	–	23.2
narrow C.C.T. 6	20.9	0.50	–	27.65
broad C.C.T. 1	31.25	0.60	24.0	26.35
broad C.C.T. 2	29.2	0.70	26.63	30.2
broad C.C.T. 3	26.95	0.70	29.38	–
broad C.C.T. 4	24.5	0.62	26.38	21.5
broad C.C.T. 5	23.05	0.69	32.63	24.55
broad C.C.T. 6	20.7	0.65	31.89	20.15
small 3 p bend 1	11.5	–	–	57.2
small 3 p bend 2	8.9	0.53	38.8	50.7
small 3 p bend 3	8.0	0.625	47.0	46.05
small 3 p bend 4	7.3	0.50	38.8	38.3
small 3 p bend 5	6.4	0.52	59.6	50.2
Charpy 1	6.4	0.60	37.5	41.1
Charpy 2	6.2	0.75	46.8	42.1
Charpy 3	5.85	0.61	43.8	39.95
Charpy 4	4.9	0.74	50.2	41.55
Charpy 5	4.95	0.79	53.8	45.1
Charpy 6	6.0	0.70	47.2	41.9
Charpy 7	4.65	0.85	58.3	49.5

Table 7   Results for stainless steel specimen tests

Specimen	Ligament length (mm)	C.O.D. initiation (mm)	J.initiation by ref 8 (kg mm/mm^2)
A25-2-4p	4.35	0.05	1.5
A25-3-4p	3.7	0.05	1.25
A25-4-3p	4.4	0.1	3.4
A25-5-3p	4.45	0.1	3.0
B25-1-4p	9.0	0.08	2.25
B25-2-4p	9.5	0.1	2.25
B25-3-4p	10.0	0.095	2.35
		0.1	2.85
C25-1-4p	14.5	0.1	4.8
C25-2-4p	13.6	0.16	4.2
C25-3-4p	14.2	0.17	3.35
C25-4-3p	16.0	0.14	4.45 ?
C25-5-4p	14.8	0.1	215
D25-1-4p	17.5	0.085	2.0
D25-2-4p	18.0	0.08	1.9
D25-3-4p	19.5	0.1	2.3
D25-4-3p	19.3	0.1	2.5
E25-1-4p	30.0	0.095	2.2
E25-2-4p	29.4	0.14	2.8
E25-3-4p	28.2	0.12	2.45
A10-1-4p	4.6	0.075	1.6
A10-2-3p	4.6	0.07	2.05
B10-1-4p	10.25	0.1	216
B10-213p	9.7	0.075	2.2
C10-1-4p	15.1	0.11	2.8
C10-2-3p	13.6	0.075	2.82
D10-1-3p	18.9	0.075	1.7
D10-2-4p	20.5	0.105	2.42

Table 8   Results for aluminium alloy specimen tests.

584

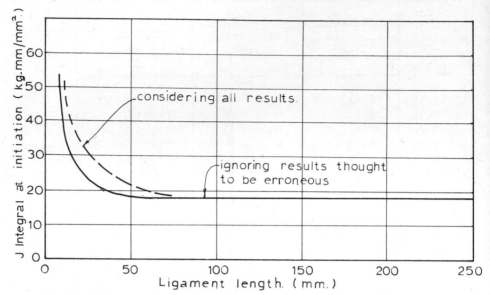

Fig.4　J-integral at crack growth initiation versus
　　　ligament length for stainless steel specimens.

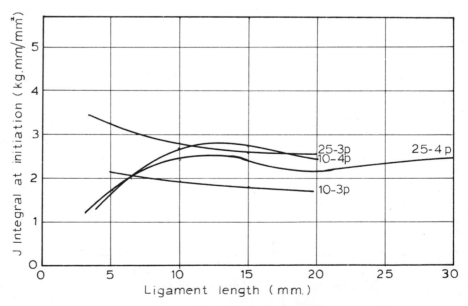

Fig.5　J-integral at crack growth initiation versus
　　　ligament length for alluminium alloy specimens.

Figure 5, relating J-integral at crack growth initiation to ligament length for the various aluminium alloy specimens does not exhibit the same dramatic variations as that for the stainless steel. There is, though, some change in J-integral at or below a ligament length of about 8 mm. Using the equation of Begley and Landes above, this yields a value for α of about 60 for this material. It is also noted that any change of J-integral for ligament lengths below 8 mm seems to be influenced by the test configuration, three point bend tests tend to have increasing J-integral for short ligaments while four point bend tests seem to lead to decreasing J-integral for short ligaments. Three point bend tests by Landes and Begley[10] on a rotor steel revealed that for a value of α of less than about 25, J-integral values at initiation of crack growth were noted to fall slightly.

There is obviously a specimen size effect in operation, but it would seem too that test configuration for small specimens plays a large part in determining J-integral values. All the small specimens of stainless steel were loaded in three point bending and yielded high J-integral values. This agrees with the results for the aluminium alloy specimens. Obviously, too, material differences have a big influence but, in all cases, it is seen that there is a ligament length limitation as described, which may rule out the use of very small specimens for J-integral testing.

8 REFERENCES

1. Rice, J.R., "A Path Independent Integral and the Approximate Analysis of Strain Concentration by Notches and Cracks.", Transactions of A.S.M.E. Journal of App. Mech., June 1968, pp 379-386.

2. Rice, J.R., "Mathematical Analysis in the Mechanics of Fracture.", Fracture, Vol.2, (H.Leibowitz. ed.) Academic Press, New York, 1968, pp 191-311.

3. A.S.T.M. "Progress in Measuring Fracture Toughness and Using Fracture Mechanics: Fifth Report of a special A.S.T.M. Committee." A.S.T.M. Materials Research and Standards, March 1964, pp 107-109.

4. Montgomery, F.R. "A Fracture Mechanics Test for Failure in Ductile Metals" Ph.D. Thesis, Queen's University of Belfast 1975.

5. Bucci, R.J., Paris, P.C., Landes, J.D. and Rice J.R., "J-Integral Estimation Procedures", Fracture Toughness, Proceedings of the 1971 National Symposium on Fracture Mechanics, Part 2, A.S.T.M.-S.T.P.514, American Soc. for Testing and Materials, 1972, pp 40-69.

6. Montgomery F.R. "J-Integral Measurements on Various Types of Specimens in AISI 304 S.S." Nuclear Science and Technology Report EUR 5521e, Commission of the European Communities, Luxembourg, 1976.

7. Begley, J.A., and Landes, J.D., "The J-Integral as a Fracture Criterion", American Soc. for Testing and Materials, Special Technical Publication STP 514, 1972, pp 1-20.

8. Rice, J.R., Paris, P.C., and Merkle, J.G., "Some Further Results of J-Integral Analysis and Estimates.", American Soc. for Testing and Materials, Special Technical Publication STP 536, 1973, pp 231-245.

9. "Crack Opening Displacement (C.O.D.) Testing", British Standards Institution, Draft for Development DD19:1972 (superceeded by BS5762:1979).

10. Landes, J.D., and Begley, J.A., "The Effect of Specimen Geometry on J_{IC}," American Society for Testing and Materials, Special Technical Publication STP 514, 1972, pp 24-39.

SECTION 6
DESIGN AND PERFORMANCE OF COMPONENTS AND STRUCTURES

REPRESENTATION OF INELASTIC BEHAVIOUR

E.T. Onat

Department of Engineering and Applied Science, Yale University,

New Haven, Connecticut 06520, U.S.A.

SUMMARY

In the presence of elevated temperatures and/or stress most materials exhibit mechanical behavior that is nonlinear and hereditary (inelastic). When this kind of behavior is present the strain of a material element will depend (nonlinearly) not only on the current stress and temperature but also on the temperatures and stress previously applied to it. The present paper is concerned with the problem of mathematical representation of the relationship that exists between histories of stress and isothermal deformations. For simplicity we shall confine ourselves to the case of small deformations and rotations. The thesis of the paper is that the study of geometric and global aspects of representations of mechanical behavior based on state variables and differential equations provides a unified point of view that enables one to compare existing representations and more importantly offers rational ways of improving a given representation or of constructing a new one.

1. INTRODUCTION

Often it is of vital importance that the engineer be able to predict the behavior of a structure (say, a component of a nuclear reactor) under extreme operating conditions. With this aim in mind a vast amount of data are produced and collected on the mechanical behavior of the materials used in the construction of the structure of interest. The data comes from phenomenological experiments and more recently from various kinds of microscopy (that shed a welcome light on the physical processes that accompany inelastic deformations). Next the data must be combined and studied to develop mathematical models that reproduce, with various levels of accuracy the observed relationship between histories of stress and deformation for a given material. The engineer will then choose from this set, a model that is appropriate to the task at hand and will use the model, together with the usual conservation laws of mechanics and appropriate numerical techniques to solve the structural problem of interest. We emphasize again that the choice of model depends on the nature of the structural problem. If, for instance, dynamic effects are expected to be important, the representation chosen must be capable of handling high rates of strain.

It is our opinion that the task of developing reasonably truthful representations of behavior with the help of experimental evidence is still poorly understood. The reason for this is that the general problem of identification and representation of nonlinear systems is a difficult one and this problem did not receive any systematic attention until very recent times.

Nevertheless during the past decade it became clear that representations based on systems of differential equations have definite advantages. Indeed most of the recent work on representation of mechanical behavior (in the presence of small isothermal deformations) can be put in the following common (canonical) form based on differential equations and on the concept of internal state (cf. [1,2])

$$\sigma(t) = f(S(t)) \qquad\qquad f: \Sigma \to T_2^S, \; \Sigma \subset R^N \qquad (a)$$
$$\frac{dS}{dt}(t) = g(S(t), \frac{d\varepsilon}{dt}(t)) \qquad g: \Sigma \times T_2^S \to R^N \qquad (b)$$

$$(1.1)$$

where $\sigma(t)$ and $\frac{d\varepsilon(t)}{dt}$ are, respectively, the tensors of stress and infinitesimal strain rate. T_2^S is the space of second rank symmetric tensors.

It is useful to think of the coordinates of the vector $S(t) \in \Sigma \subset R^N$ in (1.1) as N parameters that measure certain statistical aspects of the arrangement of the atoms or molecules within the material element at time t. It is assumed that the future mechanical behavior of the element depends (to within a given approximation) only on these N parameters and on the stimulus applied to it in the future. It is expected that N will be less than ten, say. This expectation comes from the knowledge that many fine details of internal structure cannot affect "coarse" aspects of mechanical behavior. We may say that S represents the internal state and orientation of the material element as far as an approximate description of mechanical behavior is concerned.

The first equation in (1.1) simply states that the current stress is a function of the internal state and orientation. The second system of equations in (1.1) constitutes the growth laws for S. It states that the rate of change of S is a function only of the current state and orientation and the rate of strain applied to the element. Thus if $\varepsilon(t)$ is given on $(0,\infty)$ and if the initial state S_0 is known (1.1b) will enable us to obtain $S(t)$ by integration. We can then read the stress $\sigma(t)$ from (1.1a).

Often very different-looking representations are offered by authors of different backgrounds for the mechanical behavior of a given material. One reason for this is that any one-to-one and smooth map F on R^N, s = F(S), could be used to replace S in (1.1) by $F^{-1}(s)$ and this would give rise to a seemingly new representation of the type (1.1) with new functions $\hat{f}$ and $\hat{g}$ that depend on s and $\varepsilon\cdot$. Another reason for variety is that different aspects of behavior may have been represented in different works. It is therefore highly desirable to look for those basic aspects of the representation (1.1) that is invarient under coordinate transformations and under changes of purpose, taste, background and authorship. The thesis of the present paper is that the study of geometric and global aspects of the representation (1.1) provides a unified point of view that enables one to compare various modes of representations and (more importantly) offers rational ways of improving a given representation or of constructing a new one.

2. GLOBAL AND GEOMETRICAL ASPECTS OF REPRESENTATION

In this section we shall work with a particular version of (1.1) which is appropriate to the study of metals. Moreover, for simplicity, we shall consider only the case of uniaxial isothermal stressing. We start by decomposing the state S into two major components: the current stress σ and the inelastic state q: S = (q,σ). In general q will be composed of a number of parameters, say, $(q_1,\ldots,q_n)$. Next we adopt the following special form for the laws of evolution of the state.

$$\frac{d\sigma}{dt}(t) = g_\sigma(q(t),\sigma(t)) + E\varepsilon\cdot(t) \qquad g_\sigma: \Sigma \to R; \ \Sigma \subset R^{n+1}$$

$$\frac{dq}{dt}(t) = g_q(q(t),\sigma(t)) \qquad\qquad g_q: \Sigma \to R^n \tag{2.1}$$

where E denotes the constant modulus of elasticity. We let the state of a virgin element be represented by the origin $(0,0)$ of R^{n+1}, and we add to (2.1) the initial conditions

$$g_\sigma(0,0) = 0; \qquad g_q(0,0) = 0 \ . \tag{2.2}$$

The above conditions imply that a virgin element does not change its initial state in the absence of stresses (and temperature changes). This is equivalent to saying that the virgin material is non-aging.

We note that the growth laws (2.1) are special in that the strain rate $\varepsilon\cdot$ appears linearly and only in the first equation. This in turn implies that a sudden change $\Delta\varepsilon$ in strain causes no change in the inelastic state q but increases the stress according to Hooke's law: $\Delta\sigma = E\Delta\varepsilon$. This growth law is supported by many experiments. A crucial example: A bar is deformed plastically and carries the current yield stress. A small amplitude loading wave is passed through the bar and it is observed that the wave travels with the speed based on the elasticity modulus E and not on the tangent modulus E_t. This and other tests show that a "fast" enough change in the dimension of the material element is not accompanied by a sudden change in the arrangement and distribution of dislocations.

It will be seen in Section 4 that the representation (2.1) is also capable of modeling plastic behavior that metals exhibit at certain rates of strain and even at elevated temperatures.

Next we observe once more that given a differentiable strain history $\varepsilon(t)$ (or stress history $\sigma(t)$) on $(0,\infty)$ (2.1) provides a system of ordinary differential equations for determining q(t) and $\sigma(t)$ (or $\varepsilon(t)$). It is desirable to see this in a geometric setting. Fig. 1 constitutes a schematic

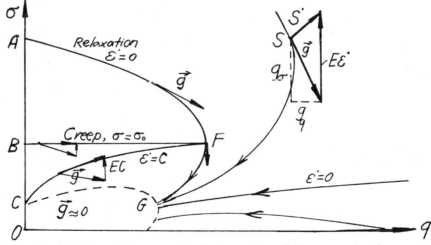

Fig. 1 Global and Geometrical Aspects of Representation.

representation of the state space. The vertical axis is for the stress and the horizontal axis represents symbolically a n-dimensional space. The set of inelastic states is a subset of this space. At a given state $S = (q,\sigma)$ the rate of change of state $\frac{dS}{dt} = S^{\cdot} = (q^{\cdot},\sigma^{\cdot})$ is found by the vector diagram shown in Fig. 1. The vector g has components (g_q,g_σ) and when the vector $(0, E\varepsilon^{\cdot})$ is added to it, produces the desired vector $(q^{\cdot},\sigma^{\cdot})$. It is important to remember that g is specified at each point of interest (q,σ) and that the vector field thus created constitutes the heart of the representation.

We see from Fig. 1 that when $E\varepsilon^{\cdot}$ is much larger than g in magnitude, the state point moves essentially in the σ-direction. Thus lines parallel to the σ-axis are of <u>infinitely fast deformations</u> in the state space.

<u>Relaxation tests.</u> In a relaxation test ε is kept constant ($\varepsilon^{\cdot}$ = 0) and therefore the state point moves in the direction of the local g-vector. Thus the curves which accept the g-vector as tangents are trajectories created by relaxation tests (relaxation lines). In Fig. 1, OAG represents the image of an ideal relaxation test. In the first step of the test (OA) the element is stressed with infinite speed. Beyond A the strain is kept constant and the state point moves along the relaxation line AFG. In the early part of this episode the stress is high so that inelastic processes such as the movement and even the production of dislocations could take place in spite of the fact that the total strain ε is kept constant. But these processes will relax the stress and this, in turn, will slow down the movement of the state point in the "positive"q direction. Eventually thermally activated events may become prominent and the inelastic state q may start to move towards the distant origin of the state space. The question arises: Will the state point reach the origin eventually? It is well known that at temperatures used in structural applications complete recovery is impossible. Therefore in the representation (2.1) g must be chosen in such a way that the motion of the state point on a relaxation line "stops" before the origin is reached. It follows then that g $\approx$ 0 in a region surrounding the origin.

<u>Creep tests.</u> In a creep test the stress σ is kept constant. The corresponding motion of the state point takes place on the σ = σ_o hyperplane of the state space according to the system of differential equations

$$q^{\cdot} = g_q(q, \sigma_o) \qquad (2.3)$$

while the first equation in (2.1) furnishes the creep strain rate:

$$\varepsilon^{\cdot} = -\frac{1}{E} g_\sigma(q, \sigma_o). \qquad (2.4)$$

In Fig. 2 we used a two-dimensional space of inelastic states (q = (q_1, q_2)) to illustrate the geometric image of a creep test (OB: sudden loading, BF: creep). If the creep trajectories on the σ = σ_o level have a fixed point F where

$$g_q(q, \sigma_o) = 0 \qquad (2.5)$$

the state point will move towards F. At F, q will remain fixed and the material will creep at the steady rate given by (2.4). This means that the material is undergoing shape change without altering its internal state. In more physical terms we may say that in steady creep the statistics of dislocation distribution within the element remains fixed but the passage of dislocations through the element provides the shape change. There is some H.E.V. microscopy that supports this picture in certain materials. However if one also wants to describe the tertiary stage of creep then one must not allow the creep lines in Fig. 2 to have a fixed point. What one needs, instead, is creep lines that take the state point to a set on which creep rupture occurs. (cf. [3]).

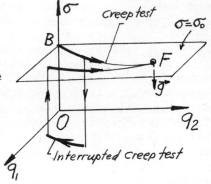

Fig. 2 Creep; interrupted creep

Constant strain rate tests.

When $\varepsilon^{\cdot}$ = const = C, the state point moves along a curve which can be obtained by a step by step application of the vector diagram shown in Fig. 1. In this figure we show the image of a particular constant strain rate test where at point F (where $g_q = 0$) the state point comes to a halt. This happens because at F we have $-\frac{1}{E}g_\sigma(F) = \varepsilon^{\cdot} = C$. We observe that a creep test performed at the appropriate level of stress, $\sigma = \sigma(F)$, will also terminate at the point F. These observations show once more that the points where g_q vanishes are of exceptional importance in the representation of mechanical behavior.

3. EXPLORATION OF THE STATE SPACE BY PHENOMENOLOGICAL EXPERIMENTS.

We have just seen that, once the g-field is known, the response of the material to various kinds of stimuli applied to it can be discussed, in geometrical terms, by following the corresponding evolutions of the state point. Conversely in order to discover the nature of the g-field, the state point must be taken by appropriate experiments to all points of interest of the state set Σ. We wish to show in this brief section that each classical mode of testing enables us to reach only certain parts of the state space. It follows then that a judicious combination of various types of tests would be needed to explore the entire state set Σ.

Let us make these points clearer by examples. We start by considering a sequence of relaxation tests performed by Hart [4]. The sequence is made up of segments of fast straining followed by episodes of relaxation. The image of such a sequence in the state-space would be composed of line segments OA, AB, BC, etc. (Fig. 3). The portion of the state space that can be explored this way is the shaded domain shown in the figure. Next we turn to multistep loading tests of the engineer where a number of sudden changes in stress level is followed by episodes of creep. It is seen from Fig. 3 that these tests can provide information on the g-field in a

larger portion of the state
space. This figure contains
the image of a test corresponding to the sequential applica-
tion of the stresses $(\sigma_0, 0, \sigma_0)$.
Of course when the dimension of
the inelastic state space is
two or more Fig. 3 is inadequate. When n = 2, the image
of the test may be like the one
shown in Fig. 2. Comparison
of the Figs. 2 and 3 brings out
this important point: When
n = 1 (Fig. 3) a portion of the
creep trajectory at the σ_0
level will be traversed twice.
This means that the strain
rates observed over a time re-

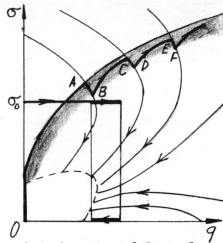

Fig. 3 Exploration of State Space

versal in the third stage of the test will be identical with
those observed on some time interval of the first stage. For
the case of n = 2 however the creep behavior may be markedly
different in the first and third stages of the test because of
the fact that the image point may be moving on two different
creep trajectories on the $\sigma = \sigma_0$ plane. These observations show
that the tests of step loading could be used to advantage to
estimate the dimension of the state space. It will be seen in
Section 5 that the movements that take place in the state space
during a <u>cyclic straining</u> test are complicated and cover a
large portion of the state space.

We have seen in Section 2 that considerations of the
physics of inelastic processes that accompany the deformation
enabled us to establish certain global aspects of the g-field.
Moreover we have just observed that the dimension of the state
set, namely the minimal number of parameters needed to describe the observed behavior is an issue of importance. We now
start the search for state variables (such as yield strength)
that can be obtained from phenomenological experiments.

4. CREEP AND PLASTICITY

In this section we study in some detail an aspect of mechanical behavior that is common to most metals and metal
alloys (e.g. Steel 304). It is well known that in the presence
of certain strain rates, Steel 304 and many other materials behave, even at elevated temperatures, like <u>plastic</u> solids. Thus
the stress-strain curves obtained with this material at strain
rates that lie between .30/hr and 30/hr and at 650°C differ
from each other only a little [5]. Moreover at these speeds
and temperature the material behaves elastically during unloading and it exhibits yield phenomenon upon reloading. Even
the intricate phenomenon of cyclic hardening observed in this

594

material is nearly independent of speed at the above-mentioned
rates of straining. It is natural to ask whether the representation (2.1) discussed in previous sections is capable of
reproducing this particular aspect of mechanical behavior. To
see that the answer to this query is yes let us consider the
images of the two tests performed at .30/hr and 30/hr in the
state space. We assume, with some justification derived from
microscopic observations that the state space images of these
two tests will also be close to each other. [This is equivalent to saying that at these speeds of straining there is ample time for the dislocations to move and to multiply under
current levels of stress, but there is not enough time to accumulate thermally activated changes in the dislocation arrangement. Thus the current inelastic state is essentially a
function of the current stress. However at much lower speeds
of testing there will be enough time for the thermally activated changes to become significant so that the inelastic
state will no longer be a function of the current stress alone
but it will also depend on time. (the conjectured image of a
very slow test ($\varepsilon^{\cdot} = 10^{-5}$/hr) is shown in Fig. 4. When the

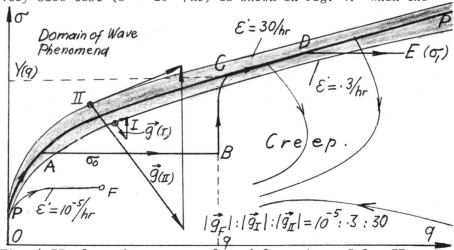

Fig. 4 PP: fast, but not too fast deformations. Below PP creep.

state reaches the point F the material will creep steadily at
the applied rate of straining). On the other hand, at speeds
of testing much higher than 30/hr the development of dislocation distribution will lag behind the stress. Thus at very
high strain rates elastic behavior will be observed.] Close-
ness to each other of the images of tests performed at .3/hr
and 30/hr imply in view of the representation (2.1) that the
g-vector must increase 100 times in magnitude as one traverses
the thin shaded region of the state space (Fig. 4) from I to
II.

This remarkable increase in $|g|$ permits us to introduce
the following simplifications in Fig. 4. One replaces the
thin shaded region with a line (surface if n > 1) PP and in-
sists that under fast enough deformations the state point

moves along PP as would be the case with plastic solids
(Fig. 4). But if at a point on PP, $\varepsilon^{\bullet}$ is such that the construction given in Fig. 1 takes the state point below PP, then
the representation (2.1) becomes applicable. (For details,
see, [6]). One could say that the happenings below PP correspond to creep like phenomenon. It is instructive to study
the image of a multistep test (σ_o, σ_1) in Fig. 4. It will be
seen that the initial fast loading moves the state point along
OPA. Then as the stress is maintained at the σ_o level, creep
takes place and the state point moves along AB. The second
sudden increase in stress moves the state point along BCD. BC
is nearly vertical; at C we reach the line PP and the stress
becomes equal to the yield value Y corresponding to the inelastic state q. If $\sigma_1 > Y$, as we assume here, the image point
next moves along CD. And finally as the stress remains at the
σ_1 level, creep takes place along DE.

We see from the above remarks that a fast loading performed with a material element which is in the inelastic state
q, produces a yield stress Y. We can therefore associate with
each q a yield stress Y(q). (The line PP of Fig. (4) will
nearly be the graph of the function Y(q)). Conversely in the
case where the space of inelastic states is one dimensional
and all stresses of interest are nonnegative, we can associate
with each value of Y an inelastic state q. Thus in this case
we can use the current yield strength of the material as a
measure of its inelastic state. The total state S of the
material will then be composed of the current stress σ and the
current yield strength Y:

$$S = (Y, \sigma)$$

It must be noted that in the work of Hart and others, the
above composition of the state is used tacitly. The advantage
of using Y as a measure of inelastic state of the material is
that the line PP and the g-field shown in Fig. 4 can be constructed, in principle, by appropriately designed experiments,
provided that n = 1 and the interest is focussed on non-nega-
tive stresses. In the next section we will further exploit
these ideas and their extensions.

5. YIELD STRENGTH AS A MEASURE OF INELASTIC STATE. CYCLIC
STRAINING

It is easy to convince ourselves that the intricate behavior observed under cyclic straining (cyclic hardening,
passage from one saturation cycle to another one, the effect
of early creep deformation on saturation cycle etc. [5]) cannot be reproduced by a model where q lives in a one-dimension-
al space. If one looks for two measurable inelastic state
variables, it is natural to follow the ideas of the previous
section and to consider the tensile and compressive yield
strengths Y_1 and Y_2 of an inelastic state. Let us be more
precise: at a given time t a deforming specimen may be in

the inelastic state q and it may carry the stress σ and the total strain ε. Now by subjecting this specimen to fast straining we can measure the tensile yield strength Y_1 ($\varepsilon^{\cdot}$ = C > 0) and the "compressive" yield strength Y_2($\varepsilon^{\cdot}$ = − C < 0) of the state q. It must be noted right away that the exact definitions of Y_1 and Y_2 (especially Y_2, in the case shown in Fig. 5) would require the choice of an offset

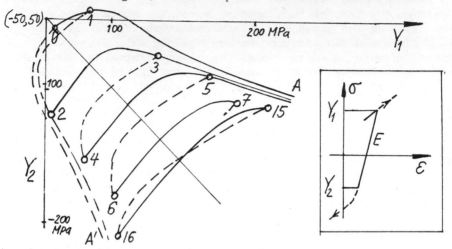

Fig. 5 Image of a cyclic straining test. Steel 304, Δε=1%,650°C

strain. This makes Y_1 and Y_2 a little fuzzy but we can live with this amount of fuzziness.

Thus we consider the inelastic state q as composed of Y_1 and Y_2 and some other additional variables

$$q = (Y_1, Y_2, q_3, \ldots, q_n)$$

We hope that for some materials taking q as composed solely of Y_1 and Y_2 would provide a crude but workable minimal representation. In any case we can, in principle, measure the first two state variables (Y_1, Y_2) of an inelastic state by fast tests that seek limits of the elastic range. This is easier said than done but F. A. Leckie is attempting to do just that.

We now propose to follow a deformation process, by means of its image in the (Y_1, Y_2) plane. We consider the cyclic straining performed with Steel 304 at 650°C using Δε = 1% and $\varepsilon^{\cdot}$ = ± .30/hr. [5] The point 0 in Fig. 5 corresponds to initial yielding (it is assumed that Y_2 = − Y_1 for the virgin state). The point 1 corresponds to the beginning of second loading, etc. The solid lines are conjectural and they correspond to episodes with $\varepsilon^{\cdot}$ = + C. Similarly the dashed lines are associated with $\varepsilon^{\cdot}$ = − C. OA and OA' represent the images of simple tension and compression tests respectively. It is seen from this figure that the elastic range (Y_1-Y_2) of the material increases as the data points move towards saturation. At saturation (Y_1-Y_2) changes a little during a cycle, but an increase in (Y_1-Y_2) is compensated by a subsequent decrease. On the other hand during saturation, the midpoint of the elastic range $Y_1 + Y_2/2$, oscillates between two extreme values. As we remarked earlier, all the curves in this figure can, in principle, be determined experimentally.

Now if we assume that there are only two inelastic state variables, the curves in Fig. 5 along which the state point moves during fast deformations will not change with time and/ or additional deformation. During yielding in tension the rate of state $(Y_1^{\cdot},Y_2^{\cdot})$ will then be related by the vector h_1 to the rate of strain $\dot{\varepsilon}$ in the following way:

$$(Y_1^{\cdot},Y_2^{\cdot}) = h_1 \varepsilon^{\cdot}; \qquad (\varepsilon^{\cdot} > 0 \text{ and large enough}) \qquad (4.1)$$

$h_1(Y_1,Y_2)$ is tangential to the appropriate trajectory and decreases in magnitude as the state point moves along a solid line towards a distant fixed point F. A similar equation will have to be written for the case of $\varepsilon^{\cdot} = - C$. In [6] we used the two families of curves in the (Y_1,Y_2) space together with the vector fields h_1 and h_2 to "predict" the outcome of typical cyclic straining tests. We have shown that for each value of strain range there exists a butterfly shaped limit cycle in the (Y_1,Y_2) plane. We have also shown that if the strain range is abruptly changed, say, from $\Delta\varepsilon_1$ to $\Delta\varepsilon_2$ then the state point will leave one limit cycle and approach the one appropriate to $\Delta\varepsilon_2$. We have also considered the motions in the (Y_1,Y_2) plane during cyclic creep tests. These qualitative comparisons of the predictions of the model (cf. (4.1) and Fig. 5) with experiments indicate that uniaxial behavior of Steel 304 can be described with some satisfaction by using Y_1 and Y_2 as the sole variables of inelastic state.

However, the situation is different for 2 1/4 Cr - 1 Mo Steel. In Fig. 6 we show the results of two cyclic tests for this material. In the first test a periodic strain history is applied ($\Delta\varepsilon$ = .41%, θ = 510°C) and the resulting response plotted in the manner of Fig. 5 using the round black data points derived from Fig. 4 of ORNL/TM-5226. In the second test the stress in a virgin specimen is increased to 170 MPa (A → B) and a creep strain of 0.048% is accumulated in 23 hours and at 510°C. During this episode the image point

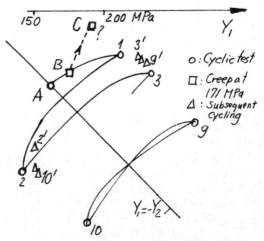

Fig. 6 2¼Cr-1 Mo Steel $\Delta\varepsilon$=.41%, 510°C.

in the (Y_1, Y_2) plane moves from B to C. The point C must lie on the dotted line corresponding to $Y_1 = 185$ MPa, but the exact position of the point is not known (the associated Y_2 value is not available) and hence the point C is conjectural. After this episode the material is unloaded to zero stress and a cyclic test is performed with $\Delta\varepsilon = .41\%$. The triangular data points that represent this stage of the test on Fig. 6 are taken from the $\sigma-\varepsilon$ diagram shown in Fig. 5 of ORNL/TM-5226.

It is seen from Fig. 6 that in the second test the material is near saturation around the tenth cycle and at the maximum stress level of 232MPa, whereas in the first test saturation starts to occur at $\sigma_{max} = 267$MPa. Next we show that if there were only two inelastic state variables, or equivalently if the lines of fast deformation were fixed in the Y_1, Y_2 plane then the above situation could not occur: To see this, note that the triangular data point $9'$ is close to the round point 3. Therefore, if the $\varepsilon^\cdot = \pm$ C trajectories were fixed, both points would move to the same saturation cycle.

Since this does not happen, we decide that for a satisfactory description of mechanical behavior of this material one needs to work at least with <u>three</u> inelastic state variables. This raises the question as to what additional <u>measurable</u> quantity could be used as the third variable. Discussion of this question is left to a future paper.

Fig. 5 illustrates that during cyclic straining a combination of kinematic and isotropic hardening takes place. It must be noted that there are theories of plasticity that allow for both types of hardening (for instance, the work of Dafalias, Eisenberg and Miller. See also experiments of Lamba and Sidebottom [7] where references to the above-mentioned work can be found). There is also the work of J. Morrow, G. M. Sinclair and others on computer based simulation of cyclic behavior [8]. Finally there is the work of Halphen [9] concerning the existence of limit cycles in the stress-strain space for some ideal solids. It is our belief that the state space representation of mechanical behavior is especially suitable for the study of material response to periodic stimuli. We hope that this Section and [6] provides support for this view.

6. ANISOTROPY THAT DEVELOPS DURING THE COURSE OF DEFORMATION. TENSORIAL NATURE OF STATE VARIABLES.

When one considers general (but for simplicity still small and isothermal) deformations of a material, the questions of material symmetry, both in virgin and deforming material, become an important issue. Anisotropy that may develop during the course of deformation often produces clues as to the nature of internal processes. It is also important from the point of view of mathematical analysis to know the strength of anisotropy that is present within the material.

To study whether a given material element which is in the state and orientation S (cf. (1.1)) is anisotropic one would apply a "sudden" rigid body rotation Q to the element $(Q \in 0^+(3)$: The group of rigid body rotations in R^3). This would rotate the arrangement of atoms without changing their relative positions. Thus it can be said that the rotated element is in the same state but in a different orientation when compared with the replica of an unrotated element. Let S' denote the state and orientation of the rotated element. If $S \neq S'$ for some Q, we say that the material is anisotropic.

Note that whether $S = S'$ or not could, in principle, be ascertained by applying the same future deformations to the rotated and unrotated elements and by studying the resulting behavior. If these elements exhibit the same behavior in such pairs of tests then in view of (1.1) $S = S' = P_Q S$ and the material element possesses internal symmetry with respect to the rotation Q. If these elements behave differently then $S \neq S'$ and the material element is anisotropic.

It must be clear from the foregoing remarks that S',, the state and orientation of the rotated element can depend only on S and Q:

$$S' = P_Q S \qquad P_Q: \Sigma \rightarrow \Sigma \qquad (6.1)$$

where P_Q is the transformation induced on Σ by $Q \in 0^+(3)$. It is important to note that in the case of initially isotropic solids the following interpretation of P_Q is possible. Let S be the state and orientation created at time t by the strain history $\varepsilon(\tau)$ applied to a virgin element on $0 \leq \tau \leq t$. Then the strain history $\varepsilon'(\tau) = Q\varepsilon(\tau)Q^T$ on $[0,t]$ applied to another virgin element would produce the state and orientation, $P_Q S$. Here Q^T stands for the transpose of the 3 x 3 matrix Q.

Next we observe that the set of all P_Q, $Q \in 0(3^+)$ defines a transformation group G on Σ. It is important to know the precise nature of the transformation P_Q. We were able to show in [1] that under certain restrictions, P_Q can be taken to be the restriction to Σ of a <u>linear</u> orthogonal transformation on R^N. Thus with this extended meaning the group $G = \{P_Q\}$ becomes a representation of $0^+(3)$ in R^N. The theory of group representation and an observation concerning the role of coordinate inversions can now be used to specify the action of P_Q in R^N as follows:

S can be thought of as composed of a number of <u>irreducible tensors of even rank</u>:

$$S = (q_1, \ldots, q_m) \qquad (6.2)$$

where each q_i defines, through its components, an element of an invariant subspace of R^N under G. Moreover

$$P_Q S = (P_Q q_1, \ldots, P_Q q_m) \qquad (6.3)$$

where $P_Q q_i$ has the meaning of an ordinary tensor transformation appropriate to the rank of q_i. It may be useful to remember that <u>antisymmetric</u> and <u>symmetric traceless</u> second rank tensors are irreducible. A completely symmetric and traceless fourth rank tensor is also irreducible and has 9 independent components in general, etc.

We end this formal account of our representation (1.1), (6.1)-(6.3) by adding the following invariance requirements on f and g that originate from a consideration of superimposed rigid body rotations:

$$f(P_Q S) = Qf(S)Q^T,$$

$$g(P_Q S, Q\varepsilon^{\cdot} Q^T) = P_Q g(S, \varepsilon^{\cdot}).$$

(6.4)

A more detailed account of this representation and its generalization to finite deformations can be found in [1] and ORNL Report 4783, 1972 by Onat and Fardshisheh.

<u>Example</u>. <u>Representation of internal damage</u>. We now illustrate the foregoing abstract remarks with a simple example that concerns the tertiary creep of metals. It is well-known that the tertiary phase of creep and the ensuing creep rupture is caused by the development of intergranular voids. It is also known that in some materials the growth of voids on grain boundaries perpendicular to the direction of maximum tensile stress is stronger than in grain boundaries of other orientation. In such a case voids caused by a given stress history will, in general, be distributed anisotropically amongst the grain boundaries and this could give rise to anisotropy in mechanical behavior. Thus it would be necessary to account for this kind of anisotropy in the representation of internal state. It is clear that this line of reasoning would lead one to the statistics of the distributions of the size, shape and the location of voids. For the sake of brevity we consider here only the representation of the distribution of total void volume amongst the grain boundaries. We consider a spherical material element that contains a large number of grains with planar interfaces. (Usually the size of such an element will be much smaller than the size of the structure of interest.) Next we consider the unit-sphere S of directions. Each point of this sphere defines a space direction $\underset{\sim}{n}(|\underset{\sim}{n}| = 1)$ by its position vector. An infinitesimal area $dA(\underset{\sim}{n})$ on this surface about the point $\underset{\sim}{n}$ represents a bundle of directions. Now consider the grain boundaries that are orthogonal to the directions present in this bundle. Let dV denote the total volume of voids (per unit mass of material) found on these grain boundaries. We have

$$dV = V(\underset{\sim}{n}) dA(\underset{\sim}{n})$$

(6.5)

where V is the density of void volume distribution in the material element. In general V will depend on $\underset{\sim}{n}$. In other words, $V: S \rightarrow R$. By definition V has the following property

$$V(\underset{\sim}{n}) = V(-\underset{\sim}{n}). \tag{6.6}$$

Also note that

$$V_o = \frac{1}{2} \int_S V(\underset{\sim}{n}) \, dA(\underset{\sim}{n}) \tag{6.7}$$

where V_o is the total volume of voids found per unit mass of material. The function $V(\underset{\sim}{n})$ is a measure of the internal state of the element. Is it feasible to represent the important aspects of this function by a few parameters? If $V(\underset{\sim}{n})$ had a constant value, V_o would be the only parameter needed. If $V(\underset{\sim}{n})$ is nearly constant (the case of weak anisotropy) and smooth we can represent it (to within a given error and conveniently) by a finite polynomial sum that contains only even powers of n_i (cf. (6.6), the components of $\underset{\sim}{n}$ in a fixed coordinate frame:

$$V \overset{\sim}{=} a_o + \sum_{i,j=1}^{3} a_{ij} n_i n_j + \sum a_{ijk\ell} n_i n_j n_k n_\ell + \cdots . \tag{6.8}$$

The above coefficients a_o, a_{ij}, $a_{ijk\ell}, \ldots$ (which can be taken to be completely symmetric with respect to their indices) would constitute a set of parameters to define $V: S \to R$. It is easily seen that these parameters would transform as tensors of appropriate even rank under rigid body rotations of the material element or under orthogonal coordinate transformations. A more satisfactory representation of $V(\underset{\sim}{n})$ by even rank tensors is obtained by considering the moments of the distribution $V(\underset{\sim}{n})$. The reader is referred to [3] where this point is pursued further.

7. REFERENCES

1. GEARY, J. and ONAT, E.T. - "Representation of Nonlinear Hereditary Mechanical Behavior", Oak Ridge National Laboratories, ORNL-TM-4525, 1974.

2. ONAT, E.T. and FARDSHISHEH, E. - "Representation of Creep, Rate Sensitivity and Plasticity," SIAM J. Appl. Math. 25, 522-538, 1973.

3. LECKIE, F.A. and ONAT, E.T. - "Tensorial Nature of Damage Measuring Internal Variables", to appear in Proc. IUTAM Symposium on Physical Nonlinearities in Structural Analysis, Senlis, France 1980.

4. HART, E.W. - "Constitutive Relations for Nonelastic Deformation of Metals", Journal of Engineering Materials and Technology, 193-202, 1976.

5. GREENSTREET, W.L., CORUM, J.M., PUGH, C.E. - "High-Tem-

perature Structural Design Methods for LMFBR Components", ORNL-TM-4058, 1972.

6.   ONAT, E.T. - "Representation of Inelastic Behavior", Yale University Report for ORNL-SUB-3863-2, 1976.

7.   LAMBA, H.S., SIDEBOTTOM, O.M. - "Cyclic Plasticity for Neoproportional Paths: Part 2", ASME Paper 77-NA/Mat-13, 1977.

8.   MARTIN, J.F., TOPPER, T.H., SINCLAIR, G.M. - "Computer Based Simulation of Cyclic Stress-Strain Behavior with Application to Fatigue", Materials Research and Standards, 11, pp 23-29, 1970.

9.   HALPHEN, B. - "Stress Accommodation in Elastic-Perfectly Plastic and Viscoplastic Structures", Mech. Res. Comm. 2, 273-278, 1975.

CREEP DAMAGE CONCEPTS AND APPLICATIONS TO DESIGN LIFE PREDICTION

D.A. Woodford

General Electric Company, Corporate Research & Development, Schenectady, New York 12308, U.S.A.

1. ABSTRACT

Techniques available for data extrapolation and life prediction are outlined. A recently developed graphical procedure of considerable flexibility is described and its predictive capability demonstrated. However, some of the complexities of the constant load test indicate inherent problems in applying these predictive techniques to component design. For remaining life assessment after some service exposure, the concept of creep damage is used here in a very broad sense. Cavitation, microstructural changes and environmental effects are identified as three categories of damage. The prospects for using measurements of these various forms of damage to assess remaining life are considered for several specific materials.

2. INTRODUCTION

This paper reviews some of the problems inherent in creep fracture life prediction of components using laboratory data based on constant load creep-to-failure tests. Although, in some applications, there is increasing interest in direct measurements of creep crack growth, major high temperature components manufactured from ductile alloys generally rely on smooth bar constant load testing for primary design analysis. Some of the complexities associated with this test, and the problems inherent in data extrapolation are described.

In addition to the primary life analysis, many situations demand a periodic appraisal of the remaining life. This requires an assessment of some change in microstructure, or measurable physical or mechanical property, which occurs systematically with high temperature creep deformation. Such changes are considered generically as creep damage. Thus, the major problems here are the selection of an appropriate creep damage parameter and the development of means to measure the state of damage and hence the remaining life after service exposure.

3. THE CONSTANT LOAD TEST

The vast majority of creep tests employ a fixed load, and hence, a changing stress on the minimum cross-section. Figure 1 shows a creep curve for pure nickel. There can be no period of constant creep rate associated with a steady state situation in such a test. Consequently, a linear region on a strain-time curve is either illusory or a result of a fortuitous balance of hardening and stress increase.

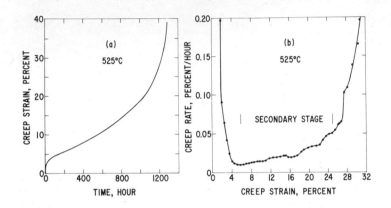

Fig. 1   Creep Curve for Nickel at 525°C and 138 MPa
   a) Strain vs. Time
   b) Strain Rate vs. Strain

When machine differentiated the creep curve shows a clear minimum in creep rate followed by a period of gradually increasing creep rate to about 26% strain at which point macroscopic necking begins.[1] Thus a period of uniform straining at increasing stress and increasing strain rate is identified as a secondary stage. Final failure is associated with both local necking and internal cavitation with a rapid increase in strain rate and a degree of triaxial deformation. The specimen clearly is subjected to a very complex strain rate history involving more than an order of magnitude variation in total extension rate and further local increases during final fracture. An effect of specimen geometry on life is to be expected and is generally observed.[1]

4. CREEP FRACTURE LIFE PREDICTION

Notwithstanding this test complexity, a considerable amount of effort has been directed towards the accurate extrapolation of available test data to design stresses.[2] Such methods include:

.   Graphical or analytical extrapolation of stress versus life.

.   The application of time-temperature parameters which may
     be used both for displaying the relative merits of different
     alloys over a range of temperature, and for data extrapolation.

• Recently developed methods which impose minimum restrictions on the form of the parameter. These include the Minimum Commitment Method (MCM)[3] and the Graphical Optimization Procedure (GOP)[4].

There are a number of problems which may be classed as intrinsic or extrinsic. Intrinsic problems include: extrapolation itself (for stress-life curves); statistical treatment of heat-to-heat variability, scatter about the mean, and use of stress or time as the dependent variable[5]; inaccuracy of time-temperature parameters, and the extensive data required to develop minimum restriction procedures. Extrinsic problems are associated with predicting life in a regime where new damage processes or failure mechanisms occur. In such circumstances the available data will be deficient in necessary information.

The recently proposed fracture mechanism maps[6] attempt to locate separate fields corresponding to different failure modes on a plot of stress versus temperature, containing contours of constant life. In principle, these maps offer an approach to life prediction in a different fracture field from that for which data are available. However, since they are constructed using creep to fracture tests, the field locations reflect the very rapid local increases in strain rate near fracture. These instabilities may have no bearing on service conditions. The extrinsic problems thus appear intractable at this time, and are generally negated in practice by conservative design.

To illustrate how an accurate representation of a reasonably complete data set may be achieved, the results on a Cr-Mo-V forging steel shown in Figure 2 are analysed using the G.O.P method.[4].

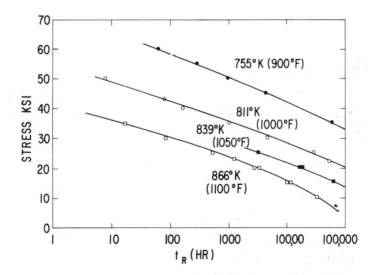

Fig. 2   Stress vs. Creep Rupture Life Curves at Various temperatures for Cr-Mo-V Steel.

Briefly, the method expresses one variable as a product of functions of the other two. Thus,

$$t_r = H(\sigma) \, Q(T) \tag{1}$$

or

$$T = H(\sigma) \, C(t_r) \tag{2}$$

where σ, T and t_r are stress, temperature and life and H, Q, C are functions. It may be shown readily that the common Orr-Sherby-Dorn[7] and Manson-Succop[8] parameters may be represented in the form of equation 1, and that the Manson-Haferd[9] and Larson-Miller[10] parameters in the form of equation 2. For example, the Larson-Miller parameter has the form -

$$H(\sigma) = T \, (K + \log t_r) \tag{3}$$

Comparing equations 2 and 3, if the L-M parameter is obeyed

$$1/C(t_r) \quad \propto \quad \log t_r \tag{4}$$

The relationships for the $H(\sigma)$ vs. σ and $C(t_r)$ vs. t_r were derived graphically as described elsewhere.[4] Curves obtained were unique and independent of temperature as shown in Figure 3, which is a plot of $1/C(t_r)$ vs. log t_r, thus confirming the general product relation of equation 2. According to equation 4, if the L-M parameter applies, Figure 3 should give a linear relationship. The curvature indicates that the optimum value of the "constant" in the L-M parameter changes as a function of time. The value of the "constant" may be obtained at any time by constructing tangents to the curve of Figure 3. The equation for these tangents may be expressed as:

$$\log t_r = \log A(t_r) + B(t_r) \; / \; C(t_r) \tag{5}$$

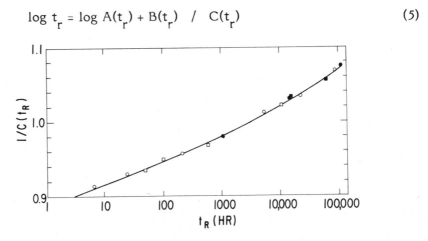

Fig. 3   Plot of $1/C(t_r)$ vs. log t_r

where A and B are functions describing the intercept and slope respectively. Rearranging -

$$C(t_r) = B(t_r) / (\log t_r - \log A (t_r))$$

(6)

An instantaneous value of the L-M "constant" is thus given by $-\log(A)t_r$ by comparison with the form of equation 3.

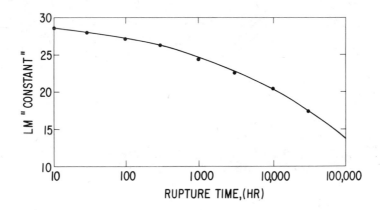

Fig. 4   Variation of Larson-Miller "Constant" as a Function of Time.

Figure 4 is a plot of this optimum value of the L-M "constant." It is apparent that K changes progressively from a value of about 28.5 at 10 hours to 13.6 at 100,000 hours. Clearly the data can be accurately described by a parameter of the form

$$H(\sigma) = T (K (t_r) + \log t_r)$$

(7)

but use of the standard parameter with K = constant (20) could give predictions which would be significantly in error.

Figure 5 shows the segmented data at each temperature when plotted using K = 20; the tagged points are taken from the failure curves at 70,000 hours. By contrast, Figure 6 shows good precision when the data are plotted using the $C(t_r)$ function shown in Figure 3 and expressed according to equation 2. There is now no separation of isothermal segments.

Clearly, a similar functional form could be generated for any of the simple parameters. By replacing the characteristic constant with a time dependent term, maximum use is made of the experimental data. Although it is not expected that the time dependence of K in the L-M parameter should be material independent, it is noteworthy that in the previous analysis of the nickel base alloy IN718, K varied from 27.1 at short lives to 20 at 10,000 hours,[4] which closely parallels that depicted in Figure 4.

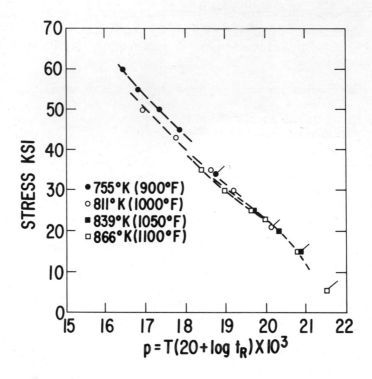

Fig. 5   Larson-Miller Parameter Plot Using a Constant of 20.

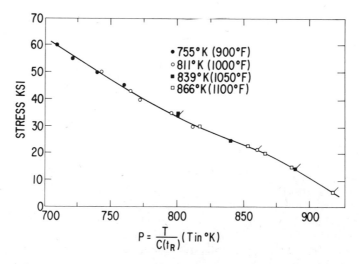

Fig. 6   Stress as a Function of Generalized Parameter $T/C(t_r)$

5.   CREEP DAMAGE

During a creep fracture test, in addition to the complex changes in stress, stress-state and strain rate, there are complex changes in the microstructure. Changes which effect the mechanical state of the material are broadly referred to as creep damaging. We identify three broad categories:

1. The nucleation and growth of grain boundary cavities.

2. Deformation enhanced microstructural changes such as precipitation and solute segregation.

3. Environmental effects such as gas metal reactions and grain boundary penetration of embrittling species.

Generally one of these categories dominates for a particular material, although clearly very complex interactions may occur. In principle, if these changes can be well characterized they may be used to determine remaining life after service exposure. We shall consider recent attempts to do this for the three types of damage.

5.1 Nucleation and Growth of Cavities

The majority of studies of nucleation and growth of cavities have been concerned with mechanisms involved rather than with detailed characterization in terms of the creep test variables. An equation was, however, proposed for copper which was consistent with a continuous increase in the number of cavities proportional to strain and an increase in the volume of an individual cavity proportional to time. The equation[11] linked total cavity volume to strain, time, stress and temperature:

$$V = \text{const.} \; \varepsilon \, t. \, \sigma^n \exp\left(\frac{-Q}{RT}\right) \tag{8}$$

Figure 7 shows this relationship for some previously published data.[12] The form of this equation has been confirmed by a number of authors.[13-15] for a variety of materials and has recently been broadened to include a grain size term.[16] However, in terms of damage assessment for remaining life calculations it has some problems. For example, as shown in Figure 8, the amount of volume change at rupture is sensitive to the test conditions.[17] Also for a given fraction of life exposure at a constant temperature the amount of damage is greater the lower the stress. Nevertheless, there have been attempts to develop life prediction schemes on the basis of metallographic observation of cavities,[18,19] and this may prove useful for some alloy classes.

The interpretation of creep cavitation mechanisms may have to be modified in view of recent work demonstrating that cavitation may be induced by thermal exposure in an oxygen containing environment, with no concomitant deformation.[20] Certainly, it is well established

610

that cavitation during creep is generally much more prevalent near the specimen surface[21] and environmental effects in general may need to be taken into account.

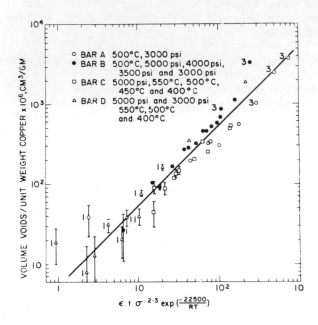

Fig. 7 Dependence of Cavity Volume on the Test Parameters for Copper.[11]

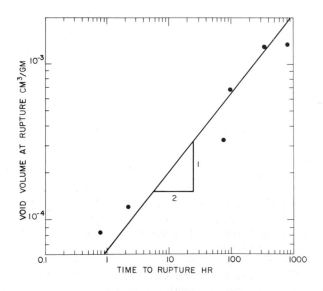

Fig. 8 Cavity Volume at Rupture as a Function of Time to Rupture in Tests on Nickel at 525°C[17]

5.2   Microstructural Changes

There have been few attempts to use microstructural changes as a direct measure of creep damage. However, post exposure mechanical property assessment has been used as an indirect measure of microstructural changes. For example, in a comprehensive study of Cr-Mo-V steel exposed at various stresses and temperatures, for different strains and times up to 60,000 hours a systematic softening was measured.[22] This softening correlated very well with fracture life in a standard test on exposed material as shown in Figure 9. For, this material, room temperature hardness provided a useful measure of damage and could in principle be used to predict remaining life after service exposure.

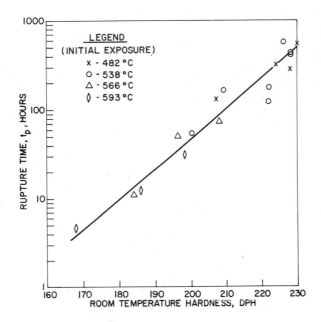

Fig. 9   Correlation Between Post-Exposure Rupture Time in the Standard Test at 538°C and 240 MPa and Room Temperature Hardness for Cr-Mo-V Steel.

An important consequence of this work was that the linear damage law was demonstrably inapplicable for stress changes, because of a loading sequence effect.[23,24] However, for temperature changes, it was quite adequate and provided a useful means for assessing remaining life of service exposed parts.[25-27] Thus, if t_s is the time in service, T_s is the life under service conditions, t_t the life at higher temprature of the exposed material and T_t the life of undamaged material at the higher temperature, then

$$\frac{t_s}{T_s} + \frac{t_t}{T_t} = 1$$

(9)

which has been shown to work well on service exposed 1 Cr - 0.5 Mo steel.[26]

On the basis of the work on Cr-Mo-V steel,[22-24] a model for life prediction under changing stress and temperature conditions and, hence, remaining life prediction has been established. The model is shown schematically on a stress vs. L-M parameter[*] plot, Figure 10, where each curve represents a state of damage or hardness. The proposed procedure for life prediction is illustrated by a stress-temperature change path ABCDE. After exposure at stress σ_3 and temperature T_1, a remaining life curve is entered at point A and the stress is reduced to σ_2 (point B) where some of the life is used to reach C. At this stage the temperature is increased to T_2 but the position on the plot is not changed, i.e. the parameter is constant. Exposure at this temperature leads us to point D where the stress is further reduced to σ_1 and the remaining life is computed from the value of the L-M parameter at point E. These constant damage curves that converge at a parameter value of about 20×10^3 i.e. 10^5 hours life at $525°C$ are entirely consistent with service experience.[28,29] In particular, a 1.25 Cr - 0.5 Mo steel, after 83,000 hours service, showed a marked reduction in high stress fracture life whereas at low stresses lives were essentially the same as new material.[30]

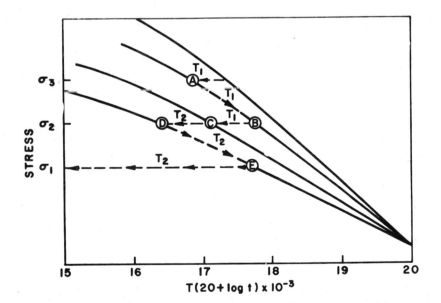

Fig. 10   Schematic Plot Illustrating Construction of Constant Damage Curves in Terms of the Larson Miller Parameter (T in $°K$, t_r in hr). The Sequence ABCDE is an Example of a Stress-Temperature Path Described in the Text.

*   For illustration purposes the L-M parameter is here considered adequate.

5.3   Environmental Effects.

A broad range of iron and nickel base alloys have been tested in environments ranging from vacuum and inert gases to hydrogen, oxygen, nitrogen, carbon monoxide, carbon dioxide and air. However, in general, these environments have rarely been shown to influence deformation rates and fracture lives by more than a factor of three.[31]   Nevertheless, it has recently been demonstrated that long exposures in air at intermediate temperatures, particularly in thin sections may lead to pronounced embrittlement and loss in life.[32] Cast nickel base superalloys are particularly susceptible as shown in Figure 11 for IN738 pre-exposed in air at 1000°C. The sensitivity to this damage increases with decreasing temperature (in the range 1000 to 700°C).   Figure 12 demonstrates that the damaging species is oxygen. It has been demonstrated that grain boundary embrittlement results from oxygen penetration to a depth at least an order of magnitude greater than oxygen penetration in the matrix.[33]   In 100 hours at 1000°C, penetration to a depth of 2.5 mm has been confirmed. Frequently, this is associated with grain boundary cavitation in the absence of applied stress in nickel[20] and nickel base alloys,[34] which is consistent with the reported profound effect of air exposure on near surface creep cavitation.[21]

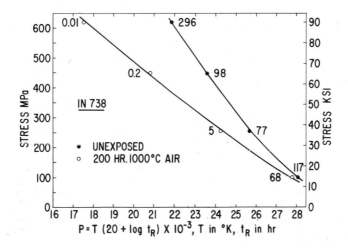

Fig. 11   Effect of Air Exposure on Rupture Life of 1N738. Individual Lives are Indicated in Hours for Tests at 700, 800, 900 and 1000°C

This damage associated with grain boundary oxygen penetration develops systematically as a function of exposure time and temperature in the high temperature materials studied so far. Although this offers the possibility of providing a basis for remaining life assessment, so far no specific attempts have been made. With nickel base alloys, employed as hot gas path components in the gas turbine, environmental damage is exacerbated by cyclic thermal straining, and the duty cycle is generally so complex that remaining life assessments are notoriously inaccurate.

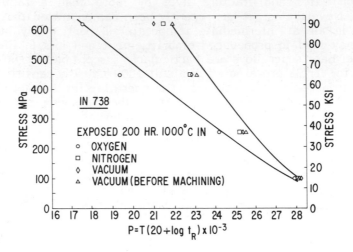

Fig. 12   Effect of Exposure Environment on Life.   Curves are Reproduced from Fig. 11.

6.   SUMMARY

Various aspects of creep fracture life prediction have been described. Although the constant load test is extremely complex and has severe limitations as a basis for life assessment of components, accurate extrapolation procedures are available. Further development of parametric schemes is probably not justifiable.

Creep damage assessment is attractive from the standpoint of predicting remaining life after service exposure, and also understanding of the possible consequences of the combined influence of temperature, stress and environment, and cyclic loadings. Three categories of creep damage are identified, viz. cavitation, microstructural changes and environmental effects. Although in many cases one of these may dominate, interactions will occur. The most complete and useful assessment of damage directed at practical problems of remaining life prediction may be the work on low alloy steels. Other areas of damage assessment need much more work. In particular, the effect of test environment and its role in intergranular cavitation may be much more important than previously realized.

7.   ACKNOWLEDGEMENT

Dr. R.H. Bricknell made a number of helpful suggestions in reviewing the manuscript.

8.   REFERENCES

1.   WOODFORD, D.A. - Trans. ASM, 1966, 59, 398.

2.   EPRI Report FP-1062, 'Development of a Standard Methodology for the Correlation and Extrapolation of Elevated Temperature Creep and Rupture Data.', 1979.

3.   MANSON, S.S. and ENSIGN, C.R. - reference 2, 299.

4.   WOODFORD, D.A. - Mat. Sci. Eng., 1974, 15, 169.

5.   HAHN, G.J. - J. Eng. Mat. Techn., 1979, 101, 344.

6.   ASHBY, M.F., GANDHI, C. and TAPLIN, D.M.R. - Acta Met., 1979, 27, 699.

7.   ORR, R.L., SHERBY, O.D., and DORN, J.E. - Trans. ASME, 1954, 46, 113.

8.   MANSON, S.S. and SUCCOP, G. - ASTM Spec. Tech. Pub. 1956, No.174, 40.

9.   MANSON, S.S. and HAFERD, A.M. - NASA-TN-2890, 1953.

10.   LARSON, F.R. and MILLER, J. - Trans. ASME, 1952, 74, 765.

11.   WOODFORD, D.A. - Metal Science Journal, 1969, 3, 50.

12.   BOETTNER, R.C. and ROBERTSON, W.D. - Trans. AIME, 1961, 221, 613.

13.   GREENWOOD, G.W. - Phil. Mag., 1969, 19, 423.

14.   GITTINS, A. - J. Mat. Sci., 1970, 5, 223.

15.   TIPLER, H.R., LINDBLOM, Y. and DAVIDSON, J.H. - High Temperature Alloys for Gas Turbines, Ed. Coutsouradis, D. et al., Applied Science Publishers, Ltd., 1979, 359.

16.   MILLER, D.A. and LANGDON, T.G. - Met. Trans. A., 1980, 11A, 955.

17.   WOODFORD, D.A. - Met. Sci. Journal, 1969, 3, 234.

616

18. DYSON, B.F. and McLEAN, D. - Met. Sci. Journal, 1972, 6, 220.

19. PIATTI, G., LUBEK, R. and MATERA, R. - Euro. Spectra, 1972, 11, No. 4, 93.

20. BRICKNELL, R.H. and WOODFORD, D.A. - Met. Trans., 1981, in press.

21. SCAIFE, E.C. and JAMES, P.L. - Metal Science Journal, 1968, 2, 217.

22. GOLDHOFF, R.M. and WOODFORD, D.A. - ASTM, 1972, STP 515, 89.

23. WOODFORD, D.A. - Int. Conf. on Creep and Fatigue in Elevated Temperature Applications, Inst. Mech. Eng., 1973-74, Paper 180.1

24. WOODFORD, D.A. - Jour. Eng. Mat. Techn., 1979, 101, 311.

25. CANE, B.J. and WILLIAMS, K.R. - Int. Conf. Mech. Behavior of Materials, 1CM3, Cambridge, England, 1979.

26. HART, R.V. - Metals Technology, Sept. 1977, 442.

27. ETIENNE, C.F., van ELST, H.C. and MEYERS, P. - Creep of Engineering Materials and Structures, Ed. Bernasoni, G. and Piatti, G., App. Sci. Publ, 1978, 149.

28. TOFT, L.H. and MARSDEN, R.A. - Iron and Steel Inst. , 1961, STP 70, 248.

29. ROBERTS, B.W., ELLIS, F.V. and BYNUM, J.E. - Jour. Eng. Mat. and Techn., 1979, 101, 331.

30. CULLEN, T.M., ROHRIG, I.A. and FREEMAN, J.W. -Trans. ASME, 1965, Paper No. 65-WA/Met. 3.

31. NIX, W.D. and FUCHS, K.P. - EPRI Reprot, 1977, ER-415.

32. WOODFORD, D.A. - Int. Conf. on Engineering Aspects of Creep, I. Mech. E., Sheffield, 1980, 55.

33. WOODFORD, D.A. and BRICKNELL, R.H. - Met. Trans - submitted for publication.

34. WOODFORD, D.A. - Met. Trans. - in press.

CRITERIA FOR PROLONGING THE SAFE OPERATION OF STRUCTURES
TROUGH THE ASSESSMENT OF THE ONSET OF CREEP DAMAGE USING
NONDESTRUCTIVE METALLOGRAPHIC MEASUREMENTS

Dr. Ing. B. Neubauer

Rheinisch-Westfälischer Technischer Überwachungs-Verein e.V...
Steubenstrasse 53, 4300 Essen, Germany

1. SUMMARY

This paper deals with microstructural changes, the onset
of tertiary creep and cavitation in structures. In analyzing
these changes, we established a nondestructive method for
measuring the creep damage in structures by metallographic
replicas.

This method is used to deal with the problem that often
premature failures occur after short exposure. These failures
are mostly caused by additional stresses from the system,
which are not capable of being properly calculated, from
uneven temperature distributions and from residual stresses
in welds.

Because the areas to be measured are small, the choice
of parts and areas must be based on careful consideration.
This will include calculations using service data and
experience of where failures most commonly occur in the parts.
Thus we can choice the most exposed spots for the measurements.

On the basis of the results of the measurements and the
loading conditions we draw up criteria for further operation.
The following three basic courses of actions are then
possible:

- further operation without service restrictions

- further operation with service restrictions

- no further operation

x

By repeated tests on the structures we can gradually establish a correlation between creep strain and creep damage of structures which will give us a safe basis for deciding on further operations or for fixing a reasonable test frequency. After about 5.000 measurements our experience has been positive and the costs were reasonable. The procedure results in a prolonged operation of components with enhanced economic performance and safety.

2. APPLICATION AND DEVELOPMENT OF THE METHOD

In analyzing the microstructural changes, the onset of tertiary creep and cavitation in structures, we established a nondestructive metallographic method for measuring the creep damage in structures.

The following method is employed: replicas taken from the structure on site contain the microstructural information, cf. Fig.1 and can be analyzed (in the same way as a specimen) by light and scanning electron microscopes. We were able to demonstrate that cavities down to a size of 0,1 µm can easily be detected [1], cf. Fig. 2. Conventional NDT-methods are less accurate by a factor of 10.000 and are therefore unable to detect these small creep cracks. Because of the small size of their test area (~ 1 cm^2), replicas can however only be used on spots.

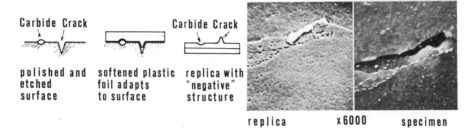

Carbide Crack | Carbide Crack

polished and etched surface | softened plastic foil adapts to surface | replica with "negative" structure

replica   x6000   specimen

Fig 1: Principle of Replication

Fig 2: Comparison between Replica and Specimen

Replica test was at short included in the german rules for boilers "TRD 508" [2]. The Replica examination of creep-loaded components is prescribed in the "Verbändeverein-barung 1978/1" [3], which was passed by the trade associations and the authorities in 1978 . W. Arnswald et.al described the state of the art and first practical experiences [4].

Like short exposure creep tests, the advantage of the replica method lies in the restlife estimation of components exhausted from the point of view of calculations (e > 1). Because of the low cost per measurement taken replicas can give information on a wide basis. This makes them an ideal complement to creep tests which require more effort and can only be used for a limited number of components. For a view of the replica method in the context of the other methods for restlife estimations see [5].

3. STRATEGY OF THE TEST AND BACKGROUND

If nondestructive structure examinations are to be employed to determine the remaining life of creep-loaded components the following approach is adopted:

a) The test is carried out on areas of highly stressed components where cracks are suspected and the smallness of the test area is thereby taken into account.

b) The structure determined on the surface of the component gives an indication of the component's creep exhaustion.

c) On the basis of the test results a conservative estimate is made of the minimum rest life.

d) After the expiry of the "minimum restlife" a new examination is carried out and, where relevant, a new minimum restlife is fixed.

3.1 Determination of crack susceptible zones in the most highly stressed components

The most highly stressed components are determined according to TRD 508, i.e. by the repeated calculation of life exhaustion "e" in the framework of internal inspection, using the service data (temperature, pressure, duration of operation) as well as the actual dimensional values, especially wall thickness. The nondestructive metallographic examinations start from an exhaustion level of e > 0.6.

The crack susceptible zones can be determined from damage analysis. Here the following locations should be mentioned:

- Parts with acting stress concentrations, especially

- welds , - T's and Forgings, - Bends.

Because of additional forces as well as metallurgically based alterations in creep resistance and undetected fabrication defects damage can occur here well before the end of the calculated restlife. Replica examinations are often carried out on these locations even before the minimum exhaustion, calculated according to TRD 508 is reached.

3.2 Correlation of the surface structure with life exhaustion of component

For creep-loaded components mainly the following steels are used:

1/2 Cr - 1/2 Mo - 1/4 V

2 1/4 Cr - 1 Mo

1 Cr - 1/2 Mo

What follows is to be understood in relation to these steels. The structural changes occuring during service in the steels mentioned can only partially be correlated with life exhaustion. The changes include:

(a) Coagulation of carbides, recrystallization

(b) Pinning of dislocations by fine dispersed carbides, dissolution and redistribution of various types of carbide, accumulation of carbide forming elements in the carbides

(c) Formation of cavities, their linkup to microcracks and finally the creep fracture of the component

3.2.a Structural changes observable by optical microscopy

The changes in structure mentioned in (a) can be observed with the aid of an optical microscope but because of the wide range of initial structures in the components they cannot for all practical purposes be correlated with exhaustion In 1976 Weber [6] did indeed demonstrate that his is in principle possible for the 1 Cr -1/2 Mo steel. However his results from annealing tests of up to 10 000 hours at 500, 550 and 600 ° C are based on the ideal condition of a hardened structure and can in no way be carried over to inspection practice owing to the range of microstructures present in components.

3.2.b Structural changes observable by electron microscopy

The microstructural changes listed in (b) can possibly be related to life exhaustion as is indicated by tests carried out by Schwaab [7] . The electron microscopic examinations are, however, so costly and time consuming that they can only be used in special cases.

3.2. c Changes in structure before creep fracture

The changes in structure listed in (c) can in contrast be measured metallographically without difficulty. As mentioned above the measurement can record cavities which are greater than 0,1 µm. The following information is referred to in order to ensure a reasonable correlation of the cavities measured with the extent of life exhaustion:

(1) Measurements taken on copper have shown that there is, during the creep test, a decrease in density which runs against the creep curve, cf. Fig. 3. The accelerated density decrease in the third creep regime can only be attributed to the parallel formation of cavities and microcracks.

(2) Since in the past structural changes were not considered when recording creep curves, we have only one work by Mints [8] to refer to, cf. Fig. 4. Mints demonstrated that the 1/2 Cr- 1/2 Mo- 1/4 V -steel shows substantial cavity formation towards the end of the second creep range. The beginning of the third creep range is characterized by the appearance of oriented micropores which at first just occur in the outside zone of the specimen. These experiments confirm that measurable cavity formation begins at the end of steady- state creep. Moreover they confirm that creep damage begins on the outside surface of components, which are accessible to replica measurements.

(3) A first evaluation of creep curves for the above mentioned steels shows that at the commencement of the third creep range a life exhaustion of $0.5 < e < 0.8$ has been reached.

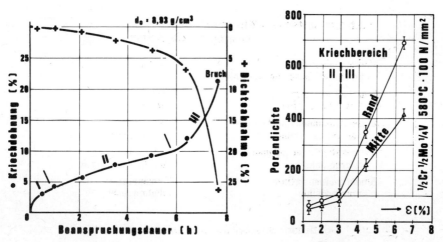

Fig 3: Density decrease in copper by creep

Fig 4: Cavity formation in steel

3.3 Conservative estimation of "minimum rest life"

Taking account of the above information it is possible to estimate a life exhaustion of $e < 0.8$ for a component with marked cavity formation. From this it can be seen that a component with a service life of 80.000 hours has a restlife of >20.000 hours (2 to 3 years)

3.4 Re-test and re-estimation of "minimum rest life"

The establishment of the minimum rest life R_{min} has the advantage that the prediction made must only cover a relatively short period of time. For each minimum rest life the microstructure of the component is obtained and with this the progressive development of creep damage, depending on material and stress. Quantitive measurements as in [4] can be fruitfully applied here. The actual rest life R up to the replacement of the exhausted component is then made up of R_{min1} + R_{min2} + R_{min3} ..... + $R_{min\ n}$. When R_{min} becomes too small then continued operation would be uneconomical and repairs would be undertaken.

4. PRACTICAL APPLICATION OF REPLICA METHOD

4.1 Classification of measurements taken

In the assessment of the replicas five ratings are employed to describe the microstructure, see table 1 left hand side. This procedure helps avoid misunderstandings and saves time. In this way the test results can generally be made available within 24 hours. Fig 5 shows these ratings for weld metal identical to 2 1/4 Cr - 1 Mo.

states of material after exposure / corresponding action

states of material after exposure	corresponding action
1 normal microstructure - precipitations beginning / going on	none
2 marked exposure - onset of cavitation	none watch
3 creep damage beginning - marked cavitation	watch fix intervals
4 late state of creep damage - microcracks	limited service quick replacement
5 nearly creep failure - cracks visible	no operation before repair

Table 1: Schematic evaluation of replica tests

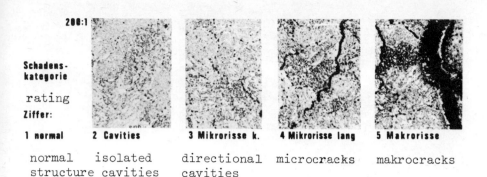

200:1

Schadens-
kategorie

rating
Ziffer:

1 normal	2 Cavities	3 Mikrorisse k.	4 Mikrorisse lang	5 Makrorisse
normal structure	isolated cavities	directional cavities	microcracks	makrocracks

<u>Fig 5</u>: Creep damage in weld 123.000 h, 535°C, 180 bar

4.2 Assessment of "minimum restlife"

This assessment is based on the number, distribution and size of the cavities and microcracks as well as on the loading conditions of the component. Changes in the microstructure are normally not assessed unless they indicate a wrong heat treatment of the component in the delivered condition. In such cases suitable steps must be taken to deal with a suspected inadequate creep resistance in the component. The actions in the power station resulting from the measurement in general are listed in .table 1 , right hand side.

For the ratings 1 and 2, no actions have to be undertaken. The "minimum restlife" for the ratings 3 to 5 varies from about 20.000 hours to zero.

In every case the "minimum restlife" depends on the extent of damage and the operating conditions. For the purpose of accumulating experience operation should always be continued, if only for a short time, before repairs are undertaken.

The following examples are concerned with welds, where early creep damage occurs with particular frequency.

4.3 Examples for assessing welded components

a. 1 Cr - 1/2 Mo steel - tube girth weld, cf. <u>Fig 6</u>

Numerous longitudinal cracks were detected in the weld metal and the adjacent base material. This shows a general exhaustion of life in the area surrounding the weld. After a short period of further operation the weld and adjacent base material were replaced.

624

B.M.   HAZ/ W.M.   W.M.

<u>Fig.6</u> : Directional cavities across weld: 106.000 h, 535°C, 180 bar

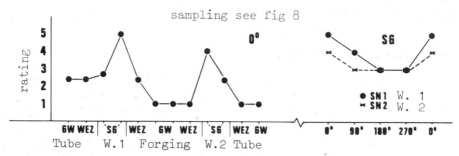

<u>Fig. 7:</u> Damage distribution by bending across and along weld: 123.000 h, 535°C, 190 bar

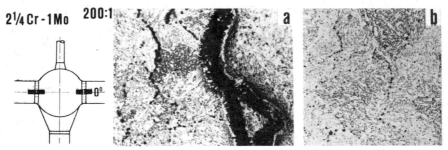

<u>Fig.8:</u> Creep damage in weld by bending: 123.000 h, 535°C, 190 bar. a) surface, b) 1,5 mm ground off

b. 2 1/4 Cr - 1 Mo steel - forging with 2 weld connections to tube, cf. <u>Figs. 7 and 8</u> .

      The creep damage concentrating in the weld metal directly adjacent to the forging is typical for welds on 2 1/4 Cr - 1 Mo steel. The circumferential damage (Fig.7 right) indicated a bending stress. The crack detected, cf. Fig 8, could be removed by grinding off by 1.5 mm. Since there was still an adequate wall thickness remaining , a continued operation until repair was possible .

c. 1/2 Cr - 1/2 Mo - 1/4 V steel - tube girth weld near forging, cf. Fig 9.

After approximately 76 000 operating hours a large creep crack was detected in the fine-grained heat affected zone of the girth weld. It ran parallel to the weld, see Fig 9 a. By grinding off by 2 mm it was possible to remove the crack leaving a few micropores, cf. Fig 9 b. After approximately 5000 h further operation it was observed that the damage had progressed, cf. Fig 9 c. By advanced ultrasonic testing a fabrication defect was detected under the cavitated area. After removing the fabrication defect by grinding, operation was continued. The component will soon be tested again.

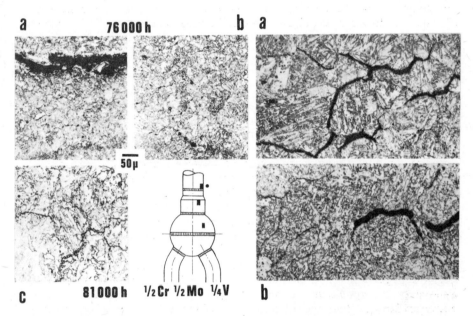

Fig.9: Continued operation
of creep damage
component
530°C, 60 bar
a: surface
b: 2 mm ground off
c: same as b

Fig.10: Creep damage from:
a: Relaxation Cracks
b: Hot cracking
1/2Cr-1/2 Mo-1/4 V - steel
80.000 h, 530°C, 60 bar

4.4 Assessment of welds with fabrication defects

Even after approximately 100 000 hours the replica examination often reveals fabrication defects such as relaxation cracks and hot cracks (cf. Fig 10). These cracks of course represent weakspots in the component. While incipient creep cracks only occur on the surface of the component, here we have defects in the body whose extension is mostly unknown. For this reason it is difficult to assess how far the component is endangered by fabrication defects.

The following procedure has developed:

If the preexisting crack has not grown by creep then it can be left, since a sudden failure after such a long period of operation is improbable. As soon as creep crack growth is detected e.g. because of pore formation at the tips of the crack, cf. Fig. 10, then the crack as a whole is classified as a creep crack. As a rule it will then be necessary to replace the weld. A particularly careful nondestructive examination of the welds in the begin-of-life condition of the components can help avoid high costs at a later date.

5. DISCUSSION

The replica method is a valuable aid in the estimation of the life expectancy of creep-loaded components in power plants. Because of the small measuring surface careful preparatory work is necessary to determine highly loaded components and zones suspected of having cracks, where the measurements can be taken. As soon as the first indications of creep damage are obtained a minimum restlife is established. There then follows a further measurement or repair work. Examples are given of the assessment of creep-damaged welds. Fabrication defects in welds can necessitate early repair.

The replica method can reveal creep damage at a very early stage. For this reason creep damaged components can generally be operated up to the point of repair without restriction of the operation. In this way the replica method increases considerably the availability of equipment with enhanced safety.

author's address:

Dr. B. Neubauer
RWTÜV e.V.
Postfach 10 32 61
D-4300 ESSEN 1

REFERENCES:

[1] NEUBAUER, B.
DVM, 8. Sitzung des AK Rasterelektronenmikroskopie am
10./12.10.1977, Berlin, S 105-111

[2] DDA
Technische Regeln für Dampfkessel, TRD 508
Beuth Verlag, Köln, Oktober 1978

[3] VdTÜV
Merkblatt Dampfkessel 451-78/1 (Verbändevereinbarung
1978/1 VdTÜV (Hrsg.) Juli 1978

[4] ARNSWALD, W., BLUM,R., NEUBAUER, B., POULSEN, U.E.
VGB Kraftwerkstechnik 59 (1979) 7, S. 581 - 593

[5] DRUCKS, G., JÄGER, P, KAUFMANN, H.R.
3R International 18 (1979) 5, S. 320 - 27

[6] FABRITIUS, H., WEBER, H.,
Chemie-Ing.- Techn. MS 383/76

[7] FABRITIUS, H., WEBER, H.,
VGB Werkstofftagung 1979, Vortragsband S 179-217

[8] MINTS, I. I., BEREZINA, T.G., KHODYKINA, L. Ye.,
Fizika metallov i metallovedenie 37 (1974) 4,
S. 823 - 831

DEFORMATION IN 2CrMo-½CrMoV PRESSURE VESSEL WELDMENTS AT ELEVATED TEMPERATURE

M.C. Coleman and J.D. Parker

Central Electricity Generating Board, Marchwood Engineering Laboratories, Marchwood, Southampton, SO4 4ZB, UK.

SUMMARY

The elastic and creep deformation occurring in low alloy ferritic steel pipe to pipe weldments has been studied in pressure vessel experiments carried out at 838 K and a range of internal steam pressures. The welds were made in heavy section ½CrMoV parent pipe, using 2CrMo weld metal and tested in either the as-welded or stress relieved condition. The results obtained are analysed in terms of the deformations that occur in the hoop and axial direction of the parent pipe and weld metals.

Elastically the parent pipe and weld metals behave identically, and both exhibit primary and steady state creep. The steady state behaviour of the parent pipe agrees with that expected from multi-axial creep deformation theory. The stress relieved welds behave similarly to the parent pipe in the hoop direction, but not in the axial direction. This is considered to be due to offloading of stress predominantly in the hoop direction. In the as-welded condition, the hoop and axial creep strains and strain rates are greater than in the parent pipe or stress relieved welds. The reasons for this are discussed in terms of welding residual stresses.

1. INTRODUCTION

The Central Electricity Generating Board's steam generating plant contains many weldments in low alloy ferritic steel pipes that are required to operate for extended periods at temperatures where creep can occur. The design of such pipework is usually based on the application of a mean diameter hoop stress to the parent material uniaxial creep properties. Previous work has shown that this approach is conservative, with respect to the deformation behaviour of plain pipes[1]. However, premature cracking of welded pipes has been experienced within the CEGB[2,3] suggesting that additional factors affect the behaviour of welds.

The manufacture of a weldment introduces microstructural and mechanical property variations that do not arise in plain pipes. Low alloy steel weldments contain metallurgical structures ranging from the mainly ferritic normalised and tempered parent pipe through refined and coarse grained bainitic regions of the heat affected zones (HAZs) to the mixed columnar and equiaxial structures of the weld metal[4]. Each of these microstructural constituents will have different mechanical properties and consequently will behave differently when subjected to plant operating conditions. In addition, the overall performance of the weldment may be further influenced by the presence of welding residual stresses.

The effect of these factors on the high temperature performance of heavy section weldments is being studied in pressure vessel research programmes at the CEGB's Marchwood Engineering Laboratories (MEL). This paper presents data obtained from these programmes on the behaviour of 2CrMo welds in $\frac{1}{2}$CrMoV parent pipe, this being the most common low alloy ferritic steel weldment encountered in plant. Results are presented on the elastic and creep behaviour of the parent pipe and of the weld metal. The relationships between the deformation in these materials is then considered and the effects of residual stress and heat treatment on the strain accumulation are discussed.

2. EXPERIMENTAL

This paper draws on information obtained from three different research programmes being carried out at MEL. These are concerned with the performance of sound weldments, weldments containing defects and weldments manufactured by different welding procedures. As a result the data presented here come from 4 pressure vessels and 5 weldments. However, in all cases the weld metal-parent material combination is 2CrMo-$\frac{1}{2}$CrMoV.

2.1 Vessel Design

A typical pressure vessel is shown in Figure 1 and consists of $\frac{1}{2}$CrMoV forged end cap and hot drawn seamless pipe sections welded together with 2CrMo manual metal arc (MMA) deposited weld metal. The pipe sections were 60 mm wall thickness and about 350 mm outside diameter, giving an outside to inside diameter ratio of $\sim$ 1.5. In each vessel the weld centre line spacings were somewhat different, but always such that interactions between the welds were negligible.

The welds were manufactured by depositing MMA electrodes in a standard "J" preparation, as shown inset in Figure 1.

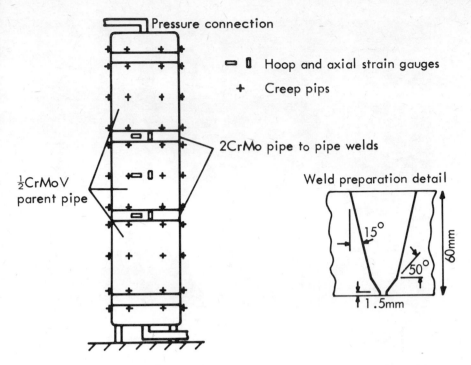

Fig.1 Typical 2CrMo-½CrMoV pressure vessel design showing weld location, monitoring positions and weld preparation detail.

The electrode diameters used ranged from 2.5 mm in the root area through 3.25 mm and 4 mm to 5 mm for the majority of each weld. Welding was carried out at a preheat of 473 K within the current and voltage range recommended by the electrode man-ufacturers. Some welds were left in the as-welded condition while others were stress relieved within the range 948 to 973 K for times between 2½ and 3 hours. A section typical of the weldments tested in the pressure vessel is shown in Figure 2, illustrating the macroscopic features of the weld metal, HAZ and parent material.

2.2 Testing Procedure

The pressure vessels were installed in a purpose built Pressure Vessel Testing Facility at MEL[5]. Heating to the test temperature of 838 K was achieved using air circulation bell furnaces. The vessels were then pressurised with steam incrementally up to the test pressure which in all cases was achieved within a period of about one hour. A short time was allowed to elapse after each pressure increment to ensure that temperature gradients in the vessels were kept to a minimum. The three research programmes involved different internal pressures; these being 350, 379 and 455 bar. At the end of each test period the vessels were returned to ambient condit-ions for non-destructive testing (NDT) by reducing first

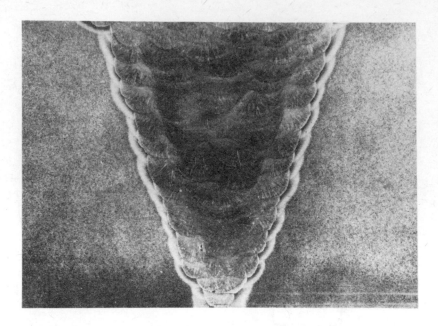

Fig.2 General macroscopic appearance of the 2CrMo-$\frac{1}{2}$CrMoV pressure vessel weldments (x1.5).

pressure then temperature.

2.3 Strain Monitoring

The deformation occurring in the vessels was monitored continuously while the vessels were at the test temperature, both during pressurisation and at testing pressure, and intermittently while the vessels were returned to ambient conditions for NDT.

Continuous monitoring was carried out using Planer high temperature capacitance gauges, Figure 3, developed at the Central Electricity Research Laboratories. These function up to ∿ 1000 K and have an effective gauge length of ∿ 20 mm[6]. Consequently both hoop and axial strains were obtained from relatively small areas of the vessels pipe material and weld centre sections, as indicated in Figure 1. In general, the gauges have a strain range of ∿ 1% and drift rates at temperature typically less than 3×10^{-8} strain per hour.

Intermittent strain data at room temperature were obtained from direct measurements across reference "creep pips". These consisted of stellite domes, welded to small cylinders compatible with the parent material, Figure 3. The creep pips were Tungsten Inert Gas welded to the vessel surface, equi-spaced

Fig.3 Detailed view of the capacitance gauges and creep pips used to monitor strain in the pressure vessels ($\sim \frac{2}{3}$ full size).

around the circumference and at different levels, Figure 1, and measurements were made between appropriate pairs. From these measurements hoop strains were calculated based on a pipe diameter gauge length of ~ 350 mm, whereas axial strains related to gauge lengths of ~ 150 mm and 100 mm for the parent material and weld metal respectively. The accuracy of this method is approximately $\pm 0.01\%$ strain.

3. RESULTS AND DISCUSSION

The results obtained from the pressure vessel experiments are considered in terms of the elastic strains occurring during pressurisation and the subsequent creep strains accumulating with time. These data presented are specific, but nevertheless typical, examples of the deformations observed in the pressure vessels pipe to pipe weldments.

3.1 Elastic Strain during Pressurisation

The hoop and axial loading strains for parent pipe, as-welded and stress relieved weld sections of the vessels subjected to 455 bar pressure are given in Figure 4. These show a linear relationship between the hoop and axial strains during pressurisation, as did all the gauges on the pressure vessels examined, and indicate that deformation during this period was

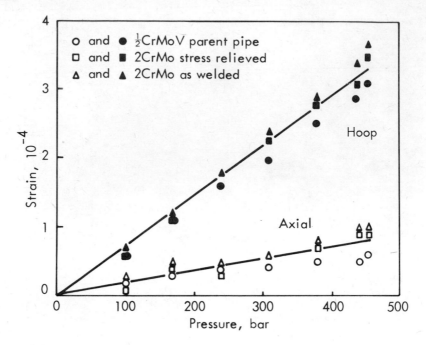

Fig.4 The hoop and axial strains measured during pressurising as welded and stress relieved 2CrMo-½CrMoV weldments at 838K.

elastic. In every case the hoop strain was approximately four times that observed in the axial direction, as previously reported for plain pipes[1]. In addition, hoop and axial strains were the same in the parent pipe and weld metal loc- ations monitored, indicating that a uniform elastic stress distribution is established during pressurisation.

These data allow the accuracy of the capacitance strain gauge monitoring system to be appraised. In a multi-axial situation the elastic hoop strain is given by

$$\varepsilon_h = \frac{1}{E} (\sigma_h - \nu(\sigma_a + \sigma_r)) \qquad (1)$$

where ε is strain, σ stress, ν Poissons ratio, E the elastic modulus and h, a and r denote the hoop, axial and radial dir- ections respectively. The hoop to axial strain ratio on the outer surface of an elastically loaded cylinder, where σ_r is zero, is given by

$$\frac{\varepsilon_h}{\varepsilon_a} = \frac{\sigma_h - \nu\sigma_a}{\sigma_a - \nu\sigma_h} \qquad (2)$$

Using the data from Figure 4 in equation (2) together with $\nu = 0.3$, gives a hoop to axial stress ratio on the pipe surface

of 2:1, as expected from theoretical considerations[7]. This
approach was adopted during every pressurisation to confirm
the reliability of the strain monitoring.

3.2   Creep Deformation in the ½CrMoV Parent Pipe

The deformation behaviour observed in the parent material
of each vessel is shown in Figure 5. In general, the creep
pips and capacitance strain gauges indicate a similar pattern
of behaviour. In each case the hoop strains show a definite
primary stage through which the strain rate progressively decreases towards a steady state creep rate. It is apparent
that as the internal pressure is increased the magnitude of
the strain accumulated during primary creep is increased and
the time in primary decreased. A similar transient period
was observed in the axial strains, but in all vessels these
were very small.

The primary and steady state creep behaviour are dependent
on the uniaxial creep properties of the parent pipe. In a
uniaxial test, increasing the stress results in an increase in
the primary creep strain, a decrease in the time in primary
and a higher steady state rate. In addition, increasing the
stress also increases the creep rate:stress exponent, n, which,

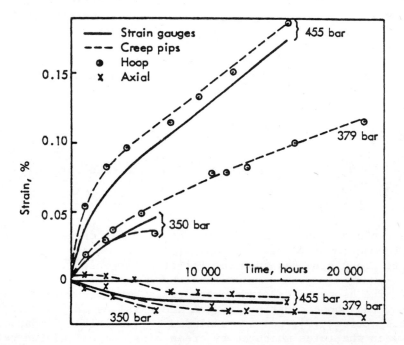

Fig.5 The variation of hoop and axial strain with time for
½CrMoV parent pipe subjected to different pressures at 838K.

in a multi-axial situation, will increase the rate at which the elastic stresses can redistribute to the steady state values, further decreasing the time in primary. These phenomena are clearly reflected in the primary hoop deformation behaviour observed in the pressure vessels. Increasing the pressure over the range 350 to 455 bar effectively doubles the primary strain from about 0.04% to 0.08% and decreases the time in primary from $\sim$ 8000 to 4000 hours.

The steady state strain rates monitored in the hoop and axial directions using the creep pips and strain gauges are given in Table 1. The rates are similar to those reported

Press- ure bar	Strain Gauge Rates		Creep Pip Rates		Theoretical	
	Hoop	Axial	Hoop	Axial	$\sigma*$ MNm^{-2}	$\dot{\varepsilon}_{Hoop}$
350	6.8	0	2.0	0	65	4.9
379	8.0	0	3.5	0	70	6.7
455	6.9	0	6.7	0	84	14

All rates x 10^{-8} $\varepsilon.h^{-1}$

Theoretical data based on A = 3.2 x 10^{-15}, n=4.

TABLE 1: The steady state strain rates measured in $\frac{1}{2}$CrMoV parent pipe at 838 K and various pressures

previously for $\frac{1}{2}$CrMoV parent pipe in a thick walled pressure vessel tested at 455 bar and 838 K[1]. In this case it was shown that the steady state creep rate could best be described using the von Mises equivalent stress, $\sigma*$, where

$$\sigma* = \frac{1}{\sqrt{2}} \left[(\sigma_1 - \sigma_2)^2 + (\sigma_2 - \sigma_3)^2 + (\sigma_3 - \sigma_1)^2 \right] \qquad (3)$$

The strain rates in a given direction are then obtained from an equation of the form

$$\dot{\varepsilon}_1 = A\sigma_*^{n-1} \left[\sigma_1 - \frac{1}{2}(\sigma_2 + \sigma_3) \right] \qquad (4)$$

where A and n are obtained from the uniaxial creep properties of the appropriate material and σ_1, σ_2 and σ_3 are the steady state creep stresses from the Bailey equations[8].

The hoop steady state creep rates have been calculated using the above equations with A and n values of 3.2 x 10^{-15} and 4 respectively[9], and are included in Table 1. Comparison between the rates measured by creep pips and calculated rates show that the predictions are generally higher than those actually obtained. However, the general trend is correct and the creep pip values are within a factor of $\sim$ 2 of the calculated rates. From the strain gauges the rates are similar, so that the predicted decrease with decreasing pressure is not found.   However, at the lowest pressure the minimum, steady state, rate may not yet have been reached and, from the creep pip data, the parent pipe appears to be deforming at a rate approaching the drift rate of the gauges.   Nevertheless, while a consistent trend is not indicated the measured rates, $6.8 - 8 \times 10^{-8}$ strain per hour, are again, within a factor of ~ 2 of the predicted rates.

For the axial case, on the surface of a pressurised pipe the hoop stress is twice the axial stress and the radial stress is zero.   As a consequence rewriting equation (4) to predict the axial steady state strain rate, $\dot{\varepsilon}_2$, the deviatoric component $[\sigma_2 - \frac{1}{2}(\sigma_3 - \sigma_1)]$ is zero and the strain rate is zero. Hence the axial steady state rates measured on all the vessels agree with the theoretical predictions.

3.3   Creep Behaviour of Stress Relieved Welds

The hoop and axial strain time behaviour for the stress relieved 2CrMo welds are shown in Figures 6 and 7 respectively, for the pressure vessels tested at 455 bar.   In each case the parent material data measured at the same test pressure is included for comparison purposes.   Significant primary and steady state creep is observed in each case and the creep pip measurements show similar hoop strain behaviour in the parent pipe and stress relieved welds.

For the stress relieved welds the hoop steady state creep rates determined from the creep pip measurements are in agreement with those obtained from the strain gauge data, Table 2, and are similar to those in the parent pipe, Figure 6.   This

Pressure bar	Strain Gauge Rates				Creep Pip Rates			
	Stress Relieved		As-welded		Stress Relieved		As-Welded	
	Hoop	Axial	Hoop	Axial	Hoop	Axial	Hoop	Axial
350	5.5	5.7	5.8	8.7	3.9	6.8	4.2	9.5
379	6.4	5	–	–	3.8	2	–	–
455	5	4	30	16	7.5	5	40	12.5

All rates $\times 10^{-8}$ $\varepsilon.h^{-1}$

TABLE 2:   The steady state strain rates measured in stress relieved and as-welded 2CrMo welds at 838 K and various pressures

agreement is somewhat unexpected since, as shown in Figure 3, the creep pips were positioned outside the HAZ's associated with each weld while the strain gauges were entirely within

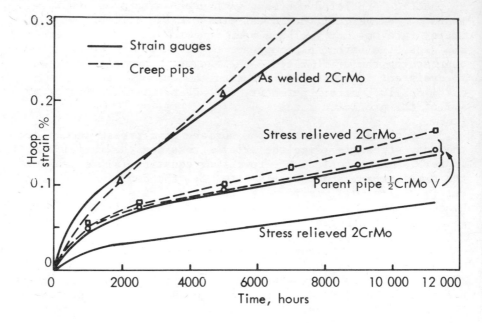

Fig.6 The variation of hoop strain with time for stress relieved and as welded 2CrMo welds at 455 bar and 838K.

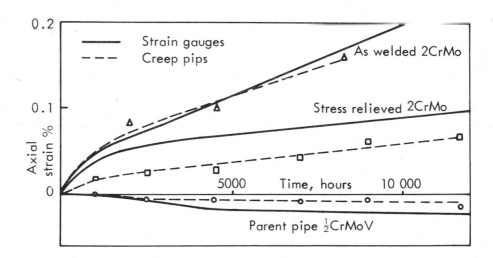

Fig.7 The variation of axial strain with time for stress relieved and as welded 2CrMo welds at 455 bar and 838K.

the weld metal. Nevertheless, at a given pressure, similar
hoop strain rates are observed in the parent pipe remote from
the weld, in the parent pipe adjacent to the stress relieved
weld and in the stress relieved weld metal centre. It appears
therefore that under the present conditions hoop deformation
occurs at the same strain rate along the length of the vessel.
Finite element analysis of the stationary state stresses present within weldments has shown that in the hoop direction,
stress is offloaded from materials weaker in creep to those
that are stronger[10]. The present work suggests that this
stress redistribution continues until a compatibility of
strain rate is achieved across the various structures comprising the weldment.

While the hoop steady state creep rates indicated by the
creep pips and the strain gauges agree for stress relieved
welds, hoop strain accumulation during primary creep is not
the same. From Figure 6, it can be seen that more primary
strain is measured at the creep pip locations than at the
centre line of the corresponding stress relieved weld metal.
This behaviour appears to be another consequence of the redistribution of stresses within a weldment. It was shown in
section 3.1 that a similar elastic stress distribution was
established in the vessels independent of material location.
If these stress levels remained unaltered for any significant
time, strain would accumulate at a higher rate within the weld
than in the parent pipe, since the 2CrMo weld metal is known to
be weaker in creep than the $\frac{1}{2}$CrMoV parent at 838 K[9]. The
reduced strain accumulation in the stress relieved weld metal
therefore suggests that redistribution of the elastic pressure
stresses in a weldment occurs early in the life of the component. This results in a weld metal stationary stress level
below that in the parent pipe and consequently less primary
strain is observed in the stress relieved weld than in the pipe.

The axial strain time behaviour is illustrated in Figure
7 for the pressure vessels tested at 455 bar. In this orientation the pattern of deformation observed in the parent pipe
differs appreciably from that of the welds. Positive axial
strain accumulation is measured in the stress relieved welds
and the creep rate decreases with time to a positive steady
state value in all the pressure vessels, as given in Table 2.
This contrasts with the parent pipe where axial steady state
strain rates are zero, Table 1. While the detailed strain/
time behaviour varies, similar steady state creep rates are
again indicated by both creep pips and strain gauges.

The large positive axial creep rates observed immediately
following pressurisation support the view that redistribution
of the elastic pressure stresses in a weldment occur rapidly.
It has been shown that, in agreement with theory, a hoop to
axial steady state stress ratio of 2:1 on the surface of a
pipe results in a zero axial creep rate. Elastically, the

surface stress ratio is 2:1 and therefore a significant change
in the surface stresses must have taken place for a positive
axial rate to occur.

Changes in the stationary state stress distribution resulting from the materials property variations introduced by a
weldment are greater in the hoop direction than the axial[10].
Stress redistribution can also occur in the axial direction,
but only to a limited extent. The degree of stress redistribution may be assessed by examination of equation (4). Using
this approach, the ratio of the steady state hoop rate, $\dot{\varepsilon}_h$, to
the steady state axial rate, $\dot{\varepsilon}_a$, on the surface of a pipe can
be expressed as

$$\frac{\dot{\varepsilon}_h}{\dot{\varepsilon}_a} = \frac{(\sigma_h/\sigma_a) - \frac{1}{2}}{1 - \frac{1}{2}(\sigma_h/\sigma_a)} \tag{5}$$

Substitution of the observed values of $\dot{\varepsilon}_h$ and $\dot{\varepsilon}_a$ shows that
for the case of the stress relieved 2CrMo welds, the stationary
state stress ratio σ_h/σ_a, is ~ 1.

3.4 Creep Behaviour of As-Welded Welds

The deformation of the as-welded welds shown in Figures 6
and 7, is clearly much more rapid than in stress relieved
weldments or the plain pipe sections. The magnitude of the
primary strain is greater in the as-welded material, particularly in the hoop direction, where at $\sim 0.1\%$ it is about twice
that observed in the stress relieved weld or parent pipe. The
time spent in primary also appears somewhat shorter for the
as-welded case. More significantlly, however, the hoop and
axial steady state creep rates are always greater in the as-
welded material, Table 2, by factors of up to ~ 4, depending
on the direction, measuring technique and pressure being
considered.

These differences in behaviour do not appear to be related
simply to mechanical property differences since, uniaxial test
data showed little change in the high temperature strength of
the weld metal as the result of stress relief heat treatment
[9].   It therefore appears that the increased deformation in
the as-welded condition is due to welding residual stresses.
In addition, since the increase in strain accumulation is
observed both in the weld metal and in the pipe sections
adjacent to the as-welded weld, as indicated by the agreement
between creep pip and strain gauge data, it appears that these
residual welding stress effects influence the deformation of
the parent material.

The build up of welding residual stresses in 2CrMo MMA
heavy section weldments and their distribution before and
after heat treatment, have been the subject of extensive work
at MEL[11,12].   However, the manner in which residual

stresses combine with pressure to produce deformation in a heavy section multi-axially stressed weldment is still uncertain. Nevertheless, a preliminary analysis of the present results, pre-supposing that the principal residual and pressure stresses are simply additive, indicates that only very low residual stresses, of the order of $10 - 20$ MNm^{-2}, are necessary to generate the strain rates observed in the as-welded welds.

Stress relaxation data have shown that stresses of this magnitude can be sustained under the conditions of the present tests[13], and measurements on the pressure vessels, not reported here, indicate that such stresses are indeed present. Furthermore, uniaxial work shows that creep cavitational damage is produced under these conditions of stress relaxation and can lead to low ductility failures[14]. Thus the present preliminary finding of accelerated creep in as-welded 2CrMo is consistent with the occurrence of cracking in inadequately stress relieved weldments.

4. CONCLUSIONS

1. The elastic strain behaviour of as-welded and stress relieved weld metal in pressurised heavy section weldments is identical to that observed in the parent pipe. This agrees with theory, and indicates that a uniform elastic stress field is established on the surface of a pipe weldment immediately after pressurisation.

2. In parent pipe, the hoop creep strain varies with time giving a primary stage, during which the creep rate decays, followed by a steady state creep condition. The extent of primary creep and the value of the steady state creep rate increases with increasing internal pressure. Thus, the uni-axial behaviour of materials is reflected in a multi-axially stressed component.

In the parent pipe axial direction, the creep rate decays to a steady state value of zero at all pressures, again in agreement with theory.

3. In the stress relieved welds the hoop steady state creep rates were similar to those of the parent pipe tested at the same pressure. This is consistent with redistribution of the steady state stresses to produce equivalence of strain rate in the hoop direction throughout the weldment.

In the axial direction, the steady state creep rates were positive, not zero as in the parent pipe, and approximately equal to the hoop strain rates. This indicates that under steady state conditions an equi-biaxial stress distribution exists on the surface of the stress relieved welds.

Y

4. The as-welded weldment showed enhanced strain accumulation in both hoop and axial directions compared to the parent pipe or stress relieved welds. Preliminary analysis suggests that this can be explained by the contribution of the welding residual stresses present in the as-welded material.

5. ACKNOWLEDGEMENT

This paper is published by permission of the Central Electricity Generating Board.

6. REFERENCES

1. COLEMAN, M.C., PARKER, J.D., WALTERS, D.J. and WILLIAMS, J.A. - 'The deformation behaviour of thick walled pipe at elevated temperatures'. Third International Conference on Mechanical Behaviour of Materials, Cambridge, 1979, 2, 193.

2. TOFT, L.H. and YELDHAM, D.E. - 'Weld performance in high pressure steam generating plant in the Midlands Region, CEGB'. International Conference on Welding Research related to Power Plant, Southampton, 1972, 5.

3. CHEETHAM, D., FIDLER, R., JAGGER, M. and WILLIAMS, J.A. - 'Relationships between laboratory data and service experience in the cracking of CrMoV weldments'. International Conference on Residual Stresses in Welded Construction and Their Effects, London, 1977, 145.

4. COLEMAN, M.C. - 'The structure of weldments and its relevance to high temperature failure'. Fifth Bolton Landing Conference on Weldments:Physical Metallurgy and Failure Phenomena, 1978, 409.

5. COLEMAN, M.C., FIDLER, R. and WILLIAMS, J.A. - 'Crack growth monitoring in pressure vessels under creep conditions'. In Detection and Measurement of Cracks, Welding Institute, Cambridge, 1976, 40.

6. NOLTINGK, B.E., McLACHLAN, D.F.A., OWEN, C.K.V. and O'NEILL, P.C. - 'High stability capacitance strain gauges for use at elevated temperatures'. Proc. Inst. Elec. Eng., 1972, 119, 897.

7. LAMÉ, G. - 'Lecons sur la théore... de l'élasticité'. Gauthier-Villars, Paris, 1852.

8. BAILEY, R.W. - 'The utilisation of creep test data in engineering design'. Proc. Inst. Mech. Engrs., 1935, 131, 131.

9. CANE, B.J. and MIDDLETON, C.J. Private Communication.

10. WALTERS, D.J. - 'The stress analysis of cylindrical
    butt-welds under creep conditions'. CEGB, 1976,
    Note No. RD/B/N3716.

11. FIDLER, R. - 'The complete distribution of residual
    stress in a $\frac{1}{2}$CrMoV/2CrMo main steam pipe weld in the
    as-welded condition'. CEGB, 1977, Report No.R/M/R261.

12. FIDLER, R. - 'A finite element analysis for the stress
    relief of a $\frac{1}{2}$CrMoV/2CrMo main steam pipe weld'. CEGB,
    1978, Report No. R/M/R270.

13. MYERS, J. and RILEY, J.F. - 'Uniaxial stress relaxation
    testing of welds'. CEGB, 1973, Note No. R/M/N669.

14. WOLSTENHOLME, D.A. - 'Transverse cracking and creep
    ductility of 2CrMo weld metals'. International
    Conference on Trends in Steels and Consumables for
    Welding, London, 1978, 541.

CREEP FAILURE ANALYSIS OF BUTT WELDED TUBES

R.J. Browne*, B.J. Cane**, J.D. Parker*** and D.J. Walters****.

*   Central Electricity Generating Board, South Eastern Region, Canal Road, Gravesend, Kent.

**   B.J. Cane, Central Electricity Research Laboratories, Kelvin Avenue, Leatherhead, Surrey.

***   J.D. Parker, Marchwood Engineering Laboratories, Marchwood, Southampton, Hants.

****   D.J. Walters, Berkeley Nuclear Laboratories, Berkeley, Glos.

SUMMARY

As part of a major research programme to investigate the influence of butt welds on the life expectancy of tubular components, a series of internal-pressure, stress-rupture tests have been carried out. Thick walled $\frac{1}{2}Cr\frac{1}{2}Mo\frac{1}{4}V$ tube specimens were welded with mild steel, $1Cr\frac{1}{2}Mo$ steel, $2\frac{1}{4}Cr1Mo$ steel or nominally matching $\frac{1}{2}Cr\frac{1}{2}Mo\frac{1}{4}V$ steel to give a wide range of weld metal creep strengths relative to the parent tube. The weldments were tested at $565^{\circ}C$ at two values of internal pressure, and gave failure lives of up to 44,000 hrs.

Finite element techniques have been used to determine the stationary state stress distribution in the weldment which was represented by a three material model. Significant stress redistribution was indicated and these results enabled the position and orientation of cracking and the rupture life to be predicted. The theoretical and experimental results have been used to highlight the limitations of current design methods which are based on the application of the mean diameter hoop stress to the parent material stress rupture data.

1. INTRODUCTION

Design methods for pipework and tubular components operating at temperatures where creep deformation can occur are specified in British Standards 806 and 1113, respectively. It is assumed that the pipe or tube operates at constant known pressure and temperature and that the creep life of the component is related to uniaxial stress rupture data through the mean diameter hoop stress, σ_{mdh}, such that

$$\sigma_{mdh} = \frac{P(OD - t)}{2t} \tag{1}$$

where P is the operating pressure, OD is the outside diameter and t the wall thickness of the pipe. For the required design life, the operating stress is obtained by applying a safety factor to the mean stress to rupture, and the component wall thickness calculated using equation (1)

It may be seen therefore that in the design process no direct account is taken of the presence of butt welds; it is merely specified that weld metal properties should match the parent pipe. In practice, however, the presence of welds will lead to

(a) variable metallurgical structures in the weld metal, across the heat affected zones (HAZ) and into the parent material.

(b) residual welding stresses which decay at high temperature by creep.

Since previous theoretical investigations [1] have demonstrated that these metallurgical effects will cause significant variations in the stress distribution within the weldment, there is a requirement to examine current design approaches. Recent needs to undertake remanent life assessments of power plant have also added importance to the development of accurate and conservative life prediction methods.

In order to investigate the influence of the variation of mechanical properties within a weldment, an extensive collaborative research programme has been established within the CEGB [2]. This comprises stress analysis, uniaxial and torsion creep testing, residual stress measurement and internal pressure creep testing of tubes and pipes containing welds. This report presents some preliminary results of the stress analysis and small vessel testing programme and examines methods of life prediction.

2. EXPERIMENTAL METHODS AND RESULTS

The creep deformation and stress rupture properties of each of the constituent materials employed in the fabrication of the tube specimens are required for use with the stress analysis programme and subsequently in the life predictions. These data were generated in a comprehensive programme of uniaxial creep tests carried out at the Central Electricity

Research Laboratories (CERL) and at the Electrical Research
Association (ERA) [3].

The relationships between rupture life t_{RUP}, and minimum
creep rate, $\dot{\varepsilon}_{min}$, with applied stress, σ, for each of the
materials are given in Figures 1 and 2 respectively. Figure
2 indicates that the minimum creep rates are adequately
described by Norton's Law,

$$\dot{\varepsilon}_{min} = A\sigma^n \tag{2}$$

with the stress exponent n $\sim$ 4 for the parent and weld
materials at applied stresses less than 100 MN/m^2. Thus the
difference in creep strain rate between parent and weld
material may be expressed as a ratio of the constant A, equation (2). These ratios were determined as 1690, 14, 5 and 1
for mild steel, 1CrMo, 2$\frac{1}{4}$CrMo and $\frac{1}{2}$CrMoV welds respectively.

Since the collaborative programme was designed to examine
the high temperature behaviour of main steam pipes, the tube
welds were manufactured using a specialized welding procedure
[4]. This ensured that the weldments contained coarse
grained bainitic HAZ's typical of thick section components.
To obtain the creep and rupture properties of this HAZ material
uniaxial samples were subjected to a programme of thermal
simulation prior to testing [3]. The results of these tests
are also included in Figures 1 and 2. As noted earlier,

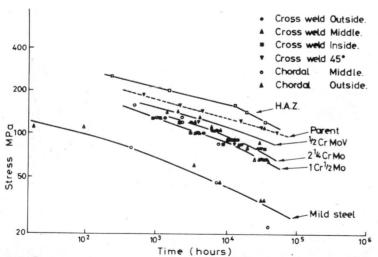

FIG. 1.   Stress Rupture Data for the Weld Metals, Parent Material
and Simulated H.A.Z.

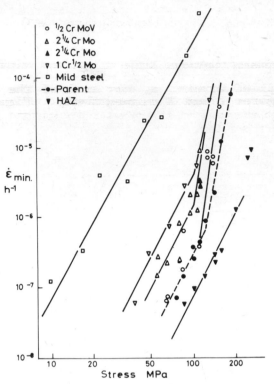

FIG. 2. Stress Dependence of the Minimum Creep Rate for the Weld. Parent and Simulated H.A.Z. Material.

the stress dependence of the minimum creep rate for the parent material and the weld metals is accurately represented by equation 2 with n=4. Although not a precise fit to the data, the same value of stress index has been used for the HAZ data as this simplifies the stress analysis (one solution suffices for different imposed pressures). Fig. 2 shows that a reasonable fit over the stress range of interest is obtained by selecting a HAZ/parent material creep rate ratio of 0.25.

In order to apply uniaxial rupture data to pressurised components a multiaxial stress rupture criterion has been developed. Previous work [5] has established that for $2\frac{1}{4}$Cr1Mo material this is given by the general form

$$t_{RUP} = \frac{c}{\sigma_1{}^q \, \bar{\sigma}^{(r-q)}} \qquad \ldots\ldots \quad (3)$$

where σ_1 is the maximum principal stress, $\bar{\sigma}$ is the von Mises equivalent stress, c is a constant, r is the slope of the uniaxial stress-rupture plot and q is the principal stress exponent of rupture. On this basis a representative rupture stress σ_{RUP}, can be defined which when applied to uniaxial data will predict the life of the structure, so that

$$\sigma_{RUP} = \sigma_1^{q/r} \bar{\sigma}^{(r-q)/r} \qquad (4)$$

Tensile and torsion rupture tests on the thermally simulated, coarse grained bainitic HAZ material, Table 1, indicated that the appropriate values of q and r in equation (4) were 3 and 8.5 respectively.

Test Method	von Mises Equivalent Stress, $\bar{\sigma}$	Max. Principal Stress, σ_1	Rupture Life
Torsion	250	144.3	1344
Tension	250	250	239
Torsion	200	115.5	6,600
Tension	200	200	1,511

Table 1. Comparison of the Rupture Lives of Simulated Coarse Grained Bainitic $\frac{1}{2}$CrMoV Sampler Tested in Torsion and Tension at 565°C

The tube specimens, Figure 3, were of $\frac{1}{2}$Cr$\frac{1}{2}$Mo$\frac{1}{4}$V parent steel and each contained two test welds of either mild steel, 1CrMo, 2$\frac{1}{4}$CrMo or $\frac{1}{2}$Cr$\frac{1}{2}$Mo$\frac{1}{4}$V weld metal. Each specimen was post weld heat treated for 3 hours at 700°C to reduce residual welding stresses, and testing was carried out at 565°C under constant internal pressure. Two values of internal pressure were employed and the test programme was undertaken at ERA.

All dimensions in mm.

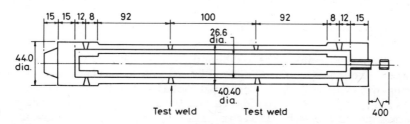

FIG.3.   Geometry of the Tube Burst Specimens showing the Location of the Test Welds.

The results of these tests together with details of the failure mode are given in Table 2. In every case the tube failure occurred by means of an axial HAZ crack propagating through the tube wall to cause a steam leak. While a decrease in rupture life with weld metal creep strength was found at both internal pressures, the duration of the tests with $\frac{1}{2}$CrMoV weld metal were significantly greater than those for the other weld metal. However, reducing the weld metal creep resistance below that of $2\frac{1}{4}$CrMo had only a second order effect upon the test life for internally pressurised tubes.

Weld Metal	Internal Pressure MPa	Rupture Life h	Failure Mode	Subsidiary Cracking	
				Failed Weld	Other Weld
Mild Steel	45.3	14691	Axial HAZ Cracking	Axial HAZ Cracking Axial and Circumferential Weld Metal Cracking	Axial HAZ Cracking Circumferential Weld Metal Cracking
1CrMo Steel	45.3	18774	Axial HAZ Cracking	Axial HAZ Cracking Axial and Circumferential Weld Metal Cracking	Axial HAZ Cracking Circumferential Weld Metal Cracking.
$2\frac{1}{4}$CrMo Steel	45.3	22039	Axial HAZ Cracking	Axial HAZ Cracking Circumferential Weld Metal Cracking	Axial HAZ Cracking Circumferential Weld Metal Cracking
$\frac{1}{2}$CrMoV Steel	45.3	43871	Axial HAZ Cracking	Axial HAZ Cracking Axial Weld Metal Cracking	Axial HAZ Cracking Axial Weld Metal Cracking
Mild Steel	51.5	7752	Axial HAZ Cracking	Axial HAZ Cracking Axial and Circumferential Weld Metal Cracking	Axial HAZ Cracking
1CrMo Steel	51.5	7712	Axial HAZ Cracking	Axial HAZ Cracking	Axial HAZ Cracking
$2\frac{1}{4}$CrMo Steel	51.5	9285	Axial HAZ Cracking	Axial HAZ Cracking. 'Crazed' Weld Metal Cracking	Axial HAZ Cracking
$\frac{1}{2}$CrMoV Steel	51.5	16511	Axial HAZ Cracking	Axial HAZ Cracking. Axial Parent Metal Cracking	Axial HAZ Cracking

Table 2 Results of Tube Burst Testing

3. STRESS ANALYSIS RESULTS

The stress analysis was conducted using the TESS finite element programme [6]. The weldment was idealised as a three component system comprising weld metal, H.A.Z. and parent material, Figure 4. Boundaries between each region were assumed to be smooth straight lines giving abrupt changes in material properties. Since the size, shape and distribution of finite elements can all affect the numerical results obtained, a brief description of the generation of a satisfactory mesh and the associated assumptions has been included as an Appendix.

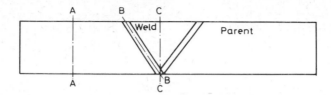

FIG. 4.   Schematic Representation of Weld.

Having determined an acceptable distribution of elements
within the mesh, combinations of material properties were
investigated. These systematically covered the range of creep
rate ratios indicated by the uniaxial testing. For each test
case, positional scans of the stress distributions were carried
out. These were made along the inside and outside surfaces of
the weldment, along the weld and HAZ centre lines and in the
pipe material remote from the weld. The positions are identified in Figure 4, and typical stationary state stress distributions are given in Figures 5a-c for the weld metal : parent
metal: HAZ creep rate ratios of 10:1:0:25. By way of example,
for each material the variation of peak stress as a function
of weld metal/parent metal creep rate ratio is given in Figure
6. Clearly as the stresses increase in the HAZ a complementary
decrease occurs in the weld metal, while stresses within the
parent material remain close to those obtained in a plain tube.
This stress redistribution is similar to that reported
previously [1] and is typical of a weldment containing two or
more materials exhibiting different creep characteristics.
However, the present results indicate that when creep resistant
HAZ structures are present, the stress redistribution in the
weldment is more sensitive to the HAZ/parent pipe creep rate
ratio than to the weld metal/parent pipe creep rate ratio.
Indeed, from Figure 6, for a tube containing a weld with creep
properties matching those of the parent, high stress levels
are still present in the HAZ material.

4. LIFE PREDICTIONS

For the parametric plot of the variation of peak stresses
with weld metal/parent metal creep rate ratio, Figure 6, the
maximum principal and equivalent stresses may be determined in
the weld, parent and H.A.Z. respectively, Table 3. A minimum
estimate of life may then be obtained for each material by
applying the larger of the two stresses to the appropriate
uniaxial stress rupture data, Figure 2. It is assumed throughout that the shortest life predicted in one material determines
the position and failure time for the complete weldment. The
orientation of cracking is assumed to occur in a plane perpendicular to the maximum principal stress. Using this approach

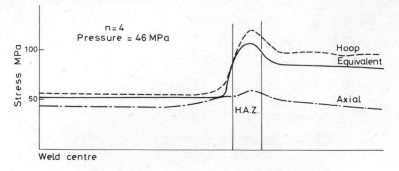

FIG. 5a.   Stress Distribution along Outer Surface B.C.

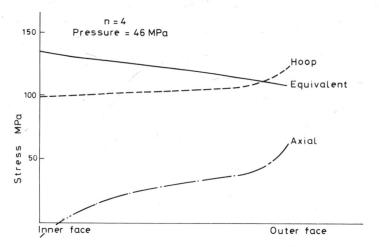

FIG. 5b.   Stress Distribution on H.A.Z. Centre Line B.B.

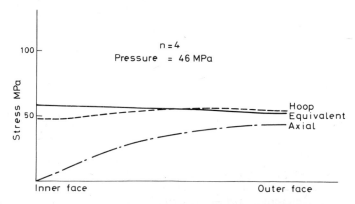

FIG. 5c.   Stress Distribution on Weld Centre Line C.C.

Fig. 5   Steady State Creep Stress Distribution

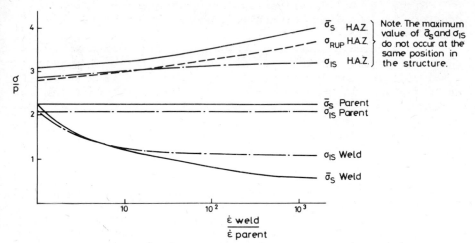

FIG.6. Variation of Peak Stresses with Weld Metal/Parent Metal Creep Rate Ratio.

Weld Metal	Internal Pressure MPa	σ_{MDH} MPa	Max. Equivalent Stress $\bar{\sigma}_s$			Max. Principal Stress σ_{1s}			Max.* HAZ Rupture Stress MPa
			Weld Metal	HAZ	Parent	Weld Metal	HAZ	Parent	
Mild Steel	45.3	110	26	185	104	51 Axial	145 Hoop	95 Hoop.	170
1 CrMo Steel	45.3	110	54	152	104	55 Hoop	138 Hoop	95 Hoop	147
2 ¼ CrMo Steel	45.3	110	68	145	104	67 Hoop	132 Hoop	95 Hoop	140
½ CrMoV Steel	45.3	110	97	140	104	94 Hoop	129 Hoop	95 Hoop	136
Mild Steel	51.5	125	29.5	209	118	59 Axial	165 Hoop	108 Hoop	192
1 CrMo Steel	51.5	125	62	172	118	62 Hoop	156 Hoop	108 Hoop	159
2 ¼ CrMo Steel	51.5	125	79	163	118	78 Hoop	150 Hoop	108 Hoop	157
½ CrMoV Steel	51.5	125	110	159	118	105 Hoop	146 Hoop	108 Hoop	153

* HAZ Rupture Stress $= \sigma_{1s}^{0.35} \cdot \bar{\sigma}_s^{0.65} = [\sigma_{RUP}]$

Table 3. Maximum Stationary State Stress Computed by Finite Element Analysis

the minimum life predicted invariably occurs in the H.A.Z. material, Table 4.

Since in this location significant variations were consistently observed between the maximum values of σ_1 and $\bar{\sigma}$, the multiaxial failure criterion was examined. Using equation 4 calculations of σ_{RUP} were made for each position through the wall thickness of the HAZ from the local σ_1 and $\bar{\sigma}$ values. The maximum value of σ_{RUP} determined is also given in Table 3 and this value was then applied to the uniaxial stress rupture curve to give a modified failure estimate. The results are included in Table 4. In each case while the times to failure are increased, the life of each tube specimen still appears to be determined by the life of the HAZ material. It should be noted that if this multiaxial criterion is applied to calculate σ_{RUP} for weld metal and parent material, the lives predicted are greater than those obtained for the HAZ.

Weld Metal	Internal Pressure MPa	Observed Rupture Life h.	Life Prediction,h						Maximum Rupture Stress HAZ	Maximum Principal Stress HAZ
			Maximum Equivalent Stress			Mean Diameter Hoop Stress				
			Weld Metal	HAZ	Parent	σ_{mdh}	$1.25\,\sigma_{mdh}$	$1.5\,\sigma_{mdh}$		
Mild Steel	45.3	14,691	23,000	5,000	40,000	31,000	5,200	1,500	8,000	17,000
1CrMn	45.3	18,774	50,000	14,000	40,000	31,000	5,200	1,500	17,000	28,000
2½CrMo	45.3	22,039	50,000	18,000	40,000	31,000	5,200	1,500	18,000	26,000
½CrMoV	45.3	43,871	40,000	21,000	40,000	31,000	5,2001	1,500	24,000	24,000
Mild Steel	51.5	7,752	18,000	2,000	18,000	11,000	1,900	300	3,500	10,000
1CrMo	51.5	7,712	36,000	7,500	18,000	11,000	1,900	300	11,000	13,000
2½CrMo	51.5	9,285	28,000	10,000	18,000	11,000	1,900	300	12,500	15,000
½CrMoV	51.5	16,511	18,000	12,000	18,000	11,000	1,900	300	14,500	18,000

Table 4. Comparison of experimental rupture lives with predictions based on various stress criteria

The above failure predictions from the finite element analysis together with those of the mean diameter formula, equation 1, are compared with the results of the tube burst tests in Table 4. Consideration of this Table indicates that the finite element analysis correctly predicts the material structure in which the failure is located, the orientation being perpendicular to the maximum principal stress direction. The rupture lives predicted by the approach are mainly under-estimates which would be expected as the peak stress in the H.A.Z. has been used to determine failure. Nonetheless the predictions lie within the scatter to be expected in materials properties.

5. CLOSING REMARKS

This paper reports preliminary results of one part of an extensive research programme which is currently in progress in the CEGB to examine design and life prediction methods for butt welded pipes operating in the creep range. The present work is concerned with thick-walled tube specimens which were designed to model high temperature steam pipes. As such, the weldment microstructures are not those normally associated with welded tubes but are typical of butt welds in steam pipes [4].

Design and remanent life estimation methods currently utilise the mean diameter hoop stress, equation 1, and parent material rupture data. Life predictions for the present tube tests using that approach are included in Table 4 and it is clear that under the conditions employed, apart from the tubes containing ½CrMoV weld metal, the mean diameter predict-ions overestimate specimen lives. However, if the safety factors commonly employed in design, 1.5 and in remanent life estimation 1.25, are applied to the mean diameter stresses, predicted lives are markedly less than experimental lives.

While the mean diameter formula takes no account of the variations with creep lives observed with the different weld metals, the finite element calculations offer a more realistic approach to failure prediction. The stationary state stress distributions indicate that the introduction of a butt weld into a thick walled pipe results in a significant perturbation in the stress distribution. Weld materials which creep more rapidly than the parent material, off-load onto the material which has higher creep strength. Conversely heat affected zone material which creeps at a slower rate than parent material experiences an increased stress level which coupled with its reduced ductility leads to premature failure of the weldment.

The life prediction method of associating the stationary state stress levels with the rupture properties of the constituent materials of the weldment is a simplification. Clearly this stress distribution will be modified by the onset of tertiary creep or cracking within the weldment.

656

Consequently for ductile materials or structures exhibiting a discontinuous fracture process the stress distribution at failure may differ significantly from the steady state conditions. In such a situation the full history of the weldment will require consideration. However in the present application where the heat affected zone comprises a high percentage of coarse-grained bainite (∿50%) which has low ductility and shows little tertiary creep, the need for such detailed calculations is avoided. This is indicated by the comparisons given in Table 4 which show that although the life predictions are in general underestimates of the observed life, they are acceptable and lie within the basic materials scatter.

The paper has indicated the influence of the variation in mechanical properties within weldments. It has shown that heat affected zones which contain large proportions of coarse grained bainite are deleterious and lead to premature failure. Weld materials which creep more rapidly than the parent material are not necessarily detrimental provided they are ductile and the joint is subjected to pressure loading alone. However it is important to note that if additional axial loading is considered, there is not scope for significant stress redistribution as axial equilibrium must be satisfied. Thus if axial loading were to equal the hoop value as permitted by BS806, the life of the vessel with the mild steel weld would reduce from the observed 14,000 hrs. to approximately 100 hrs. Further work is in hand to examine the effect of superimposed end loading on the life of internally pressurised tubes and the application of the present analysis to full sized component pressure vessels.

6.   ACKNOWLEDGMENT

This paper is published by permission of the Central Electricity Generating Board.

7.   REFERENCES

1.   WALTERS, D.J. and COCKROFT, R.D.H., 'A Stress Analysis and Failure Criteria for High Temperature Butt-Welds', International Institute of Welding, Toronto, 1972.

2.   ROWLEY, T. and COLEMAN, M.C. 'Collaborative Programme on the Correlation of Test Data for the Design of Welded Steam Pipes', CEGB, 1973, Report No. RD/M/N.710.

3.   CANE, B.J. and MIDDLETON, C.J. - to be published.

4.   COLEMAN, M.C. - to be published.

5. CANE, B.J., 'Creep Cavitation and Rupture in $2\frac{1}{4}$CrlMo Steel under Uniaxial and Multiaxial Stresses', Third International Conference on Mechanical Behaviour of Materials, Cambridge, 1979, 2, 173.

6. ECCLESTONE,M.J. and LEACH, A., 1969, CEGB Report No. RD/C/N363.

7. COLEMAN, M.C., PARKER, J.D., WALTERS, D.J. and WILLIAMS, J.A., 'The Deformation Behaviour of Thick Walled Pipe at Elevated Temperatures', Third International Conference on Mechanical Behaviour of Materials, Cambridge, 1979, 2, 193.

8. FINNIE, I. and HELLER, W.R., Creep of Engineering Materials, Chapter 7, McGraw-Hill, London, 1959.

Appendix

METHOD OF ANALYSIS

In the present studies the weldment was first considered to be a homogeneous structure. The analytical solution for this steady-state stress distribution is well established by Finnie and Heller and direct comparison of the calculated through thickness stress distributions was made with the TESS finite element solution. Refinement of the finite element mesh was continued until calculated values were within 2% of the values determined by Finnie and Heller [8]. Throughout the theoretical work, it was ensured that the membrane state stress was obtained at a point in the weldment remote from the weld/parent interface. The final and most critical area of the mesh generation occurred at the boundary between the weld and parent material. Since the magnitude of stresses and strains in this region were to be used in the life assessment methods, further detailed refinement was undertaken. The final development of the finite element mesh which incorporated a total of 1100 elements is shown in Figure 7. All numerical results presented in this paper are obtained from the finite element distribution.

ASSUMPTIONS MADE IN THE ANALYSIS

(a) The boundary between the pipe and weld was assumed a smooth straight line. Previous analytical studies [7] have shown that the detailed stress distributions are sensitive to boundary changes. However, providing the mean boundary angle and relative volumes of material do not change, the peak stresses are essentially constant.

(b) Material properties were assumed constant throughout their designated areas. This assumption, enforced by the computer program, places great emphasis upon the initial disposition of finite elements. Even though this will not be the case in a test weldment, it is known that the most significant properties controlling the peak steady state stresses are the maximum and minimum creep strain rates occurring within the component. Therefore provided that peak stresses are used the comparison remains reasonable.

(c) The von Mises equivalent stress $\bar{\sigma}$ is assumed to control creep strain deformation in a multiaxial structure where

$$\bar{\sigma} = \frac{1}{\sqrt{2}} \left[(\sigma_1 - \sigma_2)^2 + (\sigma_2 - \sigma_3)^2 + (\sigma_3 - \sigma_1)^2 \right]^{\frac{1}{2}} \quad (5)$$

Previous comparisons of experiment and theory [7] for the
creep of pipework and tubing indicate that the von Mises
equivalent stress accurately represents steady state deformation behaviour of ferritic materials.

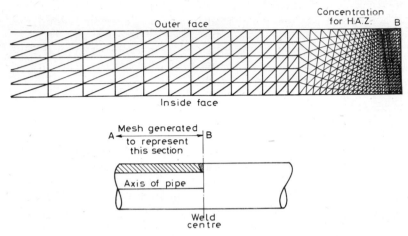

FIG. 7.   Finite Element Mesh Generated for Weld Studies.

CREEP OF BUTT-WELDED TUBES

B.G.Ivarsson and R.Sandström

Swedish Institute for Metals Research,

Stockholm, Sweden

SUMMARY

In this paper the investigations on creep in butt-welded AISI 316 tubes performed by the Swedish Creep Committee are reviewed. Uniaxial creep properties of cold worked and annealed material, weld metal, and HAZ structures have been determined and used in computer simulations of tube bursting tests. The calculations resulted in accurate predictions of time to rupture and the position and orientation of cracks. A fast, simple, and accurate method of estimating rupture times has been developed.

The expression for creep deformation has been refined to include primary creep resulting in an excellent agreement between measured and calculated strains.

1. INTRODUCTION

Failures in plants operating at high temperatures where creep is the dominating deformation mechanism often occur in or close to welded joints even if the weldments have no physical defects. Since many welds are performed in connection with dimensional changes these failures can to some extent be attributed to stress concentrations due to the geometrical inhomogeneities.

But it has been shown,(1,2,3) that even for butt welds in straight tubes, where no dimensional variations occur, the life of the tubes is appreciably reduced. These reductions in rupture strength are due to stress redistributions in the welded joints caused by different creep properties in different zones of the weldments.

In most pressure vessel codes used today welded joints are treated schematically and little or no attention is paid to the detrimental effect of the variations in creep properties.

However, with the use of computerized finite elements and finite differences calculational methods the creep deformation and rupture behaviour of components with spatially varying creep properties (and dimensions) can be predicted in detail. Thus more extensive stress-strain analyses can therefore be anticipated to be required in new and revised codes and have in fact in one case (4) already been prescribed.

These thorough analyses demand a more extensive knowledge of the creep behaviour than is requested in present codes where at most stresses to rupture and 1% deformation in 10000 and 100000 hours are needed. For that reason the Swedish Creep Committee has initiated a research programme in which the representation and application of a more complete description of the creep behaviour are investigated in several projects. In the most fundamental of those welded joints in AISI 316 steel have been studied. The results will be summarized and commented upon in this paper.

2.   DETERMINATION OF UNIAXIAL CREEP PROPERTIES

To attain a high accuracy the creep specimens used must have long gauge lengths in order to ensure uniaxiality. This is easily achieved for the parent metal and as a rule also for the weld metal. No single actual HAZ structure can, however, be tested due to the rather steep structure gradients, but these structures have to be simulated.

Annealed and weld simulated AISI 316 specimens had creep properties intermediate to those of annealed metal and weld metal, see Figs 1a and b. The lengths of the zones with homogeneous structure were, however, very small (6 mm) which made it impossible to achieve a purely uniaxial stress state. Furthermore the fast cooling rates measured in thin walled tubes could not be reproduced.

For these reasons furnace heat treatments have been used for the simulations since they result in a homogeneous structure in the entire gauge lengths of the specimens used. To determine an appropriate heat treatment, comparisons were made between structure, grain size, and micro hardness of HAZ and simulated specimens. The creep properties in the HAZ between the simulated structure were interpolated using the measured hardnesses as "weight functions". For further details, see (1).

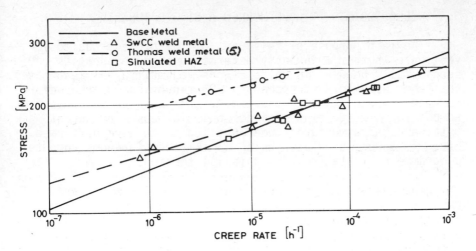

Fig 1a. Stress dependence of creep rates for parent metal and two weld metals. Creep test results for weld simulated specimens.

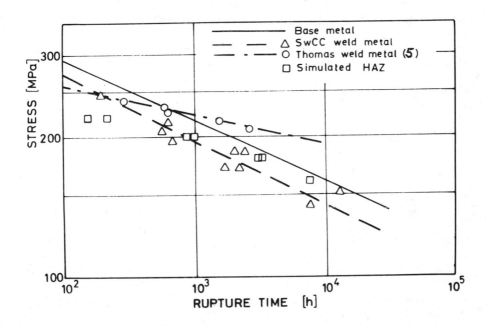

Fig 1b. Stress dependence of rupture times for parent metal and two weld metals. Creep test results for weld simulated specimens.

3. WELDED JOINTS IN COLD WORKED AISI 316 TUBES

The main purpose of this part of the project was to investigate whether uniaxially determined creep properties could successfully be used in calculations for multiaxial load cases. The tubes were cold stretched to 20% reduction in area before welding in order to increase the variation in creep properties across the welded joint. Besides uniaxial tests of parent metal and HAZ structures tube bursting tests were performed. These tests were simulated using a finite differences computer programme and data from the uniaxial testing.

The results of the experiments and calculations are presented in detail in (1) but some of the more important findings are presented here. In Fig 2 the calculated stress distribution along the tube for different times is shown. A stress peak is built up in the HAZ and, with the assumption of maximum stress (hoop stress) controlled rupture, failure is predicted to occur as longitudinal cracks about 7 mm from the centre of the weld in accordance with the results for the burst tested tubes. The agreement between experimental and calculated rupture times was also excellent. The creep life was reduced by a factor 15-30 (depending on stress level) as compared to homogeneous cold worked tubes.

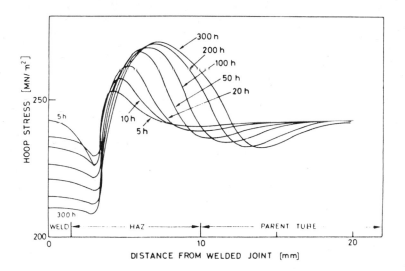

Fig 2.   Hoop stress at different times as a function of distance from the centre of the welded joint. Internal pressure 24.5 MPa.

4. CALCULATIONS FOR WELDED JOINTS IN ANNEALED TUBES

Since the calculated rupture behaviour of cold worked
tubes were accordant with experimental data similar calculations were performed for welded annealed tubes. Own data were
used for the parent tube. Two different sets of weld metal
data were used - one own and one obtained from Thomas (5), see
Figs 1a and b. No comparative tube bursting tests were performed. A complete presentation of the results of the calculations is given in (6).

From these calculations it is obvious that, in order to
decrease the loss in strength caused by the presence of a weld,
the creep deformation and rupture properties of the different
zones should be as alike as possible. If a stronger weld metal
is chosen, which is often recommended, it will experience a
stress increase due to the higher creep deformation resistance.
As a rule this increased stress will not be outbalanced by the
higher rupture strength. A weaker weld metal, on the other
hand, will shed load to the HAZ and the parent tube causing a
premature failure. Rupture can of course occur in the weld
metal if its rupture strength is much poorer than those of
the adjacent zones.

The calculations also indicate that a smooth transition
in creep properties between the weld metal and the parent
tube will improve the life of the tube. Such a transition
could be achieved by a careful choice of welding procedure.

5. A SIMPLIFIED GRAPHICAL METHOD FOR PREDICTION OF RUPTURE TIMES

Though the above - mentioned calculations resulted in
excellent descriptions of the rupture behaviour they are too
complicated and time-consuming to be used in practical construction work. Therefore a simpler method has been developed
in which the graphical representations of creep deformation
and rupture properties of the weld and parent metals are used.
The method is described mathematically in the Appendix of
Ref.(6) and graphically in Fig 3.

The method is based on the fact that the stress redistribution results in equal creep rates in the weld and
parent metals at the fusion boundary. Due to its smaller
extension about two thirds of the stress difference is distributed to the weld metal. Hence the maximum stresses in weld
metal and parent tube can be estimated. The corresponding
rupture times and thus the time and location of rupture can
then easily be evaluated.

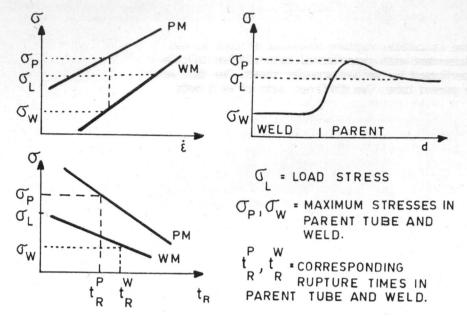

Fig 3. A graphical description of the simplified method for rupture time estimation.

Comparisons with the computer calculations show that this method gives a good though somewhat conservative estimation of the rupture time.

6. INFLUENCE OF PRIMARY CREEP

In the previously reported calculations the uniaxial creep deformation behaviour was expressed in terms of secondary creep rates. Although this led to accurate predictions of where and when rupture occurred the calculated strains underestimated the actual ones. Since the plastic strains on loading were calculated to be rather small the differences in strain were attributed to the neglect of primary creep. For that reason the uniaxial creep test data were reanalyzed and fitted to the following time-hardening constitutive equation

$$\dot{\varepsilon} = 10^a \cdot t^b \cdot \sigma^{(c+d \cdot \log t)} \tag{1}$$

which is easily intergrated to

$$\varepsilon = \frac{\dot{\varepsilon} \cdot t}{b+1+d \cdot \log \sigma} \tag{2}$$

Further details are given in (7).

The reasons for choosing this expression are, besides a

good fit to experimental data for primary creep, the facts
that the constants a, b, c, and d are easy to determine and
that the equation is easy to include in the computer programme.

But Eq.(1) also has some drawbacks. As a rule the constant
b turns out to be negative which leads to an infinite initial
creep rate. This deficiency is, however, easily eliminated in
the calculations. Since the expression is chosen to give an
accurate representation of primary creep the description of
the secondary and tertiary creep is less satisfactory, see
Fig 4. This was, however, regarded as less important since the
actual strains in the tubes were only about 1%.

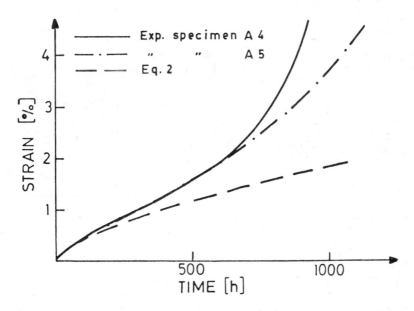

Fig 4. Fit of Eq 2 to experimental creep curves. Cold worked
        material, 255 MPa, two specimens.

It has been stated (8) that equations where a strain
dependence of the creep rate is included would give a better
estimation of the creep strains (for a complex load history).
This improvement is, however, rather small when the scatter
in experimental data is considered. Since the introduction of
strain dependence also would complicate the calculations it
has not been considered.

In Fig 5 the results of calculations using primary and
secondary creep behaviour respectively are compared to the
actual strains. Besides leading to a much improved prediction
of the strains in the tubes the new representation of creep
deformation had virtually no effect on the previous excellent
predictions of time to and position of rupture.

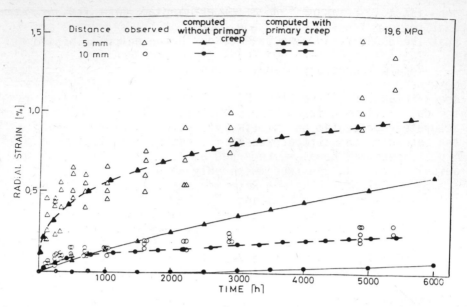

Fig 5. Hoop strains at 5 and 10 mm from the centre of the
welded joints. Comparison of experimental values to
values calculated using primary and secondary creep
representation respectively. Internal pressure 19.6 MPa.

7. CONCLUSIONS

Experiments and calculations performed by the Swedish
Creep Committee show that an excellent estimate of rupture
behaviour for a multiaxial load case can be achieved with the
use of uniaxially determined creep properties.

Further improvements of the calculations were accomplished
by the introduction of primary creep in the creep deformation
representation. The calculated strains were in full accordance with the experimentally determined ones.

ACKNOWLEDGEMENTS

This work has been financed by the Swedish Creep
Committee which is gratefully acknowledged. Permission to
publish from the Swedish Creep Committee and the Swedish
Institute for Metals Research is appreciated.

REFERENCES

1. IVARSSON, B. and SANDSTRÖM, R. -'Creep Deformation and Rupture of Butt-Welded Tubes of Cold Worked 316 Steel'. Metals Technology, 1980, 7 , 440.

2. WALTERS, D.J. and COCKROFT, R.D.H. - 'A Stress Analysis and Failure Criteria for High Temperature Butt Welds'. Paper presented at the IIW Colloquium on Creep Behaviour of Welds in Boilers, Pressure Vessels, and Pipelines Toronto, 1972.

3. UDOGUCHI, T. - 'Resent Research and Developments in High Temperature Design in Japan'. Nominated Lecture at the International Conference 'Engineering Aspects of Creep'   Sheffield, 1980.

4. ASME Boiler and Pressure Vessel Code Case N-47 (1592) Class 1 Components in Elevated Temperature Service, Section III Division 1. Am. Soc.of. Mech. Eng., New York, 1977.

5. THOMAS, R.G. - 'The Effect of δ-ferrite on the Creep Rupture Properties of Austenitic Weld Metals'. Weld. J., 1978, 57 , 81.

6. IVARSSON, B. and SANDSTRÖM, R. -'Calculational Methods of Determining the Rupture Life of Butt-Welded Tubes. Contribution to the International Conference'Engineering Aspects of Creep ' Sept 15-19, 1980 in Sheffield.

7. IVARSSON, B. - ' Creep of Butt-Welded AISI 316 TUBES. The Role of Plastic Strain and Primary Creep'. SwCC Internal Report VHK 125. To be published.

8. PENNY, R.K. and MARRIOT, D.L. - Design for Creep. Mc Graw-Hill, Maidenhead, 1971.

STRATHCLYDE UNIVERSITY LIBRARY

30125 00087573 1

ML

ANDERSONIAN LIBRARY
★
WITHDRAWN
FROM
LIBRARY
STOCK
★
UNIVERSITY OF STRATHCLYDE